Kinetics and Mechanisms of the Gas-Phase Reactions of the Hydroxyl Radical with Organic Compounds

Journal of
Physical and Chemical Reference Data

David R. Lide, Jr., Editor

The Journal of Physical and Chemical Reference Data (ISSN 0047-2689) is published quarterly by the American Chemical Society (1155 16th St., N. W., Washington, DC 20036-9976) and the American Institute of Physics (335 E. 45th St., New York, NY 10017-3483) for the National Institute of Standards and Technology. Second-class postage paid at Washington, DC and additional mailing offices. POSTMASTER: Send address changes to Journal of Physical and Chemical Reference Data, Membership and Subscription Services, P. O. Box 3337, Columbus, Ohio 43210.

The objective of the Journal is to provide critically evaluated physical and chemical property data, fully documented as to the original sources and the criteria used for evaluation. Critical reviews of measurement techniques, whose aim is to assess the accuracy of available data in a given technical area, are also included. The Journal is not intended as a publication outlet for original experimental measurements such as those that are normally reported in the primary research literature, nor for review articles of a descriptive or primarily theoretical nature.

Supplements to the Journal are published at irregular intervals and are not included in subscriptions to the Journal. They contain compilations which are too lengthy for a journal format.

The Editor welcomes appropriate manuscripts for consideration by the Editorial Board. Potential contributors who are interested in preparing a compilation are invited to submit an outline of the nature and scope of the proposed compilation, with criteria for evaluation of the data and other pertinent factors, to:

David R. Lide, Jr., Editor
J. Phys. Chem. Ref. Data
National Institute of Standards and Technology
Gaithersburg, MD 20899

One source of contributions to the Journal is The National Standard Reference Data System (NSRDS), which was established in 1963 as a means of coordinating on a national scale the production and dissemination of critically evaluated reference data in the physical sciences. Under the Standard Reference Data Act (Public Law 90-396) the National Institute of Standards and Technology of the U.S. Department of Commerce has the primary responsibility in the Federal Government for providing reliable scientific and technical reference data. The Office of Standard Reference Data of NIST coordinates a complex of data evaluation centers, located in university, industrial, and other Government laboratories as well as within the National Institute of Standards and Technology, which are engaged in the compilation and critical evaluation of numerical data on physical and chemical properties retrieved from the world scientific literature. The participants in this NIST-sponsored program, together with similar groups under private or other Government support which are pursuing the same ends, comprise the National Standard Reference Data System.

The primary focus of the NSRDS is on well-defined physical and chemical properties of well-characterized materials or systems. An effort is made to assess the accuracy of data reported in the primary research literature and to prepare compilations of critically evaluated data which will serve as reliable and convenient reference sources for the scientific and technical community.

Information for Contributors

Manuscripts submitted for publication must be prepared in accordance with Instructions for Preparation of Manuscripts for the Journal of Physical and Chemical Reference Data, available on request from the Editor.

New and renewal subscriptions should be sent with payment to the Office of the Controller at the American Chemical Society, 1155 Sixteenth Street, N.W., Washington, DC 20036-9976. Address changes, with at least six weeks advance notice, should be sent to Journal of Physical and Chemical Reference Data, Membership and Subscription Services, American Chemical Society, P.O. Box 3337, Columbus, OH 43210. Changes of address must include both old and new addresses and ZIP codes and, if possible, the address label from the mailing wrapper of a recent issue. Claims for missing numbers will not be allowed: if loss was due to failure of the change-of-address notice to be received in the time specified; if claim is dated (a) North America: more than 90 days beyond issue date, (b) all other foreign: more than one year beyond issue date.

Members of AIP member and affiliate societies requesting member subscription rates should direct subscriptions, renewals, and address changes to American Institute of Physics, Dept. S/F, 335 E. 45th St., NY 10017-3483.

Subscription Prices (1989)
(not including supplements)

	U.S.A.	Foreign (surface mail)	Optional air freight Europe Mideast N. Africa	Asia and Oceania
Members (of ACS, AIP, or affiliated society)	$ 65.00	$ 75.00	$ 85.00	$ 85.00
Regular rate	$290.00	$300.00	$310.00	$310.00

Rates above do not apply to nonmember subscribers in Japan, who must enter subscription orders with Maruzen Company Ltd., 3-10 Nihonbashi 2-chome, Chuo-ku, Tokyo 103, Japan. Tel: (03) 272-7211.

Back numbers are available at a cost of $80 per single copy and $320 per volume.

Orders for reprints, supplements, and back numbers should be addressed to the American Chemical Society, 1155 Sixteenth Street, N. W., Washington, DC 20036-9976. Prices for reprints and supplements are listed at the end of this issue.

Copying Fees: The code that appears on the first page of articles in this journal gives the fee for each copy of the article made beyond the free copying permitted by AIP. (See statement under "Copyright" elsewhere in this journal.) If no code appears, no fee applies. The fee for pre-1978 articles is $0.25 per copy. With the exception of copying for advertising and promotional purposes, the express permission of AIP is not required provided the fee is paid through the Copyright Clearance Center, Inc. (CCC), 27 Congress Street, Salem, MA 01970. Contact the CCC for information on how to report copying and remit payment.

Microfilm subscriptions of the Journal of Physical and Chemical Reference Data are available on 16 mm and 35 mm. This journal also appears in Sec. I of Current Physics Microform (CPM) along with 26 other journals published by the American Institute of Physics and its member societies. A Microfilm Catalog is available on request.

Journal of
**Physical and
Chemical
Reference Data**

Monograph No. 1

Kinetics and Mechanisms of the Gas-Phase Reactions of the Hydroxyl Radical with Organic Compounds

Roger Atkinson

Statewide Air Pollution Research Center, University of California, Riverside, CA 92521

Published by the **American Chemical Society**
and the **American Institute of Physics** for
the **National Institute of Standards and Technology**

Library of Congress Catalog Card Number 89-82387

International Standard Book Number
0-88318-720-5

American Institute of Physics, Inc.

335 East 45th Street

New York, New York 10017-3483

Printed in the United States of America

Foreword

The *Journal of Physical and Chemical Reference Data* is published jointly by the American Institute of Physics and the American Chemical Society for the National Institute of Standards and Technology. Its objective is to provide critically evaluated physical and chemical property data, fully documented as to the original sources and the criteria used for evaluation. One of the principal sources of material for the journal is the National Standard Reference Data System (NSRDS), a program coordinated by NIST for the purpose of promoting the compilation and critical evaluation of property data.

The regular issues of the *Journal of Physical and Chemical Reference Data* are published bimonthly and contain compilations and critical data reviews of moderate length. Longer works, volumes of collected tables, and other material unsuited to a periodical format have previously been published as *Supplements to the Journal*. Beginning in 1989 the generic title of these works has been changed to *Monograph*, which reflects their character as independent publications. This volume, "Kinetics and Mechanisms of the Gas-Phase Reactions of the Hydroxyl Radical with Organic Compounds" by Roger Atkinson, is presented as *Monograph No. 1* of the *Journal of Physical and Chemical Reference Data*.

David R. Lide, Jr., Editor
Journal of Physical and Chemical Reference Data

Kinetics and Mechanisms of the Gas-Phase Reactions of the Hydroxyl Radical with Organic Compounds

Roger Atkinson

Statewide Air Pollution Research Center, University of California, Riverside, CA 92521

Received September 7, 1988; revised manuscript received April 13, 1989

The literature kinetic and mechanistic data for the gas-phase reactions of the OH radical with organic compounds (through 1988) have been tabulated, reviewed and evaluated over the entire temperature ranges for which data are available.

Key words: hydroxyl radical; organic compounds; reaction kinetics; reaction mechanisms.

1

J. Phys. Chem. Ref. Data, Monograph 1 (1989)

Contents

List of Tables

List of Figures

1. Introduction

The hydroxyl (OH) radical is a key reactive intermediate in both combustion[1-9] and atmospheric chemistry.[10-12] Thus, recent computer modeling studies of the combustion of organic compounds show that the major reaction routes of these organics are by reaction with $O(^3P)$ and H atoms and OH radicals, with the relative importance of these reaction pathways depending on the particular combustion conditions, for example, on the fuel/oxygen ratio.[2-5,9] At combustion temperatures these reactive species are interconverted, at least in part, through the H_2-O_2 reactions.

$$O(^3P) + H_2 \rightleftharpoons OH + H$$

$$OH + H_2 \rightleftharpoons H_2O + H$$

$$H + O_2 \rightleftharpoons OH + O(^3P)$$

$$OH + OH \rightleftharpoons H_2O + O(^3P)$$

which are common to all high temperature organic combustion systems.[5] Measurements of OH radical concentrations, and of their temporal profiles, are now routinely carried out in flames and other combustion media.[13-17]

The hydroxyl radical has also been shown to play a pivotal role in the chemistry of the atmosphere,[10,12] and apparently reliable ambient OH radical concentrations in the troposphere[18-25] and stratosphere[26-28] are now available from *in situ* measurements, in addition to the global tropospheric average OH radical concentration derived from halocarbon lifetime measurements.[29] In the clean troposphere, hydroxyl radicals are generated through the reaction sequence,

$$O_3 + h\nu(\lambda < 319 \text{ nm}) \rightarrow O(^1D) + O_2(^1\Delta_g)$$

$$O(^1D) + H_2O \rightarrow 2\,OH$$

$$O(^1D) + M \rightarrow O(^3P) + M \ (M = \text{air})$$

with additional formation processes occurring in polluted urban areas.[12]

For the majority of organic compounds emitted into the troposphere from either biogenic or anthropogenic sources, reaction with the OH radical is their major, if not sole, chemical loss process.[12] Indeed, in the troposphere the chemical reactions responsible for the degradation of organic compounds comprise, in essence, a low-temperature combustion system. For these chemicals reaction with the OH radical leads to their removal from the atmosphere and limits their atmospheric concentrations, and this removal process (or the lack of it) is critical to the ongoing discussion of the effects of present and future anthropogenic halocarbons on the stratospheric ozone layer.

Clearly, OH radical kinetics and reaction mechanisms need to be known under the temperature, pressure and third-body conditions encountered in combustion sys-

tems and in the atmosphere. During the past several years major research efforts have been carried out to obtain these necessary experimental data. As an example, at room temperature reaction rate constants have been measured for over 350 organic compounds,[12] while at the higher temperatures (>1000 K) representative of combustion chemistry flash photolysis and pulsed radiolysis techniques are now providing absolute kinetic data, obtained in many cases over large temperature ranges.[30-40] This accelerating acquisition of kinetic and mechanistic data will surely continue, especially at combustion temperatures.

However, in order to provide a consistent and integrated overview and to allow the available data base to be effectively used by other scientists (for example, chemical modelers), this data base must be reviewed and critically evaluated on an ongoing basis. This is crucial to both combustion chemistry as well as to the chemistry of the troposphere and stratosphere. In addition to providing a recommended set of kinetic data, such reviews and evaluations provide an up-to-date status of the kinetic and mechanistic information available and are the most reliable source of data for the development of techniques for rate constant and mechanism estimations. As the information base continues to grow, these ongoing evaluations become increasingly necessary, and they must be viewed as an integral part of the experimental and theoretical research efforts in combustion and atmospheric chemistry.

Unfortunately, to date this has not been the case. Thus, while the reactions of OH radicals with the inorganic reactants of atmospheric importance and certain C_1-C_3 organics (mainly the haloalkanes) are now being included in the NASA[41] and CODATA/IUPAC[42] evaluations, there is no ongoing review and evaluation of the literature kinetic and mechanistic data for the reactions of the OH radical with the other organic compounds which comprise the vast majority of combustion fuels and atmospheric emissions. Indeed, for the reactions of the OH radical with organic compounds, few reviews have been carried out during the past decade, despite the relevance of these reactions to combustion and atmospheric chemistry.

For atmospheric purposes, Atkinson *et al.*[43] compiled and reviewed the literature kinetic and mechanistic data (through mid-1978) for organic compounds at temperatures $\leqslant 500$ K. No recommendations were given, and the emphasis was on atmospheric chemistry. In 1981, Baulch and Campbell[44] published a review covering the literature for the period 1972 through October 1979, and, because of the Atkinson *et al.* article,[43] dealt in most detail with inorganic compounds and those aspects of the reactions with organic compounds, such as high temperature data, which Atkinson *et al.*[43] did not cover. In addition, Cohen and Westberg[45] included the reactions of the OH radical with alkanes in their evaluation of the kinetics of selected reactions involved in combustion chemistry.

More recently, two evaluations and reviews have appeared which deal with organic compounds; one by Baulch *et al.*[46] dealing with the kinetics of the gas-phase reactions of the OH radical with alkanes over the temperature range ~200–2000 K, covering the literature data through October 1984, and the other by Atkinson[12] dealing with the kinetics and mechanisms of the gas-phase reactions of the OH radical with organic compounds under atmospheric conditions, with the literature being covered through late 1985/early 1986.. Although the review of Atkinson[12] focused on OH radical reactions under atmospheric conditions and, in general, only kinetic and mechanistic data below 500 K were reviewed, data at higher temperatures (up to ~1500 K) were included if these data were obtained in studies extending to temperatures below 500 K.[12] An analogous review of the gas-phase reactions of the OH radical with inorganic compounds has been carried out by Paraskevopoulos and Singleton.[47]

In this article, the review of Atkinson[12] has been updated and extended to cover the entire temperature ranges for which kinetic and mechanistic data are available. The present article deals with the kinetics and mechanisms of the initial OH radical reactions with organic compounds, and the subsequent reactions of the initially-formed product species are not dealt with. In the remainder of this section, the major experimental techniques which have been used to obtain kinetic data are briefly discussed, together with the methods of presentation of the kinetic data in Sec. 2.

1.1. Experimental Techniques Used

Two general experimental approaches, namely absolute and relative rate constant measurement methods, have been used to determine rate constants for the reactions of the OH radical with organic compounds. In the absolute technique, for the bimolecular reaction,

$$A + B \rightarrow products$$

either the psuedo-first order decay of one species is measured in the presence of a known excess concentration of the other reactant, with

$$-d\ln[A]/dt = k[B],$$

or the concentrations of both species are measured and the rate constant k derived from the equation

$$\frac{d[product]}{dt} = \frac{-d[A]}{dt} = \frac{-d[B]}{dt} = k[A][B]$$

At lower temperatures, typically $\lesssim 1000$ K, the discharge flow and flash (or laser) photolysis techniques interfaced to a variety of detection systems have been, and continue to be, widely used. To date, these detection systems for OH radical reactions have included mass spectrometry,[48] electron paramagnetic resonance,[49] laser magnetic resonance,[50] resonance absorption[51–54] (including laser absorption[55]) and resonance fluorescence[56,57] (including laser induced fluorescence[58,59]).

Relevant to this article was the determination by Kaufman and co-workers[51–53] that the initially-used electric discharge in water vapor in a fast flow system[60,61] was subject to secondary reactions regenerating OH radicals, hence yielding erroneous kinetic data. This finding[51–53] then invalidates any data obtained in this manner, and such studies are not included in this evaluation. All more recent discharge flow studies have used the reaction of H atoms with NO_2

$$H + NO_2 \rightarrow OH + NO$$

as a clean source of electronically and vibrationally ground state OH radicals.[62] The characteristics of these discharge flow and flash (or laser) photolysis methods are discussed by Howard,[63] Michael and Lee,[64] Kaufman[65] and Atkinson,[12] and these articles should be consulted for more details.

At the higher temperatures characteristic of flames, OH radical concentrations have been measured by mass-spectrometry,[66] laser induced fluorescence[67] and resonance absorption,[68] and have also been calculated from equilibrium considerations.[69]

In the relative rate method, the rate constant of interest is determined relative to that for another reaction, normally a reaction of the OH radical with a second, or reference, species. Generally, the decay rates of two or more compounds are monitored in the presence of OH radicals, with other loss processes (chemical or physical) of these reactants being either quantitatively known or minimized.[12] Hydroxyl radicals have been generated by numerous methods, including photolysis of NO-NO_2-organic-air,[70] $HONO$-NO-air,[71,72] and CH_3ONO-NO-air[73] mixtures, photolysis of H_2O_2,[74,75] the dark reaction of N_2H_4 with O_3,[76] the thermal decomposition of H_2O_2[77] (at elevated temperatures) or $HOONO_2$,[78] and the heterogeneous reaction of H_2O_2 with NO_2.[79] Detection methods for the reactant organic and the reference species have included gas chromatography,[70,73,75] Fourier transform infrared absorption spectroscopy[72,76] and differential optical absorption spectroscopy.[80] The methods utilized at around room temperature for the determination of OH radical reaction rate constants using relative rate techniques have been discussed in detail by Atkinson.[12]

At elevated temperatures, relative rate constants have been derived from studies utilizing the thermal decomposition of H_2O_2 as a source of OH radicals[77] and from the effects of small amounts of added organic to the H_2-O_2 reaction system.[81] In flames, numerous studies (see, for example, Refs. 82 and 83) have been carried out in which the decay rate of the organic, presumed to be due to reaction with the OH radical, was measured and the OH radical concentration calculated from the formation rate of CO_2, produced from the reaction.

$$OH + CO \rightarrow H + CO_2$$

In these studies the rate constants derived were thus relative to the rate constant for the reaction of OH radicals with CO.

J. Phys. Chem. Ref. Data, Monograph 1 (1989)

1.2. Presentation of Kinetic Data

In Sec. 2, the kinetic data for the reactions of OH and OD radicals with the various classes of organic compounds [alkanes (including cycloalkanes), haloalkanes, alkenes (including di- and tri-alkenes and cycloalkenes), haloalkenes, alkynes, oxygen-, sulfur-, nitrogen-, phosphorus- and silicon-containing organics, aromatics, organometallics and organic radicals] are presented and discussed. As far as possible, the initial reaction mechanisms are discussed in conjunction with the compilations and evaluations of the available rate constant data. Data from relative rate constant studies have been reevaluated on the basis of the recommended rate constants for the reference reactions at the temperatures employed in the relative rate studies.

Three OH radical reactions with inorganic compounds have been used in more than one study as the reference reaction; namely the reactions of OH radicals with H_2, CO, and HONO. The rate constant for the reaction of OH radicals with H_2 has been reviewed and evaluated by Cohen and Westberg,[45] and their recommendation (also accepted by the recent evaluation of Tsang and Hampson[84]) of

$$k(OH + H_2) = 1.06$$

$$\times 10^{-17} T^2 e^{-1490/T} cm^3 \ molecule^{-1} \ s^{-1}$$

over the temperature range 240–2400 K is utilized in this evaluation.

For the reaction of OH radicals with CO, the "low pressure" rate constant recommended in the evaluation of Baulch et al.[85] of

$$k(OH + CO) = 1.12$$

$$\times 10^{-13} e^{0.000907T} cm^3 \ molecule^{-1} \ s^{-1}$$

is employed. This low total pressure rate constant is consistent with the most recent NASA[41] and CODATA/IUPAC[42] room temperature evaluations and agrees with the recent temperature-dependent study of Ravishankara and Thompson.[86] At around room temperature this rate constant is pressure dependent for the more effective third body gases such as N_2, O_2, CF_4 and SF_6,[87–92] and at 298 K the most recent NASA[41] and IUPAC[42] evaluations recommend that for M = O_2, N_2 and air

$$k(OH + CO) = 1.5$$

$$\times 10^{-13} (1 + 0.6 P) cm^3 \ molecule^{-1} \ s^{-1},$$

where P is the total pressure in atmospheres. This pressure dependence is essentially independent of temperature for temperatures <300 K,[92] although Hynes et al.[92] have observed that the pressure dependence of the rate constant for this reaction for M = air at 371 K is somewhat less than at 299 K or 262 K. Golden and co-workers[93] have also carried out a theoretical study of this reaction, showing that, as expected, the pressure-dependent portion of this reaction becomes less important at elevated temperatures.

In this review article, a rate constant for the reaction of OH radicals with CO of

$$k(OH + CO) = 1.12 \times 10^{-13} e^{0.000907T}$$

$$\times [1 + 2.4 \times 10^{-20} [M](T/298)^{-1}] cm^3 \ molecule^{-1} \ s^{-1}$$

is used for M = O_2 and/or N_2, with the term in square brackets being an empirical relationship to take into account the pressure dependence of this rate constant over the temperature range ~290–1000 K. This expression essentially reproduces the NASA[41] and IUPAC[42] evaluations at 298 K and the data reported by Hynes et al.,[92] and exhibits the behavior predicted by Golden and co-workers.[93] It should be noted, however, that the pressure-dependence of the rate constant at temperatures below ~298 K calculated from this equation appears to be more temperature dependent than shown by the data cited by Hynes et al.[92] [This is of no real consequence to the present evaluation since the reaction of OH radicals with CO has not been used as a reference reaction in relative rate studies carried out at temperatures below ~290 K.]

As noted above, the pressure dependence of this rate constant for the reaction of OH radicals with CO is dependent on the third body M,[87–92] with no pressure dependence being observed for M = He or Ar at around room temperature.[88,89,91] This pressure dependence of the rate constant for the reaction of OH radicals with CO, and its dependence on the temperature and the particular third-body or third-bodies employed, introduces added uncertainties into the derivation of rate constants from relative rate studies employing CO as the reference compound. Accordingly, the rate constants derived from these relative rate studies utilizing CO as the reference compound are given a lower weight in the evaluations.

For the reaction of OH radicals with HONO, the recent data of Jenkin and Cox[94] are used, with

$$k(OH + HONO) = 1.80 \times 10^{-11} e^{-390/T} cm^3 \ molecule^{-1} \ s^{-1}.$$

In the data tabulations in Sec. 2, the experimental techniques used are denoted by abbreviations such as (for example) DF-RF, where the first letters denote: DF, discharge flow; FP, flash photolysis; LP, laser photolysis; LH, laser heating; SH, shock heating; MPS, modulation-phase shift; PR, pulsed radiolysis; and the second set of letters denote the detection technique; MS, mass-spectroscopy (including photoionization-mass spectroscopy); EPR, electron paramagnetic resonance; KS, kinetic spectroscopy; LMR, laser magnetic resonance; RA, resonance absorption; RF, resonance fluorescence; and LIF, laser induced fluorescence. Relative rate studies are denoted by the abbreviation RR, and the reference compound and the OH radical reaction rate constant used for the reference reaction are given.

The tables list, whenever available, the rate constants obtained at the various temperatures studied. Throughout this article, cm^3 molecule^{-1} s^{-1} units are used for bimolecular reactions, and pressures are expressed in Torr (1 Torr = 133.3 Pa). The cited Arrhenius preexponential factors A and temperature dependent parameters B in the expression $k = A e^{-B/T}$ are also listed, where B is in K. In some studies covering wide temperature ranges, the simple Arrhenius expression has, as expected, been shown not to hold, with pronounced curvature in the Arrhenius plots being observed.[12,30,33,95] In these cases a three-parameter expression of the form

$$k = AT^n e^{-B/T} \qquad \text{(I)}$$

has been used (where $n = 0$ for the Arrhenius expression) and the reported values of A, B and n are tabulated. Since to date most of the available OH radical rate constant data have been obtained over relatively limited temperature ranges, the simple Arrhenius expression, although obviously too simplistic, is often adequate and convenient for expressing these experimental data over the limited temperature ranges studied.

In those cases where data are available over only limited temperature ranges (for example, at temperatures $\lesssim 500$ K) and no obvious non-Arrhenius behavior of the data is evident, recommendations are given in the form of the Arrhenius equation

$$k = A e^{-B/T} \qquad \text{(II)}$$

For organic compounds for which reliable data exist covering large temperature ranges, for example, from $\lesssim 300$ to $\gtrsim 1000$ K, or for which their Arrhenius plots exhibit obvious curvature, a more realistic expression is used for the recommendations. The expression

$$k = CT^2 e^{-D/T} \qquad \text{(III)}$$

has been chosen in these evaluations since this has been used in the recent NASA evaluation[41] and is consistent with the experimental data. Furthermore, values of $n \sim 2$ in the above three parameter expression have been derived from previous experimental studies[30,33] and theoretical evaluations[96] of these reactions over wide temperature ranges. It should be noted, however, that Cohen and Benson[97,98] have used transition state theory to calculate values of $n = 1.1-1.8$ in Eq. (I) for the reactions of the OH radical with a series of halomethanes and haloethanes, although the differences between these formulations of the three-parameter expression, i.e., with $n = 1$ or $n = 2$, are likely to be within the uncertainties of the experimental data. The expression $k = CT^n e^{-D/T}$ can be transformed into an Arrhenius expression, $k = A e^{-B/T}$, centered at a temperature T with $A = C(eT)^n$ and $B = D + nT$.

In the rate constant data tables, the error limits cited are those reported. In many cases these are one or two least-squares standard deviations and in others they are the estimated overall error limits. While for relative rate constant studies the use of two least-squares standard deviations may be a realistic estimation of the overall error

limits with respect to the reference reaction rate constant, with additional systematic uncertainties being associated with the rate constant used for the reference reaction, the overall error limits for the absolute rate constant determinations are expected to be of the order of $\sim 10-15\%$, except for some of the most recent studies for which the overall error limits may have been reduced to $\sim 6-10\%$.

For the alkenes, haloalkenes, alkynes and aromatics, which react with the OH radical at around room temperature, at least partially, by initial OH radical addition to the C=C and C≡C bonds or to the aromatic ring(s), the measured rate constants are often in the fall-off regime between second- and third-order kinetics. For these classes of organic compounds, in general only the data obtained (or thought to have been obtained) at, or close to, the high-pressure limit are tabulated, and data which were obtained (or now realized to have been obtained) in the fall-off region are not explicitly given. However, the pressure ranges at which the high-pressure region are (effectively) attained are discussed and, where sufficient experimental data are available, the parameters in the Troe fall-off expression,[99-101]

$$k = \left[\frac{k_o[M]}{1 + k_o[M]/k_\infty} \right] F^z$$

$$z = \left[1 + \{\log(k_o[M]/k_\infty)\}^2 \right]^{-1}$$

where k_o and k_∞ are the limiting low pressure third-order and high pressure second-order rate constants, respectively, are derived (mainly for M = N_2, O_2, air or Ar). [These rate constants k_o and k_∞ are given in units of cm^6 molecule^{-2} s^{-1} and cm^3 molecule^{-1} s^{-1}, respectively, in this article]. The broadening coefficient, F, is also temperature dependent, and can be approximately represented by[42,100,101]

$$F = e^{-T/T^*} + e^{-4T^*/T}$$

where T^* is a constant (in K) for a given reactant. This treatment then allows the effects of fall-off behavior, especially at elevated temperatures, to be taken into account.

References

[1]C. K. Westbrook and F. L. Dryer, Combust. Sci. Technol. **27**, 31 (1981).

[2]C. K. Westbrook, Combust. Flame **46**, 191 (1982).

[3]C. K. Westbrook, F. L. Dryer, and K. P. Schug, 19th International Symposium on Combustion, 1982; The Combustion Institute, Pittsburgh, PA, 1982, p. 153.

[4]J. A. Miller, R. E. Mitchell, M. D. Smooke, and R. J. Kee, 19th International Symposium on Combustion, 1982; The Combustion Institute, Pittsburgh, PA, 1982, p. 181.

[5]J. Warnatz, H. Bockhorn, A. Möser, and H. W. Wenz, 19th International Symposium on Combustion, 1982; The Combustion Institute, Pittsburgh, PA, 1982, p. 197.

[6]J. D. Bittner and J. B. Howard, 19th International Symposium on Combustion, 1982; The Combustion Institute, Pittsburgh, PA, 1982, p. 211.

[7]J. Warnatz, Ber. Bunsenges Phys. Chem. **87**, 1008 (1983).

[8]J. Warnatz, in *Combustion Chemistry*, W. C. Gardiner, Jr., Editor (Springer-Verlag, New York, NY, 1984), p. 197.

[9]J. Warnatz, 20th International Symposium on Combustion, 1984; The Combustion Institute, Pittsburgh, PA, 1985, p. 845.

[10]J. A. Logan, M. J. Prather, S. C. Wofsy, and M. B. McElroy, J. Geophys. Res. **86**, 7210 (1981).

[11]R. Atkinson and A. C. Lloyd, J. Phys. Chem. Ref. Data **13**, 315 (1984).

[12]R. Atkinson, Chem. Rev. **86**, 69 (1986).

[13]G. P. Smith and D. R. Crosley, 18th International Symposium on Combustion, 1980; The Combustion Institute, Pittsburgh, PA, 1981, p. 1511.

[14]M. Aldén, H. Edner, G. Holmstedt, S. Svanberg, and T. Högberg, Appl. Optics **21**, 1236 (1982).

[15]D. Klick and E. W. Kaiser, Appl. Optics **23**, 4184 (1984).

[16]R. J. Cattolica and S. R. Vosen, 20th International Symposium on Combustion, 1984; The Combustion Institute, Pittsburgh, PA, 1985, p. 1273.

[17]K. Kohse-Höinghaus, P. Koczar, and Th. Just, 21st International Symposium on Combustion, 1986; The Combustion Institute, Pittsburgh, PA, 1988, p. 1719.

[18]G. Hübler, D. Perner, U. Platt, A. Tönnissen, and D. H. Ehhalt, J. Geophys. Res. **89**, 1309 (1984).

[19]T. M. Hard, R. J. O'Brien, C. Y. Chan, and A. A. Mehrabzadeh, Environ. Sci. Technol. **18**, 768 (1984).

[20]T. M. Hard, C. Y. Chan, A. A. Mehrabzadeh, W. H. Pan, and R. J. O'Brien, Nature **322**, 617 (1986).

[21]B. Shirinzadeh, C. C. Wang, and D. Q. Deng, Geophys. Res. Lett. **14**, 123 (1987).

[22]L. I. Davis, Jr., J. V. James, C. C. Wang, C. Guo, P. T. Morris, and J. Fishman, J. Geophys. Res. **92**, 2020 (1987).

[23]D. Perner, U. Platt, M. Trainer, G. Hübler, J. Drummond, W. Junkermann, J. Rudolph, B. Schubert, A. Volz, D. H. Ehhalt, K. J. Rumpel, and G. Helas, J. Atmos. Chem. **5**, 185 (1987).

[24]U. Platt, M. Rateike, W. Junkermann, J. Rudolph, and D. H. Ehhalt, J. Geophys. Res. **93**, 5159 (1988).

[25]C. C. Felton, J. C. Sheppard, and M. J. Campbell, Nature **335**, 53 (1988).

[26]J. G. Anderson, Geophys. Res. Lett. **3**, 165 (1976).

[27]W. S. Heaps and T. J. McGee, J. Geophys. Res. **88**, 5281 (1983).

[28]W. S. Heaps and T. J. McGee, J. Geophys. Res. **90**, 7913 (1985).

[29]R. Prinn, D. Cunnold, R. Rasmussen, P. Simmonds, F. Alyea, A. Crawford, P. Fraser, and R. Rosen, Science **238**, 945 (1987).

[30]F. P. Tully and A. R. Ravishankara, J. Phys. Chem. **84**, 3126 (1980).

[31]G. P. Smith, P. W. Fairchild, and D. R. Crosley, J. Chem. Phys. **81**, 2667 (1984).

[32]C. D. Jonah, W. A. Mulac, and P. Zeglinski, J. Phys. Chem. **88**, 4100 (1984).

[33]S. Madronich and W. Felder, 20th International Symposium on Combustion, 1984; The Combustion Institute, Pittsburgh, PA, 1985, p. 703.

[34]G. P. Smith, P. W. Fairchild, J. B. Jeffries, and D. R. Crosley, J. Phys. Chem. **89**, 1269 (1985).

[35]W. Felder and S. Madronich, Combust. Sci. Technol. **50**, 135 (1986).

[36]A.-D. Liu, W. A. Mulac, and C. D. Jonah, Int. J. Chem. Kinet. **19**, 25 (1987).

[37]A. Liu, W. A. Mulac, and C. D. Jonah, J. Phys. Chem. **92**, 131 (1988).

[38]F. P. Tully, Chem. Phys. Lett. **143**, 510 (1988).

[39]A. Liu, W. A. Mulac, and C. D. Jonah, J. Phys. Chem. **92**, 3828 (1988).

[40]A. Liu, W. A. Mulac, and C. D. Jonah, J. Phys. Chem. **92**, 5942 (1988).

[41]W. B. DeMore, M. J. Molina, S. P. Sander, D. M. Golden, R. F. Hampson, M. J. Kurylo, C. J. Howard, and A. R. Ravishankara, Evaluation No. 8, NASA Panel for Data Evaluation, JPL Publication 87-41, Jet Propulsion Laboratory, Pasadena, CA, September 15, 1987.

[42]R. Atkinson, D. L. Baulch, R. A. Cox, R. F. Hampson, Jr., J. A. Kerr, and J. Troe, J. Phys. Chem. Ref. Data, **18**, 881 (1989).

[43]R. Atkinson, K. R. Darnall, A. C. Lloyd, A. M. Winer, and J. N. Pitts, Jr., Adv. Photochem. **11**, 375 (1979).

[44]D. L. Baulch and I. M. Campbell, Gas Kinetics and Energy Transfer **4**, 137 (1981).

[45]N. Cohen and K. R. Westberg, J. Phys. Chem. Ref. Data **12**, 531 (1983).

[46]D. L. Baulch, M. Bowers, D. G. Malcolm, and R. T. Tuckerman, J. Phys. Chem. Ref. Data **15**, 465 (1986).

[47]G. Paraskevopoulos and D. L. Singleton, Rev. Chem. Intermed. **10**, 139 (1988).

[48]E. D. Morris, Jr., D. H. Stedman, and H. Niki, J. Amer. Chem. Soc. **93**, 3570 (1971).

[49]A. A. Westenberg, J. Chem. Phys. **43**, 1544 (1965).

[50]C. J. Howard and K. M. Evenson, J. Chem. Phys. **61**, 1943 (1974).

[51]F. Kaufman and F. P. Del Greco, J. Chem. Phys. **35**, 1895 (1961).

[52]F. P. Del Greco and F. Kaufman, Disc. Faraday Soc. **33**, 128 (1962).

[53]F. Kaufman and F. P. Del Greco, 9th International Symposium on Combustion (Academic Press, New York, NY, 1963), p. 659.

[54]N. R. Greiner, J. Chem. Phys. **45**, 99 (1966).

[55]A. Wahner and C. Zetzsch, 8th International Symposium on Gas Kinetics, Univ. of Nottingham, Nottingham, U.K., July 15–20, 1984.

[56]J. G. Anderson and F. Kaufman, Chem. Phys. Lett. **16**, 375 (1972).

[57]F. Stuhl and H. Niki, J. Chem. Phys. **57**, 3671 (1972).

[58]J. S. Robertshaw and I. W. M. Smith, J. Phys. Chem. **86**, 785 (1982).

[59]F. P. Tully, Chem. Phys. Lett. **96**, 148 (1983).

[60]O. Oldenberg, J. Chem. Phys. **3**, 266 (1935).

[61]L. I. Avramenko and R. V. Lorenzo, Dokl. Akad. Nauk. SSR **67**, 867 (1949).

[62]C. J. Howard, J. Chem. Phys. **65**, 4771 (1976).

[63]C. J. Howard, J. Phys. Chem. **83**, 3 (1979).

[64]J. V. Michael and J. H. Lee, J. Phys. Chem. **83**, 10 (1979).

[65]F. Kaufman, J. Phys. Chem. **88**, 4909 (1984).

[66]J. Peeters and G. Mahnen, 14th International Symposium on Combustion, 1972; The Combustion Institute, Pittsburgh, PA, 1973, p. 133.

[67]C. Morley, 18th International Symposium on Combustion, 1980; The Combustion Institute, Pittsburgh, PA, 1981, p. 23.

[68]J. Ernst, H. Gg. Wagner, and R. Zellner, Ber. Bunsenges Phys. Chem. **82**, 409 (1978).

[69]C. P. Fenimore and G. W. Jones, J. Chem. Phys. **41**, 1887 (1964).

[70]G. J. Doyle, A. C. Lloyd, K. R. Darnall, A. M. Winer, and J. N. Pitts, Jr., Environ. Sci. Technol. **9**, 237 (1975).

[71]R. A. Cox, J. Photochem. **3**, 291 (1974/75).

[72]H. Niki, P. D. Maker, C. M. Savage, and L. P. Breitenbach, J. Phys. Chem. **82**, 132 (1978).

[73]R. Atkinson, W. P. L. Carter, A. M. Winer, and J. N. Pitts, Jr., J. Air Pollut. Contr. Assoc. **31**, 1090 (1981).

[74]R. A. Gorse and D. H. Volman, J. Photochem. **1**, 1 (1972).

[75]T. Ohta, J. Phys. Chem. **87**, 1209 (1983).

[76]E. C. Tuazon, W. P. L. Carter, R. Atkinson, and J. N. Pitts, Jr., Int. J. Chem. Kinet. **15**, 619 (1983).

[77]D. E. Hoare, Nature **194**, 283 (1962).

[78]I. Barnes, V. Bastian, K. H. Becker, E. H. Fink, and F. Zabel, Atmos. Environ. **16**, 545 (1982).

[79]I. M. Campbell, B. J. Handy, and R. M. Kirby, J. Chem. Soc. Faraday Trans. 1, **71**, 867 (1975).

[80]H. W. Biermann, H. Mac Leod, R. Atkinson, A. M. Winer, and J. N. Pitts, Jr., Environ. Sci. Technol. **19**, 244 (1985).

[81]R. R. Baldwin and R. W. Walker, J. Chem. Soc. Faraday Trans. 1, **75**, 140 (1979), and references therein.

[82]A. A. Westenberg and R. M. Fristrom, J. Phys. Chem. **65**, 591 (1961).

[83]C. P. Fenimore and G. W. Jones, J. Phys. Chem. **65**, 2200 (1961).

[84]W. Tsang and R. F. Hampson, J. Phys. Chem. Ref. Data **15**, 1087 (1986).

[85]D. L. Baulch, D. D. Drysdale, J. Duxbury, and S. Grant, "Evaluated Kinetic Data for High Temperature Reactions," Vol. 3, *Homogeneous Gas Phase Reaction of the O_2-O_3 System, the CO-O_2-H_2 System, and of Sulphur Containing Species* (Butterworths, London, 1976).

[86]A. R. Ravishankara and R. L. Thompson, Chem. Phys. Lett. **99**, 377 (1983).

[87]R. Overend and G. Paraskevopoulos, Chem. Phys. Lett. **49**, 109 (1977).

[88]R. A. Perry, R. Atkinson, and J. N. Pitts, Jr., J. Chem. Phys. **67**, 5577 (1977).

[89]G. Paraskevopoulos and R. S. Irwin, J. Chem. Phys. **80**, 259 (1984).

[90]A. Hofzumahaus and F. Stuhl, Ber. Bunsenges Phys. Chem. **88**, 557 (1984).

[91]W. B. DeMore, Int. J. Chem. Kinet. **16**, 1187 (1984).

[92]A. J. Hynes, P. H. Wine, and A. R. Ravishankara, J. Geophys. Res. **91**, 11815 (1986).

[93]C. W. Larson, P. H. Stewart, and D. M. Golden, Int. J. Chem. Kinet. **20**, 27 (1988).

[94]M. E. Jenkin and R. A. Cox, Chem. Phys. Lett. **137**, 548 (1987).

[95]F. P. Tully, A. T. Droege, M. L. Koszykowski, and C. F. Melius, J. Phys. Chem. **90**, 691 (1986).

[96]R. Zellner, J. Phys. Chem. **83**, 18 (1979).

[97]N. Cohen and S. W. Benson, J. Phys. Chem. **91**, 162 (1987).

[98]N. Cohen and S. W. Benson, J. Phys. Chem. **91**, 171 (1987).

[99]J. Troe, J. Chem. Phys. **66**, 4758 (1977).

[100]J. Troe, J. Phys. Chem. **83**, 114 (1979).

[101]R. G. Gilbert, K. Luther, and J. Troe, Ber. Bunsenges Phys. Chem. **87**, 169 (1983).

2. Kinetic and Mechanistic Data

In this section, the kinetics and mechanisms of the reactions of the OH and OD radical with the various classes of organic compounds (alkanes, haloalkanes, alkenes, haloalkenes, alkynes, oxygen-, sulfur-, nitrogen-, phosphorus- and silicon-containing organics, aromatics, organometallics and organic radicals) are dealt with separately. Only the gas-phase reactions of $OH(X^2\Pi_i)_{v=0}$ radicals are dealt with in these sections, since few kinetic data exist for the reactions of vibrationally excited OH radicals,[1-6] and these measurements are mainly for vibrational quenching[1,3-5] rather than for chemical reaction. Indeed, only for the reaction of the $OH(X^2\Pi_i)_{v=1}$ radical with CH_4 is a rate constant for chemical reaction available, with an upper limit to the rate constant of $\leqslant 3 \times 10^{-14}$ cm^3 molecule^{-1} s^{-1} being reported by Spencer *et al.* at 295 ± 2 K.[2]

As far as possible, the initial reaction mechanisms are discussed together with the tabulations and evaluations of the available rate constant data. As noted above, for the relative rate studies the data have been reevaluated on the basis of the recommended rate constants for the reference reactions at the temperatures and, if necessary, the pressures employed in those relative rate studies. If such a reevaluation was not possible, then the data from these relative rate studies are not tabulated or considered in the evaluations. As also noted above, those relative rate studies employing the reaction of OH radicals with CO as the reference reaction are subject to additional uncertainties due to the dependence of the rate constant for this reference reaction on the total pressure and the diluent gas(es) present, especially at temperatures <500 K.[7] This introduces additional uncertainties into the derivation of rate constants from relative rate studies employing CO, and accordingly, rate constants from these studies are given a lower weight in the evaluations.

References

[1]S. D. Worley, R. N. Coltharp, and A. E. Potter, Jr., J. Phys. Chem. **76**, 1511 (1972).

[2]J. E. Spencer, H. Endo, and G. P. Glass, 16th International Sympo-sium on Combustion, 1976; The Combustion Institute, Pittsburgh, PA, 1977, p. 829.

[3]G. P. Glass, H. Endo, and B. K. Chaturvedi, J. Chem. Phys. **77**, 5450 (1982).

[4]B. J. Finlayson-Pitts, D. W. Toohey, and M. J. Ezell, Int. J. Chem. Kinet. **15**, 151 (1983).

[5]B. J. Finlayson-Pitts, D. W. Toohey, and M. J. Ezell, Int. J. Chem. Kinet. **17**, 613 (1985).

[6]U. Meier, H. H. Grotheer, G. Riekert and Th. Just, Ber. Bunsenges Phys. Chem. **89**, 325 (1985).

[7]C. W. Larson, P. H. Stewart, and D. M. Golden, Int. J. Chem. Kinet. **20**, 27 (1988).

2.1. Alkanes
a. Kinetics and Mechanisms

The literature rate constant data for the reactions of the OH radical with the alkanes are given in Tables 1 (acyclic alkanes) and 2 (cycloalkanes). The available rate constants for the reactions of the OD radical with alkanes are given in Table 3. In these tables, the rate constants given are those for the overall reactions. These OH radical reactions with the alkanes and cycloalkanes proceed by H-atom abstraction from the C—H bonds,[119-121]

$$OH + RH \rightarrow H_2O + R$$

and hence in general a variety of alkyl radicals are formed with differing rate constants. Only for propane, *n*-butane and 2-methylpropane are sufficient experimental data available to allow the rate constants for the formation of the differing alkyl radicals to be derived in any direct manner.

In the evaluations of the rate constants for the individual alkanes and cycloalkanes, the previous reviews of Atkinson *et al.*[119] and Atkinson[120] are utilized to aid in the assessment of those studies which are judged (possibly subjectively) to be free of systematic errors and are hence used for the evaluations. The kinetic data for the individual alkanes and cycloalkanes are discussed below. For methane and ethane a sufficient number of absolute rate constant data are available over a large temperature range that the recommended rate expressions can be derived solely from these absolute data, and the reliability of the relative rate studies for these alkanes can be assessed. For propane, *n*-butane, 2-methylpropane and the higher alkanes, relative rate data judged to be reliable (for example, from identical or related studies to those which agreed with the recommended absolute rate constant data for methane and ethane) were utilized together with absolute rate data in the evaluations.

(1) Methane

The available rate constant data are tabulated in Table 1. A large number of kinetic studies have been carried out for methane using both absolute and, especially at elevated temperatures, relative rate methods. In view of the large number of absolute rate data available, covering the temperature range from 240 to 1900 K, the recommended rate constant expression for methane is based solely upon the absolute rate data.

TABLE 1. Rate constants k and temperature-dependent parameters for the gas-phase reactions of the OH radical with alkanes

Alkane	$10^{12} \times A$ (cm³ molecule⁻¹ s⁻¹)	n	B (K)	$10^{12} \times k$ (cm³ molecule⁻¹ s⁻¹)	at T (K)	Technique	Reference	Temperature range covered (K)
Methane				17	1650–1840	RR [relative to k(CO) $= 1.12 \times 10^{-13}e^{0.0009077T}$][a]	Westenberg and Fristrom[1]	1650–1840
				9.8	1445	RR [relative to k(CO) $= 1.12 \times 10^{-13}e^{0.0009077T}$][a]	Fenimore and Jones[2]	1225–1800
				10	1560			
				8.7	1580			
				10	1690			
				13	1800			
				0.171	673	RR [relative to k(CO) $= 1.12 \times 10^{-13}e^{0.0009077T}$][a]	Hoare[3]	673–923
				0.238	723			
				0.49	798			
				0.67	873			
				0.88	923			
				22	1370–1680	RR [relative to k(CO) $= 1.12 \times 10^{-13}e^{0.0009077T}$][a]	Fristrom[4]	1370–1680
				0.41 ± 0.21	773	RR [relative to k(CO) $= 2.26 \times 10^{-13}$][a]	Blundell et al.[5]	
				0.175	673	RR [relative to k(CO) $= 1.12 \times 10^{-13}e^{0.0009077T}$][a]	Hoare[6]	673–923
				0.270	723			
				0.49	798			
				0.69	873			
				0.93	923			
				0.53	798	RR [relative to k(CO) $= 2.31 \times 10^{-13}$][a]	Hoare and Peacock[7]	
				0.92 ± 0.19	773	RR [relative to k(H₂) $= 9.22 \times 10^{-13}$][a]	Baldwin et al.[8]	
				5.0 ± 1.7	1285	Flame - RA	Dixon-Lewis and Williams[9]	
				0.0108 ± 0.0025	300	DF-EPR	Wilson and Westenberg[10]	
				0.00880 ± 0.00033	301 ± 1	FP-KS	Greiner[11]	
	83		2516	0.0179	298	FP-KS	Horne and Norrish[12]	298–423
				14 ± 3	1750–2000	RR [relative to k(CO) $= 1.12 \times 10^{-13}e^{0.0009077T}$][a]	Wilson et al.[13]	1750–2000
				0.00848 ± 0.00071	295	FP-KS	Greiner[14]	295–498
				0.00953 ± 0.00028	295			
				0.0106 ± 0.00025	296			
				0.0103 ± 0.00053	296			
				0.00804 ± 0.00020	301			
				0.00805 ± 0.00041	301			
				0.00903 ± 0.00088	302			
				0.0154 ± 0.0006	333			
				0.0352 ± 0.0007	370			
				0.0611 ± 0.0023	424			
				0.121 ± 0.004	492			
				0.121 ± 0.003	493			
				0.120 ± 0.003	493			
				0.113 ± 0.002	497			
	$5.5^{+0.8}_{-0.6}$		1898 ± 51	0.122 ± 0.003	498			

TABLE 1. Rate constants k and temperature-dependent parameters for the gas-phase reactions of the OH radical with alkanes — Continued

Alkane	$10^{12} \times A$ (cm³ molecule⁻¹ s⁻¹)	n	B (K)	$10^{12} \times k$ (cm³ molecule⁻¹ s⁻¹)	at T (K)	Technique	Reference	Temperature range covered (K)
				1.0 ± 0.1	773	RR [relative to $k(H_2)$ = 9.22×10^{-13}]ᵃ	Baldwin et al.[15]	
				$0.18^{+0.18}_{-0.09}$	548	RR [relative to $k(CO)$ = 1.84×10^{-13}]ᵃ	Simonaitis et al.[16]	
	50		3020	6.7	1500	Flame - MS	Peeters and Mahnen[17]	1100–1900
				0.00204 ± 0.00036^b	240	FP-RF	Davis et al.[18]	240–373
				0.00508 ± 0.00020^b	276			
				0.00775 ± 0.00063^b	298			
	2.36 ± 0.21		1711 ± 88	0.0242 ± 0.0037^b	373			
				0.00715 ± 0.00042	293	DF-RF	Margitan et al.[19]	293–427
				0.0212 ± 0.0004	359			
				0.0306 ± 0.0001	384			
				0.0422 ± 0.0018	407			
	3.83 ± 0.20		1842 ± 20	0.0521 ± 0.0016	427			
				0.0261 ± 0.0027	381	PR-RA	Gordon and Mulac[20]	381–416
				0.0548 ± 0.0017	416			
				0.00651 ± 0.00027	295 ± 2	FP-RA	Overend et al.[21]	
				0.0095 ± 0.0014	296	DF-LMR	Howard and Evenson[22]	
				0.0088 ± 0.0007	298	FP-RA	Zellner and Steinert[23]	298–892
				0.0148	330			
				0.020	358			
				0.028	381			
				0.061	444			
				0.070	453			
				0.113	498			
				0.174	525			
				0.257	564			
				0.251 ± 0.033	576			
				0.276 ± 0.033	584			
				0.335	622			
				0.551	629			
				0.822	671			
				0.830	680			
				1.12	738			
				1.21	756			
				1.51	776			
	$5.76^{+1.17}_{-0.98} \times 10^{-9}$	3.08	1010	2.71	892			
			1804 ± 120 (300–500 K)					
				0.0063 ± 0.0008	296 ± 2	RR [relative to $k(H_2)$ = 6.05×10^{-15}]ᵃ	Cox et al.[24]	
				9.6	1300	RR [relative to $k(H_2)$ = 5.69×10^{-12}]ᵃ	Bradley et al.[25]	
				3.82	1140	SH/FP-RA	Ernst et al.[26]	1140–1505
				3.82	1160			
				4.48	1165			
				3.49	1188			
				3.49	1192			
				4.82	1203			
				4.15	1220			
				3.82	1245			
				3.99	1260			

TABLE 1. Rate constants k and temperature-dependent parameters for the gas-phase reactions of the OH radical with alkanes — Continued

Alkane	$10^{12} \times A$ (cm^3 molecule^{-1} s^{-1})	n	B (K)	$10^{12} \times k$ (cm^3 molecule^{-1} s^{-1})	at T (K)	Technique	Reference	Temperature range covered (K)
				4.15	1260			
				3.49	1265			
				4.48	1270			
				4.65	1270			
				3.82	1275			
				3.99	1275			
				3.99	1303			
				4.32	1313			
				5.65	1335			
				5.31	1404			
				5.15	1410			
				4.48	1415			
				5.65	1500			
				5.31	1505			
				0.0070 ± 0.00067	296	FP-RA (of CH$_3$) with computer modeling	Sworski et al.[27]	
				0.00750 ± 0.00060	298	FP-RF	Tully and Ravishankara[28]	298–1020
				0.0473 ± 0.0045	398			
				0.081 ± 0.011	448			
				0.145 ± 0.012	511			
				0.167 ± 0.006	529			
				0.314 ± 0.040	600			
				0.275 ± 0.044	619			
				0.578 ± 0.058	696			
				0.84 ± 0.15	772			
				1.50 ± 0.15	915			
	1.32×10^{-5}	1.92	1355	2.00 ± 0.20	1020			
				0.00766 ± 0.00064	300	FP-RF	Husain et al.[29]	
				1.25 ± 0.45	830 ± 50	LH-LIF	Fairchild et al.[30]	830–1400
				1.3 ± 0.4	1030 ± 50			
				4.3 ± 1.0	1400 ± 50			
				0.00557 ± 0.00054	269	DF-RF	Jeong and Kaufman[31,32]	269–473
				0.00789 ± 0.00049	297			
				0.0178 ± 0.0012	339			
				0.0347 ± 0.0023	389			
				0.0549 ± 0.0035	419			
	1.28×10^{-12}	4.23	453 ± 775	0.102 ± 0.007	473			
	5.26 ± 0.88		1917 ± 60					
				0.0392 ± 0.0033	413	RR [relative to k(CO) $= 1.12 \times 10^{-13}$ e$^{0.000907T}$][a]	Baulch et al.[33]	413–693
				0.0555 ± 0.0033	417			
				0.0369 ± 0.0030	422			
				0.0654 ± 0.0066	443			
				0.0792 ± 0.0073	471			
				0.0981 ± 0.0063	505			
				0.103 ± 0.018	517			
				0.0936 ± 0.0179	521			
				0.112 ± 0.006	546			
				0.165 ± 0.004	553			
				0.267 ± 0.015	603			
				0.349 ± 0.020	663			
				0.589 ± 0.060	693			
				0.00650	298	PR-RA	Jonah et al.[34]	298–1229
				0.00846	298			
				0.0189	348			
				0.0351	373			
				0.106	398			
				0.0938	415			

TABLE 1. Rate constants k and temperature-dependent parameters for the gas-phase reactions of the OH radical with alkanes — Continued

Alkane	$10^{12} \times A$ (cm^3 molecule^{-1} s^{-1})	n	B (K)	$10^{12} \times k$ (cm^3 molecule^{-1} s^{-1})	at T (K)	Technique	Reference	Temperature range covered (K)
				0.175	424			
				0.124	450			
				0.209	483			
				0.174	483			
				0.336	543			
				0.584	571			
				0.545	613			
				0.760	667			
				0.893	709			
				0.991	712			
				1.04	769			
				1.52	858			
				2.48	873			
				2.27	974			
				2.01	974			
				4.50	1071			
				3.22	1125			
				4.19	1125			
				4.93	1229			
				0.0085 ± 0.0006	298 ± 3	FP-RF	Madronich and Felder[35]	298–1512
				0.0228 ± 0.0043	362 ± 10			
				0.0463 ± 0.0034	407 ± 5			
				0.0629 ± 0.009	410 ± 14			
				0.154 ± 0.014	510 ± 10			
				0.177 ± 0.017	525 ± 10			
				0.202 ± 0.010	546 ± 5			
				0.439 ± 0.038	626 ± 16			
				0.478 ± 0.07	698 ± 22			
				1.48 ± 0.08	900 ± 12			
				2.12 ± 0.23	967 ± 35			
				2.16 ± 0.11	1005 ± 15			
				2.72 ± 0.15	1103 ± 17			
				3.34 ± 0.20	1164 ± 17			
				2.41 ± 0.22	1174 ± 22			
				3.18 ± 0.17	1176 ± 17			
				3.89 ± 0.25	1196 ± 17			
				4.26 ± 0.39	1196 ± 37			
				3.77 ± 0.32	1238 ± 18			
				3.68 ± 0.23	1244 ± 17			
				3.58 ± 0.27	1261 ± 23			
				3.80 ± 0.37	1261 ± 18			
				4.74 ± 0.20	1300 ± 18			
				4.20 ± 0.21	1307 ± 18			
				4.84 ± 0.50	1314 ± 23			
				5.32 ± 0.31	1345 ± 18			
				5.61 ± 0.34	1365 ± 19			
				6.44 ± 0.53	1396 ± 19			
				5.98 ± 0.69	1455 ± 20			
				6.52 ± 1.15	1510 ± 20			
	$2.6^{+9.7}_{-2.1} \times 10^{-5}$	1.83	1396 ± 134	6.74 ± 0.35	1512 ± 20			
				3	1220	SH-RA	Cohen and Bott[36]	
				1.25 ± 0.6	830	LH-LIF	Smith et al.[37]	830–1412
				2.25 ± 1.0	870			
				1.55 ± 0.7	930			
				2.0 ± 1.0	966			
				2.1 ± 1.2	975			
				1.33 ± 0.5	1030			
				3.6 ± 0.9	1120			
				1.7 ± 0.7	1150			
				2.35 ± 0.7	1176			
				3.3 ± 1.0	1200 ·			

TABLE 1. Rate constants k and temperature-dependent parameters for the gas-phase reactions of the OH radical with alkanes — Continued

Alkane	$10^{12} \times A$ (cm³ molecule⁻¹ s⁻¹)	n	B (K)	$10^{12} \times k$ (cm³ molecule⁻¹ s⁻¹)	at T (K)	Technique	Reference	Temperature range covered (K)
				2.9 ± 0.35	1240			
				4.4 ± 0.8	1400			
				4.2 ± 0.8	1412			
Methane-^{13}C				0.00833	c	RR [relative to $k(CH_4)$ $= 8.36 \times 10^{-15}$]^d	Rust and Stevens[38]	
				0.00810 ± 0.00006	297 ± 3	RR [relative to $k(CH_4)$ $= 8.18 \times 10^{-15}$]^d	Davidson et al.[39]	
Methane-d_1				0.0365 ± 0.0017	416	PR-RA	Gordon and Mulac[20]	
Methane-d_2				0.0299 ± 0.0017	416	PR-RA	Gordon and Mulac[20]	
Methane-d_3				0.0111 ± 0.0005	416	PR-RA	Gordon and Mulac[20]	
Methane-d_4				0.0050 ± 0.0002	416	PR-RA	Gordon and Mulac[20]	
Ethane				14.0	813	RR [relative to $k(H_2)$ $= 1.12 \times 10^{-12}$]^a	Baldwin and Simmons[40]	
				11.9	1420	RR [relative to $k(CO)$ $= 1.12 \times 10^{-13}e^{0.0009077}$]^a	Fenimore and Jones[41]	1420–1610
				13.4	1440			
				16.3	1600			
				16.3	1600			
				19.3	1610			
				~4.5	1300–1550	RR [relative to $k(CO)$ $= 1.12 \times 10^{-13}e^{0.0009077}$]^a	Westenberg and Fristrom[42]	1300–1550
	210^{+840}_{-170}		1812 ± 302	0.478	298	FP-KS	Horne and Norrish[12]	298–423
				0.292 ± 0.038	302 ± 2	FP-KS	Greiner[43]	
				$\leqslant 1.0$	300	DF-EPR	Wilson and Westenberg[10]	
				6.8	734	RR [relative to $k(CH_4)$ $= 6.95 \times 10^{-18}T^2$ $e^{-1282/T}$]^d	Hoare and Patel[44]	734–798
				7.9	773			
				10.7	798			
				5.26	773	RR [relative to $k(H_2)$ $= 9.22 \times 10^{-13}$]^a	Baldwin et al.,[45] Baldwin and Walker[46]	
				0.310 ± 0.007	297	FP-KS	Greiner[14]	297–493
				0.340 ± 0.010	298			
				0.282 ± 0.007	299			
				0.239 ± 0.013	299			
				0.304 ± 0.035	300			
				0.224 ± 0.042	301			
				0.457 ± 0.010	335			
				0.750 ± 0.050	369			
				0.936 ± 0.058	424			
	$18.6^{+3.3}_{-2.7}$		1232 ± 53	1.55 ± 0.033	493			
				0.664 ± 0.033	381	PR-RA	Gordon and Mulac[20]	381–416
				0.797 ± 0.050	416			
				3.74 ± 1.2	653	RR [relative to $k(CH_4)$ $= 4.16 \times 10^{-13}$]^d	Hucknall et al.[47]	
				0.264 ± 0.017	295 ± 2	FP-RA	Overend et al.[21]	
				0.290 ± 0.060	296	DF-LMR	Howard and Evenson[48]	

TABLE 1. Rate constants k and temperature-dependent parameters for the gas-phase reactions of the OH radical with alkanes — Continued

Alkane	$10^{12} \times A$ (cm³ molecule⁻¹ s⁻¹)	n	B (K)	$10^{12} \times k$ (cm³ molecule⁻¹ s⁻¹)	at T (K)	Technique	Reference	Temperature range covered (K)
				27.8	1300	RR [relative to $k(\mathrm{H_2})$ $= 5.69 \times 10^{-12}$][a]	Bradley et al.[25]	
				0.26 ± 0.04	298	DF-RF	Leu[49]	
				0.112 ± 0.018	250	DF-RF	Anderson and Stephens[50]	250–364
				0.176 ± 0.022	275			
				0.257 ± 0.031	298			
				0.349 ± 0.051	322			
	16.4 ± 2.6		1245 ± 46	0.526 ± 0.080	364			
				0.231 ± 0.040	295 ± 1	DF-RF	Lee and Tang[51]	
				0.080	238	LP-RF	Margitan and Watson[52]	
				0.259 ± 0.021	297	FP-RF	Tully et al.[53]	297–800
				0.771 ± 0.076	400			
				1.58 ± 0.10	499			
				2.61 ± 0.33	609			
				3.65 ± 0.25	697			
	1.43×10^{-2}	1.05	911	5.07 ± 0.34	800			
				0.679 ± 0.048	403	RR [relative to $k(\mathrm{CO})$ $= 1.12 \times 10^{-13}\mathrm{e}^{0.0009077}$][a]	Baulch et al.[33]	403–683
				1.21 ± 0.12	443			
				1.30 ± 0.09	493			
				2.51 ± 0.18	561			
				2.26 ± 0.25	595			
				4.47 ± 0.51	683			
				0.196 ± 0.013	248	DF-RF	Jeong et al.[32]	248–472
				0.228 ± 0.014	273			
				0.310 ± 0.020	294			
				0.306 ± 0.021	298			
				0.426 ± 0.027	333			
				0.403 ± 0.027	333			
				0.538 ± 0.035	375			
				0.529 ± 0.034	375			
				0.799 ± 0.054	428			
				0.770 ± 0.048	429			
				0.993 ± 0.068	464			
	3.87×10^{-9}	3.09	–171 ± 342	1.03 ± 0.067	472			
	6.11 ± 0.60		886 ± 35					
	16.1		1173	0.324	300	PR-RA	Nielsen et al.[54]	~300–400
				0.105 ± 0.004	240	FP-RF	Smith et al.[55]	240–295
				0.137 ± 0.006	251			
				0.205 ± 0.009	273			
	18.0 ± 2.5		1240 ± 110	0.263 ± 0.010	295			
				0.275	295	DF-RF	Devolder et al.[56]	
				0.22 ± 0.03	295	LP-LIF	Schmidt et al.[57]	
				0.267 ± 0.040	295 ± 2	DF-RF	Baulch et al.[58]	
				0.239 ± 0.010	292.5	LP-LIF	Tully et al.[59]	293–705
				0.407 ± 0.017	340			
				0.651 ± 0.027	396			
				1.15 ± 0.048	478			
				1.23 ± 0.051	484			
				2.01 ± 0.083	577			
				2.11 ± 0.088	586			
	8.51×10^{-6}	2.06	430	3.48 ± 0.144	705			

TABLE 1. Rate constants k and temperature-dependent parameters for the gas-phase reactions of the OH radical with alkanes — Continued

Alkane	$10^{12} \times A$ (cm³ mole-cule⁻¹ s⁻¹)	n	B (K)	$10^{12} \times k$ (cm³ molecule⁻¹ s⁻¹)	at T (K)	Technique	Reference	Tempera-ture range covered (K)
				0.298 ± 0.021	295	PR-RA	Nielsen et al. [60]	
				0.127 ± 0.008	248	LP-RA	Stachnik et al. [61]	248–297
				0.129 ± 0.009	248			
				0.251 ± 0.006	297			
				0.250 ± 0.006	297			
				0.32 ± 0.06	296	RR [relative to k(pro-pane) = 1.13×10^{-12}]ᵈ	Edney et al. [62]	
				0.277 ± 0.03	296 ± 2	DF-RF	Bourmada et al. [63]	
				0.088 ± 0.013	226	FP-RF	Wallington et al. [64]	226–363
				0.107 ± 0.010	241			
				0.162 ± 0.018	261			
				0.230 ± 0.026	296			
	8.4 ± 3.1		1050 ± 100	0.487 ± 0.055	363			
				0.261 ± 0.013	296	LP-LIF	Zabarnick et al. [65]	
Ethane-d_3 (CH₃CD₃)				0.142 ± 0.007	293	LP-LIF	Tully et al. [59]	293–705
				0.250 ± 0.011	338			
				0.419 ± 0.018	396			
				0.794 ± 0.033	478			
				1.52 ± 0.063	586			
	7.65×10^{-7}	2.38	411	2.65 ± 0.110	705			
Ethane-d_6 (C₂D₆)				0.0523 ± 0.0060	293	LP-LIF	Tully et al. [59]	293–705
				0.105 ± 0.007	339.5			
				0.199 ± 0.010	396			
				0.435 ± 0.020	478			
				0.965 ± 0.041	586			
	2.43×10^{-7}	2.56	663	1.83 ± 0.077	705			
Propane				27.5	793	RR [relative to k(H₂) = 1.02×10^{-12}]ᵃ	Baldwin[66]	
				1.37 ± 0.21	298 ± 1	FP-KS	Greiner[43]	
				8.23	753	RR [relative to k(H₂) = 8.31×10^{-13}]ᵃ	Baker et al.,[67] Baldwin and Walker[46]	
				1.21 ± 0.08	296	FP-KS	Greiner[14]	296–497
				1.26 ± 0.14	298			
				1.19 ± 0.04	298			
				1.01 ± 0.03	299			
				1.10 ± 0.05	299			
				1.30 ± 0.02	299			
				1.30 ± 0.13	299			
				1.44 ± 0.04	335			
				1.91 ± 0.05	375			
				2.19 ± 0.07	423			
				2.92 ± 0.12	497			
				3.19 ± 0.15	497			
				3.15 ± 0.07	497			
				2.97 ± 0.15	497			
	$12.0^{+1.5}_{-1.3}$		679 ± 38	3.39 ± 0.15	497			
				0.83 ± 0.17	300	DF-EPR	Bradley et al. [68]	
				2.1 ± 0.6	298	RR [relative to k(CO) = 1.49×10^{-13}]ᵃ	Gorse and Volman[69]	

TABLE 1. Rate constants k and temperature-dependent parameters for the gas-phase reactions of the OH radical with alkanes   —   Continued

Alkane	$10^{12} \times A$ (cm³ mole-cule^{-1}s^{-1})	n	B (K)	$10^{12} \times k$ (cm³ molecule^{-1}s^{-1})	at T (K)	Technique	Reference	Temperature range covered (K)
				2.16 ± 0.10	381	PR-RA	Gordon and Mulac[20]	381–416
				1.91 ± 0.08	416			
				5.3 ± 0.8	613	RR [relative to k(ethane) = 1.42 $\times 10^{-17}T^2e^{-462/T}$]^d	Hucknall et al.[47]	613–653
				6.5	653			
				2.02 ± 0.11	295 ± 2	FP-RA	Overend et al.[21]	
				1.98 ± 0.08	329 ± 5	MPS	Harker and Burton[70]	
				1.49 ± 0.21	300 ± 1	RR [relative to k(n-butane) = 2.56 $\times 10^{-12}$]^d	Darnall et al.[71]	
				2.0	300	RR [relative to k(ethene) = 8.44 $\times 10^{-12}$]^d	Cox et al.[72]	
				0.686 ± 0.107	253	DF-RF	Anderson and Stephens[50]	253–365
				0.879 ± 0.123	273			
				0.929 ± 0.121	297			
				1.126 ± 0.163	329			
	6.21 ± 2.37		552 ± 113	1.409 ± 0.195	365			
				1.21 ± 0.05	299 ± 2	RR [relative to k(n-butane) = 2.55 $\times 10^{-12}$]^d	Atkinson et al.[73]	
				1.05 ± 0.04	297	FP-RF	Tully et al.[53]	297–690
				1.48 ± 0.06	326			
				2.51 ± 0.20	378			
				3.37 ± 0.23	469			
				4.78 ± 0.34	554			
	1.59×10^{-3}	1.40	428	8.78 ± 0.97	690			
				1.91 ± 0.15	428	RR [relative to k(CO) = $1.12 \times 10^{-13}e^{0.0009077}$]^a	Baulch et al.[33]	428–696
				2.81 ± 0.23	489			
				2.84 ± 0.08	538			
				4.02 ± 0.22	589			
				4.77 ± 0.51	641			
				7.11 ± 0.68	696			
				26.2 ± 6.7	1220 ± 15	SH-RA	Bott and Cohen[74]	
				21.9 ± 6.0	1074	LH-LIF	Smith et al.[37]	
				1.0 ± 0.2	295	LP-LIF	Schmidt et al.[57]	
				1.20 ± 0.18	295 ± 2	DF-RF	Baulch et al.[58]	
				1.10 ± 0.04	293	LP-LIF	Droege and Tully[75]	293–854
				1.52 ± 0.06	342			
				1.61 ± 0.07	351.5			
				2.14 ± 0.09	401			
				2.49 ± 0.10	428			
				3.24 ± 0.13	491			
				3.36 ± 0.14	501.5			
				3.34 ± 0.14	505			
				4.84 ± 0.20	602			
				4.84 ± 0.20	603			
				7.28 ± 0.30	732			
	1.04×10^{-4}	1.72	145	9.31 ± 0.38	854			

TABLE 1. Rate constants k and temperature-dependent parameters for the gas-phase reactions of the OH radical with alkanes — Continued

Alkane	$10^{12} \times A$ (cm³ mole-cule⁻¹ s⁻¹)	n	B (K)	$10^{12} \times k$ (cm³ molecule⁻¹ s⁻¹)	at T (K)	Technique	Reference	Temperature range covered (K)
				1.14 ± 0.15	296	RR [relative to $k(n\text{-butane}) = 2.51 \times 10^{-12}$]^d	Edney et al.[62]	
				1.38	300 ± 3	RR [relative to $k(n\text{-butane}) = 2.56 \times 10^{-12}$]^d	Behnke et al.[76]	
				1.27 ± 0.11	295 ± 2	PR-RA	Nielsen et al.[77]	
				1.27 ± 0.09	300	RR [relative to k(series of organics)]^e	Behnke et al.[78]	
Propane-d_2 (CH₃CD₂CH₃)				0.610 ± 0.028	295	LP-LIF	Droege and Tully[75]	295–854
				0.802 ± 0.034	328.5			
				1.20 ± 0.05	376.5			
				1.72 ± 0.07	437.2			
				2.47 ± 0.10	503.5			
				3.79 ± 0.16	603			
				5.92 ± 0.24	732			
	2.02×10^{-4}	1.63	383	7.86 ± 0.32	854			
Propane-d_3 (CH₃CH₂CD₃)				0.984 ± 0.050	295	LP-LIF	Droege and Tully[75]	295–854
				1.28 ± 0.06	328.5			
				1.62 ± 0.07	376.5			
				2.17 ± 0.09	437.2			
				2.88 ± 0.12	503.5			
				4.19 ± 0.18	603			
				6.20 ± 0.26	728			
	2.26×10^{-5}	1.90	40	8.06 ± 0.34	854			
Propane-d_5 (CH₃CD₂CD₃)				0.478 ± 0.021	295	LP-LIF	Droege and Tully[75]	295–840
				0.621 ± 0.026	328.5			
				0.950 ± 0.040	376.5			
				1.38 ± 0.06	437.2			
				1.96 ± 0.08	503.5			
				3.13 ± 0.13	603			
				4.93 ± 0.20	728			
	2.59×10^{-5}	1.91	303	6.60 ± 0.27	840			
Propane-d_6 (CD₃CH₂CD₃)				0.826 ± 0.040	295	LP-LIF	Droege and Tully[75]	295–840
				0.999 ± 0.045	328.5			
				1.37 ± 0.06	376.5			
				1.79 ± 0.08	437.2			
				2.46 ± 0.10	503.5			
				3.55 ± 0.15	603			
				5.31 ± 0.22	728			
	1.03×10^{-5}	2.00	23	6.78 ± 0.28	840			
Propane-d_8 (C₃D₈)				0.408 ± 0.045	295	LP-LIF	Droege and Tully[75]	295–854
				0.527 ± 0.043	328.5			
				0.746 ± 0.047	376.5			
				1.09 ± 0.06	437.2			
				1.50 ± 0.07	503.5			
				2.55 ± 0.11	603			
				4.25 ± 0.18	732			
	2.36×10^{-7}	2.53	15	5.88 ± 0.25	854			
n-Butane				36.7	793	RR [relative to $k(\text{H}_2)$ $= 1.02 \times 10^{-12}$]^a	Baldwin and Walker[79]	
				11.0	753	RR [relative to $k(\text{H}_2)$ $= 8.31 \times 10^{-13}$]^a	Baker et al.,[67] Baldwin and Walker[46]	

TABLE 1. Rate constants k and temperature-dependent parameters for the gas-phase reactions of the OH radical with alkanes — Continued

Alkane	$10^{12} \times A$ (cm^3 molecule^{-1} s^{-1})	n	B (K)	$10^{12} \times k$ (cm^3 molecule^{-1} s^{-1})	at T (K)	Technique	Reference	Temperature range covered (K)
				2.56 ± 0.08	298	FP-KS	Greiner[14]	298–495
				2.59 ± 0.22	301			
				2.79 ± 0.32	336			
				2.96 ± 0.10	373			
				4.85 ± 0.18	425			
				4.12 ± 0.15	428			
	$14.1^{+4.1}_{-3.1}$		524 ± 93	4.90 ± 0.17	495			
				4.1	298	DF-MS	Morris and Niki[80]	
				2.35 ± 0.35	298	FP-RF	Stuhl[81]	
				2.9 ± 0.7	298	RR [relative to k(CO) $= 1.49 \times 10^{-13}$][a]	Gorse and Volman[69]	
				4.22 ± 0.17	298	PR-RA	Gordon and Mulac[20]	298–416
				4.15 ± 0.17	381			
				4.98 ± 0.17	416			
				9.21 ± 0.78	653	RR [relative to k(propane) $= 5.98 \times 10^{-12}$][d]	Hucknall et al.[47]	
				2.34 ± 0.15	292 ± 2	RR [relative to k(CO) $= 1.58 \times 10^{-13}$][a]	Campbell et al.[82]	
				2.72 ± 0.27	297.7	FP-RF	Perry et al.[83]	298–420
				3.54 ± 0.35	351.0			
	17.6		559 ± 151	4.69 ± 0.47	419.6			
				2.67 ± 0.22	297 ± 2	FP-RA	Paraskevopoulos and Nip[84]	
				2.52 ± 0.25	299 ± 2	RR [relative to k(propene) $= 2.62 \times 10^{-11}$][d]	Atkinson et al.[85]	
				1.46 ± 0.22	250	DF-RF	Anderson and Stephens[50]	250–365
				1.63 ± 0.21	274			
				1.68 ± 0.23	297			
				2.10 ± 0.34	329			
	8.17 ± 4.03		443 ± 143	2.57 ± 0.38	365			
				2.71 ± 0.32	295 ± 1	RR [relative to k(propene) $= 2.68 \times 10^{-11}$][d]	Atkinson and Aschmann[86]	
				2.3 ± 0.3	295	LP-LIF	Schmidt et al.[57]	
				2.42 ± 0.10	294	LP-LIF	Droege and Tully[87]	294–509
				2.95 ± 0.12	332			
				3.53 ± 0.15	377			
				4.56 ± 0.19	439			
	2.34×10^{-5}	1.95	-134	5.84 ± 0.25	509			
				2.70 ± 0.34	300	RR [relative to k(ethene) $= 8.44 \times 10^{-12}$][d]	Barnes et al.[88]	
				2.53 ± 0.04	300	RR [relative to k(n-octane) $= 8.76 \times 10^{-12}$][d]	Behnke et al.[89]	
n-Butane-d_{10}				0.697 ± 0.068	297 ± 2	FP-RA	Paraskevopoulos and Nip[84]	

TABLE 1. Rate constants k and temperature-dependent parameters for the gas-phase reactions of the OH radical with alkanes — Continued

Alkane	$10^{12} \times A$ (cm³ molecule⁻¹s⁻¹)	n	B (K)	$10^{12} \times k$ (cm³ molecule⁻¹s⁻¹)	at T (K)	Technique	Reference	Temperature range covered (K)
				0.893 ± 0.037	294	LP-LIF	Droege and Tully[87]	294–599
				1.13 ± 0.05	332			
				1.49 ± 0.06	377			
				2.07 ± 0.09	439			
				2.87 ± 0.12	509			
	2.92×10^{-6}	2.20	-33	3.98 ± 0.17	599			
2-Methylpropane				20.4	793	RR [relative to $k(H_2)$ $= 1.02 \times 10^{-12}$][a]	Baldwin and Walker[79]	
				2.13 ± 0.12	297 ± 1	FP-KS	Greiner[43]	
				10.5	753	RR [relative to $k(H_2)$ $= 8.31 \times 10^{-13}$][a]	Baker et al.,[67] Baldwin and Walker[46]	
				2.14 ± 0.12	297	FP-KS	Greiner[14]	297–498
				2.22 ± 0.05	297			
				2.67 ± 0.17	298			
				2.56 ± 0.05	304			
				2.69 ± 0.15	305			
				3.01 ± 0.07	338			
				2.87 ± 0.07	371			
				3.04 ± 0.13	374			
				3.57 ± 0.15	425			
	$8.7^{+1.8}_{-1.5}$		387 ± 63	4.25 ± 0.22	498			
				3.5 ± 0.9	298	RR [relative to $k(CO)$ $= 1.49 \times 10^{-13}$][a]	Gorse and Volman[69,90]	
				7.65 ± 0.42	653	RR [relative to k(propane) $= 5.98 \times 10^{-12}$][d]	Hucknall et al.[47]	
				2.2	303	RR [relative to k(cis-2-butene) $= 5.49 \times 10^{-11}$][d]	Wu et al.[91]	
				2.2^f	305	RR [relative to $k(CO)$ $= 1.59 \times 10^{-13}$][a]	Butler et al.[92]	
				2.36 ± 0.05	300 ± 1	RR [relative to k(n-butane) $= 2.56 \times 10^{-12}$][d]	Darnall et al.[71]	
				1.31 ± 0.19	251	DF-RF	Anderson and Stephens[50]	251–360
				1.46 ± 0.19	274			
				1.73 ± 0.25	299			
				1.95 ± 0.25	326			
	7.67 ± 1.12		448 ± 42	2.21 ± 0.39	360			
				2.70 ± 0.20	267	DF-RF	Trevor et al.[93]	267–324
				3.6	298			
				3.62 ± 0.40	324			
				2.24 ± 0.06	297 ± 2	RR [relative to k(n-butane) $= 2.53 \times 10^{-12}$][d]	Atkinson et al.[94]	
				1.83 ± 0.34	296	LP-LMR	Böhland et al.[95]	
				1.9 ± 0.3	295	LP-LIF	Schmidt et al.[57]	
				2.19 ± 0.11	293	LP-LIF	Tully et al.[96]	293–864
				2.59 ± 0.13	342			

TABLE 1. Rate constants k and temperature-dependent parameters for the gas-phase reactions of the OH radical with alkanes — Continued

Alkane	$10^{12} \times A$ (cm^3 mole-cule^{-1} s^{-1})	n	B (K)	$10^{12} \times k$ (cm^3 molecule^{-1} s^{-1})	at T (K)	Technique	Reference	Temperature range covered (K)
				3.21 ± 0.16	403			
				3.49 ± 0.17	424			
				4.03 ± 0.20	470			
				4.58 ± 0.23	509.5			
				5.49 ± 0.27	574			
				7.40 ± 0.37	705			
	4.31×10^{-5}	1.80	-175	10.13 ± 0.51	864			
				2.35 ± 0.34	298	RR [relative to $k(n\text{-butane}) = 2.54 \times 10^{-12}$]^d	Edney et al.[62]	
2-Methylpropane-d_9 [(CD$_3$)$_3$CH]				1.70 ± 0.09	293.5	LP-LIF	Tully et al.[96]	294–864
				1.91 ± 0.10	343			
				2.27 ± 0.11	403			
				2.81 ± 0.14	471			
				3.64 ± 0.18	574			
				5.28 ± 0.26	705			
	1.08×10^{-7}	2.57	-569	7.61 ± 0.38	864			
2-Methyl-propane-d_1 [(CH$_3$)$_3$CD]				1.36 ± 0.07	293.5	LP-LIF	Tully et al.[96]	294–864
				1.81 ± 0.09	344			
				2.44 ± 0.12	403			
				3.35 ± 0.17	473			
				4.84 ± 0.24	574			
				7.12 ± 0.36	705			
	1.20×10^{-4}	1.69	85	9.90 ± 0.49	864			
2-Methyl-propane-d_{10} [(CD$_3$)$_3$CD]				0.956 ± 0.067	293.5	LP-LIF	Tully et al.[96]	294–864
				1.21 ± 0.08	340.5			
				1.58 ± 0.09	403			
				2.10 ± 0.12	473			
				3.09 ± 0.15	574			
				4.92 ± 0.25	705			
	9.12×10^{-8}	2.63	-352	7.30 ± 0.37	864			
n-Pentane				15.0	753	RR [relative to $k(H_2) = 8.31 \times 10^{-13}$]^a	Baldwin and Walker[46]	
				6.6	303	RR [relative to $k(cis\text{-2-butene}) = 5.49 \times 10^{-11}$]^d	Wu et al.[91]	
				3.51 ± 0.13	300 ± 1	RR [relative to $k(n\text{-butane}) = 2.56 \times 10^{-12}$]^d	Darnall et al.[71]	
				5.3	300	RR [relative to $k(\text{ethene}) = 8.44 \times 10^{-12}$]^d	Cox et al.[72]	
				4.1	300	RR [relative to $k(\text{ethene}) = 8.44 \times 10^{-12}$]^d	Barnes et al.[97]	
				4.08 ± 0.08	299 ± 2	RR [relative to $k(n\text{-butane}) = 2.55 \times 10^{-12}$]^d	Atkinson et al.[73]	
				4.16	300 ± 3	RR [relative to $k(n\text{-butane}) = 2.56 \times 10^{-12}$]^d	Behnke et al.[76]	

TABLE 1. Rate constants k and temperature-dependent parameters for the gas-phase reactions of the OH radical with alkanes — Continued

Alkane	$10^{12} \times A$ (cm^3 mole-cule^{-1} s^{-1})	n	B (K)	$10^{12} \times k$ (cm^3 molecule^{-1} s^{-1})	at T (K)	Technique	Reference	Temperature range covered (K)
				4.27 ± 0.16	312	RR [relative to $k(n\text{-heptane})$ $= 7.48 \times 10^{-12}$]^d	Nolting et al.[98]	
				4.12 ± 0.05	300	RR [relative to $k(n\text{-octane})$ $= 8.76 \times 10^{-12}$]^d	Behnke et al.[89]	
				2.88 ± 0.37	243	RR [relative to $k(2\text{-methylpropane})$ $= 1.04 \times 10^{-17}T^2e^{277/T}$]^d	Harris and Kerr[99]	243–325
				2.95 ± 0.23	263			
				3.40 ± 0.33	273			
				4.05 ± 0.19	298			
				4.34 ± 0.40	314			
				4.77 ± 0.21	325			
				3.58 ± 0.82	247	RR [relative to $k(n\text{-butane})$ $= 1.51 \times 10^{-17}T^2e^{190/T}$]^d	Harris and Kerr[99]	247–327
				3.14 ± 0.37	253			
				3.25 ± 0.28	263			
				3.37 ± 0.21	273			
				3.60 ± 0.28	275			
				3.61 ± 0.22	282			
				4.25 ± 0.15	295			
				4.22 ± 0.37	305			
				4.01 ± 0.36	314			
				4.49 ± 0.12	325			
				4.42 ± 0.35	327			
				4.09 ± 0.08	300	RR [relative to k(series of organics)]^e	Behnke et al.[78]	
2-Methylbutane				2.9 ± 0.6	305 ± 2	RR [relative to $k(n\text{-butane}) = 2.62 \times 10^{-12}$]^d	Lloyd et al.[100]	
				3.54 ± 0.07	300 ± 1	RR [relative to $k(n\text{-butane}) = 2.56 \times 10^{-12}$]^d	Darnall et al.[71]	
				3.7	300	RR [relative to k(ethene) $= 8.44 \times 10^{-12}$]^d	Cox et al.[72]	
				3.90 ± 0.11	297 ± 2	RR [relative to $k(n\text{-butane})$ $= 2.53 \times 10^{-12}$]^d	Atkinson et al.[94]	
2,2-Dimethyl-propane				13.3	753	RR [relative to $k(H_2)$ $= 8.31 \times 10^{-13}$]^a	Baker et al.[67]	
				0.740 ± 0.020	292	FP-KS	Greiner[14]	292–493
				0.858 ± 0.038	292			
				0.875 ± 0.025	298			
				1.16 ± 0.08	335			
				1.41 ± 0.04	370			
				2.11 ± 0.10	424			
	$14.1^{+2.1}_{-1.8}$		844 ± 44	2.54 ± 0.08	493			
				8.48	753	RR [relative to $k(H_2)$ $= 8.31 \times 10^{-13}$]^a	Baker et al.,[101] Baldwin and Walker[46]	
				0.98 ± 0.16	300 ± 1	RR [relative to $k(n\text{-butane})$ $= 2.56 \times 10^{-12}$]^d	Darnall et al.[71]	

TABLE 1. Rate constants k and temperature-dependent parameters for the gas-phase reactions of the OH radical with alkanes — Continued

Alkane	$10^{12} \times A$ (cm^3 molecule^{-1} s^{-1})	n	B (K)	$10^{12} \times k$ (cm^3 molecule^{-1} s^{-1})	at T (K)	Technique	Reference	Temperature range covered (K)
				0.91 ± 0.10	297 ± 2	FP-RA	Paraskevopoulos and Nip[84]	
				0.76 ± 0.05	299 ± 2	RR [relative to $k(n$-butane) $= 2.55 \times 10^{-12}$]^d	Atkinson et al.[102]	
				0.414 ± 0.071	249	DF-RF	Anderson and Stephens[50]	249–364
				0.460 ± 0.089	271			
				0.533 ± 0.098	296			
				0.772 ± 0.153	327			
	6.0 ± 4.1		684 ± 187	0.987 ± 0.231	364			
				0.909 ± 0.115	287	LP-LIF	Tully et al.[59,103]	287–901
				1.27 ± 0.14	350			
				2.08 ± 0.19	431			
				3.17 ± 0.25	518			
				4.46 ± 0.38	600			
				7.02 ± 0.67	705			
				10.1 ± 1.1	812			
	8.60×10^{-9}	3.05	-340	12.5 ± 1.5	901			
				0.67 ± 0.15	300	RR [relative to k(series of organics)]^e	Behnke et al.[78]	
2,2-Dimethyl-propane-d_{12}				0.180 ± 0.012	290	LP-LIF	Tully et al.[59,103]	290–903
				0.375 ± 0.025	352			
				0.728 ± 0.048	430			
				1.30 ± 0.09	508.5			
				2.19 ± 0.17	598			
				3.94 ± 0.34	705			
				5.62 ± 0.55	812			
	1.08×10^{-7}	2.71	307	8.09 ± 0.89	903			
n-Hexane				5.5 ± 1.1	305 ± 2	RR [relative to $k(n$-butane) $= 2.62 \times 10^{-12}$]^d	Lloyd et al.[100]	
				6.0	303	RR [relative to $k(cis$-2-butene) $= 5.49 \times 10^{-11}$]^d	Wu et al.[91]	
				5.8 ± 0.4	292	RR [relative to $k(n$-butane) $= 2.47 \times 10^{-12}$]^d	Campbell et al.[82]	
				5.63 ± 0.09	299 ± 2	RR [relative to $k(n$-butane) $= 2.55 \times 10^{-12}$]^d	Atkinson et al.[102]	
				5.55 ± 0.20	298 ± 2	RR [relative to k(propene) $= 2.63 \times 10^{-11}$]^d	Atkinson et al.[104]	
				5.31 ± 0.46	295 ± 1	RR [relative to k(propene) $= 2.68 \times 10^{-11}$]^d	Atkinson and Aschmann[86]	
				5.58 ± 0.55	295	RR [relative to $k(n$-butane) $= 2.50 \times 10^{-12}$]^d	Klein et al.[105]	

TABLE 1. Rate constants k and temperature-dependent parameters for the gas-phase reactions of the OH radical with alkanes — Continued

Alkane	$10^{12} \times A$ (cm^3 molecule^{-1} s^{-1})	n	B (K)	$10^{12} \times k$ (cm^3 molecule^{-1} s^{-1})	at T (K)	Technique	Reference	Temperature range covered (K)
				6.6	300	RR [relative to k(toluene) $= 5.91 \times 10^{-12}$]^d	Klöpffer et al.[106]	
				5.91 ± 0.68	300	RR [relative to k(ethene) $= 8.44 \times 10^{-12}$]^d	Barnes et al.[88]	
				5.60	300 ± 3	RR [relative to k(n-butane) $= 2.56 \times 10^{-12}$]^d	Behnke et al.[76]	
				6.2 ± 0.6	312	RR [relative to k(n-heptane) $= 7.48 \times 10^{-12}$]^d	Nolting et al.[98]	
				5.66 ± 0.04	300	RR [relative to k(n-octane) $= 8.76 \times 10^{-12}$]^d	Behnke et al.[89]	
2-Methylpentane				4.6 ± 1.0	305 ± 2	RR [relative to k(n-butane) $= 2.62 \times 10^{-12}$]^d	Lloyd et al.[100]	
				5.3	300	RR [relative to k(ethene) $= 8.44 \times 10^{-12}$]^d	Cox et al.[72]	
				5.57 ± 0.23	297 ± 2	RR [relative to k(n-butane) $= 2.53 \times 10^{-12}$]^d	Atkinson et al.[94]	
3-Methylpentane				6.3 ± 1.3	305 ± 2	RR [relative to k(n-butane) $= 2.62 \times 10^{-12}$]^d	Lloyd et al.[100]	
				5.67 ± 0.11	297 ± 2	RR [relative to k(n-butane) $= 2.53 \times 10^{-12}$]^d	Atkinson et al.[94]	
2,2-Dimethyl-butane				2.61 ± 0.08	297 ± 2	RR [relative to k(n-butane) $= 2.53 \times 10^{-12}$]^d	Atkinson et al.[94]	
				1.28 ± 0.26	245	RR [relative to k(n-pentane) $= 2.10 \times 10^{-17} T^2 e^{233/T}$]^d	Harris and Kerr[99]	245–328
				1.47 ± 0.19	247			
				1.44 ± 0.26	253			
				1.69 ± 0.21	263			
				1.93 ± 0.36	273			
				1.89 ± 0.30	283			
				2.22 ± 0.36	299			
				2.29 ± 0.33	303			
				2.44 ± 0.25	303			
				2.62 ± 0.26	313			
				2.95 ± 0.45	326			
				2.84 ± 0.27	328			
				2.32 ± 0.06	300	RR [relative to k(series of organics)]^e	Behnke et al.[78]	
2,3-Dimethyl-butane				7.45 ± 0.22	300	FP-KS	Greiner[14]	300–498
				6.71 ± 0.22	336			
				6.81 ± 0.35	372			
				7.11 ± 0.65	424			

TABLE 1. Rate constants k and temperature-dependent parameters for the gas-phase reactions of the OH radical with alkanes — Continued

Alkane	$10^{12} \times A$ (cm^3 molecule^{-1} s^{-1})	n	B (K)	$10^{12} \times k$ (cm^3 molecule^{-1} s^{-1})	at T (K)	Technique	Reference	Temperature range covered (K)
	$4.8^{+1.1}_{-0.9}$		-129 ± 67	5.94 ± 1.25	498			
				4.9 ± 1.0	305 ± 2	RR [relative to k(2-methylpropene) $= 4.94 \times 10^{-11}$]^d	Darnall et al.[107]	
				5.32 ± 0.28	300 ± 1	RR [relative to k(n-butane) $= 2.56 \times 10^{-12}$]^d	Darnall et al.[71]	
				4.0	300	RR [relative to k(ethene) $= 8.44 \times 10^{-12}$]^d	Cox et al.[72]	
				6.18 ± 0.05	299 ± 2	RR [relative to k(n-butane) $= 2.55 \times 10^{-12}$]^d	Atkinson et al.[102]	
				6.95 ± 1.20	247	RR [relative to k(n-butane) $= 1.51 \times 10^{-17} T^2 e^{190/T}$]^d	Harris and Kerr[99]	247–327
				5.66 ± 1.21	253			
				6.02 ± 0.69	263			
				5.72 ± 0.95	273			
				5.75 ± 1.17	275			
				5.78 ± 0.36	282			
				5.90 ± 0.23	295			
				5.95 ± 0.42	305			
				4.94 ± 0.77	314			
				5.98 ± 0.12	325			
				5.75 ± 0.53	327			
n-Heptane				7.21 ± 0.16	299 ± 2	RR [relative to k(n-butane) $= 2.55 \times 10^{-12}$]^d	Atkinson et al.[73]	
				8.2	300	RR [relative to k(toluene) $= 5.91 \times 10^{-12}$]^d	Klöpffer et al.[106]	
				7.10	300 ± 3	RR [relative to k(n-butane) $= 2.56 \times 10^{-12}$]^d	Behnke et al.[76]	
				7.29 ± 0.08	300	RR [relative to k(n-octane) $= 8.76 \times 10^{-12}$)^d	Behnke et al.[89]	
				7.28 ± 0.08	300	RR [relative to k (series of organics)]^e	Behnke et al.[78]	
2,2-Dimethyl-pentane				3.37 ± 0.03	300	RR [relative to k(series of organics)]^e	Behnke et al.[78]	
2,4-Dimethyl-pentane				5.16 ± 0.11	297 ± 2	RR [relative to k(n-butane) $= 2.53 \times 10^{-12}$]^d	Atkinson et al.[94]	
2,2,3-Trimethyl-butane				5.23 ± 0.12	296	FP-KS	Greiner[14]	296–497
				4.86 ± 0.12	303			
				4.50 ± 0.33	371			
				4.60 ± 0.27	373			
	$7.9^{+2.1}_{-1.6}$		115 ± 73	6.33 ± 0.19	497			

TABLE 1. Rate constants k and temperature-dependent parameters for the gas-phase reactions of the OH radical with alkanes — Continued

Alkane	$10^{12} \times A$ (cm^3 mole-cule^{-1} s^{-1})	n	B (K)	$10^{12} \times k$ (cm^3 molecule^{-1} s^{-1})	at T (K)	Technique	Reference	Temperature range covered (K)
				3.7 ± 0.8	305 ± 2	RR [relative to k(2-methylpropene) $= 4.94 \times 10^{-11}$][d]	Darnall et al.[107]	
				10.1 ± 1.3	753	RR [relative to k(H$_2$) $= 8.31 \times 10^{-13}$][a]	Baldwin et al.[108]	
				4.12 ± 0.08	297 ± 2	RR [relative to k(n-butane) $= 2.53 \times 10^{-12}$][d]	Atkinson et al.[94]	
				3.22 ± 0.48	263	RR [relative to k(n-pentane) $= 2.10 \times 10^{-17} T^2 e^{223/T}$][d]	Harris and Kerr[99]	263–303
				4.33 ± 0.41	283			
				4.18 ± 0.25	303			
				4.02 ± 0.57	303			
				4.40 ± 0.60	243	RR [relative to k(n-hexane) $= 1.35 \times 10^{-11} e^{-262/T}$][d]	Harris and Kerr[99]	243–324
				4.17 ± 0.26	244			
				4.03 ± 0.39	253			
				4.36 ± 0.54	263			
				4.23 ± 0.76	273			
				4.27 ± 0.60	282			
				4.19 ± 0.19	295			
				4.04 ± 0.20	314			
				4.11 ± 0.50	324			
				3.96 ± 0.16	324			
n-Octane				8.42 ± 1.25	296	FP-KS	Greiner[14]	296–497
				12.0 ± 0.7	371			
				10.8 ± 0.5	371			
	$29.5^{+4.4}_{-3.8}$		364 ± 60	14.3 ± 0.4	497			
				8.89 ± 0.18	299 ± 2	RR [relative to k(n-butane) $= 2.55 \times 10^{-12}$][d]	Atkinson et al.[73]	
				8.63	300 ± 3	RR [relative to k(n-butane) $= 2.56 \times 10^{-12}$][d]	Behnke et al.[76]	
				8.8 ± 0.3	312	RR [relative to k(n-heptane) $= 7.48 \times 10^{-12}$][d]	Nolting et al.[98]	
2,2-Dimethyl-hexane				4.83 ± 0.04	300	RR [relative to k(series of organics)][e]	Behnke et al.[78]	
2,2,4-Trimethylpentane				3.90 ± 0.15	298	FP-KS	Greiner[14]	298–493
				3.55 ± 0.12	305			
				4.37 ± 0.23	339			
				5.25 ± 0.15	373			
				5.43 ± 0.13	423			
	$15.5^{+3.1}_{-2.6}$		426 ± 63	6.62 ± 0.42	493			
				3.59 ± 0.16	297 ± 2	RR [relative to k(n-butane) $= 2.53 \times 10^{-12}$][d]	Atkinson et al.[94]	
2,3,4-Trimethylpentane				9.18 ± 0.23	243	RR [relative to k(n-hexane) $= 1.35 \times 10^{-11} e^{-262/T}$][d]	Harris and Kerr[99]	243–313
				9.10 ± 0.96	253			
				7.58 ± 0.45	263			
				7.81 ± 0.42	273			
				6.99 ± 0.23	295			
				6.94 ± 0.12	303			

TABLE 1. Rate constants k and temperature-dependent parameters for the gas-phase reactions of the OH radical with alkanes — Continued

Alkane	$10^{12} \times A$ (cm³ molecule⁻¹ s⁻¹)	n	B (K)	$10^{12} \times k$ (cm³ molecule⁻¹ s⁻¹)	at T (K)	Technique	Reference	Temperature range covered (K)
				6.96 ± 0.59	313			
				7.61 ± 0.18	313			
2,2,3,3-Tetra-methylbutane				1.08 ± 0.02	294	FP-KS	Greiner[14]	294–495
				1.16 ± 0.10	301			
				1.42 ± 0.04	335			
				2.04 ± 0.08	370			
				2.21 ± 0.07	424			
	$16.2^{+3.3}_{-2.7}$		802 ± 63	3.52 ± 0.12	495			
				6.65 ± 0.83	753	RR [relative to $k(H_2)$ = 8.31 × 10⁻¹³][a]	Baldwin et al.,[109] Baldwin and Walker[46]	
				1.04 ± 0.08	297 ± 2	RR [relative to $k(n\text{-butane})$ = 2.53 × 10⁻¹²][d]	Atkinson et al.[94]	
				0.948 ± 0.020	290	LP-LIF	Tully et al.[103]	290–738
				1.48 ± 0.04	348.5			
				2.38 ± 0.03	423.5			
				3.58 ± 0.05	506			
				5.27 ± 0.09	606			
	4.75×10^{-6}	2.20	68	9.36 ± 0.35	737.5			
n-Nonane				10.5 ± 0.4	299 ± 2	RR [relative to $k(n\text{-butane})$ = 2.55 × 10⁻¹²][d]	Atkinson et al.[73]	
				10.4	300 ± 3	RR [relative to $k(n\text{-butane})$ = 2.56 × 10⁻¹²][d]	Behnke et al.[76]	
				10.1 ± 0.3	312	RR [relative to $k(n\text{-heptane})$ = 7.48 × 10⁻¹²][d]	Nolting et al.[98]	
				10.3 ± 0.2	300	RR [relative to $k(n\text{-octane})$ = 8.76 × 10⁻¹²][d]	Behnke et al.[89]	
				10.3 ± 0.2	300	RR [relative to k(series of organics)][e]	Behnke et al.[78]	
2-Methyloctane				10.1 ± 0.12	300	RR [relative to k(series of organics)][e]	Behnke et al.[78]	
4-Methyloctane				9.72 ± 0.12	300	RR [relative to k(series of organics)][e]	Behnke et al.[78]	
2,3,5-Trimethyl-hexane				7.88 ± 0.09	300	RR [relative to k(series of organics)][e]	Behnke et al.[78]	
n-Decane				11.3 ± 0.6	299 ± 2	RR [relative to $k(n\text{-butane})$ = 2.55 × 10⁻¹²][d]	Atkinson et al.[73]	
				11.6 ± 0.4	312	RR [relative to $k(n\text{-heptane})$ = 7.48 × 10⁻¹²][d]	Nolting et al.[98]	
				12.4 ± 0.2	300	RR [relative to $k(n\text{-octane})$ = 8.76 × 10⁻¹²][d]	Behnke et al.[89]	

TABLE 1. Rate constants k and temperature-dependent parameters for the gas-phase reactions of the OH radical with alkanes — Continued

Alkane	$10^{12} \times A$ (cm^3 molecule^{-1} s^{-1})	n	B (K)	$10^{12} \times k$ (cm^3 molecule^{-1} s^{-1})	at T (K)	Technique	Reference	Temperature range covered (K)
				12.4 ± 0.3	300	RR [relative to k(series of organics)][e]	Behnke et al.[78]	
n-Undecane				13.6 ± 0.3	312	RR [relative to k(n-heptane) = 7.48 × 10^{-12}][d]	Nolting et al.[98]	
				13.3 ± 0.2	300	RR [relative to k(n-octane) = 8.76 × 10^{-12}][d]	Behnke et al.[89]	
n-Dodecane				15.0 ± 0.5	312	RR [relative to k(n-heptane) = 7.48 × 10^{-12}][d]	Nolting et al.[98]	
				13.9 ± 0.2	300	RR [relative to k(n-octane) = 8.76 × 10^{-12}][d]	Behnke et al.[89]	
n-Tridecane				17.4 ± 0.6	312	RR [relative to k(n-heptane) = 7.48 × 10^{-12}][d]	Nolting et al.[98]	
				15.4 ± 0.2	300	RR [relative to k(n-octane) = 8.76 × 10^{-12}][d]	Behnke et al.[89]	
n-Tetradecane				19.2 ± 0.7	312	RR [relative to k(n-heptane) = 7.48 × 10^{-12}][d]	Nolting et al.[98]	
n-Pentadecane				22.2 ± 1.0	312	RR [relative to k(n-heptane) = 7.48 × 10^{-12}][d]	Nolting et al.[98]	
n-Hexadecane				24.9 ± 1.3	312	RR [relative to k(n-heptane) = 7.48 × 10^{-12}][d]	Nolting et al.[98]	

[a]See Introduction.
[b]Calculated by least-squares analyses of the cited first-order OH radical decay rates against the CH$_4$ concentration.
[c]Room temperature, not reported.
[d]From present recommendations, see text.
[e]The reference organics and the rate constants (in units of 10^{-12} cm^3 molecule^{-1} s^{-1}) used were: n-butane, 2.55; n-hexane, 5.63; n-octane, 8.79; 2,2,3,3-tetramethylbutane, 1.08; 2,2,4-trimethylpentane, 3.70; and hexafluorobenzene, 0.219.[78]
[f]From the data obtained at 100 Torr total pressure. Rate constants derived from the higher pressure data decrease monotonically with increasing pressure, for unknown reasons.

J. Phys. Chem. Ref. Data, Monograph 1 (1989)

TABLE 2. Rate constants k and temperature-dependent parameters for the gas-phase reactions of the OH radical with cycloalkanes

Alkane	$10^{12} \times A$ (cm³ molecule⁻¹ s⁻¹)	n	B (K)	$10^{12} \times k$ (cm³ molecule⁻¹ s⁻¹)	at T (K)	Technique	Reference	Temperature range covered (K)
Cyclopropane				0.08 ± 0.02	295	FP-RF	Zetzsch[110]	
				0.062 ± 0.014	298 ± 2	FP-RA	Jolly et al.[111]	
Isopropyl-cyclopropane				2.84 ± 0.06	298 ± 2	RR [relative to $k(n\text{-butane})$ $= 2.54 \times 10^{-12}$]^a	Atkinson and Aschmann[112]	
Cyclobutane				1.2 ± 0.3	298	RR [relative to $k(CO)$ $= 1.49 \times 10^{-13}$]^b	Gorse and Volman[69]	
Cyclopentane				6.1	298	RR [relative to $k(CO)$ $= 1.49 \times 10^{-13}$]^b	Volman[113]	
				4.43 ± 0.27	300 ± 1	RR [relative to $k(n\text{-butane})$ $= 2.56 \times 10^{-12}$]^a	Darnall et al.[71]	
				5.26 ± 0.07	299 ± 2	RR [relative to $k(n\text{-butane})$ $= 2.55 \times 10^{-12}$]^a	Atkinson et al.[102]	
				5.18 ± 0.38	298 ± 2	FP-RA	Jolly et al.[111]	
				5.02 ± 0.22	295	LP-LIF	Droege and Tully[114]	295–491
				6.12 ± 0.27	344			
				7.23 ± 0.32	402.5			
	6.04×10^{-4}	1.52	-111	9.45 ± 0.41	491			
Cyclopentane-d_{10}				1.83 ± 0.08	295	LP-LIF	Droege and Tully[114]	295–602
				2.46 ± 0.11	342			
				3.33 ± 0.15	401			
				4.81 ± 0.21	491			
	4.50×10^{-3}	1.21	257	6.75 ± 0.29	602			
Cyclohexane				7.95 ± 0.43	295	FP-KS	Greiner[14]	295–497
				8.40 ± 0.55	338			
				7.70 ± 0.72	338			
				11.8 ± 0.6	370			
				9.93 ± 0.23	373			
				10.4 ± 0.6	425			
				10.1 ± 0.6	425			
	$23.5^{+4.7}_{-4.0}$		319 ± 73	12.4 ± 0.4	497			
				6.7 ± 1.7	298	RR [relative to $k(CO)$ $= 1.49 \times 10^{-13}$]^b	Gorse and Volman[69]	
				6.6	303	RR [relative to $k(cis\text{-2-butene})$ $= 5.49 \times 10^{-11}$]^a	Wu et al.[91]	
				7.48 ± 0.05	299 ± 2	RR [relative to $k(n\text{-butane})$ $= 2.55 \times 10^{-12}$]^a	Atkinson et al.[102]	
				7.43 ± 0.26	299 ± 2			
				7.07 ± 0.42	299 ± 2	RR [relative to $k(propene)$ $= 2.62 \times 10^{-11}$]^a	Atkinson et al.[115]	
				7.38 ± 0.11	300 ± 3	RR [relative to $k(n\text{-butane})$ $= 2.56 \times 10^{-12}$]^a	Tuazon et al.[116]	

TABLE 2. Rate constants k and temperature-dependent parameters for the gas-phase reactions of the OH radical with cycloalkanes   —   Continued

Alkane	$10^{12} \times A$ (cm^3 molecule^{-1} s^{-1})	n	B (K)	$10^{12} \times k$ (cm^3 molecule^{-1} s^{-1})	at T (K)	Technique	Reference	Temperature range covered (K)
				5.24 ± 0.36	295	PR-RA	Nielsen et al.[60]	
				6.22 ± 0.45	298	RR [relative to k(n-butane) $= 2.54 \times 10^{-12}$]^a	Edney et al.[62]	
				8.6 ± 0.8	296 ± 2	DF-RF	Bourmada et al.[63]	
				7.14 ± 0.31	292	LP-LIF	Droege and Tully[114]	292–491
				8.49 ± 0.37	342			
				10.1 ± 0.44	401			
	1.09×10^{-3}	1.47	−125	12.9 ± 0.56	491			
Cyclohexane-d_{12}				2.76 ± 0.12	292	LP-LIF	Droege and Tully[114]	292–603
				3.64 ± 0.16	342			
				4.83 ± 0.21	401			
				6.94 ± 0.30	491			
	3.48×10^{-4}	1.62	56	9.78 ± 0.42	603			
Cycloheptane				13.1 ± 2.1	298 ± 2	FP-RA	Jolly et al.[111]	
				11.8 ± 0.2	300	RR [relative to k(series of organics)]^c	Behnke et al.[78]	
Methylcyclo-hexane				10.4 ± 0.3	297 ± 2	RR [relative to k(n-butane) $= 2.53 \times 10^{-12}$]^a	Atkinson et al.[94]	
Cyclooctane				13.7 ± 0.3	300	RR [relative to k(series of organics)]^c	Behnke et al.[78]	
1,1,3-Trimethyl-cyclohexane				8.73 ± 0.09	300	RR [relative to k(series of organics)]^c	Behnke et al.[78]	
Bicyclo[2.2.1]-heptane				5.49 ± 0.14	299 ± 2	RR [relative to k(cyclohexane) $= 7.51 \times 10^{-12}$]^a	Atkinson et al.[117]	
Bicyclo[2.2.2]-octane				14.7 ± 1.0	299 ± 2	RR [relative to k(cyclohexane) $= 7.51 \times 10^{-12}$]^a	Atkinson et al.[117]	
Bicyclo[3.3.0]-octane				11.0 ± 0.6	299 ± 2	RR [relative to k(cyclohexane) $= 7.51 \times 10^{-12}$]^a	Atkinson et al.[117]	
cis-Bicyclo[4.3.0]-nonane				17.2 ± 1.3	299 ± 2	RR [relative to k(cyclohexane) $= 7.51 \times 10^{-12}$]^a	Atkinson et al.[117]	
trans-Bicyclo[4.3.0]-nonane				17.6 ± 1.3	299 ± 2	RR [relative to k(cyclohexane) $= 7.51 \times 10^{-12}$]^a	Atkinson et al.[117]	
cis-Bicyclo[4.4.0]-decane				19.9 ± 1.4	299 ± 2	RR [relative to k(cyclohexane) $= 7.51 \times 10^{-12}$]^a	Atkinson et al.[117]	
trans-Bicyclo[4.4.0]-decane				20.4 ± 1.3	299 ± 2	RR [relative to k(cyclohexane) $= 7.51 \times 10^{-12}$]^a	Atkinson et al.[117]	

TABLE 2. Rate constants k and temperature-dependent parameters for the gas-phase reactions of the OH radical with cycloalkanes

Alkane	$10^{12} \times A$ (cm^3 molecule^{-1} s^{-1})	n	B (K)	$10^{12} \times k$ (cm^3 molecule^{-1} s^{-1})	at T (K)	Technique	Reference	Temperature range covered (K)
Tricyclo-[5.2.1.0^{2,6}]-decane[d]				11.3 ± 0.4	299 ± 2	RR [relative to k(cyclohexane) = 7.51 × 10^{-12}][a]	Atkinson et al.[117]	
Tricyclo-[3.3.1.1^{3,7}]-decane[e]				23.1 ± 2.1	299 ± 2	RR [relative to k(cyclohexane) = 7.51 × 10^{-12}][a]	Atkinson et al.[117]	
				22.1 ± 0.3	300	RR [relative to k(series of organics)][c]	Behnke et al.[78]	

[a]From present recommendations (see text).
[b]See Introduction.
[c]The reference organics and the rate constants (in units of 10^{-12} cm^3 molecule^{-1} s^{-1}) used were: n-butane, 2.55; n-hexane, 5.63; n-octane, 8.79; 2,2,3,3-tetramethylbutane, 1.08; 2,2,4-trimethylpentane, 3.70; and hexafluorobenzene, 0.219.[78]

d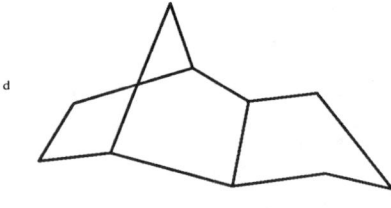

e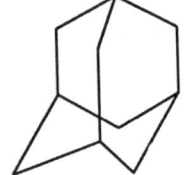

TABLE 3. Rate constants k for the gas-phase reactions of the OD radical with alkanes

Alkane	$10^{12} \times k$ (cm^3 molecule^{-1} s^{-1})	at T (K)	Technique	Reference
Methane	0.0080 ± 0.0003	300	FP-KS	Greiner[118]
Ethane	0.274 ± 0.027	300	FP-KS	Greiner[118]
n-Butane	2.76 ± 0.22	297 ± 2	FP-RA	Paraskevopoulos and Nip[84]
n-Butane-d_{10}	0.804 ± 0.063	297 ± 2	FP-RA	Paraskevopoulos and Nip[84]

The absolute rate data of Wilson and Westenberg,[10] Horne and Norrish,[12] Greiner[14] (which supersedes the earlier room temperature study of Greiner[11]), Peeters and Mahnen,[17] Davis et al.,[18] Margitan et al.,[19] Zellner and Steinert,[23] Ernst et al.,[26] Tully and Ravishankara,[28] Jeong and Kaufman,[31,32] Jonah et al.,[34] Madronich and Felder[35] and Smith et al.[37] (which supersedes the earlier preliminary data reported by Fairchild et al.[30]) are plotted in Fig. 1 (the absolute room temperature rate constants of Overend et al.,[21] Howard and Evenson,[22] Husain et al.[29] and those obtained at elevated temperatures by Dixon-Lewis and Williams,[9] Gordon and Mulac[20] and Cohen and Bott[36] are not included for reasons of clarity), while all of the available rate constants obtained at around 298 K are plotted in Fig. 2.

The rate constants obtained from the relative rate studies of Westenberg and Fristrom,[1] Fenimore and Jones,[2] Fristrom,[4] Blundell et al.,[5] Hoare[6] (which supersedes the earlier study of Hoare[3]), Wilson et al.,[13] Baldwin et al.[15] (which is judged to supersede the earlier study of Baldwin et al.[8]), Simonaitis et al.,[16] Cox et al.,[24] Bradley et al.[25] and Baulch et al.,[33] together with the absolute rate constants determined by Dixon-Lewis and Williams,[9] Gordon and Mulac[20] and Cohen and Bott,[36] are plotted in Fig. 3.

The rate constants obtained by Horne and Norrish[12] for methane (and also ethane, see below) are significantly higher than the other data, probably due to the occurrence of secondary reactions at the high initial OH radical concentrations used.[14] Otherwise, it can be seen from Figs. 1 and 2 that the data obtained from the absolute rate constant studies[9-11,14,17-23,26,28-32,34-37] are in general agreement, although there are certain areas of discrepancy. Thus, the room temperature rate constants (Fig. 2) range over a factor of 1.7, and at temperatures >625 K the rate constants obtained by Zellner and Steinert[23] are up to a factor of ~2 higher than those of Tully and Ravishankara,[28] Madronich and Felder[35] and

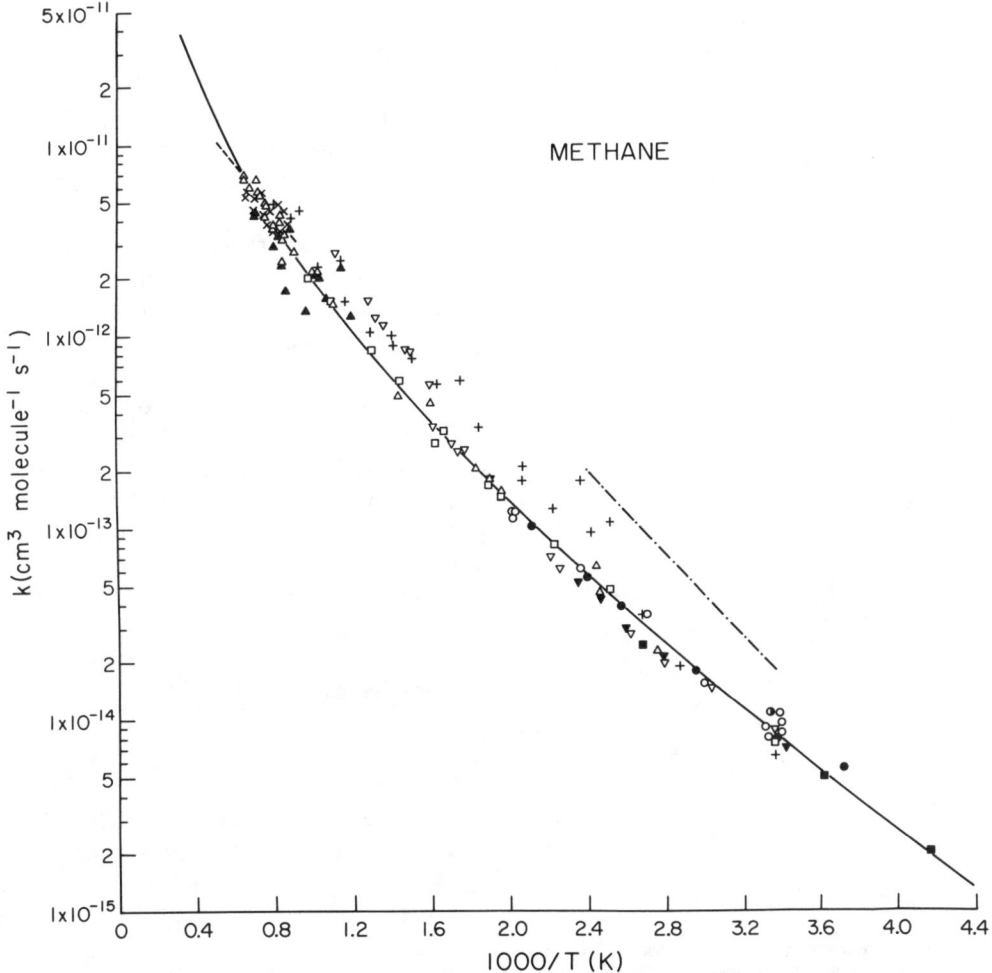

FIG. 1. Arrhenius plot of selected absolute rate constants for the reaction of the OH radical with methane. (◐) Wilson and Westenberg;[10] (–·–·–) Horne and Norrish;[12] (○) Greiner;[14] (– – –) Peeters and Mahnen;[17] (■) Davis et al.;[18] (▼) Margitan et al.;[19] (▽) Zellner and Steinert;[23] (x) Ernst et al.;[26] (□) Tully and Ravishankara;[28] (●) Jeong and Kaufman;[31,32] (+) Jonah et al.;[34] (△) Madronich and Felder;[35] (▲) Smith et al.;[37] (———) recommendation (see text).

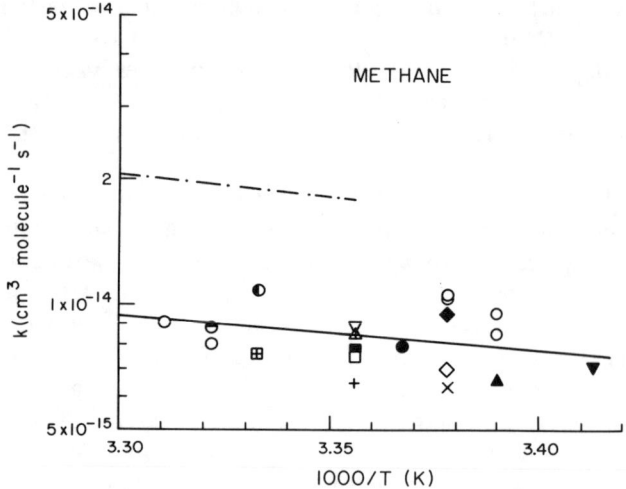

FIG. 2. Arrhenius plot of rate constants for the reaction of the OH radical with methane at around room temperature. (◑) Wilson and Westenberg;[10] (◔) Greiner;[11] (–·–·–) Horne and Norrish;[12] (○) Greiner;[14] (■) Davis et al.;[18] (▼) Margitan et al.;[19] (▲) Overend et al.;[21] (◆) Howard and Evenson;[22] (▽) Zellner and Steinert;[23] (x) Cox et al.;[24] (◇) Sworski et al.;[27] (□) Tully and Ravishankara;[28] (⊞) Husain et al.;[29] (●) Jeong and Kaufman;[31] (+) Jonah et al.;[34] (△) Madronich and Felder;[35] (——) recommendation (see text).

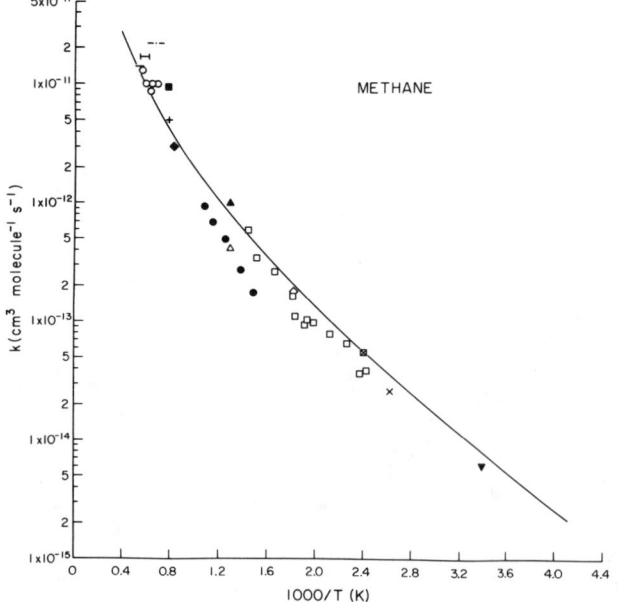

FIG. 3. Arrhenius plot of the relative and selected absolute rate constants for the reaction of the OH radical with methane. (⊢) Westenberg and Fristrom;[1] (○) Fenimore and Jones;[2] (–··–) Fristrom;[4] (△) Blundell et al.;[5] (●) Hoare;[6] (+) Dixon-Lewis and Williams;[9] (—) Wilson et al.;[13] (▲) Baldwin et al.;[15] (◇) Simonaitis et al.;[16] (x) Gordon and Mulac;[20] (▼) Cox et al.;[24] (■) Bradley et al.;[25] (□) Baulch et al.;[33] (◆) Cohen and Bott;[36] (——) recommendation (see text).

Smith et al.[37] (although the rate constants of Smith et al.[37] exhibit a relatively high degree of scatter). These higher rate constants of Zellner and Steinert[23] at temperatures >625 K are also probably due to the occurrence of secondary reactions, as discussed by Tully and Ravishankara.[28]

Furthermore, the kinetic data of Jonah et al.,[34] obtained using a pulsed radiolysis-resonance absorption technique, yield a linear Arrhenius plot over the entire temperature range studied (298–1229 K). These rate constants,[34] while in agreement with other literature data at ~300–380 K and >600 K, are significantly higher in the intermediate temperature range of ~400–600 K. Unfortunately, the reasons for these discrepancies are not presently known.

The remaining absolute rate constant data[9–11,14,17–22,26,28–32,35–37] are in good agreement, and it is apparent from these data that the Arrhenius plot of ln k vs T^{-1} exhibits a significant degree of curvature (Fig. 1). The rate constants obtained by Gordon and Mulac[20] at 381 and 416 K, while in good agreement with the other literature data for methane, exhibit significant differences from the literature rate constants for certain of the other alkanes and alkenes studied (for example, for n-butane and propene) and are hence not used in the evaluation of the rate expression. The datum of Wilson and Westenberg[10] has also been excluded from the evaluation since a stoichiometric factor was necessary to derive the rate constant for the elementary reaction from the measured rate coefficient.[10] Peeters and Mahnen[17] cited only an Arrhenius expression, not tabulating the individual rate constants, and hence their data could not be used in deriving the recommended rate expression.

Thus, the kinetic data of Dixon-Lewis and Williams,[9] Greiner,[14] Davis et al.,[18] Margitan et al.,[19] Overend et al.,[21] Howard and Evenson,[22] Ernst et al.,[26] Tully and Ravishankara,[28] Husain et al.,[29] Jeong and Kaufman,[31,32] Madronich and Felder,[35] Cohen and Bott[36] and Smith et al.[37] have been used to evaluate the rate constant for the reaction of OH radicals with methane. A unit-weighted least-squares analysis of these data,[9,14,18,19,21,22,26,28,29,31,32,35–37] using the expression $k = CT^2 e^{-D/T}$, yields the recommendation of

$$k(\text{methane}) = (6.95^{+0.45}_{-0.41})$$

$$\times 10^{-18} \, T^2 \, e^{-(1282 \pm 32)/T} \, \text{cm}^3 \, \text{molecule}^{-1} \, \text{s}^{-1}$$

over the temperature range 240–1512 K, where the error limits are two least-squares standard deviations, and

$$k(\text{methane}) = 8.36 \times 10^{-15} \, \text{cm}^3 \, \text{molecule}^{-1} \, \text{s}^{-1} \text{ at 298 K},$$

with an estimated overall uncertainty at 298 K of ±20%.

This recommendation is almost identical with that of

$$k(\text{methane}) = 6.95$$

$$\times 10^{-18} \, T^2 \, e^{-1280/T} \, \text{cm}^3 \, \text{molecule}^{-1} \, \text{s}^{-1}$$

derived by Atkinson[120] from an evaluation of the absolute rate constant data over the more restricted temperature range $\leqslant 1250$ K, and is $\sim 10\%$ higher than that of Baulch et al.[121] of

$$k(\text{methane}) = 2.5$$

$$\times\ 10^{-18}\ T^{2.13}\ e^{-1230/T}\ \text{cm}^3\ \text{molecule}^{-1}\ \text{s}^{-1}$$

over this temperature range of 240–1500 K.

As shown in Fig. 3, the rate constants derived from the relative rate studies of Westenberg and Fristrom,[1] Fenimore and Jones,[2] Wilson et al.,[13] Baldwin et al.,[15] Cox et al.[24] and Baulch et al.[33] are in reasonably good agreement with the recommended rate constant expression. This good agreement allows the related studies of these groups to be used in the evaluations of the rate constant data for those organic compounds for which less accurate absolute rate data are available or for which absolute rate data are available only for a restricted temperature range around 298 K. However, the rate constants obtained by Fristrom,[4] Hoare and co-workers,[3,6,7] Blundell et al.,[5] Simonaitis et al.[16] and Bradley et al.[25] show significant discrepancies with the present recommendation, and related studies by these groups are hence given less weight in the evaluations of the rate data for other organic compounds in this article.

(2) Methane-^{13}C, Methane-d_1, Methane-d_2 and Methane-d_3

The limited data available (Table 1) show that there is no significant isotope effect for the reaction of OH radicals with $^{13}CH_4$, when compared to $^{12}CH_4$.[38,39] The magnitude of this isotope effect is of importance with regards to the enrichment of atmospheric methane in ^{13}C relative to its sources, and the most recent determination of Davidson et al.[39] of

$$k(\text{methane-}^{12}\text{C})/k(\text{methane-}^{13}\text{C})$$

$$= 1.010 \pm 0.007 \text{ at } 297 \pm 3 \text{ K}$$

is recommended.

As expected because of the increased bond dissociation energy for C−D bonds versus C−H bonds, the rate constants for the reaction of OH radicals with methane and the deuterated methanes are observed to decrease monotonically along the series $CH_4 > CH_3D > CH_2D_2 > CHD_3 > CD_4$, by a factor of ~ 1.8 per C−D vs. C−H bond at 416 K.[20]

(3) Ethane

The available literature rate constants are given in Table 1, and are plotted in Arrhenius form in Figs. 4 and 5. As for methane, the rate constants obtained by Horne

and Norrish[12] are significantly higher than the more recent absolute rate data, presumably due to the occurrence of secondary reactions at the high initial OH radical concentrations used.[14] The absolute rate constant studies of Greiner,[14] Gordon and Mulac,[20] Overend et al.,[21] Howard and Evenson,[48] Leu,[49] Anderson and Stephens[50] (but see below), Lee and Tang,[51] Martigan and Watson,[52] Tully et al.,[53,59] Jeong et al.[32] (at temperatures $\geqslant 273$ K, see below), Nielsen et al.,[54,60] Smith et al.,[55] Devolder et al.,[56] Schmidt et al.,[57] Baulch et al.,[58] Stachnik et al.,[61] Bourmada et al.,[63] Wallington et al.[64] and Zabarnick et al.[65] are in generally good agreement. Many of these rate constant determinations were carried out to assess the reliabilities of experimental systems for the determination of OH radical reaction rate constants for other reactant molecules.[49,51,52,55,56,61,64,65]

However, somewhat disturbing is the marked disagreement at temperatures $\lesssim 250$ K between the rate constant determined by Jeong et al.[32] and those of Margitan and Watson,[52] Anderson and Stephens[50] (though it should be noted that for n-butane, 2-methylpropane, 2,2-dimethylpropane and, to a lesser extent, propane, the rate constants determined by Anderson and Stephens[50] are significantly lower than other literature data), Smith et al.,[55] Stachnik et al.[61] and Wallington et al.[64] This discrepancy at low temperatures (i.e., $\lesssim 270$ K) may suggest that erroneously high rate constants were measured in this temperature regime for methane, ethane and a series of haloalkanes by Jeong and Kaufman[31] and Jeong et al.,[32] thus leading to an exaggerated curvature in their Arrhenius plots (see also the section below dealing with the reactions of OH radicals with the haloalkanes). Clearly, further experimental data are needed for the reaction of OH radicals with ethane at temperatures $\leqslant 275$ K.

Since the rate constants obtained by Gordon and Mulac,[20] Anderson and Stephens,[50] Lee and Tang,[51] and Nielsen et al.[54,60] for certain other organic reactants do not agree with the recommendations (see below), the data from these studies have not been used in the present evaluation. Furthermore, due to a lack of experimental details, the rate constant of Schmidt et al.[57] has also been omitted from the data set used in the evaluation. Thus, the absolute kinetic data of Greiner,[14] Overend et al.,[21] Howard and Evenson,[48] Leu,[49] Margitan and Watson,[52] Tully et al.,[53,59] Jeong et al.,[32] Smith et al.,[55] Devolder et al.,[56] Baulch et al.,[58] Stachnik et al.,[61] Bourmada et al.,[63] Wallington et al.[64] and Zabarnick et al.[65] have been utilized. A unit-weighted least-squares fit of these data, using the expression $k = CT^2 e^{-D/T}$, yields the recommendation of

$$k(\text{ethane}) = (1.42^{+0.21}_{-0.18})$$

$$\times\ 10^{-17}\ T^2\ e^{-(462 \pm 43)/T} \text{cm}^3\ \text{molecule}^{-1}\ \text{s}^{-1}$$

over the temperature range 226–800 K, where the indicated error limits are two least-squares standard deviations, and

k(ethane) $= 2.68 \times 10^{-13}$ cm^3 molecule^{-1} s^{-1} at 298 K,

with an estimated overall uncertainty at 298 K of $\pm 20\%$. This recommendation is essentially identical to that of Atkinson[120] of

k(ethane) $= 1.37 \times 10^{-17} \, T^2 \, e^{-444/T}$ cm^3 molecule^{-1} s^{-1}

over the temperature range 238–800 K, but is significantly different, especially at $\gtrsim 1000$ K, from that recommended by Baulch et al.[121] of

k(ethane) $= 2.3 \times 10^{-11} \, e^{-1340/T}$ cm^3 molecule^{-1} s^{-1}.

The rate constants obtained by Fenimore and Jones,[41] Westenberg and Fristrom,[42] Hoare and Patel[44] and Bradley et al.[25] from relative rate studies exhibit a significant degree of scatter about the recommended rate expression (Fig. 4), with the rate constant derived from the study of Bradley et al.[25] being significantly dependent on which reference reaction of OH radicals (with H$_2$ or CO) is utilized. However, the relative rate data of Baldwin et al.,[45,46] Hucknall et al.,[47] Baulch et al.[33] and Edney et al.[62] are in good agreement with the recommendation, suggesting that the related studies of Baldwin and co-workers[46] and Edney et al.[62] for more complex organic compounds can be used with some confidence in the rate constant evaluations.

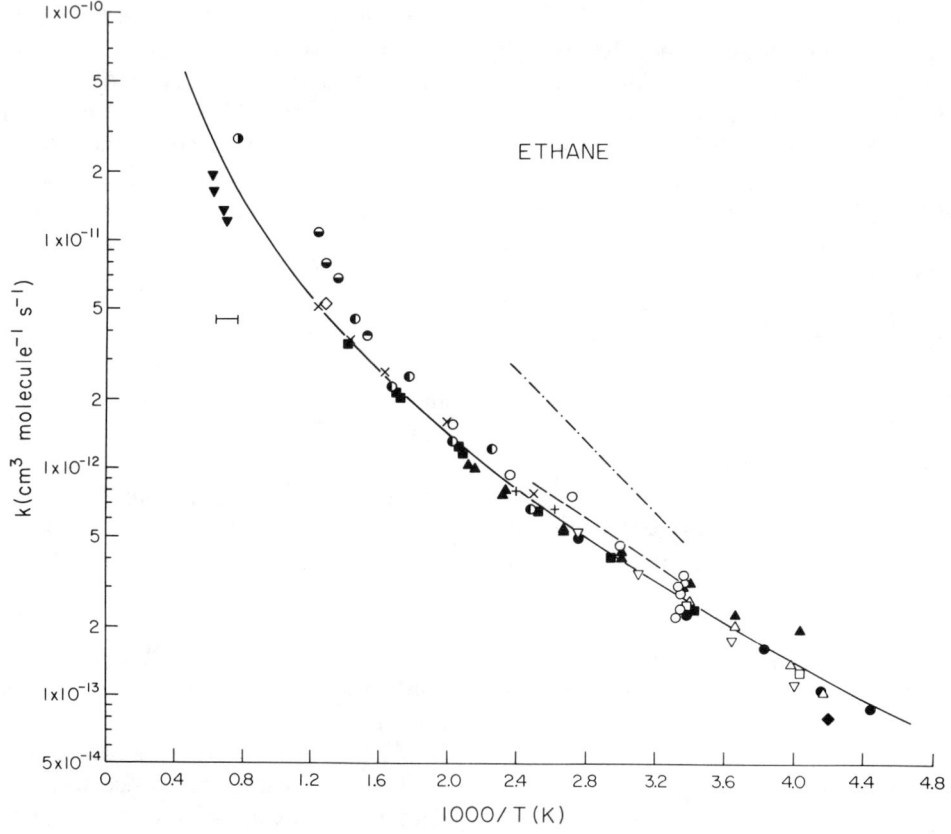

FIG. 4. Arrhenius plot of rate constants for the reaction of the OH radical with ethane. ($\blacktriangledown$) Fenimore and Jones;[41] ($\vdash$) Westenberg and Fristrom;[42] ($-\cdot-$) Horne and Norrish;[12] ($\ominus$) Hoare and Patel;[44] ($\Diamond$) Baldwin et al.;[45,46] ($\bigcirc$) Greiner;[14] ($+$) Gordon and Mulac;[20] ($\ominus$) Hucknall et al.;[47] ($\oplus$) Bradley et al.;[25] ($\triangledown$) Anderson and Stephens;[50] ($\blacklozenge$) Margitan and Watson;[52] (x) Tully et al.;[53] ($\oplus$) Baulch et al.;[33] ($\blacktriangle$) Jeong et al.;[32] ($---$) Nielsen et al.;[54] ($\triangle$) Smith et al.;[55] ($\blacksquare$) Tully et al.;[59] ($\square$) Stachnik et al.;[61] ($\bullet$) Wallington et al.;[64] (———) recommendation (see text).

J. Phys. Chem. Ref. Data, Monograph 1 (1989)

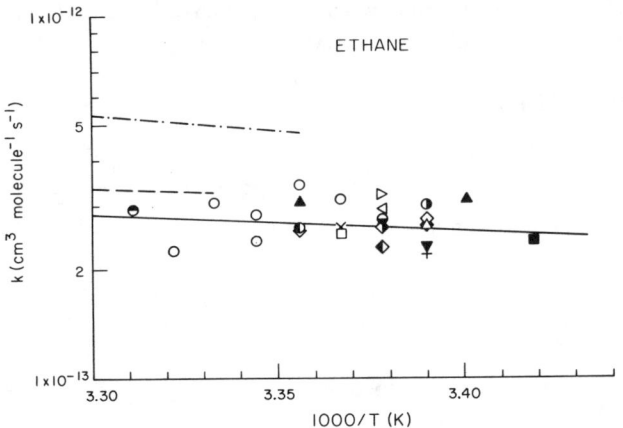

FIG. 5. Arrhenius plot of the rate constants for the reaction of the OH radical with ethane at around room temperature. (– · –) Horne and Norrish;[12] (⬤) Greiner;[43] (◯) Greiner;[14] (◆) Overend et al.,[21] Smith et al.,[55] Baulch et al.;[58] (◁) Howard and Evenson;[48] (◑) Leu;[49] (▽) Anderson and Stephens;[50] (▼) Lee and Tang;[51] (x) Tully et al.;[53] (▲) Jeong et al.;[32] (— — —) Nielsen et al.;[54] (◇) Devolder et al.;[56] (+) Schmidt et al.;[57] (■) Tully et al.;[59] (◉) Nielsen et al.;[60] (□) Stachnik et al.;[61] (▷) Edney et al.;[62] (◒) Bourmada et al.;[63] (◈) Wallington et al.;[64] (◈) Zabarnick et al.;[65] (———) recommendation (see text).

(4) Ethane-d_3 (CH$_3$CD$_3$) and Ethane-d_6

Rate constants for CH_3CD_3 and C_2D_6 (as well as C_2H_6) have been determined by Tully et al.[59] using a laser photolysis-laser induced fluorescence technique (Table 1). From these data, Tully et al.[59] observed that the rate constants for the $-CH_3$ and/or $-CD_3$ groups [$k(-CH_3)$ and $k(-CD_3)$, respectively] could be treated as being independent of the neighboring $-CH_3$ or $-CD_3$ group, and hence that for CH_3CD_3 the rate constant is given to a very good approximation by,

$$k(CH_3CD_3) = k(-CH_3) + k(-CD_3)$$

with a deuterium isotope effect of[59]

$$k(-CH_3)/k(-CD_3) = (1.01 \pm 0.06) \, e^{(456 \pm 26)/T}.$$

(5) Propane

The available kinetic data for propane, propane-d_8 and a series of partially deuterated propanes are given in Table 1, and the rate constants for propane of Baker et al.,[46,67] Greiner,[14] Bradley et al.,[68] Gorse and Volman,[69] Gordon and Mulac,[20] Hucknall et al.,[47] Overend et al.,[21] Harker and Burton,[70] Cox et al.,[72] Anderson and Stephens,[50] Atkinson et al.,[73] Baulch et al.,[33,58] Bott and Cohen,[74] Smith et al.,[37] Schmidt et al.,[57] Droege and Tully[75] and Behnke et al.[76] are plotted in Arrhenius form in Fig. 6. For reasons which are not understood, a signif-

icant amount of scatter in these rate constants is observed (up to a factor of >2 at room temperature). The absolute rate constants determined by Anderson and Stephens[50] at ⩾298 K are consistently lower, by ~20%, than those of Greiner[14] and Droege and Tully.[75] (This most recent study of Droege and Tully[75] supersedes the earlier work of Tully et al.,[53] which is believed to be in error due to a temperature calibration error[75]). Additionally, the rate constants at around room temperature of Bradley et al.,[68] Gorse and Volman,[69] Overend et al.,[21] Harker and Burton[70] and Cox et al.[72] disagree with the remaining absolute rate constant data by significant factors.

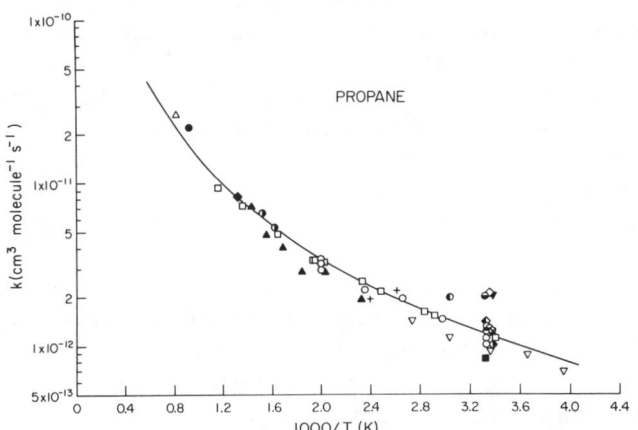

FIG. 6. Arrhenius plot of the rate constants for the reaction of the OH radical with propane. (◆) Baker et al.;[46,67] (◯) Greiner;[14] (■) Bradley et al.;[68] (◇) Gorse and Volman;[69] (+) Gordon and Mulac;[20] (◐) Hucknall et al.;[47] (▼) Overend et al.;[21] (◑) Harker and Burton;[70] (◒) Cox et al.;[72] (▽) Anderson and Stephens;[50] (◓) Atkinson et al.;[73] (▲) Baulch et al.;[33] (△) Bott and Cohen;[74] (●) Smith et al.;[37] (x) Baulch et al.;[58] (◈) Schmidt et al.;[57] (□) Droege and Tully;[75] (◈) Behnke et al.;[76] (———) recommendation (see text).

The absolute rate constants of Greiner,[14] Bott and Cohen,[74] Smith et al.,[37] Baulch et al.[58] and Droege and Tully[75] and the relative rate constants of Baker et al.[46,67] and Atkinson et al.[73] (which supersedes the study of Darnall et al.[71]) are utilized for the evaluation of the overall rate constant for this reaction. Using the expression $k = CT^2 e^{-D/T}$, a unit-weighted least-squares analysis of these data yields the recommendation of

$$k(propane) = (1.50^{+0.17}_{-0.16})$$

$$\times 10^{-17} \, T^2 \, e^{-(44 \pm 42)/T} \, cm^3 \, molecule^{-1} \, s^{-1}$$

over the temperature range 293–1220 K, where the indicated error limits are two least-squares standard deviations, and

k(propane) = 1.15

$$\times\ 10^{-12}\ \text{cm}^3\ \text{molecule}^{-1}\ \text{s}^{-1}\ \text{at}\ 298\ \text{K},$$

with an estimated uncertainty at 298 K of $\pm 30\%$. Over the temperature range $\sim$290–1200 K this recommendation is similar to those of

$$k(\text{propane}) = 1.27 \times 10^{-17}\ T^2\ e^{14/T}\ \text{cm}^3\ \text{molecule}^{-1}\ \text{s}^{-1}$$

recommended by Atkinson[120] and

$$k(\text{propane}) = 1.8 \times 10^{-20}\ T^{2.93}\ e^{390/T}\ \text{cm}^3\ \text{molecule}^{-1}\ \text{s}^{-1}$$

recommended by Baulch et al.[121]

As for methane and ethane, the relative rate constant data of Baulch et al.[33] are in reasonably good agreement with this recommendation. Furthermore, the rate constants derived from the recent relative rate studies of Edney et al.[62] and Behnke et al.[78] [which is relative to a series of organic compounds and cannot be readily reevaluated using the present recommendations (though the rate constants used[78] for the alkane reference compounds are within 2% of the present recommendations)] are in good agreement with the present recommendation.

Knox et al.[122] carried out competitive oxidations of ethane and propane over the temperature range 547–768 K, and determined that the relative disappearance rates of ethane and propane were essentially independent of temperature over this range, with a value of 0.44 $\pm$ 0.03. It is likely that the major loss process for these alkanes in the experimental system used was by reaction with the OH radical,[6] and this relative rate constant of

$$k(\text{ethane})/k(\text{propane}) = 0.44 \pm 0.03$$

is in reasonable agreement with that of 0.44 at 547 K, increasing to 0.55 at 768 K, calculated from the present recommendations for ethane and propane.

Since propane contains non-equivalent C–H bonds, the overall rate constant is the sum of the contributions from the two primary –CH_3 groups and the secondary –CH_2– group, with

$$k = 2k(-CH_3) + k(-CH_2-) = k_{\text{primary}} + k_{\text{secondary}}.$$

Using the absolute rate constants determined for C_3H_8, $CH_3CD_2CH_3$, $CH_3CH_2CD_3$, $CH_3CD_2CD_3$, $CD_3CH_2CD_3$ and C_3D_8, Droege and Tully[75] showed that the –CH_3, –CD_3, –CH_2– and –CD_2– groups could be treated as having group rate constants which were independent of the H/D isotopic nature of the neighboring group(s). Utilizing the deuterium isotope ratio of $k(-CH_3)/k(-CD_3)$ obtained from their related kinetic study of the

OH radical reactions with C_2H_6, CH_3CD_3 and C_2D_6,[59] Droege and Tully[75] obtained

$$2k(-CH_3) = 1.75$$

$$\times\ 10^{-14}\ T^{0.97}\ e^{-798/T}\ \text{cm}^3\ \text{molecule}^{-1}\ \text{s}^{-1},$$

$$k(-CH_2-) = 7.76$$

$$\times\ 10^{-17}\ T^{1.61}\ e^{18/T}\ \text{cm}^3\ \text{molecule}^{-1}\ \text{s}^{-1}$$

and

$$k(-CH_2-)/k(-CD_2-) = (1.13 \pm 0.19)\ e^{(262 \pm 78)/T}$$

Thus, from these data,[75]

$$\frac{k(-CH_2-)}{2k(-CH_3)} = \frac{k_{\text{secondary}}}{k_{\text{primary}}} = 0.00443\ T^{0.64}\ e^{816/T}$$

for propane, and hence $k_{\text{secondary}}/k_{\text{primary}} = 0.91$ at 753 K. This ratio derived from kinetic measurements is in reasonable agreement with that of 1.2 $\pm$ 0.1 obtained by Baker et al.[123] from a product study at 753 K.

Over the temperature range 250–1000 K, this ratio of $k_{\text{secondary}}/k_{\text{primary}}$ for propane of $0.00443\ T^{0.64}\ e^{816/T}$ can be well approximated by the Arrhenius expression of

$$k_{\text{secondary}}/k_{\text{primary}} = 0.39\ e^{560/T},$$

centered at 400 K. This $k_{\text{secondary}}/k_{\text{primary}}$ ratio can then be combined with the recommended expression for the overall rate constant of $(k_{\text{primary}} + k_{\text{secondary}})$ to yield the individual OH radical reaction rate constants for H-atom abstraction from the primary and secondary C–H bonds in propane at any temperature in the range $\sim$290–850 K.

(6) n-Butane and n-Butane-d_{10}

The available kinetic data for n-butane are given in Table 1, and the rate constants of Baker et al.,[46,67] Greiner,[14] Morris and Niki,[80] Stuhl,[81] Gorse and Volman,[69] Gordon and Mulac,[20] Hucknall et al.,[47] Campbell et al.,[82] Perry et al.,[83] Paraskevopoulos and Nip,[84] Atkinson et al.,[85] Anderson and Stephens,[50] Atkinson and Aschmann,[86] Schmidt et al.,[57] Droege and Tully,[87] Barnes et al.[88] and Behnke et al.[89] are plotted in Fig. 7. Unfortunately, the degree of scatter of these reported data is almost a factor of 2.5 at room temperature. The absolute rate constant data of Gordon and Mulac[20] (which also show significant discrepancies with more recent data for propene), Anderson and Stephens[50] (which are also significantly lower than other reported data for 2-methylpropane and 2,2-dimethylpropane), Schmidt et al.[57] and Morris and Niki,[80] together with the relative

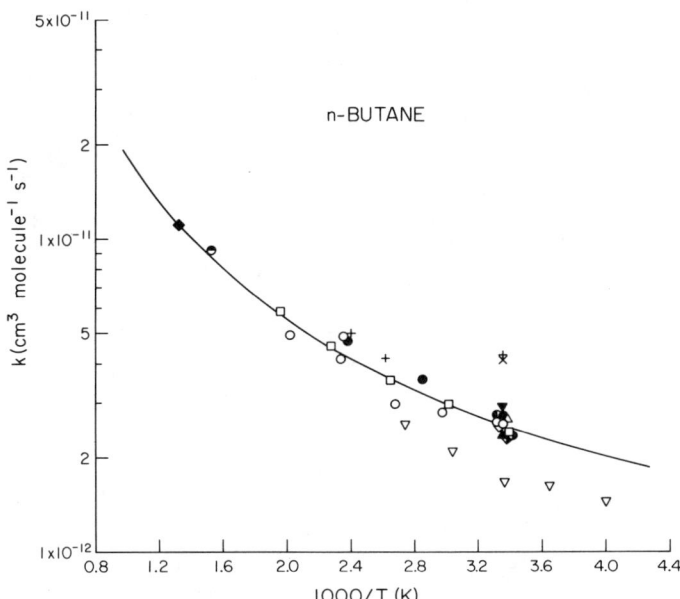

J. Phys. Chem. Ref. Data, Monograph 1 (1989)

FIG. 7. Arrhenius plot of the rate constants for the reaction of the OH radical with n-butane. ($\blacklozenge$) Baker et al.;[46,67] ($\bigcirc$) Greiner;[14] (x) Morris and Niki;[80] ($\blacktriangle$) Stuhl;[81] ($\blacktriangledown$) Gorse and Volman;[69] (+) Gordon and Mulac;[20] ($\ominus$) Hucknall et al.;[47] ($\obot$) Campbell et al.;[82] ($\bullet$) Perry et al.;[83] ($\triangle$) Paraskevopoulos and Nip,[84] Atkinson and Aschmann;[86] ($\diamondsuit$) Atkinson et al.,[85] Behnke et al.;[89] ($\triangledown$) Anderson and Stephens;[50] ($\varobar$) Schmidt et al.;[57] ($\square$) Droege and Tully;[87] ($\obar$) Barnes et al.;[88] (———) recommendation (see text).

rate constants of Gorse and Volman,[69] Hucknall et al.[47] and Campbell et al.,[82] have not been used in the evaluation. Furthermore, due to the availability of apparently reliable absolute rate constant data and the frequent use of n-butane as a reference compound in relative rate studies, the room temperature rate constants of Atkinson et al.,[85] Atkinson and Aschmann,[86] Barnes et al.[88] and Behnke et al.[89] derived from relative rate studies were also not utilized to derive the recommended rate expression.

Thus, a unit-weighted least-squares analysis of the absolute rate constants of Greiner,[14] Stuhl,[81] Perry et al.,[83] Paraskevopoulos and Nip[84] and Droege and Tully[87] and the relative rate constant of Baker et al.,[46,67] using the expression $k = CT^2 e^{-D/T}$, yields the recommendation of

$$k(n\text{-butane}) = (1.51^{+0.30}_{-0.25})$$

$$\times 10^{-17} T^2 e^{(190 \pm 64)/T} \text{ cm}^3 \text{ molecule}^{-1} \text{ s}^{-1}$$

over the temperature range 294–753 K, where the indicated error limits are two least-squares standard deviations, and

$$k(n\text{-butane}) = 2.54$$

$$\times 10^{-12} \text{ cm}^3 \text{ molecule}^{-1} \text{ s}^{-1} \text{ at 298 K,}$$

with an estimated overall uncertainty at 298 K of $\pm 20\%$. This recommendation is almost identical to the three-parameter expression of

$$k(n\text{-butane}) = 1.49 \times 10^{-17} T^2 e^{196/T} \text{ cm}^3 \text{ molecule}^{-1} \text{ s}^{-1}$$

obtained by Atkinson[120] (and is very similar over the temperature range $\sim$290–510 K to the Arrhenius expression of

$$k(n\text{-butane}) = 1.55 \times 10^{-11} e^{-540/T} \text{ cm}^3 \text{ molecule}^{-1} \text{ s}^{-1}$$

recommended by Atkinson[120]) and over the range 298–1000 K agrees to within 40% with that of

$$k(n\text{-butane}) = 1.7 \times 10^{-15} T^{1.3} \text{ cm}^3 \text{ molecule}^{-1} \text{ s}^{-1}$$

recommended by Baulch et al.[121]

The absolute rate constant of Schmidt et al.[57] is in good agreement with the recommendation, as are the rate constants obtained from the relative rate studies of Hucknall et al.,[47] Campbell et al.,[82] Atkinson et al.,[85] Atkinson and Aschmann,[86] Barnes et al.[88] and Behnke et al.[89] The rate constants reported by Morris and Niki,[80] Gordon and Mulac,[20] and Anderson and Stephens[50] exhibit significant discrepancies with the recommended rate expression.

For n-butane-d_{10}, the room temperature rate constant determined by Paraskevopoulos and Nip[84] is $\sim$20% lower than the more recent measurements of Droege and Tully.[87] From their experimental data for n-butane and n-butane-d_{10},[87] the deuterium isotope ratio obtained for ethane[59] and the rate constant ratio for H-atom abstraction from the primary and secondary C—H bonds in n-butane of

$$k_{\text{primary}}/k_{\text{secondary}} = k(-CH_3)/k(-CH_2-) = 1.035 \, e^{-536/T}$$

estimated by Atkinson,[124] Droege and Tully[87] derived the rate constants for H-atom abstraction from the $-CH_3$ and $-CH_2-$ groups in n-butane of

$$k_{\text{primary}} = 2k(-CH_3) = 6.86$$

$$\times 10^{-17} T^{1.73} e^{-379/T} \text{ cm}^3 \text{ molecule}^{-1} \text{ s}^{-1},$$

$$k_{\text{secondary}} = 2k(-CH_2-) = 1.20$$

$$\times 10^{-16} T^{1.64} e^{124/T} \text{ cm}^3 \text{ molecule}^{-1} \text{ s}^{-1}$$

and

$$k(-CH_2-)/k(-CD_2-) = (1.31 \pm 0.12) \, e^{(196 \pm 32)/T}$$

At 753 K, the ratio of $k_{\text{secondary}}/k_{\text{primary}} = 1.88$ calculated from the above expressions for k_{primary} and $k_{\text{secondary}}$ is, as it should be, similar to that of 1.97 estimated by Atkinson,[124] and is in reasonably good agreement with the ratio of $k_{secondary}/k_{primary} = 2.2$ determined by Baker et al.[125] from a product study.

As expected, these individual rate constants for H-atom abstraction from the primary and secondary C—H bonds in n-butane are totally consistent with the ratio of

$$\frac{k_{secondary}}{k_{primary}} = \frac{k(-CH_2-)}{k(-CH_3)} = 0.966\ e^{536/T}$$

estimated by Atkinson[124] for the temperature range 250–1000 K. Use of this expression, together with the recommendation for the overall OH radical reaction rate constant with n-butane, allows the individual rate constants for H-atom abstraction from the $-CH_3$ and $-CH_2-$ groups in n-butane to be calculated.

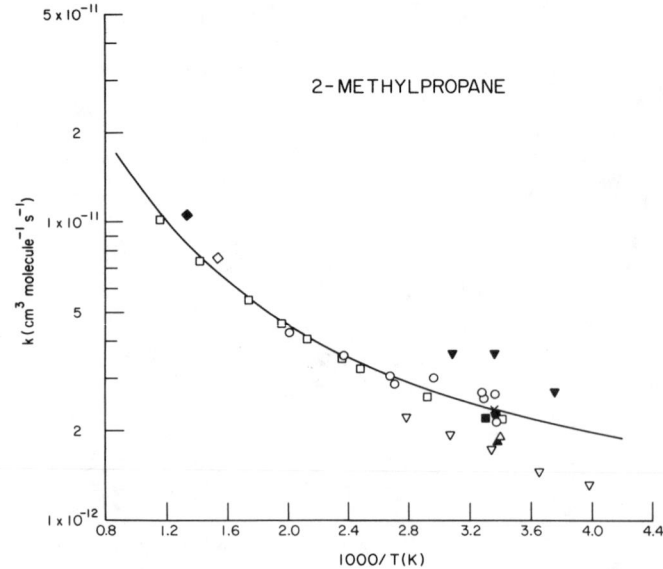

FIG. 8. Arrhenius plot of rate constants for the reaction of the OH radical with 2-methylpropane. ($\blacklozenge$) Baker et al.;[46,67] ($\bigcirc$) Greiner;[14] ($\Diamond$) Hucknall et al.;[47] ($\blacksquare$) Wu et al.;[91] ($\triangledown$) Anderson and Stephens;[50] ($\blacktriangledown$) Trevor et al.;[93] ($\bullet$) Atkinson et al.;[94] ($\blacktriangle$) Böhland et al.;[95] ($\triangle$) Schmidt et al.;[57] ($\square$) Tully et al.;[96] (x) Edney et al.;[62] (———) recommendation (see text).

(7) 2-Methylpropane and 2-Methylpropane-d_1, -d_9 and -d_{10}

The available kinetic data are given in Table 1, and the rate constants of Baker et al.,[46,67] Greiner,[14] Hucknall et al.,[47] Wu et al.,[91] Anderson and Stephens,[50] Trevor et al.,[93] Atkinson et al.,[94] Böhland et al.,[95] Schmidt et al.,[57] Tully et al.[96] and Edney et al.[62] for 2-methylpropane are plotted in Fig. 8. The relative rate constant of Butler et al.[92] is only of an approximate nature, and those of Baldwin and Walker,[79] Greiner,[43] and Darnall et al.[71] have been superseded by the more recent studies of these groups.[14,46,94] Significant discrepancies still exist, however, with the data of Anderson and Stephens[50] being a factor of ~1.5 lower, and those of Trevor et al.[93] being a factor of ~1.5 higher, than the data of Greiner,[14] Atkinson et al.[94] and Tully et al.[96]

In accordance with the criteria used to evaluate the rate constants for methane, ethane, propane and n-butane, the data of Baker et al.,[46,67] Greiner,[14] Atkinson et al.[94] and Tully et al.[96] have been used to derive the recommended rate expression. The Arrhenius plot (Fig. 8) exhibits curvature, and a unit-weighted least-squares fit of these data,[14,46,94,96] using the expression $k = CT^2e^{-D/T}$, yields the recommended expression of

$$k(2\text{-methylpropane}) = (1.04^{+0.15}_{-0.13})$$
$$\times 10^{-17}\ T^2\ e^{(277 \pm 49)/T}\ cm^3\ molecule^{-1}\ s^{-1}$$

over the temperature range 293–864 K, where the indicated error limits are two least-squares standard deviations, and

$$k(2\text{-methylpropane}) = 2.34 \times 10^{-12}\ cm^3\ molecule^{-1}\ s^{-1}$$

at 298 K, with an estimated overall uncertainty at 298 K of $\pm 25\%$. This recommended expression is essentially identical to that of

$$k(2\text{-methylpropane}) = 9.58$$
$$\times 10^{-18}\ T^2\ e^{305/T}\ cm^3\ molecule^{-1}\ s^{-1}$$

recommended by Atkinson[120] over this same temperature range, and agrees well with that of Baulch et al.[121] of

$$k(2\text{-methylpropane}) = 3.2$$
$$\times 10^{-21}\ T^{3.1}\ e^{860/T}\ cm^3\ molecule^{-1}\ s^{-1}$$

over the temperature range 298–1000 K, but diverges rapidly at temperatures below 298 K.

The room temperature absolute rate constants of Böhland et al.,[95] Schmidt et al.[57] and the rate constants derived from the relative rate studies of Hucknall et al.,[47] and Edney et al.[62] are in reasonable[57,95] or good[47,62] agreement with the present recommendation.

From the study of Falconer et al.[126] of the competitive oxidations of a series of alkanes, and assuming that the major loss process for these alkanes was by reaction with the OH radical, a rate constant ratio of

$$k(\text{propane})/k(2\text{-methylpropane}) = 0.67$$

was obtained over the temperature range 583–693 K. This rate constant ratio is in reasonable agreement with that derived from the present recommendations of 0.83 at 583 K, increasing to 0.91 at 693 K.

Tully *et al.*[96] also determined rate constants for the reactions of the OH radical with $(CH_3)_3CD$, $(CD_3)_3CH$ and $(CD_3)_3CD$. As expected, at a given temperature the rate constants for these partially or fully deuterated 2-methylpropanes are significantly lower than that for $(CH_3)_3CH$. These rate constant data were shown to be accurately expressed by

$$k = k_{primary} + k_{tertiary}$$

where $k_{primary}$ and $k_{tertiary}$ are the rate constants for H-atom abstraction from the primary C$-$H or C$-$D bonds $[= 3k(-CH_3)$ or $3k(-CD_3)]$ and the tertiary C$-$H or C$-$D bonds $[= k(>CH-)$ or $k(>CD-)]$, respectively.[96] Values of

$$3k(-CH_3) = 3.81$$
$$\times\ 10^{-16}\ T^{1.53}\ e^{-391/T}\ cm^3\ molecule^{-1}\ s^{-1},$$

$$3k(-CD_3) = 4.13$$
$$\times\ 10^{-20}\ T^{2.79}\ e^{-218/T}\ cm^3\ molecule^{-1}\ s^{-1},$$

$$k(>CH-) = 9.52$$
$$\times\ 10^{-14}\ T^{0.51}\ e^{-32/T}\ cm^3\ molecule^{-1}\ s^{-1}$$

and

$$k(>CD-) = 1.05\times 10^{-15}\ T^{1.16}\ e^{15/T}\ cm^3\ molecule^{-1}\ s^{-1}$$

were obtained.[96] For $(CH_3)_3CH$, a ratio of the rate constants for H-atom abstraction from the primary and tertiary C$-$H bonds of

$$\frac{k_{tertiary}}{k_{primary}} = \frac{k(>CH-)}{3k(-CH_3)} = 250\ T^{-1.02}\ e^{359/T}$$

was obtained, and over the temperature range 250–1000 K this is reasonably well approximated by the Arrhenius expression of

$$k_{tertiary}/k_{primary} = 0.200\ e^{767/T},$$

centered at 400 K. Use of this ratio of $k_{tertiary}/k_{primary}$, together with the recommended overall rate constant expression, allows the individual rate constants for H-atom abstraction from the primary and tertiary C$-$H bonds in 2-methylpropane to be calculated over the temperature range $\sim$290–860 K.

(8) *n*-Pentane

The available rate constant data are given in Table 1, and the rate constants obtained by Baldwin and Walker,[46] Wu *et al.*,[91] Cox *et al.*,[72] Barnes *et al.*,[97] Atkinson *et al.*[73] (which supersedes the earlier study of Darnall *et al.*[71]), Behnke *et al.*,[76,89] Nolting *et al.*[98] and Harris and Kerr[99] are plotted in Fig. 9. All of these rate constants

were obtained from relative rate studies, with those of Wu *et al.*[91] and Cox *et al.*[72] being subject to significant uncertainties. At room temperature the relative rate constants of Barnes *et al.*,[97] Atkinson *et al.*,[73] Behnke *et al.*,[76,78,89] Nolting *et al.*[98] and Harris and Kerr[99] are in excellent agreement. Furthermore, the rate constants derived by Harris and Kerr[99] relative to those for the reactions of OH radicals with *n*-butane and 2-methylpropane are in good agreement, showing that the rate constants for *n*-butane, *n*-pentane and 2-methylpropane are self-consistent.

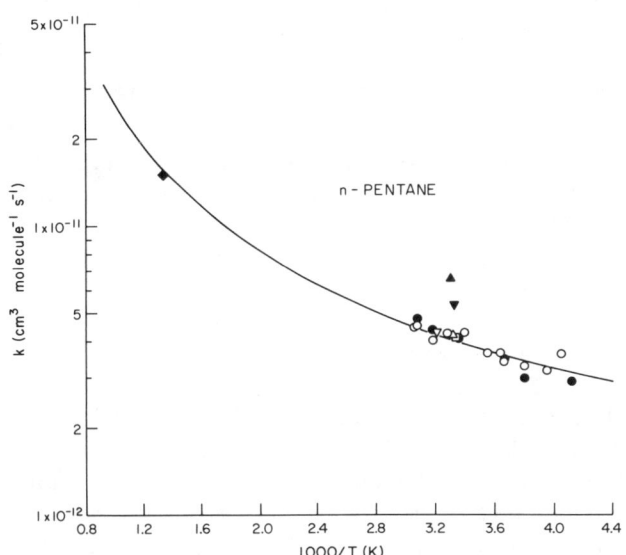

FIG. 9. Arrhenius plot of rate constants for the reaction of the OH radical with *n*-pentane. ($\blacklozenge$) Baldwin and Walker;[46] ($\blacktriangle$) Wu *et al.*;[91] ($\blacktriangledown$) Cox *et al.*;[72] ($\square$) Barnes *et al.*[97] and Atkinson *et al.*;[73] ($\triangle$) Behnke *et al.*;[76,89] ($\triangledown$) Nolting *et al.*;[98] ($\bigcirc$) Harris and Kerr,[99] relative to *n*-butane; ($\bullet$) Harris and Kerr,[99] relative to 2-methylpropane; (———) recommendation (see text).

The rate constants of Baldwin and Walker,[46] Atkinson *et al.*,[73] Behnke *et al.*,[76,89] Nolting *et al.*[98] and Harris and Kerr[99] have been used in the rate constant evaluation. A unit-weighted least-squares analysis of these data, using the equation $k = CT^2 e^{-D/T}$, yields the recommendation of

$$k(n\text{-pentane}) = (2.10^{+0.39}_{-0.33})$$
$$\times\ 10^{-17}\ T^2\ e^{(223\pm49)/T}\ cm^3\ molecule^{-1}\ s^{-1}$$

over the temperature range 243–753 K, where the indicated errors are two least-squares standard deviations, and

$$k(n\text{-pentane}) = 3.94\times 10^{-12}\ cm^3\ molecule^{-1}\ s^{-1}$$

at 298 K, with an estimated overall uncertainty at 298 K of $\pm 25\%$. Over the temperature range 250–1000 K, the rate constants given by this recommended rate expression are in good agreement (within 5%) with those calculated using the estimation method of Atkinson,[124,127] indicating that the above recommendation is self-consistent with those for the other alkanes. This point is discussed in more detail below.

(9) 2-Methylbutane

Rate constants for 2-methylbutane are only available at around room temperature (Table 1), and all are derived from relative rate studies.[71,72,94,100] The agreement is reasonable, and a rate constant of

$$k(\text{2-methylbutane}) = 3.9 \times 10^{-12} \text{ cm}^3 \text{ molecule}^{-1} \text{ s}^{-1}$$

at 298 K is recommended from the most recent study of Atkinson et al.,[94] with an estimated overall uncertainty of $\pm 40\%$.

(10) 2,2-Dimethylpropane and 2,2-Dimethylpropane-d_{12}

The available data are given in Table 1, and those of Greiner,[14] Baker et al.,[46,101] Paraskevopoulos and Nip,[84] Atkinson et al.,[102] Anderson and Stephens,[50] and Tully et al.[59] (which supersede the data reported earlier by Tully et al.[103]) for 2,2-dimethylpropane are plotted in Fig. 10. The rate constant of Darnall et al.[71] has not been included since this work has been superseded by the more recent study of Atkinson et al.[102] using a more reliable and precise technique. Consistent with the data for propane, n-butane and 2-methylpropane, the rate constants obtained by Anderson and Stephens[50] for 2,2-dimethylpropane are ~30% lower than those of Greiner,[14] Paraskevopoulos and Nip,[84] Atkinson et al.,[102] and Tully et al.,[59] all of which are in excellent agreement.

The rate constant for this reaction is evaluated from the data of Greiner,[14] Baldwin and Walker,[46] Paraskevopoulos and Nip,[84] Atkinson et al.[102] and Tully et al.[59] The Arrhenius expression clearly exhibits significant curvature (Fig. 10), and a unit-weighted least-squares analysis of these data,[14,46,59,84,102] using the expression $k = CT^2 e^{-D/T}$, yields the recommendation of

$$k(\text{2,2-dimethylpropane}) = (1.79^{+0.24}_{-0.21})$$

$$\times 10^{-17} T^2 e^{-(187 \pm 47)/T} \text{ cm}^3 \text{ molecule}^{-1} \text{ s}^{-1}$$

over the temperature range 287–901 K, where the indicated errors are two least-squares standard deviations, and

$$k(\text{2,2-dimethylpropane}) = 8.49$$

$$\times 10^{-13} \text{ cm}^3 \text{ molecule}^{-1} \text{ s}^{-1} \text{ at 298 K,}$$

with an estimated overall uncertainty at 298 K of $\pm 20\%$. This recommended expression is virtually identical to that of

$$k(\text{2,2-dimethylpropane}) = 1.75$$

$$\times 10^{-17} T^2 e^{-179/T} \text{ cm}^3 \text{ molecule}^{-1} \text{ s}^{-1}$$

recommended by Atkinson[120] over this same temperature range and is in good agreement (within ~20% over the temperature range 290–750 K) with that of

$$k(\text{2,2-dimethylpropane}) = 7.5$$

$$\times 10^{-18} T^{2.08} e^{-70/T} \text{ cm}^3 \text{ molecule}^{-1} \text{ s}^{-1}$$

recommended by Baulch et al.[121]

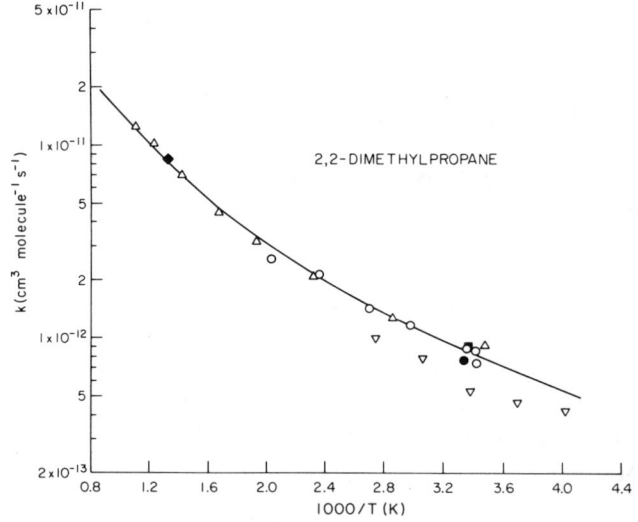

FIG. 10.   Arrhenius plot of rate constants for the reaction of the OH radical with 2,2-dimethylpropane. (◯) Greiner;[14] (◆) Baker et al.;[46,101] (■) Paraskevopoulos and Nip;[84] (●) Atkinson et al.;[102] (▽) Anderson and Stephens;[50] (△) Tully et al.;[59] (———) recommendation (see text).

From the study of Falconer et al.[126] of the competitive oxidations of a series of alkanes, and assuming that the major loss process for these alkanes was by reaction with the OH radical, rate constant ratios of

$$k(\text{ethane})/k(2,2\text{-dimethylpropane}) = 0.14 \, e^{805/T}$$

over the temperature range 601–768 K and

$$k(2,2\text{-dimethylpropane})/k(2\text{-methylpropane})$$

$$= 0.26 \, e^{503/T}$$

over the temperature range 599–765 K were obtained.[126] These rate constant ratios of

$$k(\text{ethane})/k(2,2\text{-dimethylpropane}) = 0.53 \text{ to } 0.40$$

and

$$k(2,2\text{-dimethylpropane})/k(2\text{-methylpropane})$$

$$= 0.50 \text{ to } 0.60$$

over the temperature ranges studied are in reasonably good agreement with the values of 0.50 to 0.55 and 0.79 to 0.94, respectively, calculated from the above rate constant recommendations.

As expected from the higher bond dissociation energy for $C-D$ versus $C-H$ bonds, the rate constants for the reaction of the OH radical with 2,2-dimethylpropane-d_{12} are significantly lower than those for 2,2-dimethylpropane.[59] The deuterium isotope ratio of

$$k(-CH_3)/k(-CD_3) = (0.94 \pm 0.09) \, e^{(472 \pm 46)/T}$$

is very similar to that for ethane.[59]

(11) *n*-Hexane

Rate constants for *n*-hexane are available only at around room temperature (Table 1), and all have been derived from relative rate studies. These room temperature rate constants[76,82,86,88,89,91,98,100,102,104–106] are in reasonably good agreement. The data of Atkinson and co-workers[86,102,104] and Zetzsch and co-workers[76,89,98] have been used to derive the 298 K rate constant, using a temperature dependence of $B = 262$ K (calculated from the estimation method of Atkinson[124,127] for the temperature range 250–333 K) to normalize these rate constants to 298 K. This procedure yields the recommended rate constant of

$$k(n\text{-hexane}) = 5.61 \times 10^{-12} \, cm^3 \, molecule^{-1} \, s^{-1}$$

at 298 K, with an estimated overall uncertainty of $\pm 25\%$.

This rate constant is in good agreement with the recent relative rate data of Klein *et al.*[105] and Barnes *et al.*[88] and with the less precise relative rate data of Lloyd *et al.*[100] and Wu *et al.*[91] Combining this 298 K rate constant

with the temperature dependence calculated from the estimation method of Atkinson[124,127] leads to the Arrhenius expression of

$$k(n\text{-hexane}) = 1.35 \times 10^{-11} \, e^{-262/T} \, cm^3 \, molecule^{-1} \, s^{-1},$$

which is applicable only over the restricted temperature range of ~ 250–335 K. Over a larger temperature range non-Arrhenius behavior is expected, consistent with the rate constants calculated by the estimation methods of Atkinson,[124,127] Walker[128] and Cohen.[129] The above Arrhenius expression has been used in this evaluation to place the relative rate data of Harris and Kerr[99] for 2,2,3-trimethylbutane and 2,3,4-trimethylpentane on an absolute basis.

(12) 2-Methylpentane and 3-Methylpentane

Rate constants for 2- and 3-methylpentane, all derived from relative rate studies,[72,94,100] are available only at room temperature (Table 1). Based upon the data of Atkinson *et al.*,[94] rate constants of

$$k(2\text{-methylpentane}) = 5.6 \times 10^{-12} \, cm^3 \, molecule^{-1} \, s^{-1}$$

and

$$k(3\text{-methylpentane}) = 5.7 \times 10^{-12} \, cm^3 \, molecule^{-1} \, s^{-1}$$

are recommended at 298 K, with estimated overall uncertainty limits of $\pm 30\%$.

(13) 2,2-Dimethylbutane

Rate constants for the reaction of the OH radical with 2,2-dimethylbutane have been obtained by Atkinson *et al.*,[94] Harris and Kerr[99] and Behnke *et al.*[78] from relative rate studies carried out at 297 ± 2 K, 245–328 K and 300 K, respectively. The rate constants derived from these studies are given in Table 1 and those of Atkinson *et al.*[94] and Harris and Kerr[99] are plotted in Fig. 11 (the rate constant of Behnke *et al.*[78] cannot be readily reevaluated to be consistent with the present recommendations for the reference organics used, although it is not expected to change by more than 1–2%). A unit-weighted least-squares analysis of these data,[94,99] using the Arrhenius equation, yields the recommended expression of

$$k(2,2\text{-dimethylbutane}) = (2.84^{+1.11}_{-0.80})$$

$$\times 10^{-11} \, e^{-(747 \pm 94)/T} \, cm^3 \, molecule^{-1} \, s^{-1}$$

applicable only over the temperature range 245–328 K, where the indicated errors are two least-squares standard deviations, and

k(2,2-dimethylbutane) = 2.32

$\times$ 10^{-12} cm^3 molecule^{-1} s^{-1} at 298 K,

with an estimated overall uncertainty at 298 K of $\pm$30%. The rate constant reported by Behnke et al.[78] at 300 K is in excellent agreement with this recommendation.

The temperature dependence of this rate constant appears to be somewhat high, based upon the data for other alkanes and the temperature dependence calculated from the estimation technique of Atkinson[124,127] of $B = 475$ K over this same temperature range of 250–333 K. Hence, the temperature dependence obtained from the experimental data is not recommended for use outside of the range $\sim$240–330 K.

The rate constants determined by Greiner,[14] Atkinson et al.[102] and Harris and Kerr[99] are independent of temperature, within one least-squares standard deviation, and it is recommended for the temperature range 247–498 K that

k(2,3-dimethylbutane) = 6.2

$\times$ 10^{-12} cm^3 molecule^{-1} s^{-1},

independent of temperature, with an estimated uncertainty at 298 K of $\pm$25%. This recommendation is identical to that of Atkinson,[120] although the present recommendation covers a wider temperature range (247–498 K versus 299–498 K).

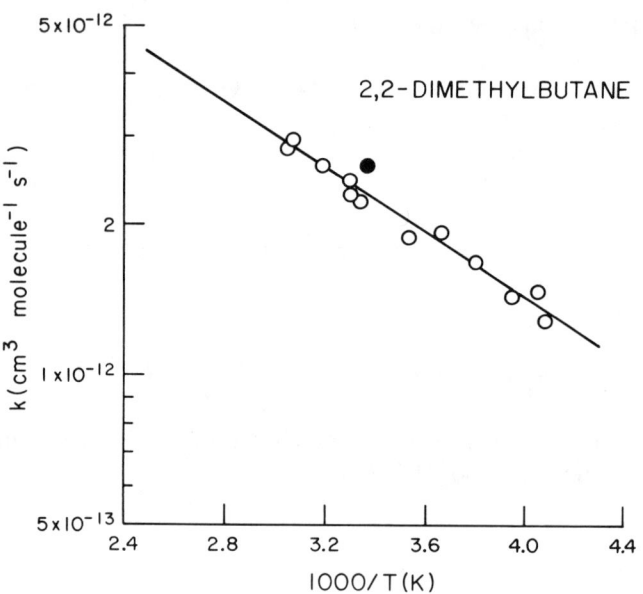

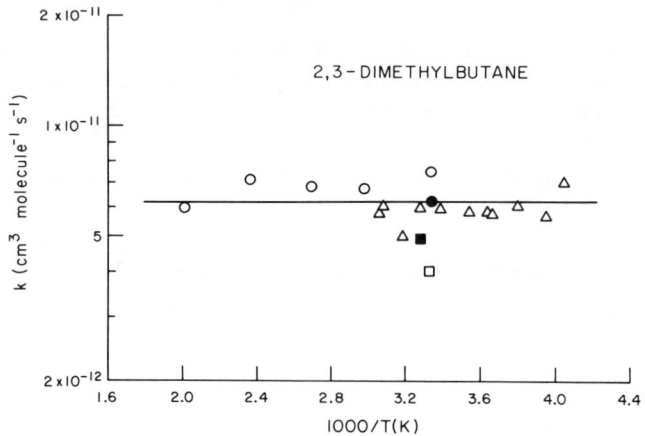

FIG. 12. Arrhenius plot of rate constants for the reaction of the OH radical with 2,3-dimethylbutane. (O) Greiner;[14] (■) Darnall et al.;[107] (□) Cox et al.;[72] (●) Atkinson et al.;[102] (Δ) Harris and Kerr;[99] (———) recommendation (see text).

FIG. 11. Arrhenius plot of rate constants for the reaction of the OH radical with 2,2-dimethylbutane. (●) Atkinson et al.;[94] (O) Harris and Kerr;[99] (———) recommendation (see text).

(15) *n*-Heptane

(14) 2,3-Dimethylbutane

The available rate constant data of Greiner,[14] Darnall et al.,[71,107] Cox et al.,[72] Atkinson et al.[102] and Harris and Kerr[99] are given in Table 1 and those of Greiner,[14] Darnall et al.,[107] Cox et al.,[72] Atkinson et al.[102] (which supersedes the earlier study of Darnall et al.[71]) and Harris and Kerr[99] are plotted in Arrhenius form in Fig. 12. The sole absolute rate constant study is that of Greiner.[14] The room temperature rate constant determined by Greiner[14] is $\sim$20% higher than those derived by Atkinson et al.[102] and Harris and Kerr,[99] which are the most recent and precise of the relative rate studies.

The available rate constants for the reaction of the OH radical with *n*-heptane are all from relative rate studies[73,76,78,89,106] carried out at room temperature. Based upon the studies of Atkinson et al.[73] and Behnke et al.[76,89] (the study of Klöpffer et al.[106] not being used in the evaluation due to a lack of details), a rate constant of

k(*n*-heptane) = 7.20 $\times$ 10^{-12} cm^3 molecule^{-1} s^{-1}

is recommended at 300 K. This recommendation is in excellent agreement with the rate constant reported from the relative rate study of Behnke et al.,[78] within the uncertainties due to experimental errors and reevaluation to be consistent with the present recommendations for the reference organics used.

J. Phys. Chem. Ref. Data, Monograph 1 (1989)

The estimation method of Atkinson[124,127] predicts that over the temperature range 290–320 K the temperature dependence of this reaction rate constant is $B \approx 300$ K, and using this temperature dependence leads to the recommendation of

$$k(n\text{-heptane}) = 7.15 \times 10^{-12} \text{ cm}^3 \text{ molecule}^{-1} \text{ s}^{-1}$$

at 298 K, with an estimated overall uncertainty of ±25%. At 312 K, the temperature at which the Nolting et al.[98] relative rate study was carried out, a rate constant of 7.48×10^{-12} cm^3 molecule^{-1} s^{-1} is then calculated and used to place their relative rate data[98] on an absolute basis.

(16) 2,2,3-Trimethylbutane

The available rate constants of Greiner,[14] Darnall et al.,[107] Baldwin et al.,[108] Atkinson et al.[94] and Harris and Kerr[99] are given in Table 1 and are plotted in Arrhenius form in Fig. 13. There is seen to be a significant degree of scatter in the reported data for temperatures <305 K. A unit-weighted least-squares analysis of the rate constant data of Greiner,[14] Baldwin et al.,[108] Atkinson et al.[94] and Harris and Kerr,[99] using the expression $k = CT^2 e^{-D/T}$, leads to the recommendation of

$$k(2,2,3\text{-trimethylbutane}) = (9.04^{+2.08}_{-1.70})$$
$$\times 10^{-18} T^2 e^{(495 \pm 63)/T} \text{ cm}^3 \text{ molecule}^{-1} \text{ s}^{-1}$$

over the temperature range 243–753 K, where the indicated errors are two least-squares standard deviations, and

$$k(2,2,3\text{-trimethylbutane}) = 4.23$$
$$\times 10^{-12} \text{ cm}^3 \text{ molecule}^{-1} \text{ s}^{-1} \text{ at 298 K,}$$

with an estimated overall uncertainty at 298 K of ±30%. In the absence of further experimental data at or below 250 K, the above expression should be used with caution at temperatures below ~275 K.

(17) n-Octane

The available kinetic data of Greiner,[14] Atkinson et al.,[73] Behnke et al.[76] and Nolting et al.[98] are given in Table 1 and are plotted in Arrhenius form in Fig. 14. These data are in excellent agreement. Since there is no evidence of curvature in the Arrhenius plot (Fig. 14), a unit-weighted least-squares analysis of the data of Greiner,[14] Atkinson et al.[73] and Behnke et al.[76] [the rate constant of Nolting et al.[98] at 312 K is relative to the less-well studied reaction of the OH radical with n-heptane (see above), and is hence not used in the evaluation] yields the recommended Arrhenius expression of

$$k(n\text{-octane}) = (3.15^{+0.74}_{-0.60})$$
$$\times 10^{-11} e^{-(384 \pm 72)/T} \text{ cm}^3 \text{ molecule}^{-1} \text{ s}^{-1}$$

over the temperature range 296–497 K, where the indicated error limits are two least-squares standard deviations, and

$$k(n\text{-octane}) = 8.68$$
$$\times 10^{-12} \text{ cm}^3 \text{ molecule}^{-1} \text{ s}^{-1} \text{ at 298 K,}$$

with an estimated overall uncertainty at 298 K of ±20%.

This recommendation is essentially identical to that of

$$k(n\text{-octane}) = 3.12 \times 10^{-11} e^{-380/T} \text{ cm}^3 \text{ molecule}^{-1} \text{ s}^{-1}$$

of Atkinson[120] over the same temperature range. Although, as seen from Fig. 14, the Arrhenius expression provides a satisfactory fit over this restricted temperature range, this is not expected to be the case over a wider temperature range extending to higher or lower temperatures, and this point is discussed below.

(18) 2,2,4-Trimethylpentane

The available rate constants[14,94] are given in Table 1 and are plotted in Arrhenius form in Fig. 14. The relative rate measurement of Atkinson et al.[94] at room temperature is in excellent agreement with that determined by Greiner[14] using flash photolysis-kinetic spectroscopy.

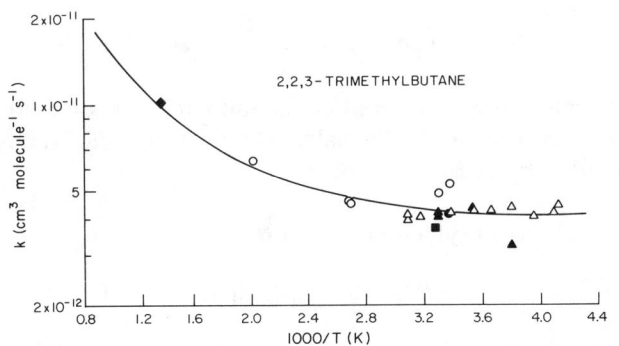

FIG. 13. Arrhenius plot of rate constants for the reaction of the OH radical with 2,2,3-trimethylbutane. (○) Greiner;[14] (■) Darnall et al.;[107] (◆) Baldwin et al.;[108] (●) Atkinson et al.;[94] (▲) Harris and Kerr,[99] relative to n-pentane; (△) Harris and Kerr,[99] relative to n-hexane; (———) recommendation (see text).

J. Phys. Chem. Ref. Data, Monograph 1 (1989)

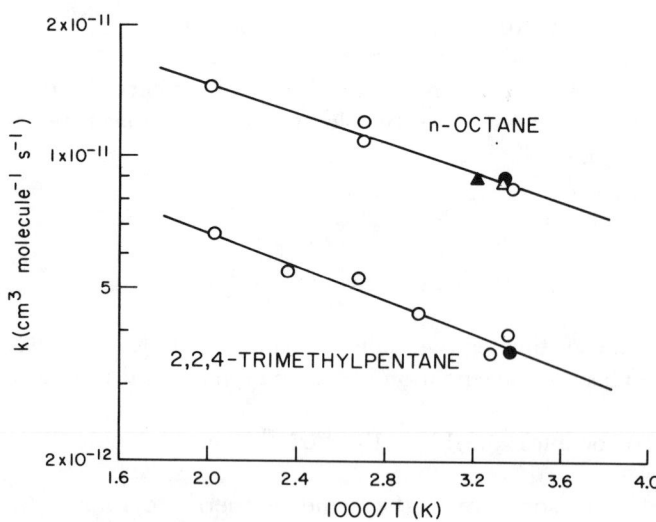

FIG. 14. Arrhenius plots of rate constants for the reactions of the OH radical with *n*-octane and 2,2,4-trimethylpentane. (○) Greiner;[14] (●) Atkinson *et al.*[73] (*n*-octane), Atkinson *et al.*[94] (2,2,4-trimethylpentane); (△) Behnke *et al.*;[76] (▲) Nolting *et al.*;[98] (———) recommendations (see text).

A unit-weighted least-squares analysis of these data[14,94] yields the recommended Arrhenius expression of

$$k(2,2,4\text{-trimethylpentane}) = (1.61^{+0.44}_{-0.35})$$

$$\times 10^{-11} e^{-(440 \pm 84)/T} \text{ cm}^3 \text{ molecule}^{-1} \text{ s}^{-1}$$

over the temperature range 297–493 K, where the indicated error limits are two least-squares standard deviations, and

$$k(2,2,4\text{-trimethylpentane}) = 3.68$$

$$\times 10^{-12} \text{ cm}^3 \text{ molecule}^{-1} \text{ s}^{-1} \text{ at 298 K},$$

with an estimated overall uncertainty at 298 K of ±20%. This expression is virtually identical to that recommended by Atkinson[120] of

$$k(2,2,4\text{-trimethylpentane}) = 1.62$$

$$\times 10^{-11} e^{-443/T} \text{ cm}^3 \text{ molecule}^{-1} \text{ s}^{-1}$$

over the same temperature range (the slight difference arising from reevaluating the relative rate constant of Atkinson *et al.*[94]).

(19) 2,2,3,3-Tetramethylbutane

The available rate constant data of Greiner,[14] Baldwin *et al.*,[46,109] Atkinson *et al.*[94] and Tully *et al.*[103] are given in

Table 1 and are plotted in Arrhenius form in Fig. 15. Again, the agreement between these kinetic studies is generally excellent. The Arrhenius plot (Fig. 15) clearly exhibits curvature and hence a unit-weighted least-squares analysis of these data of Greiner,[14] Baldwin *et al.*,[46] Atkinson *et al.*[94] and Tully *et al.*,[103] using the expression $k = CT^2 e^{-D/T}$, yields the recommendation of

$$k(2,2,3,3\text{-tetramethylbutane}) = (1.63^{+0.34}_{-0.28})$$

$$\times 10^{-17} T^2 e^{-(86 \pm 72)/T} \text{ cm}^3 \text{ molecule}^{-1} \text{ s}^{-1}$$

over the temperature range 290–753 K, where the indicated error limits are two least-squares standard deviations, and

$$k(2,2,3,3\text{-tetramethylbutane}) = 1.08$$

$$\times 10^{-12} \text{ cm}^3 \text{ molecule}^{-1} \text{ s}^{-1} \text{ at 298 K},$$

with an estimated overall uncertainty at 298 K of ±20%. This recommendation agrees to within 10% over this temperature range with that of Atkinson[120] of

$$k(2,2,3,3\text{-tetramethylbutane}) = 1.87$$

$$\times 10^{-17} T^2 e^{-133/T} \text{ cm}^3 \text{ molecule}^{-1} \text{ s}^{-1},$$

derived over the temperature range 290–738 K.

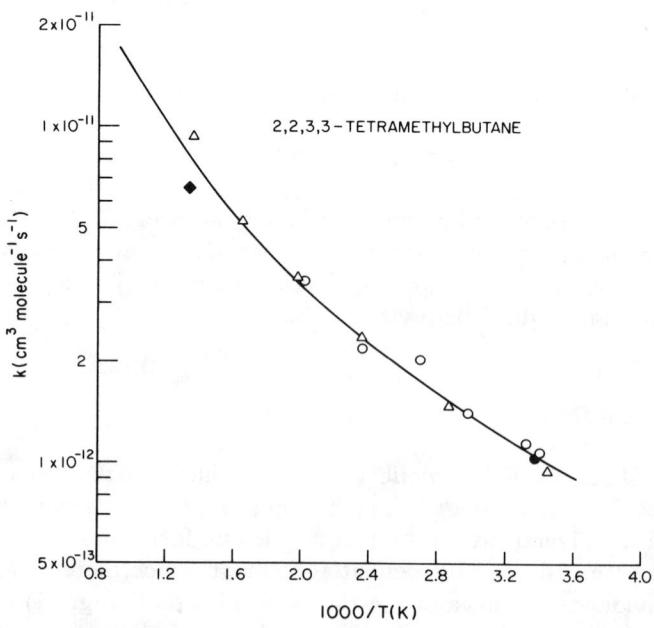

FIG. 15. Arrhenius plot of rate constants for the reaction of the OH radical with 2,2,3,3-tetramethylbutane. (○) Greiner;[14] (◆) Baldwin *et al.*;[46,109] (●) Atkinson *et al.*;[94] (△) Tully *et al.*;[103] (———) recommendation (see text).

(20) n-Nonane through n-Tridecane

For these n-alkanes, the only rate constant data available are from the relative rate studies of Zetzsch and co-workers[76,78,89,98] and, for n-nonane and n-decane, of Atkinson et al.,[73] all carried out at around room temperature. The agreement between these studies is good. For these alkanes, the temperature dependence of the rate constant around 300 K is calculated to be approximately equivalent to $B = 225$ K,[124,127] and hence the rate constants at 312 K should be ~3% higher than those at 299–300 K. From the data of Behnke et al.,[76,89] Nolting et al.[98] and Atkinson et al.,[73] the recommended rate constants at 298 K are:

$$k(n\text{-nonane}) = 1.02 \times 10^{-11} \text{ cm}^3 \text{ molecule}^{-1} \text{ s}^{-1},$$

$$k(n\text{-decane}) = 1.16 \times 10^{-11} \text{ cm}^3 \text{ molecule}^{-1} \text{ s}^{-1},$$

$$k(n\text{-undecane}) = 1.32 \times 10^{-11} \text{ cm}^3 \text{ molecule}^{-1} \text{ s}^{-1},$$

$$k(n\text{-dodecane}) = 1.42 \times 10^{-11} \text{ cm}^3 \text{ molecule}^{-1} \text{ s}^{-1}$$

and

$$k(n\text{-tridecane}) = 1.6 \times 10^{-11} \text{ cm}^3 \text{ molecule}^{-1} \text{ s}^{-1},$$

all with estimated overall uncertainties at 298 K of ±25%.

(21) Cyclopentane

The rate constant data of Volman,[113] Darnall et al.,[71] Atkinson et al.[102] (which supersedes the earlier rate constant of Darnall et al.[71]), Jolly et al.[111] and Droege and Tully[114] are given in Table 2 and are plotted in Arrhenius form in Fig. 16. No details are available concerning the rate constant obtained by Volman[113] from a relative rate study and, as noted above, the study of Darnall et al.[71] has been superseded by the more recent kinetic investigation of Atkinson et al.[102] At room temperature the absolute rate constants of Jolly et al.[111] and Droege and Tully[114] and the relative rate measurement of Atkinson et al.[102] are in excellent agreement. A unit-weighted least-squares analysis of these data,[102,111,114] using the equation $k = CT^2 e^{-D/T}$, leads to the recommendation of

$$k(\text{cyclopentane}) = (2.13^{+0.16}_{-0.14})$$

$$\times 10^{-17} T^2 e^{(299 \pm 24)/T} \text{ cm}^3 \text{ molecule}^{-1} \text{ s}^{-1}$$

over the temperature range 295–491 K, where the indicated error limits are two least-squares standard deviations, and

$$k(\text{cyclopentane}) = 5.16 \times 10^{-12} \text{ cm}^3 \text{ molecule}^{-1} \text{ s}^{-1}$$

at 298 K, with an estimated overall uncertainty at 298 K of ±20%.

Droege and Tully[114] also determined rate constants for cyclopentane-d_{10} (Table 2) and, from their data for cyclopentane and cyclopentane-d_{10}, obtained the deuterium isotope ratio of

$$\frac{k(\text{cyclopentane})}{k(\text{cyclopentane-}d_{10})} = \frac{k(-CH_2-)}{k(-CD_2-)}$$

$$= (1.16 \pm 0.10) \, e^{(254 \pm 15)/T}.$$

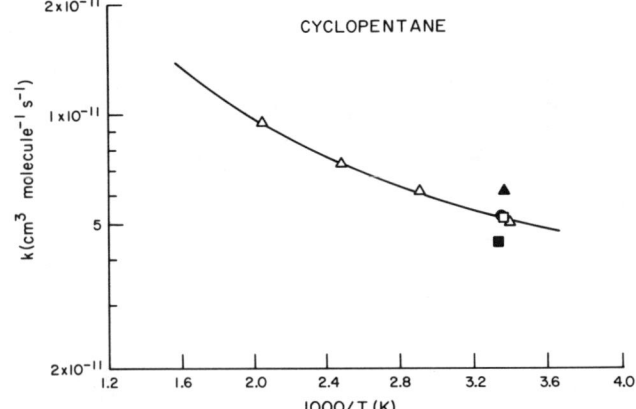

FIG. 16. Arrhenius plot of rate constants for the reaction of the OH radical with cyclopentane. (▲) Volman;[113] (■) Darnall et al.;[71] (●) Atkinson et al.;[102] (□) Jolly et al.;[111] (△) Droege and Tully;[114] (——————) recommendation (see text).

(22) Cyclohexane

The available rate constant data of Greiner,[14] Gorse and Volman,[69] Wu et al.,[91] Atkinson et al.,[102,115] Tuazon et al.,[116] Nielsen et al.,[60] Edney et al.,[62] Bourmada et al.[63] and Droege and Tully[114] are given in Table 2 and are plotted in Arrhenius form in Fig. 17. There is an appreciable degree of scatter in the rate constants determined at around room temperature. The relative rate constants of Gorse and Volman[69] and Wu et al.[91] are subject to large uncertainties (of the order of ±20–25%) and the absolute rate constant reported by Nielsen et al.[60] is significantly lower than the other room temperature data. Hence, the rate constants of Greiner,[14] Atkinson et al.,[102,115] Tuazon et al.,[116] Edney et al.,[62] Bourmada et al.[63] and Droege and Tully[114] have been used in the evaluation of this rate constant. A unit-weighted least-squares analysis of these rate constants, using the equation $k = CT^2 e^{-D/T}$, yields the recommendation of

$$k(\text{cyclohexane}) = (2.66^{+0.85}_{-0.65})$$

$$\times 10^{-17} T^2 e^{(344 \pm 95)/T} \text{ cm}^3 \text{ molecule}^{-1} \text{ s}^{-1}$$

over the temperature range 292–497 K, where the indicated error limits are two least-squares standard deviations, and

$$k(\text{cyclohexane}) = 7.49$$

$$\times 10^{-12} \text{ cm}^3 \text{ molecule}^{-1} \text{ s}^{-1} \text{ at } 298 \text{ K},$$

with an estimated overall uncertainty at 298 K of ±25%.

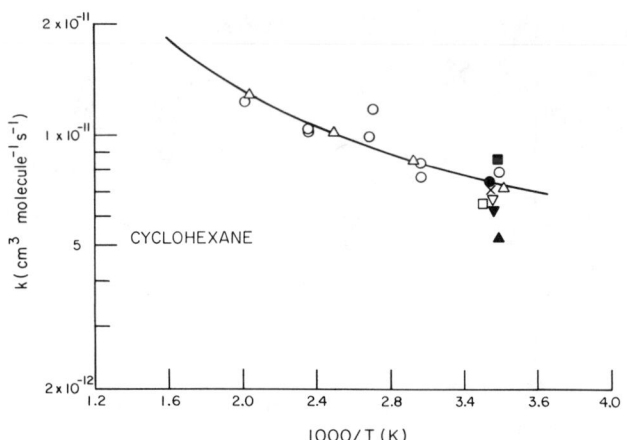

FIG. 17. Arrhenius plot of rate constants for the reaction of the OH radical with cyclohexane. (○) Greiner;[14] (▽) Gorse and Volman;[69] (□) Wu *et al.*;[91] (●) Atkinson *et al.*[102] and Tuazon *et al.*;[116] (x) Atkinson *et al.*;[115] (▲) Nielsen *et al.*;[60] (▼) Edney *et al.*;[62] (■) Bourmada *et al.*;[63] (△) Droege and Tully;[114] (————) recommendation (see text).

This recommendation yields similar rate constants over the temperature range 290–500 K to those calculated from the Arrhenius expression of

$$k(\text{cyclohexane}) = 2.73$$

$$\times 10^{-11} e^{-390/T} \text{ cm}^3 \text{ molecule}^{-1} \text{ s}^{-1}$$

recommended by Atkinson,[120] although it yields significantly different rate constants at temperatures above 500 K and below 290 K.

From their rate constants for cyclohexane and cyclohexane-d_{12}, Droege and Tully[114] derived the deuterium isotope effect of

$$\frac{k(\text{cyclohexane})}{k(\text{cyclohexane-}d_{12})} = \frac{k(-\text{CH}_2-)}{k(-\text{CD}_2-)}$$

$$= (1.16 \pm 0.06) e^{(237 \pm 9)/T}.$$

(23) Other Acyclic and Cycloalkanes

For the remaining acyclic alkanes and cycloalkanes for which rate constants are available (Tables 1 and 2), data are available only at room temperature or from only one study. No specific recommendations are made for these alkanes based upon the experimental data, although it should be noted that the data of Jolly *et al.*[111] and Behnke *et al.*[78] for cycloheptane are in good agreement, and those of Atkinson *et al.*[117] and Behnke *et al.*[78] for tricyclo[3.3.1.1^{3,7}]decane (adamantane) are in excellent agreement.

In the above evaluation and recommendation of OH radical reaction rate constants for the alkanes, data are available in many cases over only restricted temperature ranges (for example, ~300–500 K) and then often only for the overall reaction. For the majority of alkanes, the C–H bonds are non-equivalent, and hence multiple reaction pathways are operative leading to a variety of alkyl radicals. While the experimental data generally do not distinguish between these initial OH radical reaction routes, estimation methods are now available[124,127-129] which do allow the overall reaction rate constants and the distribution of alkyl radicals formed to be estimated, apparently with reasonable accuracy.

In the recent estimation technique of Atkinson,[124,127] H-atom abstraction from C–H bonds is dealt with in terms of H-atom abstraction from $-\text{CH}_3$, $-\text{CH}_2-$ and $>\text{CH}-$ groups, with:

$$k(\text{CH}_3-\text{X}) = k^\circ_{\text{prim}} F(\text{X})$$

$$k(\text{X}-\text{CH}_2-\text{Y}) = k^\circ_{\text{sec}} F(\text{X}) F(\text{Y})$$

and

$$k(\text{X-CH}\overset{\displaystyle Y}{\underset{\displaystyle Z}{\diagup}\diagdown}) = k^\circ_{\text{tert}} F(\text{X}) F(\text{Y}) F(\text{Z})$$

where

k°_{prim}, k°_{sec} and k°_{tert} are the group rate constants for H-atom abstraction from primary, secondary and tertiary groups, respectively, for a standard substituent, and $F(\text{X})$, $F(\text{Y})$ and $F(\text{Z})$ are the factors for substituent groups X, Y and Z, respectively. The standard substituent is taken to be $-\text{CH}_3$, with $F(-\text{CH}_3) = 1.00$ at all temperatures. From the previous review and evaluation of Atkinson,[120] the following parameters were obtained:[124,127]

$$k^\circ_{\text{prim}} = 4.47 \times 10^{-18} T^2 e^{-303/T} \text{ cm}^3 \text{ molecule}^{-1} \text{ s}^{-1},$$

$$k^\circ_{\text{sec}} = 4.32 \times 10^{-18} T^2 e^{233/T} \text{ cm}^3 \text{ molecule}^{-1} \text{ s}^{-1},$$

$$k^\circ_{\text{tert}} = 1.89 \times 10^{-18} T^2 e^{711/T} \text{ cm}^3 \text{ molecule}^{-1} \text{ s}^{-1},$$

and

$$F(-\text{CH}_2-) = F(>\text{CH}-) = F(>\text{C}<) = e^{76/T},$$

applicable for the temperature range $\sim 250-1000$ K. For three- to seven-membered cyclic rings, the factors F_{ring} of $F_3 = e^{-1214/T}$, $F_4 = e^{-451/T}$, $F_5 = e^{-66/T}$ and $F_6 = F_7 = 1.00$ were derived (these ring strain factors are applicable only to the C—H bonds involved in the ring structure, and not to substituent side-chains).[112] Figures 18–20 show the fits of the rate constants calculated in this manner with the recommended rate constants for those alkanes (apart from methane and ethane) for which recommendations have been made as a function of temperature. The fits are seen to be generally excellent over the temperature ranges for which experimental data and recommendations are available ($\sim 250-1000$ K). Thus, this estimation method can be used to provide the rate constants and/or temperature dependencies of the rate constants for these alkanes for which either no data are available or only room temperature rate constants are available.

In addition, this estimation technique allows the distribution of alkyl radicals formed in these OH radical reactions to be calculated at temperatures in the range $\sim 250-1000$ K. Thus, the calculated distributions of the individual reaction pathways for H-atom abstraction from the primary, secondary and tertiary C—H bonds for propane, n-butane and 2-methylpropane agree well with those derived from the kinetic studies of Tully and co-workers.[75,87,96]

Based upon the kinetic studies of Tully and co-workers for the reactions of OH radicals with ethane,[59] propane,[75] n-butane,[87] 2-methylpropane,[96] 2,2-dimethyl-propane,[59] cyclopentane[114] and cyclohexane[114] the deuterium isotope effects for H- or D-atom abstraction from primary, secondary and tertiary C—H or C—D bonds depend predominantly on whether the C—H or C—D bond is primary, secondary or tertiary, and not on the neighboring groups. Based upon the experimental data of Tully and co-workers,[59,75,87,96,114] the ratios k(abstraction from C—H bonds)/k(abstraction from C—D bonds) = k^H/k^D of

$$k^H/k^D = e^{460/T} \text{ for primary bonds,}$$

$$k^H/k^D = e^{290/T} \text{ for secondary bonds}$$

and

$$k^H/k^D \approx e^{190/T} \text{ for tertiary bonds,}$$

are applicable over the temperature range $\sim 290-800$ K. These deuterium isotope ratios can be combined with the —CH$_3$, —CH$_2$— and >CH— group rate constants discussed above to allow the OH radical reaction rate constants to be calculated for fully or partially deuterated alkanes for which no experimental data are available.

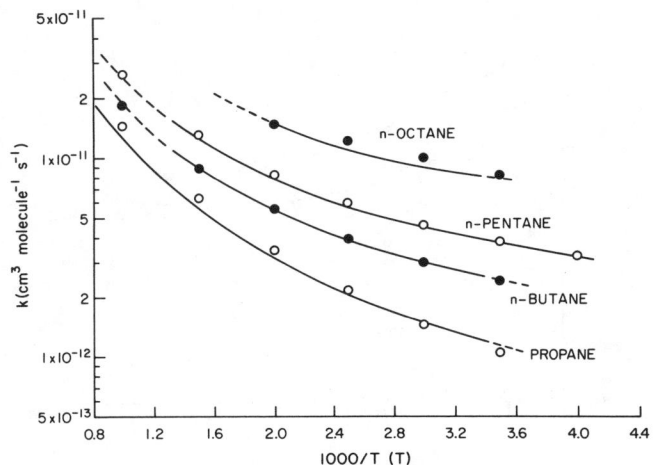

FIG. 18. Arrhenius plot of rate constants for the reactions of the OH radical with propane, n-butane, n-pentane, and n-octane. (○, ●) Recommended rate constants; (——, — — —) calculated from the estimation technique of Atkinson[124,127] (solid lines define the temperature ranges encompassed by the recommendations).

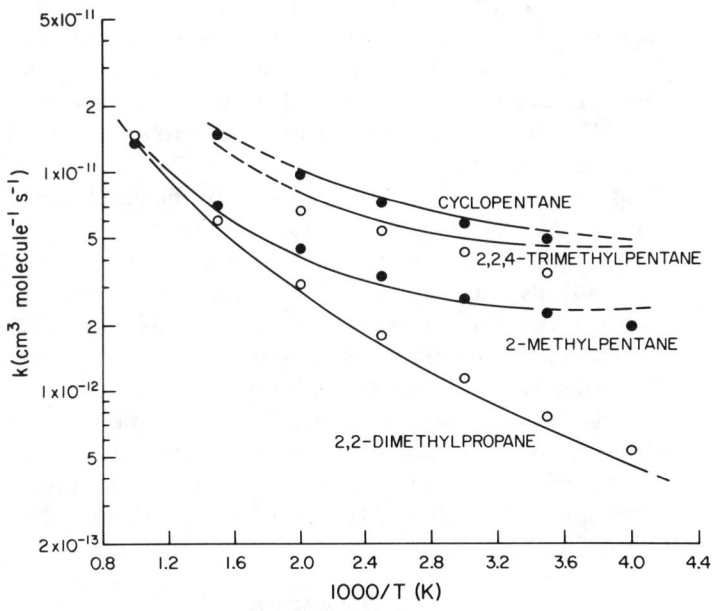

FIG. 19. Arrhenius plots of rate constants for the reactions of the OH radical with 2,2-dimethylpropane, 2-methylpentane, 2,2,4-trimethylpentane and cyclopentane. (○, ●) Recommended rate constants; (——, — — —) calculated from the estimation technique of Atkinson[124,127] (solid lines define the temperature ranges encompassed by the recommendations).

J. Phys. Chem. Ref. Data, Monograph 1 (1989)

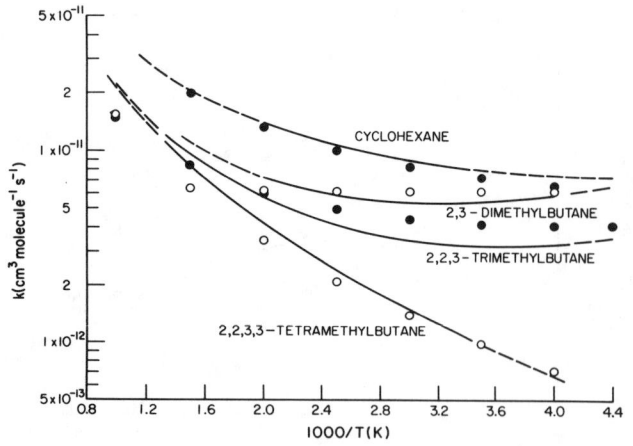

[7]D. E. Hoare and G. B. Peacock, Proc. Roy. Soc. (London) **A291**, 85 (1966).

[8]R. R. Baldwin, A. C. Norris, and R. W. Walker, 11th International Symposium on Combustion, 1966; The Combustion Institute, Pittsburgh, PA, 1967, p. 889.

[9]G. Dixon-Lewis and A. Williams, 11th International Symposium on Combustion, 1966; The Combustion Institute, Pittsburgh, PA, 1967, p. 951.

[10]W. E. Wilson and A. A. Westenberg, 11th International Symposium on Combustion, 1966; The Combustion Institute, 1967, p. 1143.

[11]N. R. Greiner, J. Chem. Phys. **46**, 2795 (1967).

[12]D. G. Horne and R. G. W. Norrish, Nature **215**, 1373 (1967).

[13]W. E. Wilson, Jr., J. T. O'Donovan, and R. M. Fristrom, 12th International Symposium on Combustion, 1968; The Combustion Institute, Pittsburgh, PA, 1969, p. 929.

[14]N. R. Greiner, J. Chem. Phys. **53**, 1070 (1970).

[15]R. R. Baldwin, D. E. Hopkins, A. C. Norris, and R. W. Walker, Combust. Flame **15**, 33 (1970).

[16]R. Simonaitis, J. Heicklen, M. M. Maguire, and R. A. Bernheim, J. Phys. Chem. **75**, 3205 (1971).

[17]J. Peeters and G. Mahnen, 14th International Symposium on Combustion, 1972; The Combustion Institute, Pittsburgh, PA, 1973, p. 133.

[18]D. D. Davis, S. Fischer, and R. Schiff, J. Chem. Phys. **61**, 2213 (1974).

[19]J. J. Margitan, F. Kaufman, and J. G. Anderson, Geophys. Res. Lett. **1**, 80 (1974).

[20]S. Gordon and W. A. Mulac, Int. J. Chem. Kinet. **Symp. 1**, 289 (1975).

[21]R. P. Overend, G. Paraskevopoulos, and R. J. Cvetanovic, Can. J. Chem. **53**, 3374 (1975).

[22]C. J. Howard and K. M. Evenson, J. Chem. Phys. **64**, 197 (1976).

[23]R. Zellner and W. Steinert, Int. J. Chem. Kinet. **8**, 397 (1976).

[24]R. A. Cox, R. G. Derwent, and P. M. Holt, J. Chem. Soc. Faraday Trans. 1, **72**, 2031 (1976).

[25]J. N. Bradley, W. D. Capey, R. W. Fair, and D. K. Pritchard, Int. J. Chem. Kinet. **8**, 549 (1976).

[26]J. Ernst, H. Gg. Wagner, and R. Zellner, Ber. Bunsenges Phys. Chem. **82**, 409 (1978).

[27]T. J. Sworski, C. J. Hochanadel, and P. J. Ogren, J. Phys. Chem. **84**, 129 (1980).

[28]F. P. Tully and A. R. Ravishankara, J. Phys. Chem. **84**, 3126 (1980).

[29]D. Husain, J. M. C. Plane, and N. K. H. Slater, J. Chem. Soc. Faraday Trans. 2, **77**, 1949 (1981).

[30]P. W. Fairchild, G. P. Smith, and D. R. Crosley, 19th International Symposium on Combustion, 1982; The Combustion Institute, Pittsburgh, PA, 1982, p. 107.

[31]K.-M. Jeong and F. Kaufman, J. Phys. Chem. **86**, 1808 (1982).

[32]K.-M. Jeong, K.-J. Hsu, J. B. Jeffries, and F. Kaufman, J. Phys. Chem. **88**, 1222 (1984).

[33]D. L. Baulch, R. J. B. Craven, M. Din, D. D. Drysdale, S. Grant, D. J. Richardson, A. Walker, and G. Watling, J. Chem. Soc. Faraday Trans. 1, **79**, 689 (1983).

[34]C. D. Jonah, W. A. Mulac, and P. Zeglinski, J. Phys. Chem. **88**, 4100 (1984).

[35]S. Madronich and W. Felder, 20th International Symposium on Combustion, 1984; The Combustion Institute, Pittsburgh, PA, 1985, p. 703.

[36]N. Cohen and J. F. Bott, 20th International Symposium on Combustion, 1984; The Combustion Institute, Pittsburgh, PA, 1985, p. 711.

[37]G. P. Smith, P. W. Fairchild, J. B. Jeffries, and D. R. Crosley, J. Phys. Chem. **89**, 1269 (1985).

[38]F. Rust and C. M. Stevens, Int. J. Chem. Kinet. **12**, 371 (1980).

[39]J. A. Davidson, C. A. Cantrell, S. C. Tyler, R. E. Shetter, R. J. Cicerone, and J. G. Calvert, J. Geophys. Res. **92**, 2195 (1987).

[40]R. R. Baldwin and R. F. Simmons, Trans. Faraday Soc. **53**, 964 (1957).

[41]C. P. Fenimore and G. W. Jones, 9th International Symposium on Combustion, 1962 (Academic Press, New York, NY, 1963), p. 597.

[42]A. A. Westenberg and R. M. Fristrom, 10th International Symposium on Combustion, 1964; The Combustion Institute, Pittsburgh, PA, 1965, p. 473.

FIG. 20.　Arrhenius plots of rate constants for the reactions of the OH radical with 2,3-dimethylbutane, 2,2,3-trimethylbutane, 2,2,3,3-tetramethylbutane and cyclohexane. (○, ●) Recommended rate constants; (——— , — — —) calculated from the estimation technique of Atkinson[124,127] (solid lines define the temperature ranges encompassed by the recommendations).

(24) Reactions of OD Radicals with Alkanes

To date, kinetic data are available (Table 3) for only four alkanes, and then only at room temperature. By comparison with the data given in Table 1, it is evident that the rate constants at room temperature for the reactions of the OD radical with methane, ethane and n-butane are essentially identical to those for the reactions of the OH radical with these alkanes. This is to be expected, since the thermochemistries of these OD radical reactions are essentially identical to those for the corresponding OH radical reactions.[130] Moreover, as with the OH radical reactions, the rate constant for the reaction of OD radicals with n-butane-d_{10}[84] is lower by a factor of ~3 than that for the reaction of OD radicals with n-butane, and is essentially identical to that for the reaction of OH radicals with n-butane-d_{10}. This is again expected on thermochemical grounds, since the abstraction of D-atoms from C−D bonds by OH or OD radicals are less exothermic by ~0.9 kcal mole^{-1} than are the corresponding abstractions of H-atoms from C−H bonds.[130]

References

[1]A. A. Westenberg and R. M. Fristrom, J. Phys. Chem. **65**, 591 (1961).

[2]C. P. Fenimore and G. W. Jones, J. Phys. Chem. **65**, 2200 (1961).

[3]D. E. Hoare, Nature **194**, 283 (1962).

[4]R. M. Fristrom, 9th International Symposium on Combustion, 1962 (Academic Press, New York, NY, 1963), p. 560.

[5]R. V. Blundell, W. G. A. Cook, D. E. Hoare, and G. S. Milne, 10th International Symposium on Combustion, 1964; The Combustion Institute, Pittsburgh, PA, 1965, p. 445.

[6]D. E. Hoare, Proc. Roy. Soc. (London) **A291**, 73 (1966).

[43]N. R. Greiner, J. Chem. Phys. **46**, 3389 (1967).

[44]D. E. Hoare and M. Patel, Trans. Faraday Soc. **65**, 1325 (1969).

[45]R. R. Baldwin, D. E. Hopkins, and R. W. Walker, Trans. Faraday Soc. **66**, 189 (1970).

[46]R. R. Baldwin and R. W. Walker, J. Chem. Soc. Faraday Trans. 1, **75**, 140 (1979).

[47]D. J. Hucknall, D. Booth, and R. J. Sampson, Int. J. Chem. Kinet., **Symp. 1**, 301 (1975).

[48]C. J. Howard and K. M. Evenson, J. Chem. Phys. **64**, 4303 (1976).

[49]M.-T. Leu, J. Chem. Phys. **70**, 1662 (1979).

[50]L. G. Anderson and R. D. Stephens, 15th International Conference on Photochemistry, Stanford, CA, June 27–July 1, 1982; Report GMR–4087, ENV #130, General Motors Research Laboratories, Warren, MI, June 29, 1982.

[51]J. H. Lee and I. N. Tang, J. Chem. Phys. **77**, 4459 (1982).

[52]J. J. Margitan and R. T. Watson, J. Phys. Chem. **86**, 3819 (1982).

[53]F. P. Tully, A. R. Ravishankara, and K. Carr, Int. J. Chem. Kinet. **15**, 1111 (1983).

[54]O. J. Nielsen, P. Pagsberg, and A. Sillesen, Proceedings, 3rd European Symposium on the Physico-Chemical Behavior of Atmospheric Pollutants, 1984; Riedel, Dordrecht, Holland, 1984, p. 283.

[55]C. A. Smith, L. T. Molina, J. J. Lamb, and M. J. Molina, Int. J. Chem. Kinet. **16**, 41 (1984).

[56]P. Devolder, M. Carlier, J. F. Pauwels, and L. R. Sochet, Chem. Phys. Lett. **111**, 94 (1984).

[57]V. Schmidt, G. Y. Zhu, K. H. Becker, and E. H. Fink, Ber. Bunsenges Phys. Chem. **89**, 321 (1985).

[58]D. L. Baulch, I. M. Campbell, and S. M. Saunders, J. Chem. Soc. Faraday Trans. 1, **81**, 259 (1985).

[59]F. P. Tully, A. T. Droege, M. L. Koszykowski, and C. F. Melius, J. Phys. Chem. **90**, 691 (1986).

[60]O. J. Nielsen, J. Munk, P. Pagsberg, and A. Sillesen, Chem. Phys. Lett. **128**, 168 (1986).

[61]R. A. Stachnik, L. T. Molina, and M. J. Molina, J. Phys. Chem. **90**, 2777 (1986).

[62]E. O. Edney, T. E. Kleindienst, and E. W. Corse, Int. J. Chem. Kinet. **18**, 1355 (1986).

[63]N. Bourmada, C. Lafage, and P. Devolder, Chem. Phys. Lett. **136**, 209 (1987).

[64]T. J. Wallington, D. M. Neuman, and M. J. Kurylo, Int. J. Chem. Kinet. **19**, 725 (1987).

[65]S. Zabarnick, J. W. Fleming, and M. C. Lin, Int. J. Chem. Kinet. **20**, 117 (1988).

[66]R. R. Baldwin, Trans. Faraday Soc. **60**, 527 (1964).

[67]R. R. Baker, R. R. Baldwin, and R. W. Walker, Trans. Faraday Soc. **66**, 2812 (1970).

[68]J. N. Bradley, W. Hack, K. Hoyermann, and H. Gg. Wagner, J. Chem. Soc. Faraday Trans. 1, **69**, 1889 (1973).

[69]R. A. Gorse and D. H. Volman, J. Photochem. **3**, 115 (1974).

[70]A. B. Harker and C. S. Burton, Int. J. Chem. Kinet. **7**, 907 (1975).

[71]K. R. Darnall, R. Atkinson, and J. N. Pitts, Jr., J. Phys. Chem. **82**, 1581 (1978).

[72]R. A. Cox, R. G. Derwent, and M. R. Williams, Environ. Sci. Technol. **14**, 57 (1980).

[73]R. Atkinson, S. M. Aschmann, W. P. L. Carter, A. M. Winer, and J. N. Pitts, Jr., Int. J. Chem. Kinet. **14**, 781 (1982).

[74]J. F. Bott and N. Cohen, Int. J. Chem. Kinet. **16**, 1557 (1984).

[75]A. T. Droege and F. P. Tully, J. Phys. Chem. **90**, 1949 (1986).

[76]W. Behnke, F. Nolting, and C. Zetzsch, J. Aeros. Sci. **18**, 65 (1987).

[77]O. J. Nielsen, H. W. Sidebottom, D. J. O'Farrell, M. Donlon, and J. Treacy, Chem. Phys. Lett. **146**, 197 (1988).

[78]W. Behnke, F. Nolting, and C. Zetzsch, 10th International Symposium on Gas Kinetics, University College of Swansea, Swansea, UK, July 24–29, 1988.

[79]R. R. Baldwin and R. W. Walker, Trans. Faraday Soc. **60**, 1236 (1964).

[80]E. D. Morris, Jr. and H. Niki, J. Phys. Chem. **75**, 3640 (1971).

[81]F. Stuhl, Z. Naturforsch. **28A**, 1383 (1973).

[82]I. M. Campbell, B. J. Handy, and R. M. Kirby, J. Chem. Soc. Faraday Trans. 1, **71**, 867 (1975).

[83]R. A. Perry, R. Atkinson, and J. N. Pitts, Jr., J. Chem. Phys. **64**, 5314 (1976).

[84]G. Paraskevopoulos and W. S. Nip, Can. J. Chem. **58**, 2146 (1980).

[85]R. Atkinson, W. P. L. Carter, A. M. Winer, and J. N. Pitts, Jr., J. Air Pollut. Contr. Assoc. **31**, 1090 (1981).

[86]R. Atkinson and S. M. Aschmann, Int. J. Chem. Kinet. **16**, 1175 (1984).

[87]A. T. Droege and F. P. Tully, J. Phys. Chem. **90**, 5937 (1986).

[88]I. Barnes, V. Bastian, K. H. Becker, E. H. Fink, and W. Nelsen, J. Atmos. Chem. **4**, 445 (1986).

[89]W. Behnke, W. Holländer, W. Koch, F. Nolting, and C. Zetzsch, Atmos. Environ. **22**, 1113 (1988).

[90]R. A. Gorse and D. H. Volman, J. Photochem. **1**, 1 (1972).

[91]C. H. Wu, S. M. Japar, and H. Niki, J. Environ. Sci. Health **A11**, 191 (1976).

[92]R. Butler, I. J. Solomon, and A. Snelson, Chem. Phys. Lett. **54**, 19 (1978).

[93]P. L. Trevor, G. Black, and J. R. Barker, J. Phys. Chem. **86**, 1661 (1982).

[94]R. Atkinson, W. P. L. Carter, S. M. Aschmann, A. M. Winer, and J. N. Pitts, Jr., Int. J. Chem. Kinet. **16**, 469 (1984).

[95]T. Böhland, F. Temps, and H. Gg. Wagner, Z. Phys. Chem. Neu. Folge **142**, S.129 (1984).

[96]F. P. Tully, J. E. M. Goldsmith, and A. T. Droege, J. Phys. Chem. **90**, 5932 (1986).

[97]I. Barnes, V. Bastian, K. H. Becker, E. H. Fink, and F. Zabel, Atmos. Environ. **16**, 545 (1982).

[98]F. Nolting, W. Behnke, and C. Zetzsch, J. Atmos. Chem. **6**, 47 (1988).

[99]S. J. Harris and J. A. Kerr, Int. J. Chem. Kinet. **20**, 939 (1988).

[100]A. C. Lloyd, K. R. Darnall, A. M. Winer, and J. N. Pitts, Jr., J. Phys. Chem. **80**, 789 (1976).

[101]R. R. Baker, R. R. Baldwin, and R. W. Walker, Combust. Flame **27**, 147 (1976).

[102]R. Atkinson, S. M. Aschmann, A. M. Winer, and J. N. Pitts, Jr., Int. J. Chem. Kinet. **14**, 507 (1982).

[103]F. P. Tully, M. L. Koszykowski, and J. S. Binkley, 20th International Symposium on Combustion, 1984; The Combustion Institute, Pittsburgh, PA, 1985, p. 715.

[104]R. Atkinson, S. M. Aschmann, and W. P. L. Carter, Int. J. Chem. Kinet. **15**, 51 (1983).

[105]Th. Klein, I. Barnes, K. H. Becker, E. H. Fink, and F. Zabel, J. Phys. Chem. **88**, 5020 (1984).

[106]W. Klöpffer, R. Frank, E.-G. Kohl, and F. Haag, Chemiker-Zeitung **110**, 57 (1986); "Methods of the Ecotoxicological Evaluation of Chemicals, Photochemical Degradation in the Gas Phase," Vol. 6, *OH Reaction Rate Constants and Tropospheric Lifetimes of Selected Environmental Chemicals*. Report 1980–1983, K. H. Becker, H. M. Biehl, P. Bruckmann, E. H. Fink, F. Führ, W. Klöpffer, R. Zellner, and C. Zetzsch, Editors, Kernforschungsanlage Jülich GmbH, November 1984.

[107]K. R. Darnall, A. M. Winer, A. C. Lloyd, and J. N. Pitts, Jr., Chem. Phys. Lett. **44**, 415 (1976).

[108]R. R. Baldwin, R. W. Walker, and R. W. Walker, J. Chem. Soc. Faraday Trans. 1, **77**, 2157 (1981).

[109]R. R. Baldwin, R. W. Walker, and R. W. Walker, J. Chem. Soc. Faraday Trans. 1, **75**, 1447 (1979).

[110]C. Zetzsch, presented at Bunsen Colloquium, Göttingen, W. Germany, October 9, 1980; private communication, 1985.

[111]G. S. Jolly, G. Paraskevopoulos, and D. L. Singleton, Int. J. Chem. Kinet. **17**, 1 (1985).

[112]R. Atkinson and S. M. Aschmann, Int. J. Chem. Kinet. **20**, 339 (1988).

[113]D. H. Volman, Int. J. Chem. Kinet. **Symp. 1**, 358 (1975).

[114]A. T. Droege and F. P. Tully, J. Phys. Chem. **91**, 1222 (1987).

[115]R. Atkinson, S. M. Aschmann, and J. N. Pitts, Jr., Int. J. Chem. Kinet. **15**, 75 (1983).

[116]E. C. Tuazon, W. P. L. Carter, R. Atkinson, and J. N. Pitts, Jr., Int. J. Chem. Kinet. **15**, 619 (1983).

[117]R. Atkinson, S. M. Aschmann, and W. P. L. Carter, Int. J. Chem. Kinet. **15**, 37 (1983).

[118]N. R. Greiner, J. Chem. Phys. **48**, 1413 (1968).

[119]R. Atkinson, K. R. Darnall, A. C. Lloyd, A. M. Winer, and J. N. Pitts, Jr., Adv. Photochem. **11**, 375 (1979).

[120]R. Atkinson, Chem. Rev. **86**, 69 (1986).

[121]D. L. Baulch, M. Bowers, D. G. Malcolm, and R. T. Tuckerman, J. Phys. Chem. Ref. Data **15**, 465 (1986).

[122]J. H. Knox, R. F. Smith, and A. F. Trotman-Dickenson, Trans. Faraday Soc. **54**, 1509 (1958).

[123]R. R. Baker, R. R. Baldwin, and R. W. Walker, Trans. Faraday Soc. **66**, 3016 (1970).

[124]R. Atkinson, Int. J. Chem. Kinet. **18**, 555 (1986).

[125]R. R. Baker, R. R. Baldwin, A. R. Fuller, and R. W. Walker, J. Chem. Soc. Faraday Trans. 1, **71**, 736 (1975).

[126]W. E. Falconer, J. H. Knox, and A. F. Trotman-Dickenson, J. Chem. Soc. 782 (1961).

[127]R. Atkinson, Int. J. Chem. Kinet. **19**, 799 (1987).

[128]R. W. Walker, Int. J. Chem. Kinet. **17**, 573 (1985).

[129]N. Cohen, Int. J. Chem. Kinet. **14**, 1339 (1982).

[130]S. W. Benson, *Thermochemical Kinetics*, 2nd Ed. (Wiley, New York, NY, 1976).

2.2. Haloalkanes

a. Kinetics

The available rate constant data are listed in Table 4. The relative rate constants reported by Butler et al.[34] are not included, since these were derived from a complex expression which cannot be reevaluated using the more recent rate constants for the reference reactions. It should also be noted that the rate constants derived from the study of Cox et al.[7] have a stated accuracy of approximately a factor of two, due to uncertainties in the number of molecules of NO oxidized per OH radical reacted.[7] It can then be seen that essentially all of the data listed in Table 4 for the C_1 and C_2 haloalkanes have been determined from absolute rate constant studies.

As discussed below for the individual haloalkanes, apart from CH_3CCl_3 for which significant discrepancies appear to have arisen in all but the most recent studies[4,28,29] due to problems associated with the presence of reactive impurities, these absolute rate data are in general agreement, with the exception of the rate constants determined from the studies of Clyne and Holt[14] and Nielsen et al.[11] As noted in previous evaluations,[35-37] for several of the haloalkanes studied by Clyne and Holt[14] the room temperature rate constants and Arrhenius activation energies are significantly higher than the other absolute literature values. Furthermore, in many cases the derived Arrhenius preexponential factors[14] (Table 4) appear to be unreasonably high. Thus, these data of Clyne and Holt[14] have not been used in the evaluations of the recommended rate expressions for the haloalkanes.

It is apparent that for most of these haloalkanes the Arrhenius plots exhibit distinct curvature. In accordance with the recent evaluations of Atkinson[36] and DeMore et al.,[37] in most cases least-squares analyses of the rate constant data for these haloalkanes have been carried out

using the expression $k = CT^2 e^{-D/T}$, and the recommendations are generally in this form. The use of this expression yields good fits to the experimental data over the temperature ranges studied (i.e., ~240–500 K), although Cohen and Benson[38,39] used transition state calculations to obtain values of n in the three parameter equation $k = AT^n e^{-B/T}$ of ~1.1–1.8 for a series of halomethanes and haloethanes.

The kinetic data for the individual haloalkanes are discussed below.

(1) CH_3F

The available rate constants[1-5] are listed in Table 4 and plotted in Arrhenius form in Fig. 21. These rate constants of Howard and Evenson,[1] Nip et al.,[2] Jeong and Kaufman[3,4] and Bera and Hanrahan[5] are in reasonably good agreement at room temperature. However, since secondary reactions of the OH radical with CH_2F radicals and other radical species were expected[5] to occur in the pulsed radiolysis study of Bera and Hanrahan,[5] the rate constant determined by Bera and Hanrahan[5] was not used in the rate constant evaluation. (This was also the case for the CH_2F_2 reaction.) A unit-weighted least-squares analysis of the data of Howard and Evenson,[1] Nip et al.[2] and Jeong and Kaufman[3,4] yields the recommended expression of

$$k(CH_3F) = (5.51^{+3.36}_{-2.09})$$
$$\times 10^{-18} T^2 e^{-(1005 \pm 168)/T} cm^3 \ molecule^{-1} \ s^{-1}$$

over the temperature range 292–480 K, where the indicated error limits are two least-squares standard deviations, and

$$k(CH_3F) = 1.68 \times 10^{-14} cm^3 \ molecule^{-1} \ s^{-1} \ at \ 298 \ K,$$

with an estimated overall uncertainty at 298 K of ±30%. This recommendation is identical to that of Atkinson.[36]

(2) CH_3Cl

The available rate constants of Wilson et al.,[6] Howard and Evenson,[1] Cox et al.[7] (which, as noted above, is uncertain by a factor of ~2), Perry et al.,[8] Davis et al.,[9] Paraskevopoulos et al.,[10] Jeong and Kaufman,[3] Nielsen et al.[11] and Taylor et al.[12] are given in Table 4 and are plotted in Arrhenius form in Fig. 22. It can be seen that the room temperature rate constants of Howard and Evenson,[1] Perry et al.,[8] Davis et al.,[9] Paraskevopoulos et al.,[10] Jeong and Kaufman[3] and Taylor et al.[12] are in good agreement. The rate constants obtained by Nielsen et al.[11] over the temperature range ~300–400 K are uniformly higher by a factor of ~1.7 than those of Howard and Evenson,[1] Perry et al.,[8] Davis et al.,[9] Paraskevopoulos et al.,[10] Jeong and Kaufman[3] and Taylor et al.,[12] probably due to fragmentation of the CH_3Cl reactant by the radiation beam, leading to enhanced OH radical reaction.[11] Furthermore, the rate constants reported from the recent laser photolysis-laser induced fluorescence study of Taylor et al.[12] are significantly higher than those of Perry et al.[8] and Jeong and Kaufman[3] at temperatures ~420–485 K. Incorporation of these data of Taylor et al.[12] into the evaluation leads to a rate expression which predicts rate constants at ~250 K which are ~30% lower than the measured rate constants of Davis et al.[9] and Jeong and Kaufman.[3] Accordingly, the rate constant data of Taylor et al.[12] have not been used in the evaluation of the rate constant for this reaction.

A unit-weighted least-squares analysis of the rate constant data of Howard and Evenson,[1] Perry et al.,[8] Davis et al.,[9] Paraskevopoulos et al.[10] and Jeong and Kaufman,[3] using the expression $k = CT^2e^{-D/T}$, yields the recommendation of

$$k(CH_3Cl) = (3.50^{+0.71}_{-0.58})$$
$$\times 10^{-18} \ T^2 \ e^{-(585 \pm 59)/T} \ cm^3 \ molecule^{-1} \ s^{-1}$$

over the temperature range 250–483 K, where the indicated error limits are two least-squares standard deviations, and

$$k(CH_3Cl) = 4.36 \times 10^{-14} \ cm^3 \ molecule^{-1} \ s^{-1} \ at \ 298 \ K,$$

with an estimated overall uncertainty at 298 K of ±20%. This recommendation is identical to that of Atkinson[36] and essentially identical to that recommended in the recent NASA evaluation.[37]

The rate constants calculated from the recommended expression at 1850–2100 K are in good agreement with those derived from the relative rate study of Wilson et al.,[6] and this observation allows the recommended expression to be used with some degree of confidence up to ~2000 K. The rate constant of Wilson[13] for CH_3Br obtained from a related relative rate study can then be used, in conjunction with the absolute rate constants determined over the temperature range 244–350 K, for the evaluation of the OH radical reaction with CH_3Br (see below).

TABLE 4. Rate constants k and temperature-dependent parameters for the gas-phase reactions of the OH radical with haloalkanes

Haloalkane	$10^{12} \times A$ (cm³ molecule⁻¹ s⁻¹)	n	B (K)	$10^{14} \times k$ (cm³ molecule⁻¹ s⁻¹)	at T (K)	Technique	Reference	Temperature range covered (K)
CH₃F				1.6 ± 0.35	296 ± 2	DF-LMR	Howard and Evenson[1]	
				2.17 ± 0.18	297 ± 2	FP-RA	Nip et al.[2]	
				1.40 ± 0.09	292	DF-RF	Jeong and Kaufman[3,4]	292–480
				2.50 ± 0.18	330			
				3.86 ± 0.33	356			
				4.76 ± 0.31	368			
				5.48 ± 0.66	385			
				8.56 ± 0.66	416			
				13.1 ± 1.1	455			
	7.96 × 10⁻¹³	4.32	277 ± 730	17.1 ± 1.1	480			
	8.11 ± 1.35		1887 ± 60					
				1.71 ± 0.24	308	PR-RA	Bera and Hanrahan[5]	
CH₃Cl				1200	1850–2100	RR [relative to $k(CO) = 1.12$ $\times 10^{-13}e^{0.000907T}$][a]	Wilson et al.[6]	1850–2100
				3.6 ± 0.8	296 ± 2	DF-LMR	Howard and Evenson[1]	
				10.2	298	RR [relative to $k(methane)$ $= 8.36 \times 10^{-15}$][b]	Cox et al.[7]	
				4.4 ± 0.5	298.4	FP-RF	Perry et al.[8]	298–423
				8.1 ± 0.8	349.3			
	4.1		1359 ± 151	16.8 ± 1.7	422.6			
				2.38 ± 0.14	250	FP-RF	Davis et al.[9]	250–350
				3.26 ± 0.06	273			

TABLE 4. Rate constants k and temperature-dependent parameters for the gas-phase reactions of the OH radical with haloalkanes — Continued

Haloalkane	$10^{12} \times A$ (cm³ molecule⁻¹ s⁻¹)	n	B (K)	$10^{14} \times k$ (cm³ molecule⁻¹ s⁻¹)	at T (K)	Technique	Reference	Temperature range covered (K)
				4.29 ± 0.21	298			
	1.84 ± 0.18		1098 ± 35	8.28 ± 0.28	350			
				4.10 ± 0.68	297	FP-RA	Paraskevopoulos et al.[10]	
				2.03 ± 0.15	247	DF-RF	Jeong and Kaufman[3,4]	247–483
				3.95 ± 0.26	293			
				6.68 ± 0.46	332			
				8.74 ± 0.58	363			
				12.8 ± 0.9	401			
				16.3 ± 1.3	426			
	2.21×10^{-9}	3.08	232 ± 423	25.4 ± 2.0	483			
	3.04 ± 0.43		1263 ± 45					
	5.31		1263	7.14	300	PR-RA	Nielsen et al.[11]	300–400
				4.9 ± 0.6	295	LP-LIF	Taylor et al.[12]	295–800
				7.0 ± 0.6	335			
				10.3 ± 1.9	375			
				10.1 ± 1.1	378			
				20.1 ± 1.9	428			
				29.3 ± 6.6	473			
				31.8 ± 3.5	475			
				53.3 ± 4.7	524			
				48.2 ± 1.4	525			
				71.0 ± 3.7	575			
				80.8 ± 3.4	615			
				103 ± 3.5	655			
				109 ± 9.7	667			
				111 ± 9.6	667			
				130 ± 12.0	695			
				137 ± 10.9	695			
				168 ± 20.1	735			
	8.38×10^{-4}	1.38	1202 ± 72	185 ± 6.8	800			
CH₃Br				760	1800–2000	RR [relative to $k(CH_4) = 6.95 \times 10^{-18}T^2 e^{-1282/T}$][b]	Wilson[13]	1800–2000
				3.5 ± 0.8	296 ± 2	DF-LMR	Howard and Evenson[1]	
				2.01 ± 0.12	244	FP-RF	Davis et al.[9]	244–350
				3.16 ± 0.15	273			
				4.14 ± 0.43	298			
	0.793 ± 0.079		889 ± 58	6.08 ± 0.4	350			
CH₂F₂				0.78 ± 0.12	296 ± 2	DF-LMR	Howard and Evenson[1]	
				0.58 ± 0.03	293	DF-RF	Clyne and Holt[14]	293–429
				1.61 ± 0.50	327			
				2.41 ± 0.35	368			
	$7.4^{+7.4}_{-3.7}$		2100 ± 200	6.03 ± 0.40	429			
				1.17 ± 0.14	297 ± 2	FP-RA	Nip et al.[2]	
				0.429 ± 0.038	250	DF-RF	Jeong and Kaufman[3,4]	250–492
				1.12 ± 0.075	298			
				2.10 ± 0.14	336			
				4.34 ± 0.27	384			
				7.27 ± 0.46	432			
				9.51 ± 0.66	464			
	2.52×10^{-9}	3.09	679 ± 458	14.1 ± 1.2	492			
	4.37 ± 0.58		1766 ± 50					

TABLE 4. Rate constants k and temperature-dependent parameters for the gas-phase reactions of the OH radical with haloalkanes — Continued

Haloalkane	$10^{12} \times A$ (cm³ molecule⁻¹ s⁻¹)	n	B (K)	$10^{14} \times k$ (cm³ molecule⁻¹ s⁻¹)	at T (K)	Technique	Reference	Temperature range covered (K)
				0.88 ± 0.14	308	PR-RA	Bera and Hanrahan[5]	
CH₂FCl				3.7 ± 0.6	296 ± 2	DF-LMR	Howard and Evenson[1]	
				1.65 ± 0.36	245	FP-RF	Watson et al.[15]	245–375
				4.21 ± 0.41	298			
	2.84 ± 0.3		1259 ± 50	9.80 ± 0.34	375			
				2.8 ± 0.5	273	FP-RA	Handwerk and Zellner[16]	273–373
				3.5 ± 0.7	293			
	3.1 ± 0.9		1320 ± 100	11 ± 2	373			
				4.45 ± 0.66	297	FP-RA	Paraskevopoulos et al.[10]	
				2.76 ± 0.18	250	DF-RF	Jeong and Kaufman[3,4]	250–486
				4.94 ± 0.30	295			
				6.60 ± 0.40	323			
				8.85 ± 0.55	348			
				14.0 ± 0.9	399			
				17.2 ± 1.1	438			
	1.57×10^{-7}	2.41	307 ± 382	25.4 ± 1.7	486			
	2.37 ± 0.29		1137 ± 40					
CH₂Cl₂				15.5 ± 3.4	296 ± 2	DF-LMR	Howard and Evenson[1]	
				12.4	298	RR [relative to k(methane) $= 8.36 \times 10^{-15}$][b]	Cox et al.[7]	
				14.5 ± 2.0	298.5	FP-RF	Perry et al.[8]	
				4.75 ± 0.57	245	FP-RF	Davis et al.[9]	245–375
				11.6 ± 0.5	298			
	4.27 ± 0.63		1094 ± 81	22.3 ± 0.5	375			
				9.59 ± 0.69	251	DF-RF	Jeong and Kaufman[3,4]	251–455
				15.3 ± 0.95	292			
				20.8 ± 1.4	323			
				27.6 ± 1.9	342			
				35.2 ± 2.4	384			
				45.0 ± 2.9	415			
	1.61×10^{-7}	2.54	186 ± 493	60.9 ± 3.8	455			
	5.57 ± 0.77		1042 ± 45					
	6.81		1117	14.6	300	PR-RA	Nielsen et al.[11]	300–400
				17.6 ± 2.0	298	LP-LIF	Taylor et al.[12]	298–775
				18.7 ± 4.7	299			
				24.8 ± 2.3	335			
				29.4 ± 2.8	376			
				43.3 ± 4.2	425			
				61.5 ± 6.9	455			
				72.1 ± 18.0	474			
				85.8 ± 5.9	495			
				97.8 ± 9.1	535			
				119 ± 7.3	575			
				151 ± 13.4	615			
				155 ± 10.9	615			
				163 ± 8.7	655			
				170 ± 7.5	655			
				202 ± 10.7	695			
				224 ± 11.7	735			
	1.52×10^{-4}	1.58	622 ± 60	257 ± 10.8	775			

TABLE 4. Rate constants k and temperature-dependent parameters for the gas-phase reactions of the OH radical with haloalkanes — Continued

Haloalkane	$10^{12} \times A$ (cm^3 mole-cule^{-1} s^{-1})	n	B (K)	$10^{14} \times k$ (cm^3 molecule^{-1} s^{-1})	at T (K)	Technique	Reference	Tempera-ture range covered (K)
CHF$_3$				$0.02^{+0.02}_{-0.015}$	296 ± 2	DF-LMR	Howard and Evenson[1]	
				183	1300	RR [relative to $k(H_2) = 5.69 \times 10^{-12}$]^a	Bradley et al.[17]	
				55	1255	SH/FP-RA	Ernst et al.[18]	1255–1445
				60	1320			
				66	1320			
				83	1345			
				70	1395			
				55	1400			
				93	1445			
				0.13 ± 0.04	296	DF-RF	Clyne and Holt[14]	296–430
				0.14 ± 0.06	430			
				0.035 ± 0.017	297 ± 2	FP-RA	Nip et al.[2]	
				0.169 ± 0.011	387	DF-RF	Jeong and Kaufman[3,4]	387–480
				0.237 ± 0.017	410			
				0.331 ± 0.027	428			
				0.448 ± 0.029	447			
				0.564 ± 0.036	465			
	2.98 ± 1.07		2909 ± 156	0.719 ± 0.045	480			
				0.23 ± 0.04	308	PR-RA	Bera and Hanrahan[5]	
CHF$_2$Cl				0.475 ± 0.048	296.9	FP-RF	Atkinson et al.[19]	297–434
				1.15 ± 0.12	348.0			
	1.21		1636 ± 151	2.71 ± 0.27	433.7			
				0.34 ± 0.07	296 ± 2	DF-LMR	Howard and Evenson[1]	
				0.170 ± 0.040	250	FP-RF	Watson et al.[15]	250–350
				0.277 ± 0.038	273			
				0.48 ± 0.046	298			
	0.925 ± 0.10		1575 ± 71	1.01 ± 0.08	350			
				0.177 ± 0.002	253	DF-RF	Chang and Kaufman[20]	253–427
				0.425 ± 0.028	296			
				1.20 ± 0.03	358			
	1.20 ± 0.16		1657 ± 39	2.49 ± 0.10	427			
				0.20	263	FP-RA	Handwerk and Zellner[16]	263–373
				0.27	273			
				0.51	283			
				0.46 ± 0.08	293			
	2.1 ± 0.6		1780 ± 150	1.7 ± 0.3	373			
				0.33 ± 0.07	294	DF-RF	Clyne and Holt[14]	294–426
				0.77 ± 0.12	321			
				1.28 ± 0.11	343			
				1.97 ± 0.07	376			
				2.77 ± 0.17	391			
	$9.5^{+1.7}_{-1.4}$		2300 ± 200	3.90 ± 0.07	426			
				0.458 ± 0.058	297	FP-RA	Paraskevopoulos et al.[10]	
				0.483 ± 0.032	293	DF-RF	Jeong and Kaufman[3,4]	293–482
				0.768 ± 0.048	327			
				1.08 ± 0.075	360			
				1.79 ± 0.14	391			
				2.75 ± 0.18	436			

TABLE 4. Rate constants k and temperature-dependent parameters for the gas-phase reactions of the OH radical with haloalkanes — Continued

Haloalkane	$10^{12} \times A$ (cm^3 mole-cule^{-1} s^{-1})	n	B (K)	$10^{14} \times k$ (cm^3 molecule^{-1} s^{-1})	at T (K)	Technique	Reference	Temperature range covered (K)
	5.03×10^{-16}	5.11	-252 ± 780	4.39 ± 0.27	482			
	1.27 ± 0.21		1661 ± 60					
CHFCl$_2$				2.6 ± 0.4	296 ± 2	DF-LMR	Howard and Evenson[1]	
				2.7 ± 0.3	298.4	FP-RF	Perry et al.[8]	298–422
				4.8 ± 0.5	349.5			
	1.75		1253 ± 151	9.1 ± 0.9	421.7			
				1.12 ± 0.12	245	FP-RF	Watson et al.[15]	245–375
				2.09 ± 0.18	273			
				2.88 ± 0.24	298			
	1.87 ± 0.2		1245 ± 26	6.68 ± 0.82	375			
				1.28 ± 0.25	241	DF-RF	Chang and Kaufman[20]	241–396
				1.73 ± 0.13	250			
				2.70 ± 0.20	288			
				3.04 ± 0.11	296			
				7.17 ± 0.16	380			
	1.16 ± 0.17		1073 ± 40	7.52 ± 0.29	396			
				3.54 ± 0.26	293	DF-RF	Clyne and Holt[14]	293–413
				6.57 ± 0.22	330			
				9.77 ± 0.38	373			
	$4.8^{+1.0}_{-0.8}$		1400 ± 100	15.2 ± 1.0	413			
				3.39 ± 0.86	297	FP-RA	Paraskevopoulos et al.[10]	
				1.88 ± 0.14	250	DF-RF	Jeong and Kaufman[3,4]	250–483
				3.37 ± 0.22	295			
				4.25 ± 0.27	315			
				5.85 ± 0.36	354			
				7.86 ± 0.48	392			
				10.5 ± 0.65	433			
	1.97×10^{-6}	1.94	382 ± 413	14.8 ± 1.0	483			
	1.19 ± 0.15		1052 ± 45					
	1.83		1787	0.515	300	PR-RA	Nielsen et al.[11]	300–400
CHCl$_3$				10.1 ± 1.5	296 ± 2	DF-LMR	Howard and Evenson[1]	
				20.0	298	RR [relative to k(methane) $= 8.36 \times 10^{-15}$][b]	Cox et al.[7]	
				4.39 ± 0.28	245	FP-RF	Davis et al.[9]	245–375
				11.4 ± 0.7	298			
	4.69 ± 0.71		1134 ± 108	21.8 ± 1.4	375			
				5.51 ± 0.41	249	DF-RF	Jeong and Kaufman[3,4]	249–487
				10.1 ± 0.65	298			
				16.0 ± 1.0	339			
				23.2 ± 1.6	370			
				30.8 ± 2.0	411			
				44.8 ± 2.7	466			
	6.91×10^{-8}	2.65	262 ± 398	55.0 ± 3.9	487			
	5.63 ± 0.68		1183 ± 45					
				29	300	RR [relative to k(toluene) $= 5.91 \times 10^{-12}$][b]	Klöpffer et al.[21]	
				10.3 ± 1.5	295	LP-LIF	Taylor et al.[12]	295–775
				11.0 ± 1.9	295			

TABLE 4. Rate constants k and temperature-dependent parameters for the gas-phase reactions of the OH radical with haloalkanes — Continued

Haloalkane	$10^{12} \times A$ (cm³ mole-cule⁻¹ s⁻¹)	n	B (K)	$10^{14} \times k$ (cm³ molecule⁻¹ s⁻¹)	at T (K)	Technique	Reference	Temperature range covered (K)
				15.5 ± 4.1	339			
				22.1 ± 2.4	383			
				32.8 ± 3.2	430			
				48.2 ± 6.4	476			
				82.8 ± 6.8	571			
				95.7 ± 10.1	626			
				108 ± 8.1	627			
				118 ± 7.9	680			
				166 ± 15.1	735			
				187 ± 11.1	772			
	1.92×10^{-8}	2.78	95 ± 60	188 ± 28.2	775			
CF₄				<0.04	296 ± 2	DF-LMR	Howard and Evenson[1]	
				<0.1	293	DF-RF	Clyne and Holt[22]	
CF₃Cl				<0.07	296 ± 2	DF-LMR	Howard and Evenson[1]	
CF₃Br				$\leqslant 0.1$	298	DF-EPR	Le Bras and Combourieu[23]	
CF₃I				12 ± 2	295	FP-RA	Garraway and Donovan[24]	
CF₂Cl₂				<0.1	297.3	FP-RF	Atkinson et al.[19]	297–424
				<0.1	342.9			
				<0.1	423.8			
				<0.04	296 ± 2	DF-LMR	Howard and Evenson[1]	
				<0.012	298	RR [relative to k(methane) $= 8.36 \times 10^{-15}$][b]	Cox et al.[7]	
				<0.06	478	DF-RF	Chang and Kaufman[25]	298–478
				<0.1	293	DF-RF	Clyne and Holt[22]	
CF₂ClBr				<0.1	293	DF-RF	Clyne and Holt[22]	
CFCl₃				<0.1	296.8	FP-RF	Atkinson et al.[19]	297–424
				<0.1	347.7			
				<0.1	423.8			
				<0.05	296 ± 2	DF-LMR	Howard and Evenson[1]	
				<0.005	298	RR [relative to k(methane) $= 8.36 \times 10^{-15}$][b]	Cox et al.[7]	
				<0.05	480	DF-RF	Chang and Kaufman[25]	381–480
				<0.1	293	DF-RF	Clyne and Holt[22]	
CCl₄				<0.4	296 ± 2	DF-LMR	Howard and Evenson[1]	
				<0.012	298	RR [relative to k(methane) $= 8.36 \times 10^{-15}$][b]	Cox et al.[7]	
				<0.1	293	DF-RF	Clyne and Holt[22]	
CH₃CH₂F				23.2 ± 3.7	297 ± 2	FP-RA	Nip et al.[2]	
CH₃CH₂Cl				39.0 ± 7.0	296	DF-LMR	Howard and Evenson[26]	

TABLE 4. Rate constants k and temperature-dependent parameters for the gas-phase reactions of the OH radical with haloalkanes — Continued

Haloalkane	$10^{12} \times A$ (cm^3 molecule^{-1} s^{-1})	n	B (K)	$10^{14} \times k$ (cm^3 molecule^{-1} s^{-1})	at T (K)	Technique	Reference	Temperature range covered (K)
				39.4 ± 5.3	297	FP-RA	Paraskevopoulos et al.[10]	
CH$_3$CHF$_2$				3.1 ± 0.7	296	DF-LMR	Howard and Evenson[26]	
				3.5 ± 0.5	293	FP-RA	Handwerk and Zellner[16]	
				4.66 ± 0.16	293	DF-RF	Clyne and Holt[14]	293–417
				7.16 ± 0.26	323			
				10.1 ± 0.8	363			
	$3.0^{+1.0}_{-0.8}$		1200 ± 100	16.4 ± 0.5	417			
				3.70 ± 0.37	297 ± 2	FP-RA	Nip et al.[2]	
CH$_2$FCH$_2$F				11.2 ± 1.2	298	FP-RA	Martin and Paraskevopoulos[27]	
CH$_3$CHCl$_2$				26.0 ± 6.0	296	DF-LMR	Howard and Evenson[26]	
CH$_2$ClCH$_2$Cl				22.0 ± 5.0	296	DF-LMR	Howard and Evenson[26]	
CH$_2$BrCH$_2$Br				25.0 ± 5.5	296	DF-LMR	Howard and Evenson[26]	
CH$_3$CF$_3$				<0.1	293	DF-RF	Clyne and Holt[14]	293–425
				0.47 ± 0.15	333			
				1.29 ± 0.35	378			
	69^{+105}_{-42}		3200 ± 500	3.84 ± 1.23	425			
				0.171 ± 0.044	298	FP-RA	Martin and Paraskevopoulos[27]	
CH$_2$FCHF$_2$				4.98 ± 0.82	293	DF-RF	Clyne and Holt[14]	293–441
				4.68 ± 0.40	294			
				6.74 ± 0.43	335			
				9.09 ± 0.42	383			
	$1.5^{+0.5}_{-0.4}$		1000 ± 100	18.9 ± 0.6	441			
				1.83 ± 0.18	298	FP-RA	Martin and Paraskevopoulos[27]	
CH$_3$CF$_2$Cl				0.283 ± 0.042	296	DF-LMR	Howard and Evenson[26]	
				0.46	298	RR [relative to k(methane) $= 8.36 \times 10^{-15}$][b]	Cox et al.[7]	
				0.192 ± 0.048	273	FP-RF	Watson et al.[15]	273–375
				0.322 ± 0.048	298			
	1.15 ± 0.15		1748 ± 30	1.09 ± 0.14	375			
				0.37 ± 0.07	293	FP-RA	Handwerk and Zellner[16]	293–373
	1.8 ± 0.5		1790 ± 150	1.4 ± 0.3	373			
				0.84 ± 0.18	293	DF-RF	Clyne and Holt[14]	293–417
				0.60 ± 0.07	293			
				1.20 ± 0.11	323			
				1.44 ± 0.37	363			
				3.09 ± 0.15	380			
	$3.3^{+4.3}_{-1.9}$		1800 ± 300	4.06 ± 0.27	417			
				0.463 ± 0.173	297	FP-RA	Paraskevopoulos et al.[10]	
CH$_3$CCl$_3$				1.5 ± 0.3	296	DF-LMR	Howard and Evenson[26]	
				3.36	298	RR [relative to k(methane) $= 8.36 \times 10^{-15}$][b]	Cox et al.[7]	

TABLE 4. Rate constants k and temperature-dependent parameters for the gas-phase reactions of the OH radical with haloalkanes — Continued

Haloalkane	$10^{12} \times A$ (cm^3 molecule^{-1} s^{-1})	n	B (K)	$10^{14} \times k$ (cm^3 molecule^{-1} s^{-1})	at T (K)	Technique	Reference	Temperature range covered (K)
				0.712 ± 0.094	260	FP-RF	Watson et al. [15]	260–375
				1.59 ± 0.16	298			
	3.72 ± 0.4		1627 ± 50	4.85 ± 0.58	375			
				1.55 ± 0.22	275	DF-RF	Chang and Kaufman[20]	275–405
				2.19 ± 0.26	298			
				3.03 ± 0.30	320			
				4.94 ± 0.48	355			
	1.95 ± 0.24		1331 ± 37	6.87 ± 0.40	405			
				1.81 ± 0.16	293	DF-RF	Clyne and Holt[22]	293–430
				2.78 ± 0.74	310			
				4.59 ± 0.56	338			
				5.73 ± 0.51	371			
				7.29 ± 0.44	399			
	$2.4^{+0.9}_{-0.7}$		1394 ± 113	8.63 ± 0.40	430			
				0.83 ± 0.07	278	DF-RF	Jeong and Kaufman[28]; Jeong et al.[4]	278–457
				1.06 ± 0.11	293			
				2.93 ± 0.19	352			
				5.52 ± 0.41	400			
	5.95×10^{-8}	2.65	858 ± 866	10.2 ± 0.65	457			
	5.04 ± 0.96		1797 ± 65					
				0.318 ± 0.095	222	FP-RF	Kurylo et al.[29]	222–363
				0.447 ± 0.135	253			
				0.540 ± 0.145	263			
				1.08 ± 0.20	296			
	5.4 ± 1.8		1810 ± 100 (253–363 K)	3.85 ± 0.75	363			
				0.87	298 ± 3	RR [relative to k(CH$_3$Cl) $= 4.36 \times 10^{-14}$][b]	Nelson et al.[30]	
CH$_2$ClCHCl$_2$				28.4 ± 2.1	277	DF-RF	Jeong and Kaufman[28]; Jeong et al.[4]	277–461
				31.8 ± 2.0	295			
				37.6 ± 2.3	322			
				43.6 ± 2.8	346			
				46.8 ± 2.9	386			
				49.2 ± 3.1	400			
				52.7 ± 3.5	424			
	6.76×10^3	-1.21	906 ± 674	57.6 ± 3.7	461			
	1.65 ± 0.27		483 ± 55					
CH$_2$FCF$_3$				0.55 ± 0.07	294	DF-RF	Clyne and Holt[14]	294–429
				1.32 ± 0.10	327			
				1.64 ± 0.31	344			
				1.92 ± 0.08	358			
				3.83 ± 0.49	393			
				4.20 ± 0.47	424			
	$3.2^{+2.3}_{-1.3}$		1800 ± 200	3.64 ± 0.38	429			
				0.515 ± 0.058	298	FP-RA	Martin and Paraskevopoulos[27]	
				0.393 ± 0.024	249	DF-RF	Jeong et al.[4]	249–473
				0.441 ± 0.040	250			
				0.552 ± 0.035	268			
				0.773 ± 0.071	291			
				0.823 ± 0.055	295			
				0.844 ± 0.073	298			
				1.54 ± 0.12	342			
				2.54 ± 0.17	380			

TABLE 4. Rate constants k and temperature-dependent parameters for the gas-phase reactions of the OH radical with haloalkanes — Continued

Haloalkane	$10^{12} \times A$ (cm³ molecule⁻¹ s⁻¹)	n	B (K)	$10^{14} \times k$ (cm³ molecule⁻¹ s⁻¹)	at T (K)	Technique	Reference	Temperature range covered (K)
				3.94 ± 0.26	430			
				4.56 ± 0.29	447			
	1.22×10^{-13}	4.36	-45 ± 388	6.44 ± 0.40	473			
	1.10 ± 0.11		1424 ± 35					
CHF$_2$CHF$_2$				0.53 ± 0.15	294	DF-RF	Clyne and Holt[14]	294–434
				1.88 ± 0.27	333			
				2.12 ± 0.41	389			
	$2.8^{+6.5}_{-2.0}$		1800 ± 400	4.82 ± 0.36	434			
CH$_2$ClCF$_3$				1.05 ± 0.23	296	DF-LMR	Howard and Evenson[26]	
				1.1 ± 0.2	263	FP-RA	Handwerk and Zellner[16]	263–373
				1.2 ± 0.2	268			
				1.2 ± 0.2	273			
				1.5 ± 0.3	283			
				1.5 ± 0.3	293			
				2.8	337			
	1.1 ± 0.3		1260 ± 60	3.6 ± 0.8	373			
				1.03 ± 0.30	294	DF-RF	Clyne and Holt[14]	294–427
				3.83 ± 0.57	322			
				3.86 ± 0.31	344			
				6.94 ± 0.33	358			
				6.58 ± 0.25	385			
				13.0 ± 1.2	407			
	39^{+46}_{-21}		2300 ± 300	15.4 ± 1.3	427			
CH$_2$ClCF$_2$Cl				0.839 ± 0.037	250	FP-RF	Watson et al.[31]	250–350
				1.9 ± 0.2	298			
	1.87 ± 0.27		1351 ± 78	3.95 ± 0.10	350			
	$\left[3^{+6}_{-1} \right.$		$\left. 1578^{+400}_{-230} \right]$c					
				1.42 ± 0.11	249	DF-RF	Jeong et al.[4]	249–473
				1.60 ± 0.10	253			
				1.91 ± 0.16	267			
				2.72 ± 0.18	295			
				2.42 ± 0.16	297			
				4.31 ± 0.28	333			
				5.95 ± 0.37	365			
				8.06 ± 0.51	383			
				10.4 ± 0.65	418			
	5.54×10^{-14}	4.58	-252 ± 377	16.0 ± 1.15	473			
	2.02 ± 0.24		1263 ± 35					
CHF$_2$CF$_3$				0.50 ± 0.22	294	DF-RF	Clyne and Holt[14]	294–441
				0.49 ± 0.14	294			
				0.62 ± 0.18	336			
				1.13 ± 0.33	378			
	$0.17^{+0.10}_{-0.06}$		1100 ± 100	1.58 ± 0.29	441			
				0.249 ± 0.028	298	FP-RA	Martin and Paraskevopoulos[27]	
CHFClCF$_3$				1.24 ± 0.19	296	DF-LMR	Howard and Evenson[26]	
				0.433 ± 0.019	250	FP-RF	Watson et al.[31]	250–375
				0.94 ± 0.03	301			
	0.613 ± 0.04		1244 ± 90	2.28 ± 0.16	375			
CHCl$_2$CF$_3$				2.84 ± 0.43	296	DF-LMR	Howard and Evenson[26]	
				1.62 ± 0.05	245	FP-RF	Watson et al.[31]	245–375

TABLE 4. Rate constants k and temperature-dependent parameters for the gas-phase reactions of the OH radical with haloalkanes — Continued

Haloalkane	$10^{12} \times A$ (cm^3 molecule^{-1} s^{-1})	n	B (K)	$10^{14} \times k$ (cm^3 molecule^{-1} s^{-1})	at T (K)	Technique	Reference	Temperature range covered (K)
				3.6 ± 0.4	298			
	1.24 ± 0.3		1056 ± 70	7.2 ± 0.35	375			
	$\left[\, 1.4 \pm 0.4 \right.$		$1102^{+157}_{-106} \left.\right]^c$					
				3.86 ± 0.19	293	DF-RF	Clyne and Holt[14]	293–429
				5.86 ± 0.15	329			
				8.01 ± 0.33	366			
	1.12 ± 0.05		1000 ± 100	11.1 ± 0.4	429			
CF_2ClCF_2Cl				<0.05	296	DF-LMR	Howard and Evenson[26]	
$CF_2ClCFCl_2$				<0.03	296	DF-LMR	Howard and Evenson[26]	
				<0.03	298	FP-RF	Watson et al.[15]	
$CH_2ClCHClCH_3$				$\leqslant 44$	~296	RR [relative to k(dimethyl ether) $= 2.96 \times 10^{-12}]^b$	Tuazon et al.[32]	
$CH_2BrCHBrCH_2Cl$				43.5 ± 5.0	296 ± 2	RR [relative to k(dimethyl ether) $= 2.96 \times 10^{-12}]^b$	Tuazon et al.[33]	

aSee Introduction.

bFrom the present recommendations.

cArrhenius expression estimated after allowance for possible contributions to the observed OH radical decay rates from the measured impurity levels present (see text).

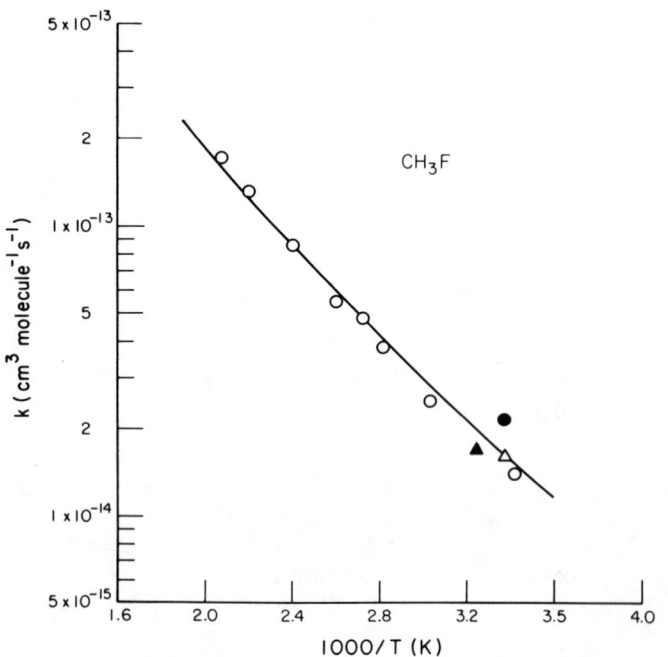

FIG. 21.   Arrhenius plot of rate constants for the reaction of the OH radical with CH$_3$F. ($\triangle$) Howard and Evenson;[1] ($\bullet$) Nip et al.;[2] ($\bigcirc$) Jeong and Kaufman;[3,4] ($\blacktriangle$) Bera and Hanrahan;[5] (———) recommendation (see text).

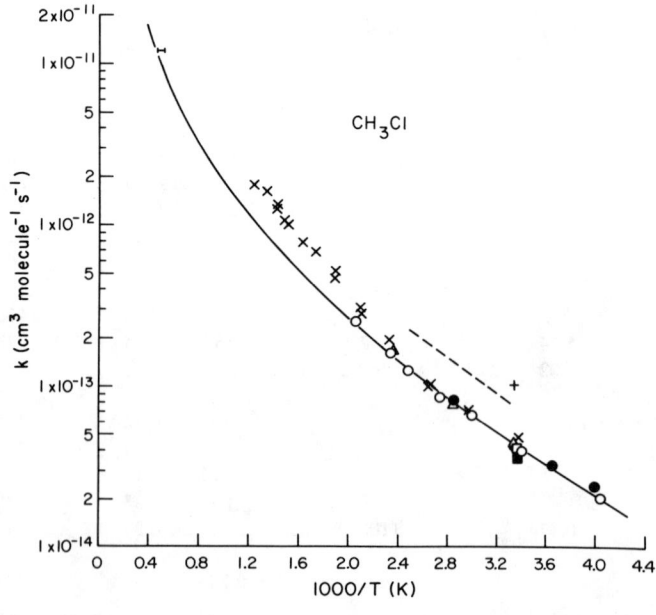

FIG. 22.   Arrhenius plot of rate constants for the reaction of the OH radical with CH$_3$Cl. ($\vdash$) Wilson et al.;[6] ($\blacksquare$) Howard and Evenson;[1] ($+$) Cox et al.;[7] ($\triangle$) Perry et al.;[8] ($\bullet$) Davis et al.;[9] ($\square$) Paraskevopoulos et al.;[10] ($\bigcirc$) Jeong and Kaufman;[3] (— — —) Nielsen et al.;[11] ($\times$) Taylor et al.;[12] (———) recommendation (see text).

(3) CH₃Br

The available rate constants of Wilson,[13] Howard and Evenson[1] and Davis *et al.*[9] are listed in Table 4 and are plotted in Arrhenius form in Fig. 23. The two absolute studies carried out[1,9] are in good agreement at room temperature. The Arrhenius plot of these absolute data[1,9] does not show any evidence of curvature over the relatively small temperature range (244–350 K) studied (Fig. 23), and a unit-weighted least-squares analysis of these data[1,9] yields the Arrhenius expression of

$$k(CH_3Br) = (7.40^{+5.32}_{-3.10})$$
$$\times 10^{-13} e^{-(875 \pm 155)/T} cm^3 \ molecule^{-1} \ s^{-1}$$

over the temperature range 244–350 K, where the indicated error limits are two least-squares standard deviations, and

$$k(CH_3Br) = 3.93 \times 10^{-14} \ cm^3 \ molecule^{-1} \ s^{-1} \ at \ 298 \ K,$$

with an estimated overall uncertainty at 298 K of ±20%. This is identical to the recommendation of Atkinson,[36] and is plotted in Fig. 23 as the dashed line.

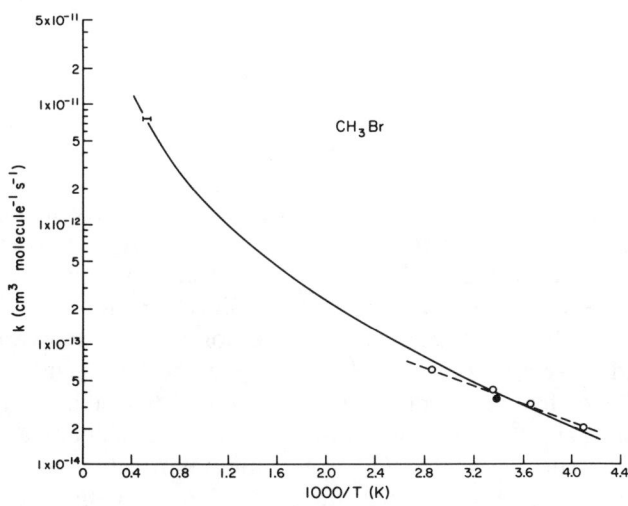

FIG. 23. Arrhenius plot of rate constants for the reaction of the OH radical with CH₃Br (⊢) Wilson;[13] (●) Howard and Evenson;[1] (○) Davis *et al.*;[9] (— — —) Arrhenius expression derived from the data of Howard and Evenson[1] and Davis *et al.*;[9] (———) recommendation (see text).

As discussed above, the rate constant obtained from the relative rate study of Wilson[13] at 1800–2000 K can be utilized to derive a recommendation applicable up to ~2000 K. Thus, using the expression $k = CT^2e^{-D/T}$, a unit-weighted least-squares analysis of the data of Wilson,[13] Howard and Evenson[1] and Davis *et al.*[9] yields the recommendation of

$$k(CH_3 \ Br) = (2.60^{+0.86}_{-0.65})$$
$$\times 10^{-18} \ T^2 \ e^{-(521 \pm 89)/T} \ cm^3 \ molecule^{-1} \ s^{-1}$$

over the temperature range 244–2000 K, where the indicated error limits are two least-squares standard deviations, and

$$k(CH_3Br) = 4.02 \times 10^{-14} \ cm^3 \ molecule^{-1} \ s^{-1} \ at \ 298 \ K,$$

with an estimated overall uncertainty at 298 K of ±20%. Over the temperature range ~240–350 K this expression yields similar rate constants to the recent NASA recommendation[37] of

$$k(CH_3Br) = 1.17 \times 10^{-18} \ T^2 \ e^{-295/T} \ cm^3 \ molecule^{-1} \ s^{-1}$$

derived from the data of Howard and Evenson[1] and Davis *et al.*[9]

(4) CH₂F₂

The available rate constants of Howard and Evenson,[1] Clyne and Holt,[14] Nip *et al.*,[2] Jeong and Kaufman[3] and Bera and Hanrahan[5] are given in Table 4 and are plotted in Arrhenius form in Fig. 24. In this case the rate constants of Clyne and Holt[14] are in reasonably good agreement with those of Howard and Evenson,[1] Nip *et al.*[2] and Jeong and Kaufman,[3] although their room temperature rate constant[14] is the lowest of those measured. In accordance with the discussion above, a unit-weighted least-squares analysis of the data of Howard and Evenson,[1] Nip *et al.*[2] and Jeong and Kaufman,[3] using the expression $k = CT^2e^{-D/T}$, yields the recommendation of

$$k(CH_2F_2) = (5.06^{+2.66}_{-1.74})$$
$$\times 10^{-18} \ T^2 \ e^{-(1107 \pm 142)/T} \ cm^3 \ molecule^{-1} \ s^{-1}$$

over the temperature range 250–492 K, where the indicated errors are two least-squares standard deviations, and

$$k(CH_2F_2) = 1.09 \times 10^{-14} \ cm^3 \ molecule^{-1} \ s^{-1} \ at \ 298 \ K,$$

with an estimated overall uncertainty at 298 K of ±30%. This recommendation is identical to that of Atkinson,[36] being based upon the same data set.

(5) CH₂FCl

The available rate constants of Howard and Evenson,[1] Watson *et al.*,[15] Handwerk and Zellner,[16] Paraskevopoulos *et al.*[10] and Jeong and Kaufman[3] are given in Table 4 and are plotted in Arrhenius form in Fig. 25. These rate constants are in reasonably good agreement, although there is a significant discrepancy between the rate constants obtained by Watson et al.[15] and Jeong and Kaufman[3] at ~250 K. Although it is not obvious from Fig. 25 whether or not the Arrhenius plot exhibits curvature, a unit-weighted least-squares analysis of these data[1,3,10,15,16] has been carried out, using the equation $k = CT^2 e^{-D/T}$, to yield the recommendation of

$$k(CH_2FCl) = (3.77^{+1.66}_{-1.16})$$

$$\times 10^{-18} T^2 e^{-(604 \pm 115)/T} cm^3 molecule^{-1} s^{-1}$$

over the temperature range 245–486 K, where the indicated error limits are two least-squares standard deviations, and

$$k(CH_2FCl) = 4.41 \times 10^{-14} cm^3 molecule^{-1} s^{-1} at 298 K,$$

with an estimated overall uncertainty at 298 K of ±20%. This recommendation is identical to those of Atkinson[36] and DeMore et al.[37]

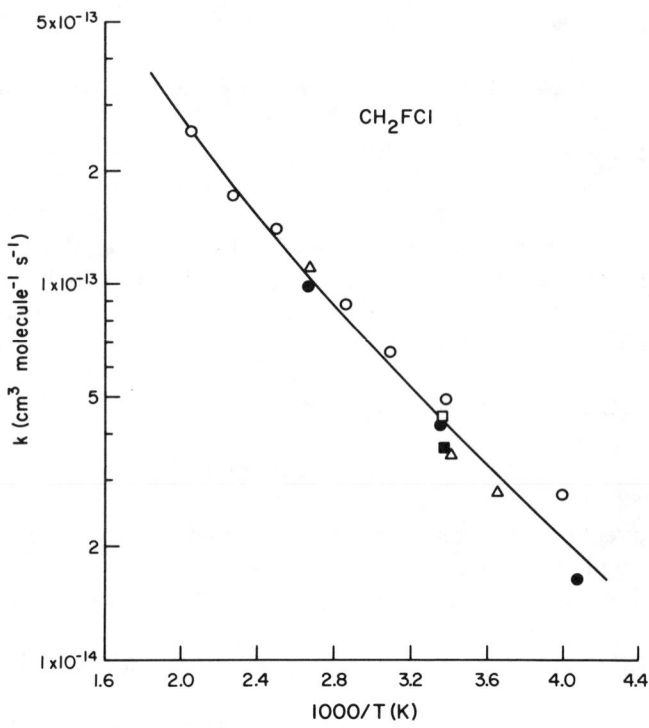

FIG. 25. Arrhenius plot of rate constants for the reaction of the OH radical with CH_2FCl. (■) Howard and Evenson;[1] (●) Watson et al.;[15] (△) Handwerk and Zellner;[16] (□) Paraskevopoulos et al.;[10] (○) Jeong and Kaufman;[3] (———) recommendation (see text).

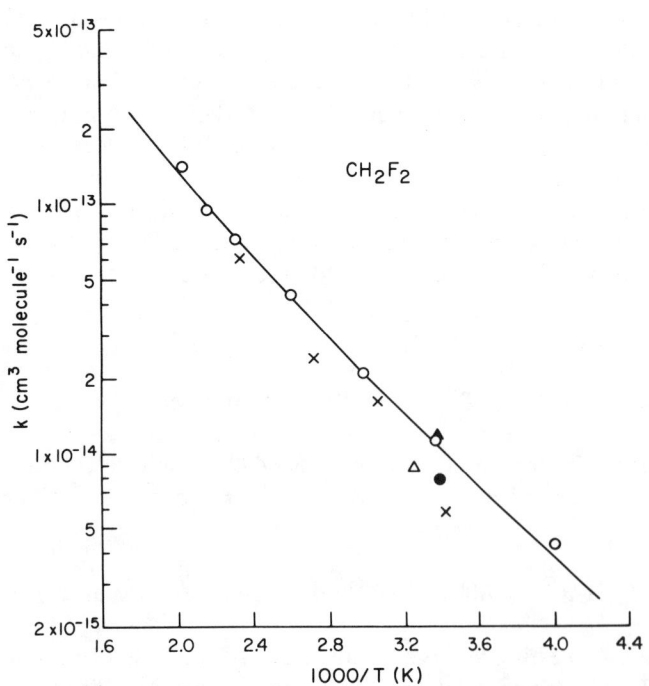

FIG. 24. Arrhenius plot of rate constants for the reaction of the OH radical with CH_2F_2. (●) Howard and Evenson;[1] (x) Clyne and Holt;[14] (▲) Nip et al.;[2] (○) Jeong and Kaufman;[3] (△) Bera and Hanrahan;[5] (———) recommendation (see text).

(6) CH_2Cl_2

The available kinetic data of Howard and Evenson,[1] Cox et al.,[7] Perry et al.,[8] Davis et al.,[9] Jeong and Kaufman,[3] Nielsen et al.[11] and Taylor et al.[12] are given in Table 4 and are plotted in Arrhenius form in Fig. 26. While the room temperature rate constants of Howard and Evenson,[1] Perry et al.[8] and Jeong and Kaufman[3] are in excellent agreement, the rate constants obtained by Davis et al.[9] are uniformly lower than those of Jeong and Kaufman[3] by ~20–40% over the temperature range common to both studies. In view of the situation concerning the data of Taylor et al.[12] for CH_3Cl discussed above, their data[12] have not been used in the evaluation of the rate constant for CH_2Cl_2. A unit-weighted least-squares analysis of the data of Howard and Evenson,[1] Perry et al.,[8] Davis et al.[9] and Jeong and Kaufman,[3] using the equation $k = CT^2 e^{-D/T}$, yields the recommendation of

$$k(CH_2Cl_2) = (8.54^{+8.18}_{-4.19})$$

$$\times 10^{-18} T^2 e^{-(500 \pm 212)/T} cm^3 molecule^{-1} s^{-1}$$

over the temperature range 245–455 K, where the indicated errors are two least-squares standard deviations, and

$$k(CH_2Cl_2) = 1.42 \times 10^{-13} \text{ cm}^3 \text{ molecule}^{-1} \text{ s}^{-1} \text{ at 298 K,}$$

with an estimated overall uncertainty at 298 K of ±25%. The rate constants measured by Taylor et al.,[12] especially for temperatures ≥350 K, are in excellent agreement with this recommended rate expression (Fig. 26). This recommendation is identical to that of Atkinson[36] and very similar to the recent NASA evaluation,[37] but with slightly higher estimated uncertainty limits at 298 K. The Arrhenius expression of Nielsen et al.[11] is in good agreement with the present recommendation over the temperature range ~300–400 K studied.[11]

since their data show no significant effect of temperature and differ by factors of >2 from the other literature data. The rate constant measured by Bera and Hanrahan[5] is clearly in error, possibly due to the presence of reactive impurities.[5]

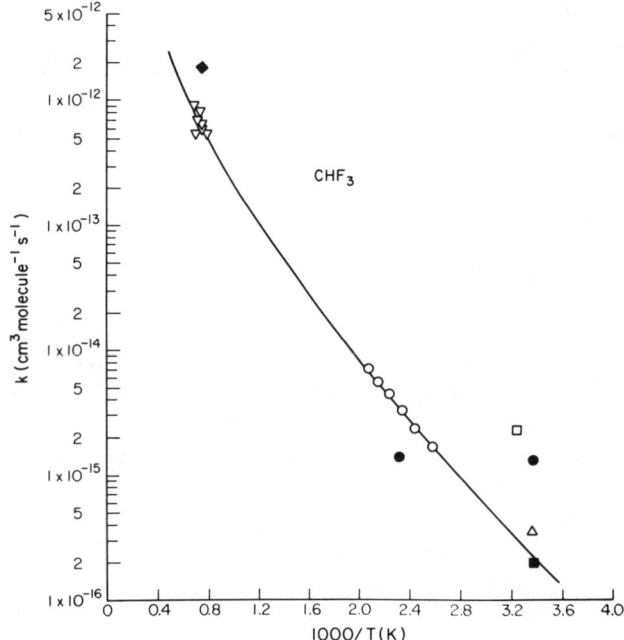

FIG. 27. Arrhenius plot of rate constants for the reaction of the OH radical with CHF₃. (■) Howard and Evenson;[1] (◆) Bradley et al.;[17] (∇) Ernst et al.;[18] (●) Clyne and Holt;[14] (△) Nip et al.;[2] (○) Jeong and Kaufman;[3,4] (□) Bera and Hanrahan;[5] (———) recommendation (see text).

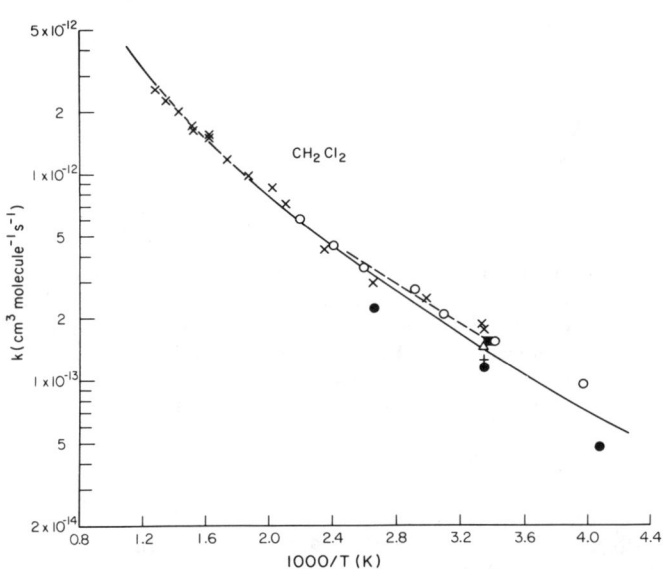

FIG. 26. Arrhenius plot of rate constants for the reaction of the OH radical with CH₂Cl₂. (■) Howard and Evenson;[1] (+) Cox et al.;[7] (△) Perry et al.;[8] (●) Davis et al.;[9] (○) Jeong and Kaufman;[3] (— — —) Nielsen et al.;[11] (x) Taylor et al.;[12] (———) recommendation (see text).

(7) CHF₃

The available rate constants of Howard and Evenson,[1] Bradley et al.,[17] Ernst et al.,[18] Clyne and Holt,[14] Nip et al.,[2] Jeong and Kaufman[3,4] and Bera and Hanrahan[5] are given in Table 4 and are plotted in Arrhenius form in Fig. 27. The reaction of the OH radical with CHF₃ is very slow at room temperature, and the rate constants determined by Howard and Evenson[1] and Nip et al.[2] are subject to large uncertainties. This appears to be also true for the rate constants reported by Clyne and Holt,[14]

In view of the significant uncertainties associated with the rate constants measured by Howard and Evenson,[1] Nip et al.[2] and Bradley et al.[17] (due to the large differences in the rate constant derived depending on whether H₂ or CO is used as the reference compound in their relative rate study),[17] a unit-weighted least-squares analysis of the data of Ernst et al.[18] and Jeong and Kaufman[3,4] was carried out, using the equation $k = CT^2 e^{-D/T}$, to yield the recommendation of

$$k(CHF_3) = (1.49^{+0.25}_{-0.21})$$

$$\times 10^{-18} T^2 e^{-(1887 \pm 92)/T} \text{ cm}^3 \text{ molecule}^{-1} \text{ s}^{-1}$$

over the temperature range 387–1445 K, where the indicated errors are two least-squares standard deviations, and

$$k(CHF_3) = 2.4 \times 10^{-16} \text{ cm}^3 \text{ molecule}^{-1} \text{ s}^{-1} \text{ at 298 K,}$$

with an estimated overall uncertainty at 298 K of ±50%.

This recommended expression yields a rate constant at 296 K in agreement, within the experimental error limits, with those measured by Howard and Evenson[1] and Nip et al.[2] Since this recommendation is based upon data obtained at temperatures $\geqslant 387$ K, it should be used with caution for temperatures $\lesssim 300$ K. This recommendation is similar to that of

$$k(CHF_3) = 2.17 \times 10^{-18} \ T^2 \ e^{-2048/T} \ cm^3 \ molecule^{-1} \ s^{-1}$$

of Atkinson,[36] derived from the data of Jeong and Kaufman[3,4] over the temperature range 387–480 K (due to a typographical error, the value of C was incorrectly cited[36] as $2.1 \times 10^{-18} \ cm^3 \ molecule^{-1} \ s^{-1}$).

(8) CHF₂Cl

The available rate constants[1,3,10,14–16,19,20] are given in Table 4 and are plotted in Arrhenius form in Fig. 28. It can be seen that the rate constants of Atkinson et al.,[19] Howard and Evenson,[1] Watson et al.,[15] Chang and Kaufman,[20] Handwerk and Zellner,[16] Paraskevopoulos et al.[10] and Jeong and Kaufman[3] are in good agreement. While the rate constants measured by Clyne and Holt[14] agree well with those studies at ~294–321 K, their rate constants at higher temperatures are increasingly higher than the consensus values from these other studies. A unit-weighted least-squares analysis of the rate constant data of Atkinson et al.,[19] Howard and Evenson,[1] Watson et al.,[15] Chang and Kaufman,[20] Handwerk and Zellner,[16] Paraskevopoulos et al.[10] and Jeong and Kaufman,[3] using the equation $k = CT^2 e^{-D/T}$, yields the recommendation of

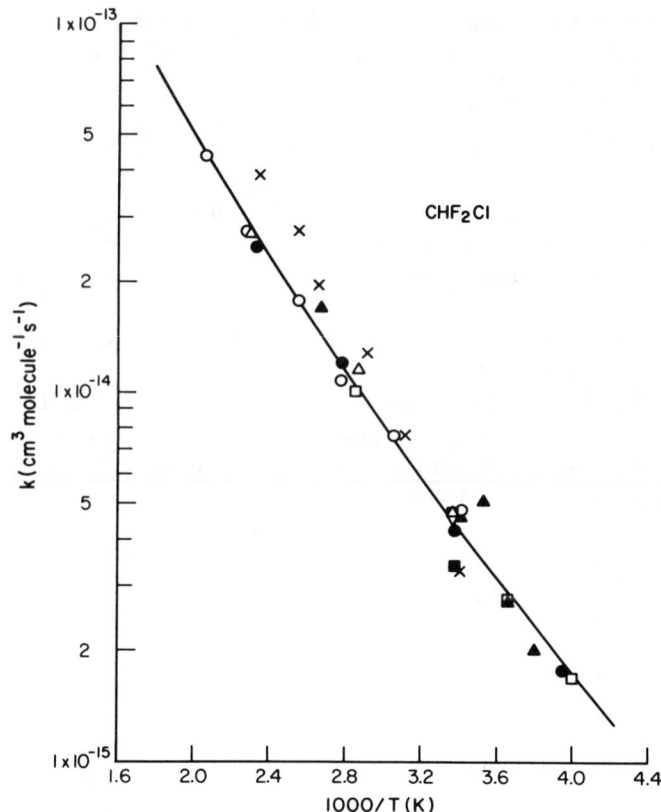

FIG. 28.   Arrhenius plot of rate constants for the reaction of the OH radical with CHF₂Cl. (△) Atkinson et al.;[19] (■) Howard and Evenson;[1] (□) Watson et al.;[15] (●) Chang and Kaufman;[20] (▲) Handwerk and Zellner;[16] (x) Clyne and Holt;[14] (▽) Paraskevopoulos et al.;[10] (○) Jeong and Kaufman;[3] (———) recommendation (see text).

$$k(CHF_2Cl) = (1.51^{+0.52}_{-0.39})$$

$$\times 10^{-18} \ T^2 \ e^{-(1000 \pm 94)/T} \ cm^3 \ molecule^{-1} \ s^{-1}$$

over the temperature range 250–482 K, where the indicated error limits are two least-squares standard deviations, and

$$k(CHF_2Cl) = 4.68 \times 10^{-15} \ cm^3 \ molecule^{-1} \ s^{-1} \ at \ 298 \ K,$$

with an estimated overall uncertainty at 298 K of ±20%. This recommendation is identical to those of Atkinson[36] and DeMore et al.[37]

(9) CHFCl₂

The available rate constants[1,3,8,10,11,14,15,20] are given in Table 4 and are plotted in Arrhenius form in Fig. 29.

Analogous to CHF₂Cl, the rate constants measured by Clyne and Holt[14] at elevated temperatures are significantly higher than those of Howard and Evenson,[1] Perry et al.,[8] Watson et al.,[15] Chang and Kaufman,[20] Paraskevopoulos et al.[10] and Jeong and Kaufman,[3] all of which are in reasonably good agreement. The data reported for this reaction by Nielsen et al.[11] are lower than those from the other studies by a factor of ~5, suggesting that the reactant studied was CHF₂Cl, and not CHFCl₂ as reported. A unit-weighted least-squares analysis of the data of Howard and Evenson,[1] Perry et al.,[8] Watson et al.,[15] Chang and Kaufman,[20] Paraskevopoulos et al.[10] and Jeong and Kaufman,[3] using the equation $k = CT^2 e^{-D/T}$, yields the recommendation of

$$k(CHFCl_2) = (1.70^{+0.47}_{-0.37})$$

$$\times 10^{-18} \ T^2 \ e^{-(479 \pm 76)/T} \ cm^3 \ molecule^{-1} \ s^{-1}$$

over the temperature range 241–483 K, where the indicated error limits are two least-squares standard deviations, and

$k(CHFCl_2) = 3.03 \times 10^{-14} \text{ cm}^3 \text{ molecule}^{-1} \text{ s}^{-1}$ at 298 K,

with an estimated overall uncertainty at 298 K of ±20%. This recommendation is identical to that of Atkinson[36] and essentially identical to that of the recent NASA evaluation.[37]

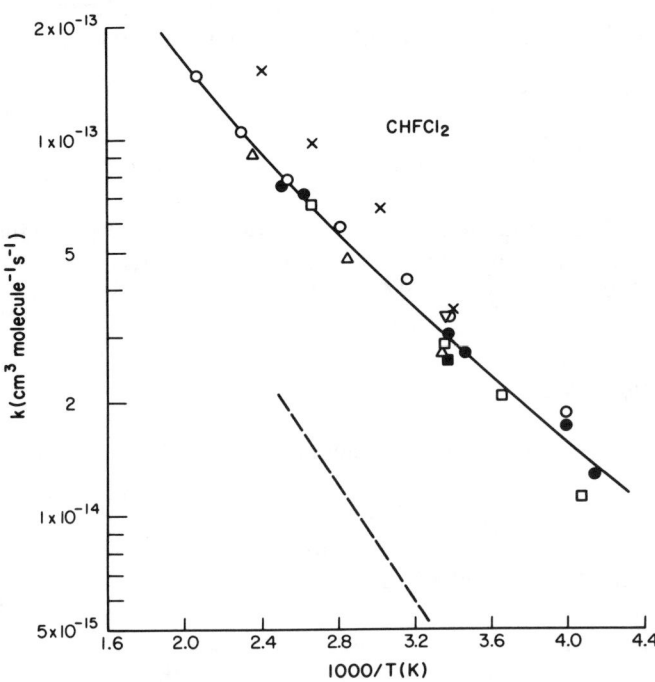

FIG. 29. Arrhenius plot of rate constants for the reaction of the OH radical with CHFCl₂. (■) Howard and Evenson;[1] (Δ) Perry et al.;[8] (□) Watson et al.;[15] (●) Chang and Kaufman;[20] (x) Clyne and Holt;[14] (∇) Paraskevopoulos et al.;[10] (○) Jeong and Kaufman;[3] (— — —) Nielsen et al.;[11] (———) recommendation (see text).

(10) CHCl₃

The available kinetic data of Howard and Evenson,[1] Cox et al.,[7] Davis et al.,[9] Jeong and Kaufman,[3] Klöpffer et al.[21] and Taylor et al.[12] are given in Table 4 and those of Howard and Evenson,[1] Cox et al.,[7] Davis et al.,[9] Jeong and Kaufman[3] and Taylor et al.[12] are plotted in Arrhenius form in Fig. 30. It can be seen that the rate constants of Howard and Evenson,[1] Davis et al.,[9] Jeong and Kaufman[3] and Taylor et al.[12] are in excellent agreement. However, consistent with the evaluations for the reactions of the OH radical with CH₃Cl and CH₂Cl₂, the rate constants of Taylor et al.[12] were not used in the derivation of the recommended rate expression for CHCl₃. Thus, a unit-weighted least-squares analysis of the data of Howard and Evenson,[1] Davis et al.[9] and Jeong and Kaufman,[3] using the equation $k = CT^2e^{-D/T}$, yields the recommendation of

$k(CHCl_3) = (6.30^{+1.18}_{-1.00})$

$\times 10^{-18} \; T^2 \; e^{-(504 \pm 56)/T} \text{ cm}^3 \text{ molecule}^{-1} \text{ s}^{-1}$

over the temperature range 245–487 K, where the indicated error limits are two least-squares standard deviations, and

$k(CHCl_3) = 1.03 \times 10^{-13} \text{ cm}^3 \text{ molecule}^{-1} \text{ s}^{-1}$ at 298 K,

with an estimated overall uncertainty at 298 K of ±20%.

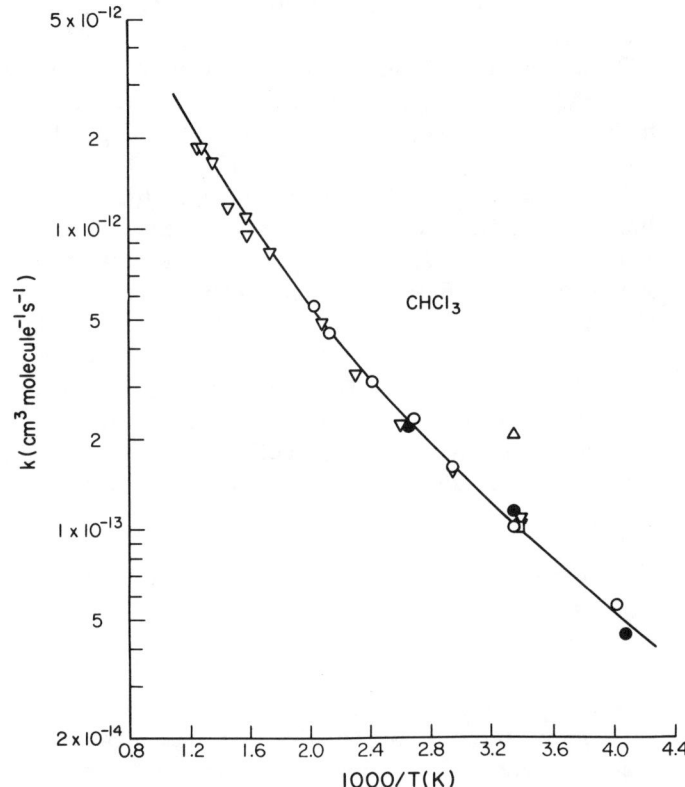

FIG. 30. Arrhenius plot of rate constants for the reaction of the OH radical with CHCl₃. (□) Howard and Evenson;[1] (Δ) Cox et al.;[7] (●) Davis et al.;[9] (○) Jeong and Kaufman;[3] (∇) Taylor et al.;[12] (———) recommendation (see text).

This recommendation is identical to those of Atkinson[36] and DeMore et al.[37]

(11) CF₄, CF₃Cl, CF₃Br, CF₂Cl₂, CF₂ClBr, CFCl₃ and CCl₄

For these halomethanes, no reaction with the OH radical has been observed. Based upon the measured room temperature upper limits to the rate constants (Table 4), the following recommendations are made for 298 K

$$k(CF_4) < 4 \times 10^{-16} \text{ cm}^3 \text{ molecule}^{-1} \text{ s}^{-1},$$

$$k(CF_3Cl) < 7 \times 10^{-16} \text{ cm}^3 \text{ molecule}^{-1} \text{ s}^{-1},$$

$$k(CF_3Br) < 1 \times 10^{-15} \text{ cm}^3 \text{ molecule}^{-1} \text{ s}^{-1},$$

$$k(CF_2ClBr) < 1 \times 10^{-15} \text{ cm}^3 \text{ molecule}^{-1} \text{ s}^{-1},$$

$$\text{and } k(CCl_4) < 5 \times 10^{-16} \text{ cm}^3 \text{ molecule}^{-1} \text{ s}^{-1}$$

These room temperature upper limits to the rate constants for CF_4, CF_3Cl, CF_3Br and CF_2ClBr are based upon the data of Howard and Evenson[1] (CF_4 and CF_3Cl), Le Bras and Combourieu[23] (CF_3Br) and Clyne and Holt[22] (CF_2ClBr). For CCl_4 the upper limit reported by Cox et al.[7] has been used, increased by a factor of 4 to take into account uncertainties in the number of NO to NO_2 conversions occurring in their relative rate study. The rate constants for these reactions at 298 K are likely to be orders of magnitude lower than the upper limits given here.

For CF_2Cl_2 and $CFCl_3$, upper limits to the rate constants for the OH radical reactions have been determined at temperatures >298 K by Chang and Kaufman[25] and Atkinson et al.[19] Based upon the upper limits to the rate constants measured by Chang and Kaufman[25] at 478–480 K and the rate expressions $k = Ae^{-B/T}$ or $k = CT^2e^{-D/T}$, with $A \geqslant 1 \times 10^{-12}$ cm^3 molecule^{-1} s^{-1} or $C \geqslant 1 \times 10^{-18}$ cm^3 molecule^{-1} s^{-1} (consistent with the recommendations for the halomethanes containing H-atoms), then at 298 K the following recommendations are made

$$k(CF_2Cl_2) < 1 \times 10^{-17} \text{ cm}^3 \text{ molecule}^{-1} \text{ s}^{-1},$$
and
$$k(CFCl_3) < 1 \times 10^{-17} \text{ cm}^3 \text{ molecule}^{-1} \text{ s}^{-1}$$

These upper limits to the 298 K reaction rate constants are somewhat more conservative than the recent NASA recommendations[37] of upper limits to the rate constants of $<6 \times 10^{-18}$ cm^3 molecule^{-1} s^{-1} for CF_2Cl_2 and $<5 \times 10^{-18}$ cm^3 molecule^{-1} s^{-1} for $CFCl_3$.

(12) CH₃CH₂Cl

The rate constants obtained by Howard and Evenson[26] and Paraskevopoulos et al.[10] at room temperature (Table 4) are in excellent agreement, and it is recommended that

$$k(CH_3CH_2Cl) = 3.9$$

$$\times 10^{-13} \text{ cm}^3 \text{ molecule}^{-1} \text{ s}^{-1} \text{ at 298 K,}$$

with an estimated uncertainty of ±35%. No temperature dependence is available.

(13) CH₃CHF₂

Rate constants have been determined for the reaction of OH radicals with CH_3CHF_2 by Howard and Even-

son,[26] Handwerk and Zellner,[16] Clyne and Holt[14] and Nip et al.[2] The rate constants of Howard and Evenson,[26] Handwerk and Zellner[16] and Nip et al.[2] are in reasonable agreement, but are significantly lower than the room temperature rate constant of Clyne and Holt.[14] Since the data of Clyne and Holt[14] are neglected in these evaluations of the OH radical reactions with the haloalkanes, a unit-weighted average of the room temperature rate constants of Howard and Evenson,[26] Handwerk and Zellner[16] and Nip et al.[2] yields the recommendation of

$$k(CH_3CHF_2) = 3.4$$

$$\times 10^{-14} \text{ cm}^3 \text{ molecule}^{-1} \text{ s}^{-1} \text{ at } \sim 295 \text{ K,}$$

with an estimated overall uncertainty of ±30%. This room temperature recommendation is identical to those of Atkinson[36] and DeMore et al.[37]

(14) CH₃CF₂Cl

The available rate constants of Howard and Evenson,[26] Cox et al.,[7] Watson et al.,[15] Handwerk and Zellner,[16] Clyne and Holt[14] and Paraskevopoulos et al.[10] are given in Table 4 and are plotted in Arrhenius form in Fig. 31. It is evident that the rate constants of Howard and Evenson,[26] Watson et al.,[15] Handwerk and Zellner[16] and Paraskevopoulos et al.[10] are in reasonably good agreement, although significantly lower than those measured by Clyne and Holt.[14] A unit-weighted least-squares analysis of these data of Howard and Evenson,[26] Watson et al.,[15] Handwerk and Zellner[16] and Paraskevopoulos et al.,[10] using the equation $k = CT^2e^{-D/T}$, yields the recommendation of

$$k(CH_3CF_2Cl) = (2.05^{+5.76}_{-1.52})$$

$$\times 10^{-18} T^2 e^{-(1171 \pm 413)/T} \text{ cm}^3 \text{ molecule}^{-1} \text{ s}^{-1}$$

over the temperature range 273–375 K, where the indicated errors are two least-squares standard deviations, and

$$k(CH_3CF_2Cl) = 3.58$$

$$\times 10^{-15} \text{ cm}^3 \text{ molecule}^{-1} \text{ s}^{-1} \text{ at 298 K,}$$

with an estimated overall uncertainty at 298 K of ±50%. This recommendation is identical to that of Atkinson,[36] obtained using the same data set.

(15) CH₃CCl₃

The available kinetic data[4,7,15,20,22,26,28–30] are given in Table 4. As discussed previously,[28,29,36] it now appears that the earlier rate constants determined by Howard and Evenson,[26] Watson et al.,[15] Chang and Kaufman[20]

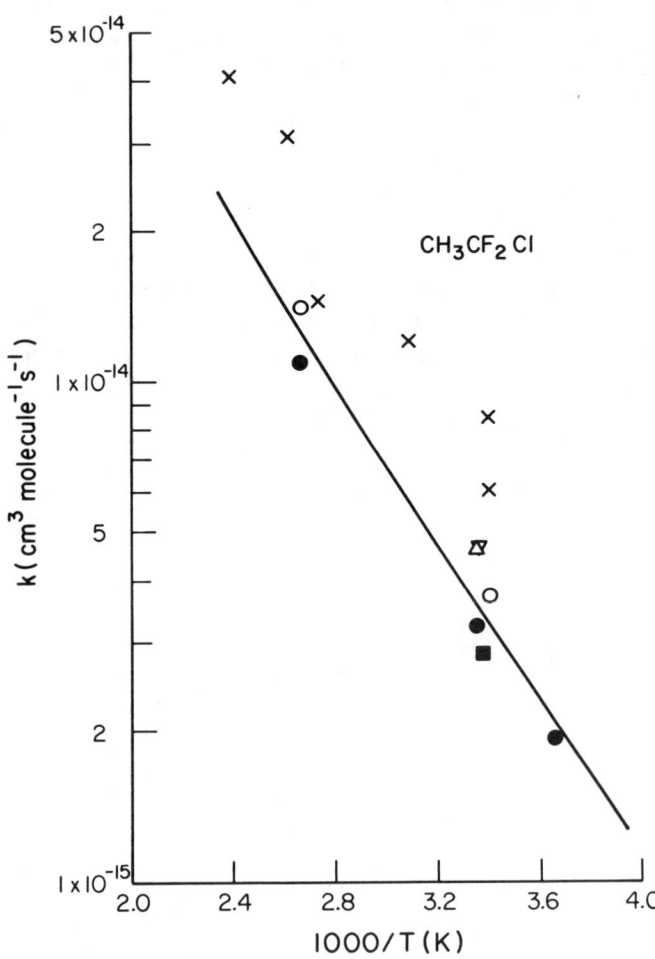

by extrapolation of the higher temperature data,[4,28,29] may still have been affected by $CH_2=CCl_2$ impurity problems.[29]

Thus, only the absolute rate constant data of Jeong and Kaufman[4,28] and those of Kurylo et al.[29] at $\geqslant 253$ K are used in the evaluation. A unit-weighted least-squares analysis of these data,[4,28,29] using the equation $k = CT^2e^{-D/T}$, yields the recommendation of

$$k(CH_3CCl_3) = (5.92^{+1.29}_{-1.05})$$

$$\times 10^{-18} T^2 e^{-(1129 \pm 62)/T} cm^3 molecule^{-1} s^{-1}$$

over the temperature range 253–457 K, where the indicated errors are two least-squares standard deviations, and

$$k(CH_3CCl_3) = 1.19$$

$$\times 10^{-14} cm^3 molecule^{-1} s^{-1} at 298 K,$$

with an estimated overall uncertainty at 298 K of $\pm 30\%$. This recommendation is identical to that of Atkinson[36] and similar to the recent NASA evaluation,[37] which, although using the same data set, utilized the simple Arrhenius expression rather than a three-parameter equation.

FIG. 31. Arrhenius plot of rate constants for the reaction of the OH radical with CH_3CF_2Cl. (■) Howard and Evenson;[26] (Δ) Cox et al.;[7] (●) Watson et al.;[15] (○) Handwerk and Zellner;[16] (x) Clyne and Holt;[14] (∇) Paraskevopoulos et al.;[10] (———) recommendation (see text).

and Clyne and Holt,[22] which yielded a room temperature rate constant of $\sim(1.5–2.2) \times 10^{-14} cm^3 molecule^{-1} s^{-1}$ and a temperature dependence of $B \sim 1300–1600$ K, were erroneously high due to contamination by small amounts of highly reactive (relative to CH_3CCl_3) $CH_2=CCl_2$ impurity. The most recent studies of Kaufman and co-workers[4,28] and Kurylo et al.,[29] in which the CH_3CCl_3 samples were extensively purified, are in excellent agreement and yield significantly lower rate constants than did these previous studies. The room temperature rate constant derived from the relative rate study of Nelson et al.[30] is in good agreement with these absolute rate constants of Jeong and Kaufman[4,28] and Kurylo et al.,[29] and these data are plotted in Arrhenius form in Fig. 32. The rate constant measured by Kurylo et al.[29] at 222 K, which is significantly higher than expected

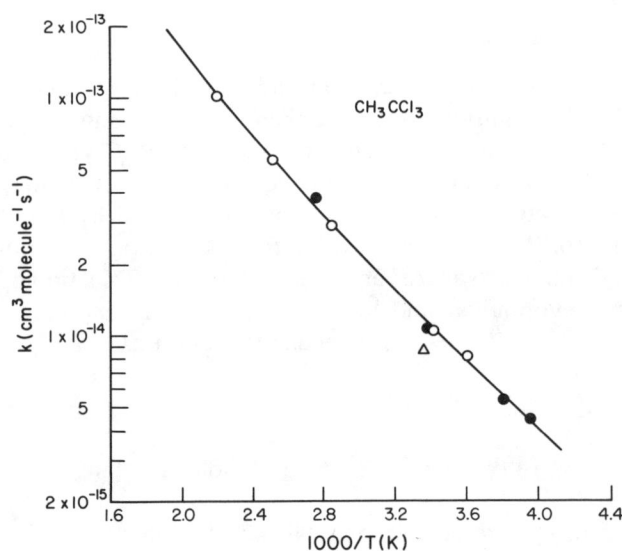

FIG. 32. Arrhenius plot of rate constants for the reaction of the OH radical with CH_3CCl_3. (○) Jeong et al.;[4,28] (●) Kurylo et al.[29] (T $\geqslant 253$ K); (Δ) Nelson et al.;[30] (———) recommendation (see text).

(16) CH₂FCF₃

The available kinetic data of Clyne and Holt,[14] Martin and Paraskevopoulos[27] and Jeong et al.[4] are given in Table 4 and are plotted in Arrhenius form in Fig. 33.

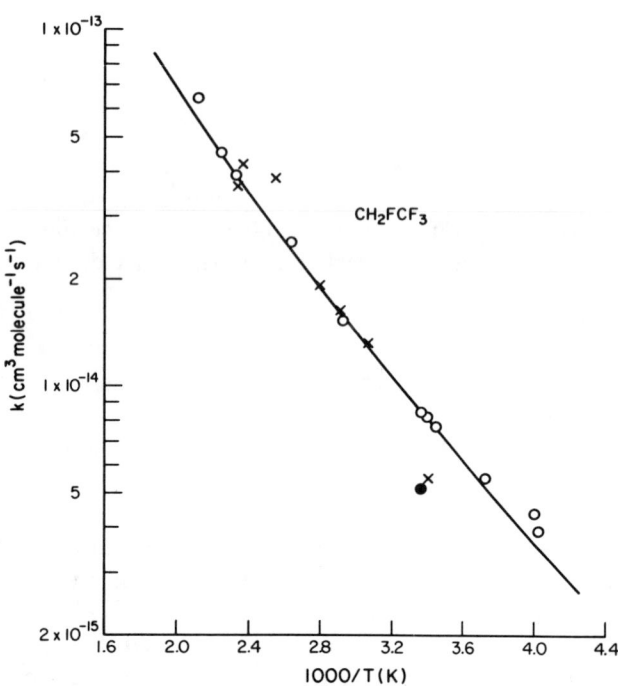

FIG. 33.   Arrhenius plot of rate constants for the reaction of the OH radical with CH₂FCF₃. (x) Clyne and Holt;[14] (●) Martin and Paraskevolpoulos;[27] (○) Jeong et al.;[4] (———) recommendation (see text).

The rate constant of Martin and Paraskevopoulos[27] at 298 K is significantly lower than that of Jeong et al.[4] (although it is in agreement with that of Clyne and Holt[14]). However, in view of the criteria for evaluating these reactions, the rate constants determined by Clyne and Holt[14] were not used in the evaluation. A unit-weighted least-squares analysis of the data of Martin and Paraskevopoulos[27] and Jeong et al.,[4] using the expression $k = CT^2 e^{-D/T}$, yields the recommendation of

$$k(\text{CH}_2\text{FCF}_3) = (1.27^{+0.87}_{-0.52})$$

$$\times \ 10^{-18} \ T^2 \ e^{-(769 \pm 163)/T} \ \text{cm}^3 \ \text{molecule}^{-1} \ \text{s}^{-1}$$

over the temperature range 249–473 K, where the indicated errors are two least-squares standard deviations, and

$$k(\text{CH}_2\text{FCF}_3) = 8.54$$

$$\times \ 10^{-15} \ \text{cm}^3 \ \text{molecule}^{-1} \ \text{s}^{-1} \ \text{at 298 K,}$$

with an estimated uncertainty at 298 K of +30%, −50%. This recommendation is identical to that of Atkinson,[36] and similar to that of NASA[37] which used the Arrhenius expression.

(17) CH₂ClCF₃

The available rate constants of Howard and Evenson,[26] Handwerk and Zellner[16] and Clyne and Holt[14] are given in Table 4 and are plotted in Arrhenius form in Fig. 34. Again, the rate constants of Clyne and Holt[14] exhibit a much higher temperature dependence than do those of Handwerk and Zellner.[16]

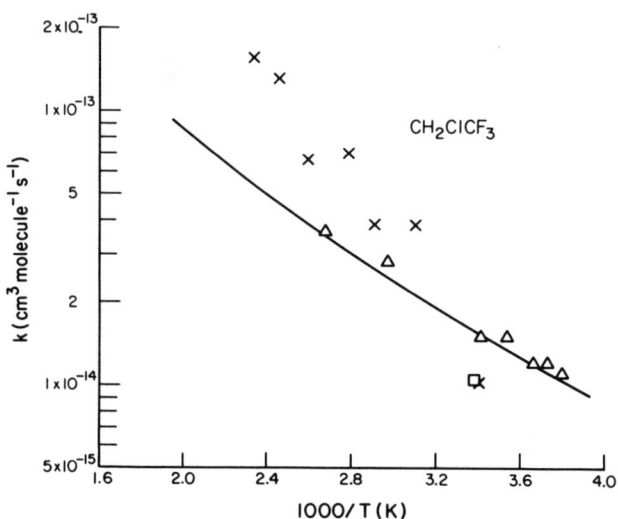

FIG. 34.   Arrhenius plot of rate constants for the reaction of the OH radical with CH₂ClCF₃. (□) Howard and Evenson;[26] (△) Handwerk and Zellner;[16] (x) Clyne and Holt;[14] (———) recommendation (see text).

A unit-weighted least-squares analysis of the rate constant data of Howard and Evenson[26] and Handwerk and Zellner,[16] using the equation $k = CT^2 e^{-D/T}$, yields the recommendation of

$$k(\text{CH}_2\text{ClCF}_3) = (8.50^{+20.75}_{-6.03})$$

$$\times \ 10^{-19} \ T^2 \ e^{-(458 \pm 362)/T} \ \text{cm}^3 \ \text{molecule}^{-1} \ \text{s}^{-1}$$

over the temperature range 263–373 K, where the indicated errors are two least-squares standard deviations, and

$$k(CH_2ClCF_3) = 1.62$$

$$\times \ 10^{-14} \text{ cm}^3 \text{ molecule}^{-1} \text{ s}^{-1} \text{ at 298 K,}$$

with an estimated uncertainty at 298 K of ± a factor of 2. This recommendation is identical to that of Atkinson.[36]

(18) CH₂ClCF₂Cl

The available rate constants of Watson et al.[31] and Jeong et al.[4] are given in Table 4 and are plotted in Arrhenius form in Fig. 35.

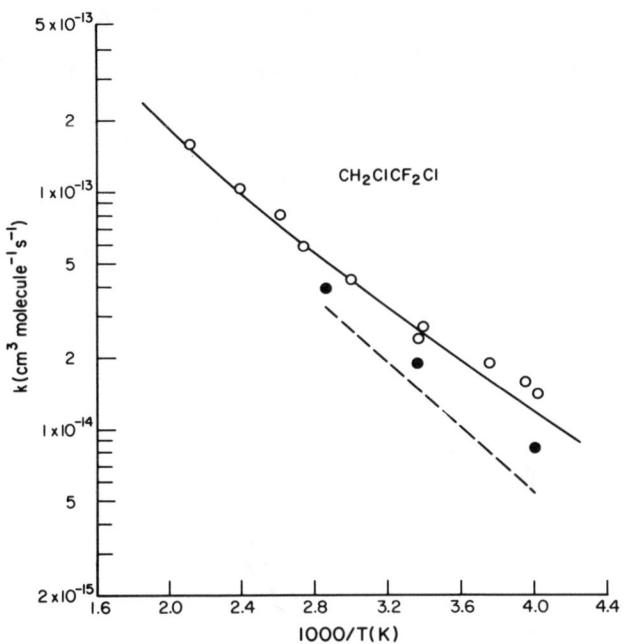

FIG. 35. Arrhenius plot of rate constants for the reaction of the OH radical with CH_2ClCF_2Cl. (●) Watson et al.,[31] measured rate constants; (— — —) Watson et al.,[31] corrected for measured impurities (see text); (○) Jeong et al.;[4] (———) recommendation (see text).

It can be seen that the measured rate constants of Watson et al.[31] are consistently lower than those of Jeong et al.,[4] especially at lower temperatures. Furthermore, Watson et al.,[31] from an analysis of the purity of the CH_2ClCF_2Cl sample used which showed the presence of ~0.045% of haloethenes, concluded that the true rate constants for this reaction were lower than those measured. Their estimated Arrhenius expression,[31] after correction for the presence of these impurities, is given in Table 4 and is shown in Fig. 35 as the dashed line. However, the CH_2ClCF_2Cl sample used by Jeong et al.[4] was stated to have a purity level of >99.999%, and hence their data should have been essentially free from any complications arising from the presence of reactive im-

purities. It should be noted that, analogous to the situation for methane and ethane, the rate constants measured by Jeong et al.[4] at temperatures ≲275 K may have been systematically high.

In the absence of further experimental data, a unit-weighted least-squares analysis of the measured rate constants of Watson et al.[31] and Jeong et al.[4] has been carried out, using the equation $k = CT^2 e^{-D/T}$, to yield the recommendation of

$$k(CH_2ClCF_2Cl) = (2.80^{+2.29}_{-1.26})$$

$$\times \ 10^{-18} \ T^2 \ e^{-(672 \pm 183)/T} \text{ cm}^3 \text{ molecule}^{-1} \text{ s}^{-1}$$

over the temperature range 249–473 K, where the indicated errors are two least-squares standard deviations (which are associated only with the measured rate constants and do not include the corrected values of Watson et al.[31]), and

$$k(CH_2ClCF_2Cl) = 2.61 \times 10^{-14} \text{ cm}^3 \text{ molecule}^{-1} \text{ s}^{-1}$$

at 298 K, with an estimated overall uncertainty at 298 K of +30%, −60%.

The NASA evaluation[37] uses the corrected Arrhenius expression of Watson et al.[31] (which is encompassed by the uncertainties associated with the above recommended 298 K rate constant). Clearly, further kinetic studies employing carefully purified CH_2ClCF_2Cl are necessary.

(19) CHFClCF₃

The rate constants of Howard and Evenson[26] and Watson et al.[31] are given in Table 4 and are plotted in Arrhenius form in Fig. 36. These two studies are in good agreement and no curvature in the Arrhenius plot is evident. Accordingly, a unit-weighted least-squares analysis of these data yields the recommended Arrhenius expression of

$$k(CHFClCF_3) = (6.38^{+18.20}_{-4.73})$$

$$\times \ 10^{-13} \ e^{-(1233 \pm 400)/T} \text{ cm}^3 \text{ molecule}^{-1} \text{ s}^{-1}$$

over the temperature range 250–375 K, where the indicated errors are two least-squares standard deviations, and

$$k(CHFClCF_3) = 1.02 \times 10^{-14} \text{ cm}^3 \text{ molecule}^{-1} \text{ s}^{-1}$$

at 298 K, with an estimated uncertainty at 298 K of ±30%. Using the expression $k = CT^2 e^{-D/T}$, a unit-weighted least-squares analysis of these data[26,31] yields

$$k(CHFClCF_3) = (9.12^{+29.27}_{-6.96})$$

$$\times \ 10^{-19} \ T^2 \ e^{-(624 \pm 416)/T} \text{ cm}^3 \text{ molecule}^{-1} \text{ s}^{-1},$$

where the indicated errors are two least-squares standard deviations, and

$$k(\text{CHFClCF}_3) = 1.00$$

$$\times\ 10^{-14}\ \text{cm}^3\ \text{molecule}^{-1}\ \text{s}^{-1}\ \text{at 298 K.}$$

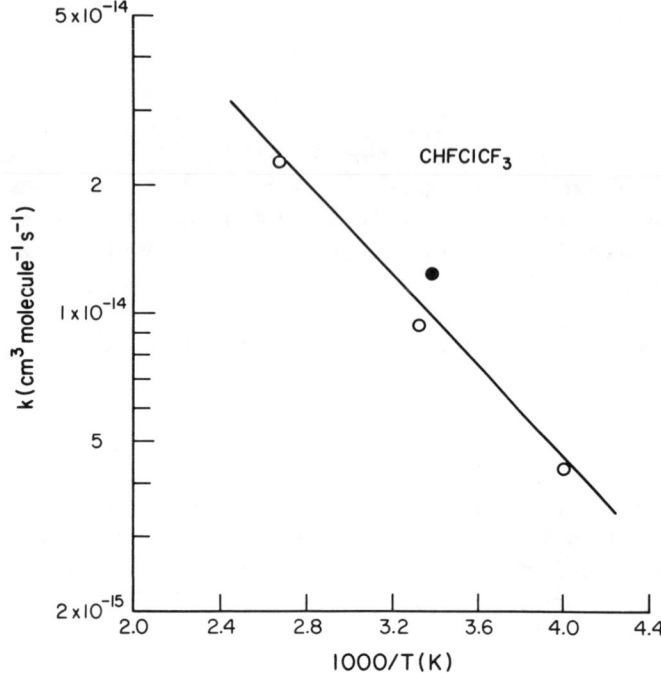

FIG. 36.   Arrhenius plot of rate constants for the reaction of the OH radical with CHFClCF₃. (●) Howard and Evenson;[26] (○) Watson et al.;[31] (⸺) recommendation (see text).

In view of the small temperature range covered (250–375 K) and the fact that these two expressions yield essentially identical (within 2%) rate constants over this range, the use of the simple Arrhenius expression (the recommended line in Fig. 36) is recommended over this temperature range.

(20) CHCl₂CF₃

The rate constants of Howard and Evenson,[26] Watson et al.[31] and Clyne and Holt[14] are given in Table 4 and are plotted in Arrhenius form in Fig. 37. Watson et al.[31] estimated that contributions of C₄ haloalkene impurities could have led to their observed rate constants being somewhat high, and estimated the corrected Arrhenius expression given in Table 4 and shown as the dashed line in Fig. 37. These estimated rate constants of Watson et al.,[31] taking into account the presence of reactive impurities, are only slightly different from their measured rate constants (which exhibit no unambiguous evidence for

curvature in the Arrhenius plot). A unit-weighted least-squares analysis of the measured rate constants of Howard and Evenson[26] and Watson et al.[31] has been carried out to yield the Arrhenius expression of

$$k(\text{CHCl}_2\text{CF}_3) = (1.16^{+1.44}_{-0.65})$$

$$\times\ 10^{-12}\ e^{-(1056\,\pm\,237)/T}\ \text{cm}^3\ \text{molecule}^{-1}\ \text{s}^{-1}$$

over the temperature range 245–375 K, where the indicated errors are two least-squares standard deviations, and

$$k(\text{CHCl}_2\text{CF}_3) = 3.35$$

$$\times\ 10^{-14}\ \text{cm}^3\ \text{molecule}^{-1}\ \text{s}^{-1}\ \text{at 298 K,}$$

with an estimated uncertainty at 298 K of +20%, −40%.

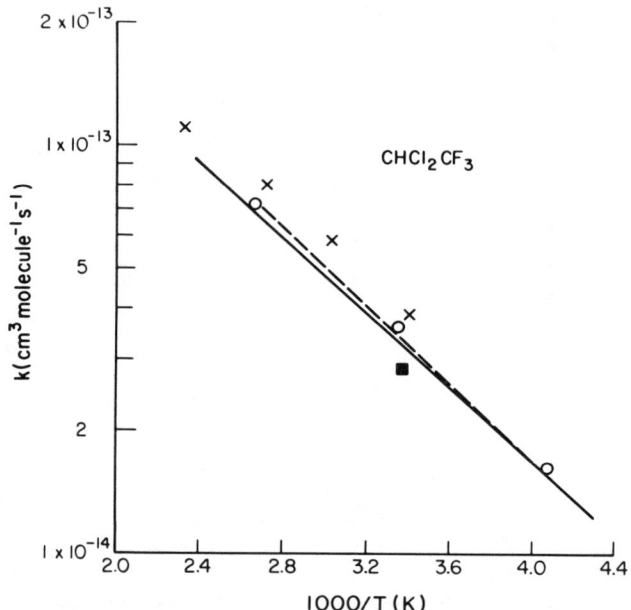

FIG. 37.   Arrhenius plot of rate constants for the reaction of the OH radical with CHCl₂CF₃. (■) Howard and Evenson;[26] (○) Watson et al.,[31] measured rate constants; (— — —) Watson et al.,[31] corrected for measured impurities (see text); (x) Clyne and Holt;[14] (⸺) recommendation (see text).

Using the expression $k = CT^2 e^{-D/T}$, a least-squares analysis of these data[26,31] leads to

$$k(\text{CHCl}_2\text{CF}_3) = (1.70^{+1.87}_{-0.89})$$

$$\times\ 10^{-18}\ T^2\ e^{-(455\,\pm\,217)/T}\ \text{cm}^3\ \text{molecule}^{-1}\ \text{s}^{-1},$$

where the indicated errors are two least-squares standard deviations, and

$k(CHCl_2CF_3) = 3.28$

$$\times \ 10^{-14} \ cm^3 \ molecule^{-1} \ s^{-1} \ at \ 298 \ K.$$

Over the temperature range 245–375 K, these two expressions yield essentially identical rate constants, and the recommended Arrhenius expression is shown in Fig. 37.

(21) Other Haloalkanes

For the remaining haloalkanes listed in Table 4, only single reliable studies have been carried out (thus, although two experimental studies were carried out for CH_3CF_3, CH_2FCHF_2 and CHF_2CF_3, the rate constant data of Clyne and Holt[14] are discounted). In the absence of further experimental data, the results of these studies (other than that of Clyne and Holt[14]) should be used, with correspondingly wide uncertainty limits.

b. Mechanism

Analogous to the alkanes, for the haloalkanes with F, Cl and Br substituents these reactions proceed via H-atom abstraction. For the reaction of the OH radical with CH_3CH_2F at 297 K, Singleton *et al.*[40] determined a rate constant ratio

$$OH + CH_3CH_2F \longrightarrow \begin{cases} H_2O + CH_3\dot{C}HF & (a) \\ H_2O + \dot{C}H_2CH_2F & (b) \end{cases}$$

of $k_a/(k_a + k_b) = 0.85 \pm 0.03$.

However, Garraway and Donovan[24] have reported a room temperature rate constant of 1.2×10^{-13} cm^3 molecule^{-1} s^{-1} for the reaction of the OH radical with CF_3I, and also reported that reaction occurs for other iodine-substituted non-hydrogen containing alkanes such as C_2F_5I and C_3F_7I. If these observations are correct, then these reactions must then occur via I-atom abstraction to yield HOI and the corresponding C_nF_{2n+1} radical.

$$OH + C_nF_{2n+1}I \rightarrow HOI + C_nF_{2n+1}$$

References

[1]C. J. Howard and K. M. Evenson, J. Chem. Phys. **64**, 197 (1976).

[2]W. S. Nip, D. L. Singleton, R. Overend, and G. Paraskevopoulos, J. Phys. Chem. **83**, 2440 (1979).

[3]K.-M. Jeong and F. Kaufman, J. Phys. Chem. **86**, 1808 (1982).

[4]K.-M. Jeong, K.-J. Hsu, J. B. Jeffries, and F. Kaufman, J. Phys. Chem. **88**, 1222 (1984).

[5]R. K. Bera and R. J. Hanrahan, Radiat. Phys. Chem., **32**, 579 (1988).

[6]W. E. Wilson, Jr., J. T. O'Donovan, and R. M. Fristrom, 12th International Symposium on Combustion, 1968; The Combustion Institute, Pittsburgh, PA, 1969, p. 929.

[7]R. A. Cox, R. G. Derwent, A. E. J. Eggleton, and J. E. Lovelock, Atmos. Environ. **10**, 305 (1976).

[8]R. A. Perry, R. Atkinson, and J. N. Pitts, Jr., J. Chem. Phys. **64**, 1618 (1976).

[9]D. D. Davis, G. Machado, B. Conaway, Y. Oh, and R. Watson, J. Chem. Phys. **65**, 1268 (1976).

[10]G. Paraskevopoulos, D. L. Singleton, and R. S. Irwin, J. Phys. Chem. **85**, 561 (1981).

[11]O. J. Nielsen, P. Pagsberg, and A. Sillesen, Proc. 3rd European Symposium on the Physico-Chemical Behavior of Atmospheric Pollutants, 1984 (Riedel Publishing Co., Dordrecht, Holland, 1984), p. 283.

[12]P. H. Taylor, J. A. D'Angelo, M. C. Martin, J. H. Kasner, and B. Dellinger, Int. J. Chem. Kinet., **21**, 829 (1989).

[13]W. E. Wilson, Jr., 10th International Symposium on Combustion, 1964; The Combustion Institute, Pittsburgh, PA, 1965; p. 47.

[14]M. A. A. Clyne and P. M. Holt, J. Chem. Soc. Faraday Trans. 2, **75**, 582 (1979).

[15]R. T. Watson, G. Machado, B. Conaway, S. Wagner, and D. D. Davis, J. Phys. Chem. **81**, 256 (1977).

[16]V. Handwerk and R. Zellner, Ber. Bunsenges Phys. Chem. **82**, 1161 (1978).

[17]J. N. Bradley, W. D. Capey, R. W. Fair, and D. K. Pritchard, Int. J. Chem. Kinet. **8**, 549 (1976).

[18]J. Ernst, H. Gg. Wagner, and R. Zellner, Ber. Bunsenges Phys. Chem. **82**, 409 (1978).

[19]R. Atkinson, D. A. Hansen, and J. N. Pitts, Jr., J. Chem. Phys. **63**, 1703 (1975).

[20]J. S. Chang and F. Kaufman, J. Chem. Phys. **66**, 4989 (1977).

[21]W. Klöpffer, R. Frank, E.-G. Kohl, and F. Haag, Chemiker-Zeitung, **110**, 57 (1986); "Methods of the Ecotoxicological Evaluation of Chemicals. Photochemical Degradation in the Gas Phase," Vol. 6, *OH Reaction Rate Constants and Tropospheric Lifetimes of Selected Environmental Chemicals*. Report 1980–1983; K. H. Becker, H. M. Biehl, P. Bruckmann, E. H. Fink, F. Führ, W. Klöpffer, R. Zellner, and C. Zetzsch, Editors, Kernforschungsanlage Jülich GmbH, November, 1984.

[22]M. A. A. Clyne and P. M. Holt, J. Chem. Soc. Faraday Trans. 2, **75**, 569 (1979).

[23]G. Le Bras and J. Combourieu, Int. J. Chem. Kinet. **10**, 1205 (1978).

[24]J. Garraway and R. J. Donovan, J. Chem. Soc., Chem. Comm. 1108 (1979).

[25]J. S. Chang and F. Kaufman, Geophys. Res. Lett. **4**, 192 (1977).

[26]C. J. Howard and K. M. Evenson, J. Chem. Phys. **64**, 4303 (1976).

[27]J.-P. Martin and G. Paraskevopoulos, Can. J. Chem. **61**, 861 (1983).

[28]K.-M. Jeong and F. Kaufman, Geophys. Res. Lett. **6**, 757 (1979).

[29]M. J. Kurylo, P. C. Anderson, and O. Klais, Geophys. Res. Lett. **6**, 760 (1979).

[30]L. Nelson, J. J. Treacy, and H. W. Sidebottom, Proc. 3rd European Symposium on the Physico-Chemical Behavior of Atmospheric Pollutants, 1984 (Riedel Publishing Co., Dordrecht, Holland, 1984), p. 258.

[31]R. T. Watson, A. R. Ravishankara, G. Machado, S. Wagner, and D. D. Davis, Int. J. Chem. Kinet. **11**, 187 (1979).

[32]E. C. Tuazon, R. Atkinson, A. M. Winer, and J. N. Pitts, Jr., Arch. Environ. Contamin. Toxicol. **13**, 691 (1984).

[33]E. C. Tuazon, R. Atkinson, S. M. Aschmann, J. Arey, A. M. Winer, and J. N. Pitts, Jr., Environ. Sci. Technol. **20**, 1043 (1986).

[34]R. Butler, I. J. Solomon, and A. Snelson, J. Air Pollut. Contr. Assoc. **28**, 1131 (1978).

[35]R. Atkinson, K. R. Darnall, A. C. Lloyd, A. M. Winer, and J. N. Pitts, Jr., Adv. Photochem. **11**, 375 (1979).

[36]R. Atkinson, Chem. Rev. **86**, 69 (1986).

[37]W. B. DeMore, M. J. Molina, S. P. Sander, D. M. Golden, R. F. Hampson, M. J. Kurylo, C. J. Howard, and A. R. Ravishankara, Evaluation No. 8, NASA Panel for Data Evaluation, JPL Publication 87–41, Jet Propulsion Laboratory, Pasadena, CA, September 15, 1987.

[38]N. Cohen and S. W. Benson, J. Phys. Chem. **91**, 162 (1987).

[39]N. Cohen and S. W. Benson, J. Phys. Chem. **91**, 171 (1987).

[40]D. L. Singleton, G. Paraskevopoulos, and R. S. Irwin, J. Phys. Chem. **84**, 2339 (1980).

2.3. Alkenes

a. Kinetics

The OH radical reaction rate constants obtained at, or close to, the limiting high-pressure second-order kinetic regime or at ~760 Torr of diluent gas are listed in Tables 5 (acyclic monoalkenes), 6 (acyclic di- and tri-alkenes), and 7 (cyclic mono-, di- and tri-alkenes). The rate constants for the OD radical reactions with monocyclic alkenes are given in Table 8. The data reported by Cox[61] from the photolysis of HONO-alkene-air mixtures at 300 K and atmospheric pressure of air have not been included, since the stoichiometric factors were not specified. However, based upon our present knowledge of the rate constants for the reactions of the OH radical with the reference compounds NO, NO_2 and HONO[62] and the reaction stoichiometries[63] for these OH-alkene reactions, these data[61] are reasonably consistent with the elementary rate constants recommended below. Simonaitis and Heicklen[64] also obtained rate constants for propene at 373 and 473 K relative to those for the reaction of OH radicals with CO at total pressures of ~400–800 Torr (mainly H_2O). Rate constant ratios of

$$k(OH + propene)/k(OH + CO) = 75 \pm 8 \text{ at } 373 \text{ K}$$

and 55 ± 6 at 473 K were determined.[64] As discussed previously,[65] while subject to significant uncertainties, mainly concerning the rate constant of the reference reaction under the experimental conditions employed, these data[64] are generally consistent with the present recommendation.

In addition, a set of rate constants at 301 ± 1 K for 2-methyl-1,3-butadiene and a series of monoterpenes can be derived from the experimental NO-photooxidation data of Grimsrud et al.[66] These data[66] must be viewed as semi-quantitative only,[28,55] since their use assumes that the OH radical concentrations were identical in the separate NO_x-organic-air irradiations carried out and that the contributions of any O_3 reactions to the monoterpene reaction rates were negligible.

As noted above, in most cases the rate constants listed in Tables 5–8 are at, or close to, the high-pressure second-order limit. However, the rate constants determined for the C_3 and C_4 alkenes using discharge flow techniques at total pressures of ~1 Torr may still be in the fall-off regime between second- and third-order kinetics. These data are not used in the evaluation of the recommended rate constants. For ethene and propene the most reliable rate constant data in the fall-off region are indicated and used to derive the fall-off parameters in the Troe fall-off expression.

(1) Ethene and Ethene-d_4

As discussed below, the available experimental data and theoretical expectations show that three reasonably distinct temperature regimes are observed for the reactions of the OH (or OD) radical with ethene and ethene-d_4. Taking the OH radical reaction with ethene as an example, these temperature regimes are characterized by the following reaction pathways: (a, —a and b) OH radical addition to ethene to form the species $HOCH_2CH_2$* (where * denotes an initially energy-rich radical), followed by stabilization and/or decomposition of this adduct radical; (c) at temperatures above ~450–550 K the thermalized $HOCH_2CH_2$ radical undergoes decomposition back to the OH radical and C_2H_4 reactants sufficiently rapidly that, unless the $HOCH_2CH_2$ radical is removed by reaction on a shorter time-scale, OH radical addition to ethene is neither observed nor of any importance; and (d) at temperatures above ~600–700 K, where the effective rate constant for OH radical addition to ethene is decreasing rapidly due to the decomposition of the thermalized $HOCH_2CH_2$ radical and/or fall-off effects, H-atom abstraction from ethene becomes the sole reaction pathway observed:

$$OH + C_2H_4 \rightleftharpoons HOCH_2CH_2^* \qquad (a, -a)$$

$$HOCH_2CH_2^* + M \rightarrow HOCH_2CH_2 + M \qquad (b)$$

$$HOCH_2CH_2 \rightarrow OH + C_2H_4 \qquad (c)$$

$$OH + C_2H_4 \rightarrow H_2O + C_2H_3 \qquad (d)$$

The bimolecular rate constants obtained for the reactions of the OH radical with ethene and ethene-d_4, with the rate constants at temperatures <650 K being supposedly at or close to the high pressure second-order limit, are given in Table 5, and these rate constants are plotted in Arrhenius form in Figs. 38 (ethene) and 39 (ethene-d_4), respectively. The analogous rate constants for the reactions of the OD radical with ethene and ethene-d_4 are given in Table 8.

Ethene. Figure 38 shows that for the reaction of OH radicals with ethene, at elevated temperatures (>700 K) the relative rate data of Westenberg and Fristrom,[1] Baldwin et al.,[2] Hoare and Patel[3] and Bradley et al.[6] exhibit significant differences from the more recent absolute rate constant data of Smith,[15] Tully,[16] and Liu et al.,[17,18] indicating that these earlier relative rate studies[1-3,6] were subject to unrecognized complexities and/or systematic errors.

Extrapolation of the elevated temperature (>650 K) kinetic data of Smith,[15] Tully[16] and Liu et al.[17,18] to room temperature leads to the conclusion that any H-atom abstraction process is totally negligible. This expectation, based upon the extrapolation of elevated temperature data, is totally consistent with (a) the thermochemistry of the H-atom abstraction reaction,[62] (b) the room temperature kinetic study of Howard[67] over the total pressure range 0.7–7 Torr of helium which showed that the rate constant extrapolates to essentially zero at zero pressure, and (c) the discharge flow-mass spectrometric study of Bartels et al.[68] which showed that the H-atom abstraction rate constant accounts for <2.5% of the overall reaction rate constant at ~2 Torr total pressure and 295 K.

In addition to the rate constants given in Table 5, Wilson and Westenberg,[69] Greiner,[70] Morris et al.[21], Smith and Zellner,[71] Bradley et al.,[22] Stuhl,[23] Pastrana and Carr,[25] Davis et al.,[72] Atkinson et al.,[7] Overend and Paraskevopoulos,[8] Howard,[67] Farquharson and Smith,[73] Tully,[10,19] Zellner and Lorenz[11] and Klein et al.[13] measured rate constants at temperatures $\lesssim 525$ K which are in the fall-off region (although this fact was not always known or appreciated[22,23,69-71]).

The experimental data obtained at temperatures $\lesssim 525$ K can be used to derive the limiting low pressure third-order and high pressure second-order rate constants k_o and k_∞, respectively. Use of the Troe fall-off equation,[74,75]

$$k = \left(\frac{k_o[M]}{1 + \frac{k_o[M]}{k_\infty}}\right) F \left\{1 + \left[\log_{10}(k_o[M]/k_\infty)\right]^2\right\}^{-1}$$

where [M] is the concentration of the diluent gas and F is the broadening factor, then allows the bimolecular OH radical addition rate constants k to be calculated as a function of temperature and pressure.

TABLE 5. Rate constants k and temperature-dependent parameters for the gas-phase reactions of the OH radical with acyclic monoalkenes at, or close to, the high pressure limit

Alkene	$10^{12} \times A$ (cm^3 molecule^{-1} s^{-1})	B (K)	$10^{12} \times k$ (cm^3 molecule^{-1} s^{-1})	at T (K)	Technique	Reference	Temperature range covered (K)
Ethene			9	1250–1400	RR [relative to k(CO) $= 1.12 \times 10^{-13} e^{0.0009077}$][a]	Westenberg and Fristrom[1]	1250–1400
			7.5	813	RR [relative to $k(H_2)$ $= 1.12 \times 10^{-12}$][a]	Baldwin et al.[2]	
			4.8	734	RR [relative to $k(CH_4)$ $= 6.95 \times 10^{-18} T^2 e^{-1282/T}$][b]	Hoare and Patel[3]	734–798
			4.1	748			
			3.9	773			
			3.8	798			
			6.23 ± 0.33	381	PR-RA	Gordon and Mulac[4]	381–416
			7.31 ± 0.33	416			
			7.55 ± 1.51	305 ± 2	RR [relative to k(n-butane) $= 2.62 \times 10^{-12}$][b]	Lloyd et al.[5]	
			22.5	1300	RR [relative to $k(H_2)$ $= 5.69 \times 10^{-12}$][a]	Bradley et al.[6]	
			7.85 ± 0.79	299.2	FP-RF	Atkinson et al.[7]	299–425
			6.76 ± 0.68	351.3			
	2.18	−388 ± 151	5.35 ± 0.54	425.1			
			10.0 ± 1.7	296	FP-RA	Overend and Paraskevopoulos[8]	
			8.38 ± 0.38	299 ± 2	RR [relative to k(n-butane) $= 2.55 \times 10^{-12}$][b]	Atkinson et al.[9]	
			8.47 ± 0.24	291	LP-LIF	Tully[10]	291–591
			6.15 ± 0.35	361.5			
			4.55 ± 0.27[c]	438			
			3.08 ± 0.13[c]	515			
			1.3[c]	591			
			8.8 ± 2.0[d]	296	LP-RF	Zellner and Lorenz[11]	296–524
	3.3 ± 1.4	−320 ± 150	$5.5^{+3.3d}_{-1.7}$	524			
			8.66 ± 0.38	295 ± 1	RR [relative to k(propene) $= 2.68 \times 10^{-11}$][b]	Atkinson and Aschmann[12]	

TABLE 5. Rate constants k and temperature-dependent parameters for the gas-phase reactions of the OH radical with acyclic monoalkenes at, or close to, the high pressure limit — Continued

Alkene	$10^{12} \times A$ (cm^3 molecule^{-1} s^{-1})	B (K)	$10^{12} \times k$ (cm^3 molecule^{-1} s^{-1})	at T (K)	Technique	Reference	Temperature range covered (K)
			8.4 ± 0.6^d	295	RR [relative to $k(n\text{-hexane})$ $= 5.55 \times 10^{-12}]^b$	Klein et al.[13]	
			7.3 ± 1.0	295	LP-LIF	Schmidt et al.[14]	
			2.5 ± 0.5	1220	LH-LIF	Smith[15]	
			0.319 ± 0.030	651	LP-LIF	Tully[16]	651–901
			0.438 ± 0.029	694			
			0.477 ± 0.030	701			
			0.615 ± 0.047	746			
			0.672 ± 0.042	757			
			0.725 ± 0.059	779			
			0.803 ± 0.057	800			
			0.899 ± 0.055	829			
			0.971 ± 0.073	849			
			1.20 ± 0.12	898			
	33.6 ± 6.4	2997 ± 144	1.16 ± 0.10	901			
			6.78	343	PR-RA	Liu et al.[17,18]	343–1173
			6.02	373			
			5.20	403			
			5.04	423			
			4.24	483			
			4.12	523			
	1.66	-479 (343–563 K)	4.01	563			
			3.14	603			
			2.06	653			
			1.58	703			
			1.29	730			
			1.47	748			
			1.70	773			
			1.65	794			
			1.51	800			
			1.85	855			
			2.32	873			
			2.15	901			
			2.30	943			
			2.60	973			
			2.51	990			
			2.92	1042			
			3.53	1087			
			3.20	1099			
			3.46	1136			
			4.28	1163			
			4.03	1173			
			7.91	295	LP-LIF	Tully[19]	295–420
			8.07	295			
			6.60	350			
			5.23	420			
Ethene-d_4			8.78 ± 0.52	298 ± 2	RR [relative to $k(\text{ethene}) =$ $8.52 \times 10^{-12}]^b$	Niki et al.[20]	
			0.132 ± 0.017	651	LP-LIF	Tully[16]	651–901
			0.209 ± 0.018	694			
			0.211 ± 0.013	701			
			0.297 ± 0.018	746			
			0.335 ± 0.026	757			

TABLE 5. Rate constants k and temperature-dependent parameters for the gas-phase reactions of the OH radical with acyclic monoalkenes at, or close to, the high pressure limit — Continued

Alkene	$10^{12} \times A$ (cm^3 molecule^{-1} s^{-1})	B (K)	$10^{12} \times k$ (cm^3 molecule^{-1} s^{-1})	at T (K)	Technique	Reference	Temperature range covered (K)
			0.403 ± 0.027	779			
			0.426 ± 0.030	800			
			0.492 ± 0.034	829			
			0.571 ± 0.029	849			
			0.614 ± 0.054	871			
	58.5 ± 15.4	3934 ± 205	0.771 ± 0.087	901			
			6.85	333	PR-RA	Liu et al.[18]	333–1123
			4.29	473			
			3.59	603			
			1.53	653			
			1.09	703			
			0.79	723			
			1.13	773			
			1.29	798			
			1.50	873			
			1.62	923			
			1.85	973			
			1.95	1023			
			2.10	1073			
			2.36	1123			
			8.49	295	LP-LIF	Tully[19]	295–420
			8.49	295			
			7.00	350			
			5.84	420			
Propene			17 ± 4	300	DF-MS	Morris et al.[21]	
			5.0 ± 1.7	300	DF-EPR	Bradley et al.[22]	
			14.5 ± 2.2	298	FP-RF	Stuhl[23]	
			13.3 ± 3.4	298	RR [relative to $k(\mathrm{CO}) = 1.49 \times 10^{-13}$]$^\mathrm{a}$	Gorse and Volman[24]	
			14.3 ± 0.7	381	PR-RA	Gordon and Mulac[4]	381–416
			20.0 ± 1.0	416			
			5 ± 1	300	DF-RA	Pastrana and Carr[25]	
			25.1 ± 2.5	297.6	FP-RF	Atkinson and Pitts[26]	298–424
			20.4 ± 2.1	345.5			
			16.4 ± 1.6	390.3			
	4.1	-544 ± 151	14.7 ± 1.5	423.6			
			25.4 ± 5.1	305 ± 2	RR [relative to $k(n\text{-butane}) = 2.62 \times 10^{-12}$]$^\mathrm{b}$	Lloyd et al.[5]	
			22.0	303	RR [relative to $k(cis\text{-2-butene}) = 5.49 \times 10^{-11}$]$^\mathrm{b}$	Wu et al.[27]	
			24.2 ± 3.6	305 ± 2	RR [relative to $k(2\text{-methylpropene}) = 4.94 \times 10^{-11}$]$^\mathrm{b}$	Winer et al.[28]	
			24.2 ± 4.9	305 ± 2	RR [relative to $k(2\text{-methylpropene}) = 4.94 \times 10^{-11}$]$^\mathrm{b}$	Winer et al.[29]	

J. Phys. Chem. Ref. Data, Monograph 1 (1989)

TABLE 5. Rate constants k and temperature-dependent parameters for the gas-phase reactions of the OH radical with acyclic monoalkenes at, or close to, the high pressure limit — Continued

Alkene	$10^{12} \times A$ (cm^3 molecule^{-1} s^{-1})	B (K)	$10^{12} \times k$ (cm^3 molecule^{-1} s^{-1})	at T (K)	Technique	Reference	Temperature range covered (K)
			26.0 ± 1.6	298	FP-RF	Ravishankara et al.[30]	
			24.6 ± 2.8	297 ± 2	FP-RA	Nip and Paraskevopoulos[31]	
			25.3	300	RR [relative to k(ethene) = 8.44×10^{-12}][b]	Cox et al.[32]	
			26.2	300	RR [relative to k(ethene) = 8.44×10^{-12}][b]	Barnes et al.[33]	
			19 ± 3	298	DF-RF	Smith[34]	
			30 ± 5[d]	297	LP-RF	Zellner and Lorenz[11]	
			46	673	RR [relative to k(2,2,3,3-tetramethyl-butane) = $1.63 \times 10^{-17}T^2e^{-86/T}$][b]	Baldwin et al.[35]	673–773
			42	713			
			47	743			
			55	773			
			8	1200	LH-LIF	Smith[36]	
			29.5 ± 2.0[d]	295	RR [relative to k(n-hexane) = 5.55×10^{-12}][b]	Klein et al.[13]	
			22 ± 4	295	LP-LIF	Schmidt et al.[14]	
			27.1 ± 0.3	293	LP-LIF	Tully and Goldsmith[37]	293–896
			21.7 ± 0.2	338.5			
			17.5 ± 0.2	400			
			15.9 ± 0.2	422			
			14.9 ± 0.3	440.5			
	4.58 ± 0.46	-524 ± 38 (293–467 K)	13.9 ± 0.2	467			
			3.79 ± 0.17	701			
			3.60 ± 0.11	705			
			4.57 ± 0.12	781			
			4.74 ± 0.08	785			
			5.44 ± 0.11	857			
	33.1 ± 7.6	1541 ± 178 (701–896 K)	5.95 ± 0.16	896			
			4.5 ± 0.7	960	LH-LIF	Smith et al.[38]	960–1210
			5.8 ± 1.1	1090			
			7.0 ± 1.0	1180			
			8.9 ± 1.2	1210			
			27.9 ± 2.6	300	RR [relative to k(ethene) = 8.44×10^{-12}][b]	Barnes et al.[39]	
			21 ± 2	298	FP-RF	Wallington[40]	
			27.0 ± 0.7	296 ± 2	RR [relative to k(cyclohexane) = 7.45×10^{-12}][b]	Atkinson and Aschmann[41]	
Propene-d_6			18.7	298	DF-MS	Morris and Niki[42]	
			16.8	298	FP-RF	Stuhl[23]	

TABLE 5. Rate constants k and temperature-dependent parameters for the gas-phase reactions of the OH radical with acyclic monoalkenes at, or close to, the high pressure limit — Continued

Alkene	$10^{12} \times A$ (cm³ molecule⁻¹ s⁻¹)	B (K)	$10^{12} \times k$ (cm³ molecule⁻¹ s⁻¹)	at T (K)	Technique	Reference	Temperature range covered (K)
			27.9 ± 0.2	293	LP-LIF	Tully and Goldsmith[37]	293–896
			22.3 ± 0.3	338			
			18.4 ± 0.3	392			
			15.7 ± 0.2	440.5			
	4.79 ± 0.51	−518 ± 39 (293–481 K)	13.7 ± 0.1	481			
			2.65 ± 0.10	701			
			2.35 ± 0.09	705			
			3.21 ± 0.16	781			
			3.29 ± 0.15	785			
			3.56 ± 0.13	857			
	18.7 ± 9.7	1403 ± 404 (701–896 K)	3.85 ± 0.12	896			
1-Butene			40.8	298	DF-MS	Morris and Niki[42]	
			15 ± 1	300	DF-RA	Pastrana and Carr[25]	
			35.3 ± 3.6	297.7	FP-RF	Atkinson and Pitts[26]	298–424
			30.0 ± 3.0	344.1			
	7.6	−468 ± 151	22.2 ± 2.2	423.7			
			28.5	303	RR [relative to k(cis-2-butene) = 5.49 × 10⁻¹¹][b]	Wu et al.[27]	
			29.5 ± 2.0	298	FP-RF	Ravishankara et al.[30]	
			33.4 ± 2.5	297 ± 2	FP-RA	Nip and Paraskevopoulos[31]	
			32.1·	300	RR [relative to k(ethene) = 8.44 × 10⁻¹²][b]	Barnes et al.[33]	
			30 ± 4	298	DF-MS	Biermann et al.[43]	
			31.3 ± 0.8	298 ± 2	RR [relative to k(propene) = 2.63 × 10⁻¹¹][b]	Ohta[44]	
			31.9 ± 1.6	295 ± 1	RR [relative to k(propene) = 2.68 × 10⁻¹¹][b]	Atkinson and Aschmann[12]	
			19.7 ± 4.2	1225	LH-LIF	Smith[15]	
			6.60 ± 0.44	650	LP-LIF	Tully[16]	650–833
			7.55 ± 0.49	691			
			8.22 ± 0.52	732			
			8.92 ± 0.64	778			
	37.4 ± 6.3	1116 ± 122	9.67 ± 0.68	833			
1-Butene-d_8			4.32 ± 0.27	650	LP-LIF	Tully[16]	650–833
			4.86 ± 0.30	691			
			5.42 ± 0.34	732			
			6.13 ± 0.45	778			
	35.6 ± 2.5	1374 ± 49	6.85 ± 0.50	833			
2-Methyl-propene			64.6	298	DF-MS	Morris and Niki[42]	
			50.7 ± 5.1	297.2	FP-RF	Atkinson and Pitts[26]	297–424
			39.0 ± 3.9	345.5			
	9.2	−503 ± 151	30.5 ± 3.1	423.9			

TABLE 5. Rate constants k and temperature-dependent parameters for the gas-phase reactions of the OH radical with acyclic monoalkenes at, or close to, the high pressure limit — Continued

Alkene	$10^{12} \times A$ (cm^3 molecule^{-1} s^{-1})	B (K)	$10^{12} \times k$ (cm^3 molecule^{-1} s^{-1})	at T (K)	Technique	Reference	Temperature range covered (K)
			50.5	303	RR [relative to k(cis-2-butene) = 5.49×10^{-11}]^b	Wu et al.[27]	
			61.6	300	RR [relative to k(ethene) = 8.44×10^{-12}]^b	Barnes et al.[33]	
			54.7 ± 0.9	298 ± 2	RR [relative to k(2-methyl-2-butene) = 8.69×10^{-11}]^b	Ohta[44]	
			52.3 ± 2.4	295 ± 1	RR [relative to k(propene) = 2.68×10^{-11}]^b	Atkinson and Aschmann[12]	
			29.6 ± 6.8	1259	LH-LIF	Smith[15]	
cis-2-Butene			61.2	298	DF-MS	Morris and Niki[42]	
	10.4	−488 ± 151	53.7 ± 5.4 43.0 ± 4.3 32.9 ± 3.3	297.6 345.7 424.9	FP-RF	Atkinson and Pitts[26]	298–425
			57.1 ± 11.5	305 ± 2	RR [relative to k(n-butane) = 2.62×10^{-12}]^b	Lloyd et al.[5]	
			60.3 ± 9.0	305 ± 2	RR [relative to k(2-methylpropene) = 4.94×10^{-11}]^b	Winer et al.[28]	
			42.6 ± 2.5	298	FP-RF	Ravishankara et al.[30]	
			54.7 ± 1.8	298 ± 2	RR [relative to k(2-methyl-2-butene) = 8.69×10^{-11}]^b	Ohta[44]	
			57.1 ± 1.4	295 ± 1	RR [relative to k(propene) = 2.68×10^{-11}]^b	Atkinson and Aschmann[12]	
trans-2-Butene			71.4	298	DF-MS	Morris and Niki[42]	
			12 ± 10	300	DF-RA	Pastrana and Carr[25]	
	11.2	−549 ± 151	69.9 ± 7.0 57.0 ± 5.7 40.3 ± 4.1	297.8 346.1 425.0	FP-RF	Atkinson and Pitts[26]	298–425
			71.4	303	RR [relative to k(cis-2-butene) = 5.49×10^{-11}]^b	Wu et al.[27]	
			59.6 ± 3.1	297 ± 2	RR [relative to k(cis-1,3-pentadiene) = 1.01×10^{-10}]^e	Ohta[45]	
			65.1 ± 1.4	295 ± 1	RR [relative to k(propene) = 2.68×10^{-11}]^b	Atkinson and Aschmann[12]	

TABLE 5. Rate constants k and temperature-dependent parameters for the gas-phase reactions of the OH radical with acyclic monoalkenes at, or close to, the high pressure limit — Continued

Alkene	$10^{12} \times A$ (cm^3 mole-cule^{-1} s^{-1})	B (K)	$10^{12} \times k$ (cm^3 molecule^{-1} s^{-1})	at T (K)	Technique	Reference	Temperature range covered (K)
			73 ± 13	297	RR [relative to k(propene) = 2.65×10^{-11}][b]	Edney et al.[46]	
			27.0 ± 3.6	1275	LH-LIF	Smith[15]	
			72.1 ± 3.8	298 ± 3	RR [relative to k(propene) = 2.63×10^{-11}][b]	Rogers[47]	
1-Pentene			42.5	298	DF-MS	Morris and Niki[42]	
			30.7	303	RR [relative to k(cis-2-butene) = 5.49×10^{-11}][b]	Wu et al.[27]	
			39.7 ± 3.8	297 ± 2	FP-RA	Nip and Paraskevopoulos[31]	
			29 ± 4	298	DF-MS	Biermann et al.[43]	
			28.7 ± 1.3	298	FP-RF	Biermann et al.[43]	
			31.9 ± 1.4	295 ± 1	RR [relative to k(propene) = 2.68×10^{-11}][b]	Atkinson and Aschmann[12]	
cis-2-Pentene			65.9	303	RR [relative to k(cis-2-butene) = 5.49×10^{-11}][b]	Wu et al.[27]	
			65.4 ± 1.7	298 ± 2	RR [relative to k(cis-2-butene) = 5.64×10^{-11}][b]	Ohta[44]	
trans-2-Pentene			66.9 ± 2.1	297 ± 2	RR [relative to k(1,3-butadiene) = 6.69×10^{-11}][b]	Ohta[45]	
2-Pentene (cis, trans mixture)			90.1	298	DF-MS	Morris and Niki[42]	
2-Methyl-1-butene			90.1	298	DF-MS	Morris and Niki[42]	
			60.4	303	RR [relative to k(cis-2-butene) = 5.49×10^{-11}][b]	Wu et al.[27]	
			60.7 ± 1.1	298 ± 2	RR [relative to k(2-methylpropene) = 5.14×10^{-11}][b]	Ohta[44]	
3-Methyl-1-butene	5.23	-533 ± 151	31.0 ± 3.1 24.0 ± 2.4 18.4 ± 1.9	299.2 349.9 423.2	FP-RF	Atkinson et al.[48]	299–423
			32.4 ± 1.1	295 ± 1	RR [relative to k(propene) = 2.68×10^{-11}][b]	Atkinson and Aschmann[12]	
2-Methyl-2-butene			119	298	DF-MS	Morris and Niki[42]	
			78 ± 8	297.7	FP-RF	Atkinson et al.[49]	298–425

TABLE 5. Rate constants k and temperature-dependent parameters for the gas-phase reactions of the OH radical with acyclic monoalkenes at, or close to, the high pressure limit — Continued

Alkene	$10^{12} \times A$ (cm³ molecule⁻¹ s⁻¹)	B (K)	$10^{12} \times k$ (cm³ molecule⁻¹ s⁻¹)	at T (K)	Technique	Reference	Temperature range covered (K)
			77 ± 8	298.0			
			67 ± 7	345.2			
			62 ± 9	421.6			
	36	−226 ± 201	62 ± 9	424.5			
			87.3 ± 8.8	299.5	FP-RF	Atkinson and Pitts[50]	299–426
			65.4 ± 6.6	356.2			
	19.1	−450 ± 151	56.0 ± 5.6	426.1			
			92 ± 7	300 ± 1	RR [relative to k(cis-2-butene) = 5.58×10^{-11}][b]	Atkinson et al.[51]	
			89.9 ± 3.4	299 ± 2	RR [relative to k(propene) = 2.62×10^{-11}][b]	Atkinson et al.[9]	
			85.0 ± 2.7	297 ± 2	RR [relative to k(1,3-butadiene) = 6.69×10^{-11}][b]	Ohta[45]	
			88.4 ± 3.5	295 ± 1	RR [relative to k(propene) = 2.68×10^{-11}][b]	Atkinson and Aschmann[12]	
1-Hexene			32.9	303	RR [relative to k(cis-2-butene) = 5.49×10^{-11}][b]	Wu et al.[27]	
			37.5 ± 1.1	295 ± 1	RR [relative to k(propene) = 2.68×10^{-11}][b]	Atkinson and Aschmann[12]	
2-Methyl-1-pentene			62.6 ± 0.9	298 ± 2	RR [relative to k(2-methyl-2-butene) = 8.69×10^{-11}][b]	Ohta[44]	
2-Methyl-2-pentene			87.8 ± 1.8	298 ± 2	RR [relative to k(2-methyl-2-butene) = 8.69×10^{-11}][b]	Ohta[44]	
			90.3 ± 1.4	298 ± 2	RR [relative to k(cis-2-pentene) = 6.54×10^{-11}][f]	Ohta[44]	
trans-4-Methyl-2-pentene			60.5 ± 0.7	298 ± 2	RR [relative to k(trans-2-pentene) = 6.65×10^{-11}][g]	Ohta[44]	
3,3-Dimethyl-1-butene			28.5	303	RR [relative to k(cis-2-butene) = 5.49×10^{-11}][b]	Wu et al.[27]	
2,3-Dimethyl-2-butene			153	298	DF-MS	Morris and Niki[42]	
			110 ± 22	298	FP-RF	Perry[52]	
			56.9 ± 1.3	298	FP-RF	Ravishankara et al.[30]	
			129 ± 9	300 ± 1	RR [relative to k(cis-2-butene) = 5.58×10^{-11}][b]	Atkinson et al.[51]	

TABLE 5. Rate constants k and temperature-dependent parameters for the gas-phase reactions of the OH radical with acyclic monoalkenes at, or close to, the high pressure limit — Continued

Alkene	$10^{12} \times A$ (cm^3 molecule^{-1} s^{-1})	B (K)	$10^{12} \times k$ (cm^3 molecule^{-1} s^{-1})	at T (K)	Technique	Reference	Temperature range covered (K)
			112 ± 6	299 ± 2	RR [relative to k(propene) = 2.62 × 10^{-11}][b]	Atkinson et al.[9]	
			115 ± 5	298 ± 2	RR [relative to k(2-methyl-1,3-butadiene) = 1.01 × 10^{-10}][b]	Atkinson et al.[53]	
			110 ± 3	294 ± 2	RR [relative to k(2-methyl-1,3-butadiene) = 1.02 × 10^{-10}][b]	Atkinson et al.[54]	
			112 ± 5	295 ± 1	RR [relative to k(propene) = 2.68 × 10^{-11}][b]	Atkinson and Aschmann[12]	
			103 ± 1	298 ± 2	RR [relative to k(2-methyl-2-butene) = 8.69 × 10^{-11}][b]	Ohta[44]	
			111 ± 3	294 ± 1	RR [relative to k(2-methyl-1,3-butadiene) = 1.02 × 10^{-10}][b]	Atkinson et al.[55]	
	37.0 ± 5.6			1237	LH-LIF	Smith[15]	
			111 ± 8	296 ± 2	RR [relative to k(2-methyl-1,3-butadiene) = 1.01 × 10^{-10}][b]	Atkinson et al.[56]	
			111 ± 3	296 ± 2	RR [relative to k(2-methyl-1,3-butadiene) = 1.01 × 10^{-10}][b]	Atkinson and Aschmann[57]	
1-Heptene			36.1 ± 7.2	305 ± 2	RR [relative to k(2-methylpropene) = 4.94 × 10^{-11}][b]	Darnall et al.[58]	
			40.5 ± 1.6	295 ± 1	RR [relative to k(propene) = 2.68 × 10^{-11}][b]	Atkinson and Aschmann[12]	
2,3-Dimethyl-2-pentene			98.2 ± 0.9	298 ± 2	RR [relative to k(2-methyl-2-butene) = 8.69 × 10^{-11}][b]	Ohta[44]	
			108 ± 2	298 ± 2	RR [relative to k(2,3-dimethyl-2-butene) = 1.10 × 10^{-10}][b]	Ohta[44]	
trans-4,4-Dimethyl-2-pentene			54.5 ± 0.7	298 ± 2	RR [relative to k(trans-2-pentene) = 6.65 × 10^{-11}][g]	Ohta[44]	

[a]See Introduction.
[b]From present recommendations (see text).
[c]Non-exponential OH radical decays observed.
[d]Extrapolated to high-pressure limit using the Troe fall-off expression.
[e]From the rate constant determined by Ohta[45] (Table 6).
[f]From the rate constant determined by Ohta.[44]
[g]From the rate constant determined by Ohta,[45] using an assumed temperature dependence of $B = -500$ K.

TABLE 6. Rate constants k and temperature-dependent parameters for the gas-phase reactions of the OH radical with acyclic di- and trialkenes at, or close to, the high pressure limit

Alkene	$10^{12} \times A$ (cm³ molecule⁻¹ s⁻¹)	B (K)	$10^{12} \times k$ (cm³ molecule⁻¹ s⁻¹)	at T (K)	Technique	Reference	Temperature range covered (K)
Propadiene			4.5 ± 2.5[a]	300	DF-EPR	Bradley *et al.*[22]	
			9.30 ± 0.93	299.0	FP-RF	Atkinson *et al.*[48]	299–421
			8.70 ± 0.87	349.7			
	5.59	-153 ± 151	8.02 ± 0.80	420.8			
			10.0 ± 1.4	297 ± 2	RR [relative to k(1,3-butadiene) = 6.69×10^{-11}][b]	Ohta[45]	
			9.84 ± 0.97	295 ± 1	RR [relative to k(propene) = 2.68×10^{-11}][b]	Atkinson and Aschmann[12]	
			9.0 ± 1.0	305	PR-RA	Liu *et al.*[59]	305–1173
			8.7 ± 0.9	373			
			8.2 ± 0.8	398			
			8.8 ± 0.9	478			
			8.0 ± 0.9	543			
	6.7 ± 0.9	-100 ± 50 (305–613 K)	7.8 ± 0.8	613			
			7.2 ± 0.7	673			
			7.3 ± 0.7	773			
			7.6 ± 0.8	808			
			7.9 ± 0.8	853			
			8.2 ± 0.8	873			
			7.8 ± 0.8	888			
			6.7 ± 0.7	973			
			6.5 ± 0.7	1073			
			5.6 ± 0.6	1173			
1,2-Butadiene			26.1 ± 2.1	297 ± 2	RR [relative to k(1,3-butadiene) = 6.69×10^{-11}][b]	Ohta[45]	
1,3-Butadiene			67.6 ± 13.6	305 ± 2	RR [relative to k(n-butane) = 2.62×10^{-12}][b]	Lloyd *et al.*[5]	
			68.5 ± 6.9	299.5	FP-RF	Atkinson *et al.*[48]	299–424
			57.2 ± 5.7	347.2			
	14.5	-468 ± 151	43.3 ± 4.4	424.0			
			65.0	300	RR [relative to k(ethene) = 8.44×10^{-12}][b]	Barnes *et al.*[33]	
			61.6 ± 1.5	297 ± 2	RR [relative to k(propene) = 2.65×10^{-11}][b]	Ohta[45]	
			68.8 ± 2.2	297 ± 2	RR [relative to k(2-methyl-2-butene) = 8.74×10^{-11}][b]	Ohta[45]	
			67.8 ± 2.2	295 ± 1	RR [relative to k(propene) = 2.68×10^{-11}][b]	Atkinson and Aschmann[12]	
			61 ± 6	313	PR-RA	Liu *et al.*[59]	313–1203
			50 ± 5	333			
			51 ± 5	338			
			46 ± 5	373			

TABLE 6. Rate constants k and temperature-dependent parameters for the gas-phase reactions of the OH radical with acyclic di- and trialkenes at, or close to, the high pressure limit — Continued

Alkene	$10^{12} \times A$ (cm³ molecule^{-1} s^{-1})	B (K)	$10^{12} \times k$ (cm³ molecule^{-1} s^{-1})	at T (K)	Technique	Reference	Temperature range covered (K)
			41 ± 4	393			
			47 ± 5	408			
			42 ± 4	438			
			35 ± 4	483			
			29 ± 3	563			
	14 ± 1	−440 ± 40 (313–623 K)	30 ± 3	623			
			30 ± 3	673			
			24 ± 3	723			
			20 ± 2	773			
			17 ± 2	873			
			15 ± 2	923			
			11 ± 1	1023			
			10 ± 1	1053			
			6.5 ± 0.6	1153			
			6.9 ± 0.7	1173			
			7.7 ± 0.8	1203			
1,2-Pentadiene			35.5 ± 1.4	297 ± 2	RR [relative to k(1,3-butadiene) = 6.69 × 10^{-11}]^b	Ohta[45]	
cis-1,3-Pentadiene			101 ± 4	297 ± 2	RR [relative to k(1,3-butadiene) = 6.69 × 10^{-11}]^b	Ohta[45]	
1,4-Pentadiene			53.3 ± 1.4	297 ± 2	RR [relative to k(propene) = 2.65 × 10^{-11}]^b	Ohta[45]	
3-Methyl-1,2-butadiene			56.9 ± 2.1	297 ± 2	RR [relative to k(1,3-butadiene) = 6.69 × 10^{-11}]^b	Ohta[45]	
2-Methyl-1,3-butadiene			78.1	300	RR [relative to k(ethene) = 8.44 × 10^{-12}]^b	Cox et al.[32]	
			99.8 ± 4.5	299 ± 2	RR [relative to k(propene) = 2.62 × 10^{-11}]^b	Atkinson et al.[9]	
			92.6 ± 15	299	FP-RF	Kleindienst et al.[60]	299–422
			76.4 ± 12	349			
	23.6	−409 ± 28	62.1 ± 8.2	422			
			99.0 ± 2.7	297 ± 2	RR [relative to k(1,3-butadiene) = 6.69 × 10^{-11}]^b	Ohta[45]	
			102 ± 4	295 ± 1	RR [relative to k(propene) = 2.68 × 10^{-11}]^b	Atkinson and Aschmann[12]	
			101 ± 2	297	RR [relative to k(propene) = 2.65 × 10^{-11}]^b	Edney et al.[46]	
trans-1,3-Hexadiene			112 ± 4	297 ± 2	RR [relative to k(1,3-butadiene) = 6.69 × 10^{-11}]^b	Ohta[45]	

TABLE 6. Rate constants k and temperature-dependent parameters for the gas-phase reactions of the OH radical with acyclic di- and trialkenes at, or close to, the high pressure limit — Continued

Alkene	$10^{12} \times A$ (cm³ molecule⁻¹ s⁻¹)	B (K)	$10^{12} \times k$ (cm³ molecule⁻¹ s⁻¹)	at T (K)	Technique	Reference	Temperature range covered (K)
trans-1,4-Hexadiene			90.3 ± 5.4	297 ± 2	RR [relative to k(1,3-butadiene) = 6.69 × 10⁻¹¹]ᵇ	Ohta[45]	
			90.9 ± 4.3	297 ± 2	RR [relative to k(propene) = 2.65 × 10⁻¹¹]ᵇ	Ohta[45]	
1,5-Hexadiene			62.2 ± 1.4	297 ± 2	RR [relative to k(1,3-butadiene) = 6.69 × 10⁻¹¹]ᵇ	Ohta[45]	
			61.7 ± 3.5	297 ± 2	RR [relative to k(propene) = 2.65 × 10⁻¹¹]ᵇ	Ohta[45]	
2,4-Hexadiene (cis + trans mixture)			134 ± 6	297 ± 2	RR [relative to k(1,3-butadiene) = 6.69 × 10⁻¹¹]ᵇ	Ohta[45]	
2-Methyl-1,4-pentadiene			78.8 ± 8.1	297 ± 2	RR [relative to k(cis-1,3-pentadiene) = 1.01 × 10⁻¹⁰]ᶜ	Ohta[45]	
3-Methyl-1,3-pentadiene			136 ± 9	297 ± 2	RR [relative to k(cis-1,3-pentadiene) = 1.01 × 10⁻¹⁰]ᶜ	Ohta[45]	
4-Methyl-1,3-pentadiene			131 ± 5	297 ± 2	RR [relative to k(cis-1,3-pentadiene) = 1.01 × 10⁻¹⁰]ᶜ	Ohta[45]	
2,3-Dimethyl-1,3-butadiene			122 ± 6	297 ± 2	RR [relative to k(1,3-butadiene) = 6.69 × 10⁻¹¹]ᵇ	Ohta[45]	
2-Methyl-1,5-hexadiene			96.1 ± 4.4	297 ± 2	RR [relative to k(1,5-hexadiene) = 6.20 × 10⁻¹¹]ᶜ	Ohta[45]	
2,5-Dimethyl-1,5-hexadiene			120 ± 2	297 ± 2	RR [relative to k(1,5-hexadiene) = 6.20 × 10⁻¹¹]ᶜ	Ohta[45]	
2,5-Dimethyl-2,4-hexadiene			210 ± 10	297 ± 2	RR [relative to k(2-methyl-1,5-hexadiene) = 9.61 × 10⁻¹¹]ᶜ	Ohta[45]	
cis-1,3,5-Hexatriene			110 ± 8	294 ± 2	RR [relative to k(2-methyl-1,3-butadiene) = 1.02 × 10⁻¹⁰]ᵇ	Atkinson et al.[54]	
trans-1,3,5-Hexatriene			111 ± 18	294 ± 2	RR [relative to k(2-methyl-1,3-butadiene) = 1.02 × 10⁻¹⁰]ᵇ	Atkinson et al.[54]	
3-Methylene-7-methyl-1,6-octadiene (Myrcene)			215 ± 16	294 ± 1	RR [relative to k(2,3-dimethyl-2-butene) = 1.13 × 10⁻¹⁰]ᵇ	Atkinson et al.[55]	

TABLE 6. Rate constants k and temperature-dependent parameters for the gas-phase reactions of the OH radical with acyclic di- and trialkenes at, or close to, the high pressure limit — Continued

Alkene	$10^{12} \times A$ (cm³ mole-cule^{-1} s^{-1})	B (K)	$10^{12} \times k$ (cm³ molecule^{-1} s^{-1})	at T (K)	Technique	Reference	Temperature range covered (K)
3,7-Dimethyl-1,3,6-octatriene (cis-, trans-Ocimene)			252 ± 20[d]	294 ± 1	RR [relative to k(2,3-dimethyl-2-butene) = 1.13×10^{-10}][b]	Atkinson et al.[55]	

[a]May not be the high pressure limit.
[b]From the present recommendations (see text).
[c]From the rate constants determined by Ohta.[45]
[d]cis- and trans-Isomers have identical rate constants within $\pm 20\%$.[55]

TABLE 7. Rate constants k and temperature-dependent parameters for the gas-phase reactions of the OH radical with cyclic mono-, di- and trialkenes

Alkene	$10^{12} \times A$ (cm³ mole-cule^{-1} s^{-1})	B (K)	$10^{12} \times k$ (cm³ molecule^{-1} s^{-1})	at T (K)	Technique	Reference	Temperature range covered (K)
Cyclopentene			67.3 ± 2.5	298 ± 2	RR [relative to k(2-methyl-1,3-butadiene) = 1.01×10^{-10}][a]	Atkinson et al.[53]	
			50.2 ± 4.0	298 ± 3	RR [relative to k(trans-2-butene) = 6.40×10^{-11}][a]	Rogers[47]	
			63.6 ± 1.7	298 ± 3	RR [relative to k(cyclohexene) = 6.77×10^{-11}][a]	Rogers[47]	
Cyclohexene			65.9	303	RR [relative to k(cis-2-butene) = 5.49×10^{-11}][a]	Wu et al.[27]	
			75.6 ± 15.2	305 ± 2	RR [relative to k(2-methylpropene) = 4.94×10^{-11}][a]	Darnall et al.[58]	
			65.4	300	RR [relative to k(ethene) = 8.44×10^{-12}][a]	Cox et al.[32]	
			67.5	300	RR [relative to k(ethene) = 8.44×10^{-12}][a]	Barnes et al.[33]	
			64.5 ± 2.5	297 ± 2	RR [relative to k(1,5-hexadiene) = 6.20×10^{-11}][b]	Ohta[45]	
			67.7 ± 1.8	298 ± 2	RR [relative to k(2-methyl-1,3-butadiene) = 1.01×10^{-10}][a]	Atkinson et al.[53]	

TABLE 7. Rate constants k and temperature-dependent parameters for the gas-phase reactions of the OH radical with cyclic mono-, di- and trialkenes — Continued

Alkene	$10^{12} \times A$ (cm^3 mole-cule^{-1} s^{-1})	B (K)	$10^{12} \times k$ (cm^3 molecule^{-1} s^{-1})	at T (K)	Technique	Reference	Temperature range covered (K)
			54.3 ± 2.4	298 ± 3	RR [relative to k(trans-2-butene) = 6.40 × 10^{-11}]^a	Rogers[47]	
1,3-Cyclo-hexadiene			164 ± 6	298 ± 2	RR [relative to k(2-methyl-1,3-buta-diene) = 1.01 × 10^{-10}]^a	Atkinson et al.[53]	
1,4-Cyclo-hexadiene			99.2 ± 3.1	297 ± 2	RR [relative to k(1,5-hexadiene) = 6.20 × 10^{-11}]^b	Ohta[45]	
			99.8 ± 4.1	298 ± 2	RR [relative to k(2-methyl-1,3-buta-diene) = 1.01 × 10^{-10}]^a	Atkinson et al.[53]	
Cycloheptene			74.4 ± 2.4	298 ± 2	RR [relative to k(2-methyl-1,3-buta-diene) = 1.01 × 10^{-10}]^a	Atkinson et al.[53]	
1,3-Cyclo-heptadiene			139 ± 5	294 ± 2	RR [relative to k(2-methyl-1,3-buta-diene) = 1.02 × 10^{-10}]^a	Atkinson et al.[54]	
1,3,5-Cyclo-heptatriene			96.9 ± 2.5	294 ± 2	RR [relative to k(2-methyl-1,3-buta-diene) = 1.02 × 10^{-10}]^a	Atkinson et al.[54]	
1-Methyl-cyclohexene			94.4 ± 18.9	305 ± 2	RR [relative to k(2-methylpropene) = 4.94 × 10^{-11}]^a	Darnall et al.[58]	
Bicyclo[2.2.1]-2-heptene			49.3 ± 4.1	298 ± 2	RR [relative to k(2-methyl-1,3-butadiene) = 1.01 × 10^{-10}]^a	Atkinson et al.[53]	
Bicyclo[2.2.1]-2,5-heptadiene			120 ± 11	298 ± 2	RR [relative to k(2-methyl-1,3-butadiene) = 1.01 × 10^{-10}]^a	Atkinson et al.[53]	
Bicyclo[2.2.2]-2-octene			40.8 ± 2.0	298 ± 2	RR [relative to k(2-methyl-1,3-butadiene) = 1.01 × 10^{-10}]^a	Atkinson et al.[53]	
α-Pinene^c			56.3 ± 8.5	305 ± 2	RR [relative to k(2-methylpropene) = 4.94 × 10^{-11}]^a	Winer et al.[28]	
			60.1 ± 8.2	298	FP-RF	Kleindienst et al.[60]	298–422
			51.0 ± 6.9	349			
	13.7	−446 ± 75	38.8 ± 5.7	422			
			55.0 ± 3.2	294 ± 1	RR [relative to k(2,3-dimethyl-2-butene) = 1.13 × 10^{-10}]^a	Atkinson et al.[55]	
β-Pinene^c			65.7 ± 9.9	305 ± 2	RR [relative to k(2-methylpropene) = 4.94 × 10^{-11}]^a	Winer et al.[28]	

TABLE 7. Rate constants k and temperature-dependent parameters for the gas-phase reactions of the OH radical with cyclic mono-, di- and trialkenes — Continued

Alkene	$10^{12} \times A$ (cm³ mole-cule⁻¹ s⁻¹)	B (K)	$10^{12} \times k$ (cm³ molecule⁻¹ s⁻¹)	at T (K)	Technique	Reference	Temperature range covered (K)
			77.6 ± 11	297	FP-RF	Kleindienst et al.[60]	297–423
			67.8 ± 11	350			
	23.6	-358 ± 58	54.2 ± 10	423			
			80.2 ± 5.2	294 ± 1	RR [relative to k(2,3-dimethyl-2-butene) = 1.13×10^{-10}][a]	Atkinson et al.[55]	
d-Limonene[c]			146 ± 22	305 ± 2	RR [relative to k(2-methylpropene) = 4.94×10^{-11}][a]	Winer et al.[28]	
			171 ± 5	294 ± 1	RR [relative to k(2,3-dimethyl-2-butene) = 1.13×10^{-10}][a]	Atkinson et al.[55]	
Δ^3-Carene[c]			87.8 ± 4.3	294 ± 1	RR [relative to k(2,3-dimethyl-2-butene) = 1.13×10^{-10}][a]	Atkinson et al.[55]	
γ-Terpinene[c]			177 ± 19	294 ± 1	RR [relative to k(2,3-dimethyl-2-butene) = 1.13×10^{-10}][a]	Atkinson et al.[55]	
α-Phellandrene[c]			313 ± 72	294 ± 1	RR [relative to k(2,3-dimethyl-2-butene) = 1.13×10^{-10}][a]	Atkinson et al.[55]	
α-Terpinene[c]			363 ± 40	294 ± 1	RR [relative to k(2,3-dimethyl-2-butene) = 1.13×10^{-10}][a]	Atkinson et al.[55]	

[a]From the present recommendations (see text).
[b]From the rate constant determined by Ohta.[45]
[c]Structures:

α-Pinene, ; β-pinene, ; d-limonene, ; Δ³-carene, ; γ-terpinene, ;

; α-phellandrene, ; α-Terpinene, .

TABLE 8. Rate constants for the gas-phase reactions of the OD radical with acyclic monoalkenes at one atmosphere total pressure of argon diluent

Alkene	$10^{12} \times k$ (cm^3 molecule^{-1} s^{-1})	at T (K)	Technique	Reference	Temperature range covered (K)
Ethene	6.14	343	PR-RA	Liu et al. [17,18]	343–1173
	4.75	373			
	4.34	403			
	3.52	483			
	3.23	563			
	2.94	603			
	1.79	653			
	1.32	703			
	1.26	748			
	1.53	773			
	1.88	873			
	2.23	973			
	2.62	1073			
	3.52	1173			
Ethene-d_4	4.91	383	PR-RA	Liu et al. [18]	383–1173
	4.28	393			
	4.23	448			
	3.20	523			
	3.10	603			
	1.47	653			
	0.84	708			
	0.88	748			
	1.08	801			
	1.13	873			
	1.30	973			
	1.86	1023			
	2.34	1173			

The broadening parameter F has been calculated to be 0.70 at 298 K,[13] and this value has been used, with the temperature dependence of F being given by,

$$F = e^{-T/T^*} + e^{-4T^*/T}$$

with $T^* = 840$ K. There are only a limited number of studies which provide reliable data concerning the third-order rate constant k_0[11,13,19] and, based upon these studies of Zellner and Lorenz,[11] Klein et al.[13] and Tully,[19] and the discussion of Klein et al.,[13] the following values of k_0 for ethene are obtained at 298 K

$$k_0(M = N_2, O_2) = 1.0 \times 10^{-28} \text{ cm}^6 \text{ molecule}^{-2} \text{ s}^{-1}$$

$$k_0(M = Ar) = 6.0 \times 10^{-29} \text{ cm}^6 \text{ molecule}^{-2} \text{ s}^{-1}$$

and

$$k_0(M = He) = 3.0 \times 10^{-29} \text{ cm}^6 \text{ molecule}^{-2} \text{ s}^{-1}$$

The data of Tully[19] for M = He at 295, 350 and 420 K allow a temperature dependence of these low pressure rate constants of T^{-3} to be estimated, leading to

$$k_0(M = N_2, O_2) = 1.0 \times 10^{-28} (T/298)^{-3} \text{ cm}^6 \text{ molecule}^{-2} \text{ s}^{-1}$$

$$k_0(M = Ar) = 6.0 \times 10^{-29} (T/298)^{-3} \text{ cm}^6 \text{ molecule}^{-2} \text{ s}^{-1}$$

and

$$k_0(M = He) = 3.0 \times 10^{-29} (T/298)^{-3} \text{ cm}^6 \text{ molecule}^{-2} \text{ s}^{-1}$$

The absolute rate constants determined by Atkinson et al.,[7] Tully[10,19] and Liu et al.[17,18] over the temperature range 291–425 K which are given in Table 5 are reasonably close to the high pressure limit, and are in good agreement (Fig. 38). A least-squares analysis of the rate constants given in Table 5 from these studies[7,10,17–19] leads to the Arrhenius expression of

$$k(\text{ethene}) = (1.85^{+0.24}_{-0.22}) \times 10^{-12} e^{(438 \pm 43)/T} \text{ cm}^3 \text{ molecule}^{-1} \text{ s}^{-1},$$

which is applicable only for the temperature range 290–425 K and a total pressure of ~760 Torr of argon diluent.

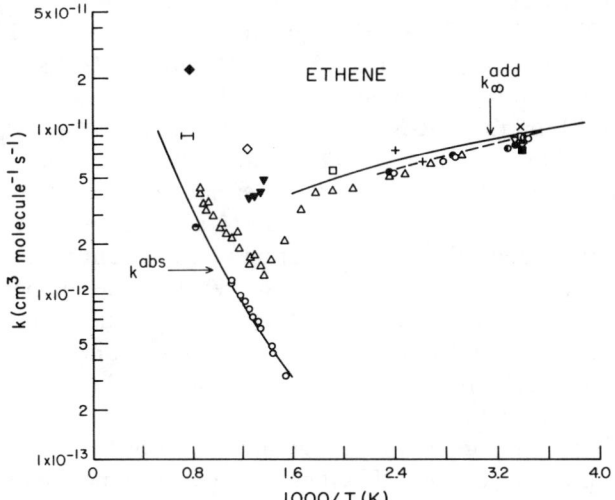

FIG. 38. Arrhenius plot of rate constants at, or close to, the high-pressure limit for the reaction of the OH radical with ethene. (⊢⊣) Westenberg and Fristrom;[1] (◇) Baldwin et al.;[2] (▼) Hoare and Patel;[3] (+) Gordon and Mulac;[4] (◐) Lloyd et al.;[5] (◆) Bradley et al.;[6] (●) Atkinson et al.;[7] (x) Overend and Paraskevopoulos;[8] (▽) Atkinson et al.;[9] (○) Tully et al.;[10,16,19] (□) Zellner and Lorenz;[11] (◑) Atkinson and Aschmann;[12] (▲) Klein et al.;[13] (■) Schmidt et al.;[14] (◖) Smith;[15] (△) Liu et al.;[17,18] (— — —) recommended Arrhenius expression applicable to 760 Torr total pressure of N₂ or air; (———) recommendations for k_∞^{add} and k^{abs} (see text).

From precise relative rate constant determinations carried out at ~740 Torr total pressure of air, Atkinson et al.[9] derived a value of

$$k(\text{ethene}) = (8.38 \pm 0.38) \times 10^{-12} \text{ cm}^3 \text{ molecule}^{-1} \text{ s}^{-1}$$

at 299 ± 2 K relative to the present recommendation for n-butane, while Atkinson and Aschmann[12] derived a value of

$$k(\text{ethene}) = 8.66$$

$$\times 10^{-12} \text{ cm}^3 \text{ molecule}^{-1} \text{ s}^{-1} \text{ at } 295 \pm 1 \text{ K},$$

relative to the atmospheric pressure of air recommendation for propene (see the discussion below concerning propene). Since this rate constant for propene was derived from a least-squares analysis of the relative rate constants for a series of alkenes and dialkenes at atmospheric pressure of air with the corresponding "high-pressure" absolute rate constant data, this rate constant of Atkinson and Aschmann[12] for ethene at 295 K has

been combined with the above temperature dependence to recommend that

$$k(\text{ethene; 760 Torr of air}) = (1.96^{+0.26}_{-0.24})$$

$$\times 10^{-12} e^{(438 \pm 43)/T} \text{ cm}^3 \text{ molecule}^{-1} \text{ s}^{-1}$$

over the restricted temperature range of 291–425 K, where the indicated error limits are equivalent to two least-squares standard deviations, and

$$k(\text{ethene}) = 8.52 \times 10^{-12} \text{ cm}^3 \text{ molecule}^{-1} \text{ s}^{-1} \text{ at } 298 \text{ K}$$

and ~760 Torr total pressure of air, with an estimated uncertainty at 298 K of ±15%. This expression is plotted in Fig. 38 as the dashed line and is used in this article to reevaluate those relative rate studies carried out at atmospheric pressure of air and utilizing ethene as the reference organic.

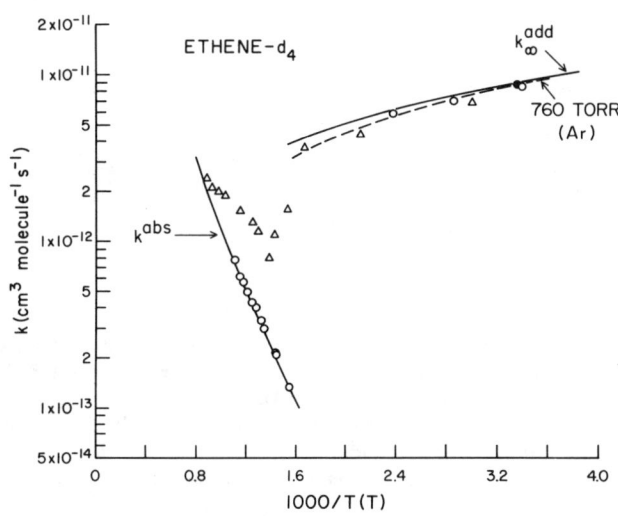

FIG. 39. Arrhenius plot of rate constants at, or close to, the high-pressure limit for the reaction of the OH radical with ethene-d₄. (●) Niki et al.;[20] (○) Tully;[16,19] (△) Liu et al.;[18] (— — —, ———) recommendations (see text).

Using the above Arrhenius expressions applicable to ~760 Torr total pressure of argon and air diluents and an assessment of the degree of fall-off (calculated from the Troe fall-off equation), a limiting high-pressure second-order rate constant of

$$k_\infty(\text{ethene}) = 9.0$$

$$\times 10^{-12} (T/298)^{-1.1} \text{ cm}^3 \text{ molecule}^{-1} \text{ s}^{-1}$$

is obtained. Use of these above values of k_o, k_∞ and F reproduce to within better than 5% the Arrhenius expressions given above which are applicable to 760 Torr total pressure of argon and air. This recommended expression for k_∞ is plotted in Fig. 38 and Fig. 40, which also shows the absolute rate constant data of Atkinson et al.,[7] Overend and Paraskevopoulos,[8] Tully,[10,16,19] Zellner and Lorenz,[11] Schmidt et al.,[14] Smith[15] and Liu et al.[17,18] Also shown in Fig. 40 are the calculated rate constants at 760 and 100 Torr total pressure of argon diluent. The absolute rate constants of Atkinson et al.,[7] Tully[10,19] and Liu et al.[17,18] obtained at temperatures $\lesssim$525 K are in good agreement with these calculations (Fig. 40).

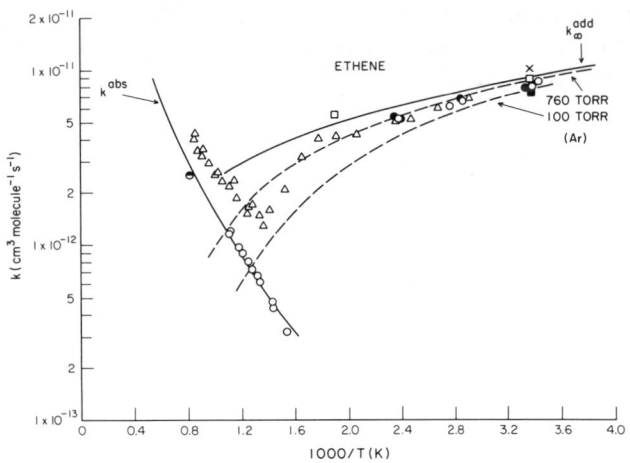

FIG. 40. Arrhenius plot of absolute rate constants obtained at, or close to, the high-pressure limit for the reaction of the OH radical with ethene. (●) Atkinson et al.;[7] (x) Overend and Paraskevopoulos;[8] (○) Tully;[10,16,19] (□) Zellner and Lorenz;[11] (■) Schmidt et al.;[14] (◕) Smith;[15] (△) Liu et al.;[17,18] (— — —, ——) recommendations (see text).

These calculations show that at 298 K the high pressure limit is not attained at 760 Torr total pressure of air, with the measured rate constant being ~5% below k_∞ under these conditions. At higher temperatures, rate data measured at 760 Torr total pressure of air, argon or helium move progressively into the fall-off region. At temperatures $\gtrsim$ 550 K, thermal decomposition of the $HOCH_2CH_2$ adduct also becomes important. The formation of the thermalized $HOCH_2CH_2$ radical from the reaction of OH radicals with ethene is calculated to be 32.1 kcal mol^{-1} exothermic[76] and hence the thermal dissociation of the thermalized $HOCH_2CH_2$ radical to reactants is expected[76] to have a high-pressure rate constant of

$$k_\infty(HOCH_2CH_2 \rightarrow OH + C_2H_4) \sim 3 \times 10^{13} \, e^{-15000/T} \, s^{-1}$$

[The thermal decomposition rates of the higher OH-alkene adducts are expected to be similar to that for the $HOCH_2CH_2$ radical; these thermal decompositions may be in the fall-off regime at the temperatures and pressures encountered.]

Thus, the decomposition rate of the thermalized $HOCH_2CH_2$ radical is calculated to be ~400 s^{-1} at 600 K (typical of the OH radical decay rates measured in the LP-LIF experiments of Tully[16]) and ~15000 s^{-1} at 700 K (the upper range of the OH radical decay rates utilized in the PR-RA experiments of Liu et al.[17,18]). Thus, from ~550 K upwards, depending upon the experimental technique used, the thermal decomposition of the $HOCH_2CH_2$ radical will lead to a rapid decrease in the measured OH radical reaction rate constant with increasing temperature. Note, however, that the temperature range at which this effect occurs is dependent upon the experimental technique, being ~100 K higher for the PR-RA technique of Liu et al.[17,18] than for the LF-LIF method of Tully.[16]

Since the temperature range in which this thermal decomposition of the OH-ethene adduct becomes important is also that in which the OH radical addition reaction rate constant is highly dependent upon the total pressure and the identity of the diluent gas (and in which the H-atom abstraction reaction is becoming significant), this temperature region from ~550 K to ~750 K is one in which the measured rate constants are dependent on the measurement method time scale, the total pressure and the identity of the diluent gas.

At temperatures $\gtrsim$600 K the only absolute rate constants available are those of Smith,[15] Tully[16] and Liu et al.,[17,18] with the rate constants of Liu et al.[17,18] being significantly higher than those of Smith[15] and Tully.[16] At least part of this difference in the measured rate constants may be due to the above mentioned effects of total pressure, diluent gas and measurement technique. Thus, the measurements of Liu et al.[17,18] were carried out with a short measurement time scale (OH radical decay rates up to ~15000 s^{-1}) at 760 Torr total pressure of argon. In contrast, the data of Tully[16] were obtained using longer measurement time scales (OH radical decay rates of up to ~750 s^{-1}) with helium as the diluent gas at total pressures of ~300–600 Torr. Thus, the study of Tully[16] was carried out at an effective pressure which was a factor of ~3 lower, and hence more into the fall-off region, than that employed in the study of Liu et al.[17,18] In addition, thermal decomposition of the $HOCH_2CH_2$ radical was more important in the study of Tully[16] due to the longer measurement time scales employed. Thus, the conditions of the study of Tully[16] were effective in suppressing the OH radical addition pathway and in isolating the H-atom abstraction route, while in the study of Liu et al.[17,18] the OH radical addition process may have contributed to the measured rate constant at temperatures $\gtrsim$750 K, leading to measured rate constants which were higher than those of the H-atom abstraction pathway up to higher temperatures than anticipated. Indeed, the rate data of Smith[15] and Tully[16] and of Liu et al.[17,18] do converge somewhat as the temperature approaches 1200 K.

Accordingly, the rate constants measured at temperatures $\geqslant 650$ K by Smith[15] and Tully[16] are taken to be those due to the H-atom abstraction pathway, and a least-squares analysis of the data from these two studies,[15,16] using the expression $k^{abs} = CT^2 e^{-D/T}$, yields the recommendation of

$$k^{abs}(\text{ethene}) = (4.87^{+1.79}_{-1.31}) \times$$

$$10^{-18} T^2 e^{-(1125 \pm 247)/T} \text{ cm}^3 \text{ molecule}^{-1} \text{ s}^{-1}$$

over the temperature range 651–1220 K, where the indicated error limits are two least-squares standard deviations. This recommended expression is plotted in Fig. 38 and Fig. 40. Extrapolation to 298 K yields an H-atom abstraction rate constant of

$$k^{abs} \sim 1 \times 10^{-14} \text{ cm}^3 \text{ molecule}^{-1} \text{ s}^{-1}, \sim 0.1\%$$

of the observed 298 K high pressure rate constant k_∞.

Interestingly, this rate constant for H-atom abstraction from ethene is very similar to the recommended rate constant for the reaction of the OH radical with methane of

$$k(\text{methane}) = 6.95$$

$$\times 10^{-18} T^2 e^{-1282/T} \text{ cm}^3 \text{ molecule}^{-1} \text{ s}^{-1}$$

$$[k^{abs}(\text{ethene})/ k(\text{methane}) = 0.70 \text{ e}^{157/T}],$$

totally consistent with the similar C—H bond energies in ethene (105.6 kcal mol^{-1}) and methane (104.8 kcal mol^{-1}).[62]

Ethene-d_4. The available rate constants measured at or close to the high pressure second-order limit at temperatures $\lesssim 525$ K[18-20] show that the rate constant k_∞ for ethene-d_4 is essentially identical to that for ethene, as expected for an addition process. Accordingly, the value of k_∞ derived above for ethene is also appropriate for ethene-d_4,

$$k_\infty(\text{ethene-}d_4) = 9.0$$

$$\times 10^{-12} (T/298)^{-1.1} \text{ cm}^3 \text{ molecule}^{-1} \text{ s}^{-1}$$

The rate constants measured in the fall-off region by Tully[19] at 295, 350 and 420 K for M = He and at 295 K for M = Ar show that at any given temperature the limiting low pressure third-order rate constant k_o is greater for ethene-d_4 than for ethene, as expected from the increased density of states in the $HOCD_2CD_2$ radical compared to the $HOCH_2CH_2$ radical.[74,75] A rate constant ratio of

$$k_o(\text{ethene-}d_4)/k_o(\text{ethene}) = 3$$

can be derived from the rate constant data of Tully.[19] Thus,

$$k_o^{He}(\text{ethene-}d_4) = 9.0$$

$$\times 10^{-29} (T/298)^{-3} \text{ cm}^6 \text{ molecule}^{-2} \text{ s}^{-1},$$

$$k_o^{Ar}(\text{ethene-}d_4) = 1.8$$

$$\times 10^{-28} (T/298)^{-3} \text{ cm}^6 \text{ molecule}^{-2} \text{ s}^{-1},$$

and

$$k_o^{N_2,O_2}(\text{ethene-}d_4) = 3.0$$

$$\times 10^{-28} (T/298)^{-3} \text{ cm}^6 \text{ molecule}^{-2} \text{ s}^{-1}.$$

The rate constant calculated for the reaction of the OH radical with ethene-d_4 at 760 Torr total pressure of argon is plotted in Fig. 39 as the dashed line. In contrast to the analogous situation for acetylene and acetylene-d_2 (see Sec. 2.5), the rate constants for the reactions of the OH radical with both ethene and ethene-d_4 at ~ 760 Torr total pressure of argon and $\lesssim 525$ K are sufficiently close to the high-pressure limit that the measured rate constants are essentially indistinguishable from k_∞ within the measurement uncertainties.

Analogous to the case for ethene, at temperatures $\geqslant 650$–750 K the measured rate constants are those for the D-atom abstraction process. By similar arguments to the ethene reaction, a least-squares analysis of the rate constant data of Tully,[16] using the equation $k = CT^2 e^{-D/T}$, yields the recommendation of

$$k^{abs}(\text{ethene-}d_4) = (1.42^{+0.45}_{-0.34})$$

$$\times 10^{-17} T^2 e^{-(2448 \pm 210)/T} \text{ cm}^3 \text{ molecule}^{-1} \text{ s}^{-1}$$

over the temperature range 651–901 K, where the indicated errors are two least-squares standard deviations. Kinetic data are required at temperatures > 900 K to ascertain whether the above three-parameter equation overestimates the D-atom abstraction rate constants above ~ 900 K.

The OD radical reactions with ethene and ethene-d_4 have been studied by Liu *et al.*,[17,18] and the rate constants measured at 760 Torr total pressure of argon diluent for these OD radical reactions (Table 8) are similar to those for the corresponding OH radical reactions, as expected. However, the rate constants over the temperature region $\lesssim 563$ K, which are those for OD radical addition to ethene and ethene-d_4, are uniformly $\sim 20\%$ lower than those for the corresponding OH radical reactions.

(2) Propene and Propene-d_6

Totally analogous to the situation for ethene, the kinetic data for the reaction of OH radicals with propene exhibit three distinct temperature regions. Below ~ 470 K the measured bimolecular rate constants are pressure dependent,[11,13,37] although this pressure dependence is much less marked than that for ethene, and the rate constants decrease with increasing temperature. At room temperature the measured rate constants are essentially independent of the total pressure and the identity of the diluent gas above ~25 Torr.[11,13,26,30] Furthermore, in this temperature region the rate constants for propene and propene-d_6 are essentially identical.[37] Above ~700 K the rate constants increase rapidly with temperature, and are independent of the pressure of the diluent gas,[37] with a significant deuterium isotope effect being observed.[37] In the intermediate temperature range of ~500–700 K non-exponential OH radical decays have been observed.[37]

As for the OH radical reaction with ethene, these data are totally consistent with the occurrence of OH radical addition and H-atom abstraction pathways, with the addition process totally dominating at temperatures ≲470 K. Above this temperature, thermal decomposition of the OH—C_3H_6 adduct occurs and the OH radical addition pathway becomes rapidly less important with increasing temperature. Above ~700 K, the measured rate constants, at least using absolute rate techniques, appear to be those for the H-atom abstraction reaction[37] (for the OH radical reaction with propene-d_6, OD radical formation arising from the OH radical addition channel was also observed at 602 K[37]).

The available rate constants, other than that of Cox[61] (as noted above), are listed in Table 5, with those at temperatures <470 K being at, or close to, the high pressure limit. The rate constants obtained from the absolute rate studies of Gordon and Mulac,[4] Atkinson and Pitts,[26] Ravishankara et al.,[30] Nip and Paraskevopoulos,[31] Zellner and Lorenz,[11] Schmidt et al.,[14] Tully and Goldsmith[37] and Smith et al.[38] (which supersedes the preliminary rate constant reported earlier by Smith[36]) and from the relative rate studies of Winer et al.,[28,29] Baldwin et al.,[35] Klein et al.,[13] Barnes et al.[39] and Atkinson and Aschmann[41] (with the rate constants at temperatures <470 K being at, or close to, the high-pressure limit) are plotted in Arrhenius form in Fig. 41.

The absolute rate constants obtained by Atkinson and Pitts,[26] Ravishankara et al.,[30] Nip and Paraskevopoulos[31] and Tully and Goldsmith[37] at ≤467 K are in excellent agreement, and a unit-weighted least-squares analysis of these data yields the Arrhenius expression of

$$k(\text{propene}, T \leqslant 467 \text{ K}) = (4.72^{+0.64}_{-0.57})$$

$$\times 10^{-12} e^{(504 \pm 45)/T} \text{ cm}^3 \text{ molecule}^{-1} \text{ s}^{-1}$$

over the temperature range 293–467 K, where the indicated errors are two least-squares standard deviations, and

$$k(\text{propene}) = 2.56 \times 10^{-11} \text{ cm}^3 \text{ molecule}^{-1} \text{ s}^{-1}$$

at 298 K and ~25–400 Torr total pressure of helium, argon or hydrogen diluent. The room temperature rate constants obtained by Lloyd et al.,[5] Wu et al.,[27] Winer et al.,[28,29] Cox et al.,[32] Barnes et al.,[33,39] Klein et al.[13] and Atkinson and Aschmann[41] from relative rate studies are

in good agreement with this expression, as are the absolute rate data of Zellner and Lorenz,[11] Schmidt et al.[14] and Wallington.[40] However, there are significant discrepancies with the rate constants reported by Morris et al.,[21] Bradley et al.,[22] Stuhl,[23] Gorse and Volman,[24] Gordon and Mulac,[4] Pastrana and Carr[25] and Smith,[34] with those of Morris et al.,[21] Bradley et al.,[22] Pastrana and Carr[25] and Smith[34] being in the fall-off region at the low total pressures (~1 Torr) employed[13] (although discrepancies still occur when the fall-off behavior is taken into account[13]). The room temperature rate constant determined by Stuhl[23] was probably low because of wall losses of propene in the static system used.

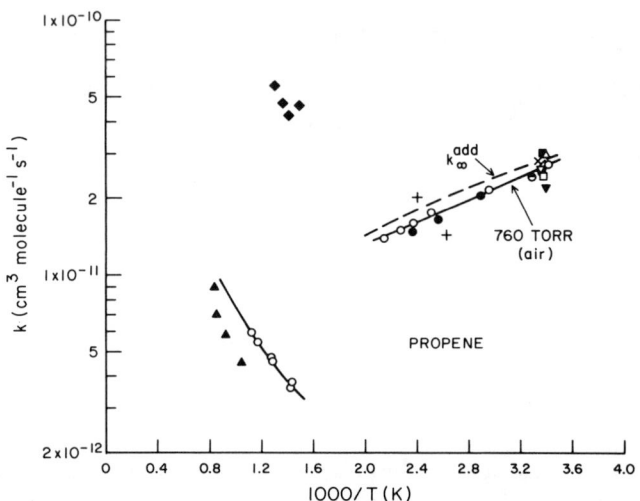

FIG. 41.   Arrhenius plot of rate constants obtained at, or close to, the high-pressure limit, for the reaction of the OH radical with propene. (+) Gordon and Mulac;[4] (●) Atkinson and Pitts;[26] (◓) Winer et al.;[28,29] (∇) Ravishankara et al.;[30] (□) Nip and Paraskevopoulos;[31] (■) Zellner and Lorenz;[11] (◆) Baldwin et al.;[35] (△) Klein et al.;[13] (▼) Schmidt et al.;[14] (○) Tully and Goldsmith;[37] (▲) Smith et al.;[38] (x) Barnes et al.;[39] (◇) Atkinson and Aschmann;[41] (— — —, ————)   recommendations (see text).

In the rate constant study of Atkinson and Aschmann,[12] relative rate constants for a series of alkenes (and n-butane and n-hexane) were obtained at 295 ± 1 K. Analogous to the procedure of Atkinson and Aschmann[12] and Atkinson,[63] a least-squares fit of these relative rate constants at 295 ± 1 K[12] to the absolute rate constants at 295 K for propene,[26,30,31,37] 1-butene,[26,30,31] 3-methyl-1-butene,[48] 2-methylpropene,[26] cis-2-butene,[26] trans-2-butene,[26] 2-methyl-2-butene,[50] propadiene[48] and 1,3-butadiene[48] (using the observed temperature dependencies or an estimated temperature dependence of B = −500 K to extrapolate or interpolate these observed absolute rate constants to 295 K) has been used to obtain a value of

k(propene) = 2.68

$$\times \ 10^{-11} \ cm^3 \ molecule^{-1} \ s^{-1} \ at \ 295 \ K,$$

with an estimated overall uncertainty of $\sim$15%.

Use of this 295 K rate constant, together with the temperature dependence derived above, yields the recommended Arrhenius expression of

$$k(propene, \ T \leqslant 467 \ K) = (4.85^{+0.66}_{-0.59})$$

$$\times \ 10^{-12} \ e^{(504 \pm 45)/T} \ cm^3 \ molecule^{-1} \ s^{-1}$$

applicable to $\sim$760 Torr total pressure of air, where the indicated error limits are two least-squares standard deviations, and

$$k(propene) = 2.63 \times 10^{-11} \ cm^3 \ molecule^{-1} \ s^{-1} \ at \ 298 \ K,$$

with an estimated uncertainty at 298 K of ±15%. This expression is $\sim$3% higher than that derived solely from the absolute rate data. In the discussions below, this recommended value of k(propene) at 295 K is used to derive rate constants at 295 K and atmospheric pressure for the other alkenes and dialkenes studied by Atkinson and Aschmann.[12]

As for ethene, values of k_o, k_∞ and F, and their temperature dependencies, are required to define the fall-off behavior for the OH radical reaction rate constant for propene. As discussed by Klein et al.,[13] this reaction approaches the high-pressure limit at relatively low total pressures, and hence few reliable data are available concerning the low pressure rate constant k_o. The recent studies of Zellner and Lorenz[11] and Klein et al.[13] have derived values of

$$k_o^{Ar} \sim 8.3 \times 10^{-28} \ cm^6 \ molecule^{-2} \ s^{-1} \ ^{[11]}$$

and

$$k_o^{Ar} = k_o^{air} \sim 8 \times 10^{-27} \ cm^6 \ molecule^{-2} \ s^{-1},^{[13]}$$

both at room temperature. The discrepancy between these derived values of k_o arises, in part, because the fall-off does not become obvious until low total pressures ($\lesssim$10 Torr), and only a small number of data points have been obtained in this low-pressure region. Indeed, at the lowest pressure studied, the rate constants measured by Zellner and Lorenz[11] and Klein et al.[13] disagree by less than a factor of 2 (and almost agree within the combined cited error limits).

Accordingly, a geometric mean of

$$k_o^{Ar} = k_o^{air} \approx 3 \times 10^{-27} \ cm^6 \ molecule^{-2} \ s^{-1}$$

at 298 K is used. By analogy with OH radical addition to ethene, a T^{-3} dependency is assumed, leading to

$$k_o^{Ar} = k_o^{air} = 3 \times 10^{-27} \ (T/298)^{-3} \ cm^6 \ molecule^{-2} \ s^{-1}$$

With $F = 0.5$ at 298 K[13] and $T^* = 430$ K, the measured rate constants at $\sim$760 Torr total pressure of argon or air are $\sim$5–6% and $\sim$8–9% lower than the limiting high-pressure rate constants k_∞ at 298 K and 420 K, respectively. Based upon the rate constant expression given above of

$$k(propene) = 4.85 \times 10^{-12} \ e^{504/T} \ cm^3 \ molecule^{-1} \ s^{-1},$$

applicable for $\sim$760 Torr total pressure of air, then

$$k_\infty = 2.8 \times 10^{-11} \ (T/298)^{-1.3} \ cm^3 \ molecule^{-1} \ s^{-1}$$

over the temperature range $\sim$290–470 K, and this expression is plotted in Fig. 41 as the dashed line.

At elevated temperatures, $\gtrsim$700 K, the reaction is expected to proceed by H-atom abstraction and, possibly, initial OH radical addition followed by rapid rearrangement and decomposition of the adduct. Rate constants for the reaction of OH radicals with propene have been measured in this temperature range by Tully and Goldsmith[37] and Smith et al.[38] using absolute methods and by Baldwin et al.[35] from a product study. The rate constants derived from the relative rate/product study of Baldwin et al.[35] are an order of magnitude higher than those of Tully and Goldsmith[37] and Smith et al.[38] and also lie above the extrapolated high pressure addition rate constant k_∞, for reasons which are not presently known. While the rate constants determined by Tully and Goldsmith[37] and Smith et al.[38] are in general agreement, there are discrepancies of the order of 50% between these studies (Fig. 41).

The data of Tully and Goldsmith[37] have been used to derive the rate constant in this temperature region. Using the expression $k = CT^2 e^{-D/T}$, a unit-weighted least-squares analysis of the data of Tully and Goldsmith[37] yields the recommendation of

$$k(propene; \ T > 700 \ K) = (7.20^{+1.98}_{-1.55})$$

$$\times \ 10^{-18} \ T^2 \ e^{(31 \pm 189)/T} \ cm^3 \ molecule^{-1} \ s^{-1}$$

over the temperature range 701–896 K, where the indicated error limits are two least-squares standard deviations. Hence the abstraction reaction will account for <3% of the observed overall high-pressure rate constant at 298 K. This estimate is totally consistent with the room temperature product data of Cvetanovic,[77] Hoyermann and Sievert[78] and Biermann et al.,[43] which showed that the abstraction reaction accounts for $\lesssim$5%,[77] <5%[78] and <2%[43] of the overall reaction rate constant under the conditions employed.

Under conditions where the rate constants were close to the high pressure limit, the rate constants measured by Tully and Goldsmith[37] for propene-d_6 at <470 K are essentially identical to those for the reaction of OH radicals with propene. This is totally consistent with the occurrence of an OH radical addition reaction under these conditions.

At >700 K, the rate constants measured by Tully and Goldsmith[37] for propene-d_6 are $\sim35\%$ lower than those for propene. However, these rate constants for propene-d_6 at temperatures $\geqslant700$ K may also include other reaction processes, such as OD radical formation, in addition to D-atom abstraction, and thus these measured rate constants for propene-d_6 at >700 K may not be solely those for the D-atom abstraction pathway (as indicated by the similar temperature dependencies of the propene and propene-d_6 reactions, despite the fact that the temperature dependence of D-atom abstraction from propene-d_6 should be higher than for H-atom abstraction from propene).

(3) 1-Butene and 1-Butene-d_8

The available rate constants are listed in Table 5. The kinetics of the OH radical reactions with 1-butene and 1-butene-d_8 are analogous to those for propene, with OH radical addition dominating at temperatures $\lesssim425$ K and OH radical addition with rapid subsequent isomerization/decomposition to products other than the initial reactants and/or H- (or D-) atom abstraction occurring at elevated temperatures ($\gtrsim600$ K). As for propene, the limiting high-pressure second-order rate constants at around room temperature are closely approached at total pressures of helium $\gtrsim20$ Torr.[30] In the lower temperature ($\lesssim425$ K) region, the most recent kinetic data of Atkinson and Pitts,[26] Wu $et\ al.$,[27] Ravishankara $et\ al.$,[30] Nip and Paraskevopoulos,[31] Barnes $et\ al.$,[33] Biermann $et\ al.$[43] (which is possibly still in the fall-off region between second- and third-order kinetics), Ohta[44] and Atkinson and Aschmann[12] are in good agreement. These rate constants of Atkinson and Pitts,[26] Wu $et\ al.$,[27] Ravishankara $et\ al.$,[30] Nip and Paraskevopoulos,[31] Barnes $et\ al.$,[33] Ohta[44] and Atkinson and Aschmann[12] and the elevated temperature rate constants of Smith[15] and Tully[16] are plotted in Arrhenius form in Fig. 42.

At temperatures $\lesssim425$ K, the sole reported temperature dependence is that of Atkinson and Pitts,[26] and hence this temperature dependence is recommended. As for ethene and propene, the rate constant derived from the best-fit analysis of the relative rate constant data of Atkinson and Aschmann[12] for a series of alkenes and di-alkenes with the available absolute data (as described above) is recommended. This yields

k(1-butene) = 3.19

$\times\ 10^{-11}$ cm^3 molecule^{-1} s^{-1} at 295 K.

This rate constant, when combined with the temperature dependence of Atkinson and Pitts,[26] leads to the recommended Arrhenius expression of

k(1-butene; $T \leqslant 425$ K) = 6.55

$\times\ 10^{-12}\ e^{467/T}$ cm^3 molecule^{-1} s^{-1}

over the temperature range 298–424 K, and

k(1-butene) = 3.14

$\times\ 10^{-11}$ cm^3 molecule^{-1} s^{-1} at 298 K,

with an estimated overall uncertainty at 298 K of $\pm20\%$.

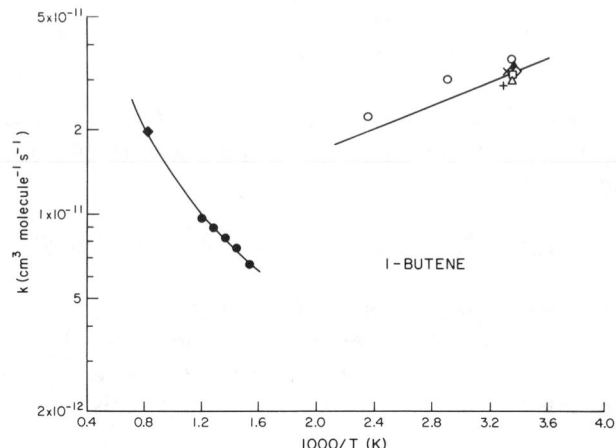

FIG. 42. Arrhenius plot of rate constants obtained at, or close to, the high-pressure limit for the reaction of the OH radical with 1-butene. ($\bigcirc$) Atkinson and Pitts;[26] (+) Wu $et\ al.$;[27] ($\triangle$) Ravishankara $et\ al.$;[30] ($\blacktriangle$) Nip and Paraskevopoulos;[31] (x) Barnes $et\ al.$;[33] ($\square$) Ohta;[44] ($\diamondsuit$) Atkinson and Aschmann;[12] ($\blacklozenge$) Smith;[15] ($\bullet$) Tully;[16] (———) recommendations, see text.

The room temperature kinetic data of Barnes $et\ al.$,[33] Biermann $et\ al.$[43] and Ohta,[44] which were not used in the evaluation, are in good agreement with this recommended rate constant.

At temperatures $\geqslant650$ K the measured rate constants[15,16] increase with increasing temperatures (but for $T <1225$ K are still lower than those at ~425 K). A unit-weighted least-squares analysis of the rate constants of Smith[15] and Tully,[16] using the expression $k = CT^2e^{-D/T}$, yields the recommendation of

k(1-butene; $T >650$ K) = $(1.04^{+0.13}_{-0.11})$

$\times\ 10^{-17}\ T^2\ e^{(273\ \pm\ 85)/T}$ cm^3 molecule^{-1} s^{-1}

over the temperature range 650–1225 K, where the indicated errors are two least-squares standard deviations [use of the rate constants of Tully[16] only leads to calculated rate constants which agree with those from the above recommendation to within $\pm5\%$ over the temperature range 650–1225 K].

Extrapolation of these expressions to room temperature indicates that H-atom abstraction will account for $<10\%$ of the measured overall high-pressure rate constant at 298 K. This estimated contribution of H-atom abstraction to the overall OH radical reaction rate constant at 298 K is in agreement with the estimates of <10% obtained from the product studies of Hoyermann and Sievert[79] and Atkinson et al.,[80] but disagrees with the percentage (20 ± 6%) measured by Biermann et al.[43] Thus, at temperatures ≲425 K the reaction of the OH radical with 1-butene proceeds mainly by OH radical addition [note, however, that extrapolation of the recommended high temperature (>650 K) rate expression predicts that the "direct" reaction channel involving decomposition of the OH-1-butene adduct to products other than the reactants and/or H-atom abstraction will account for ~20% of the overall reaction rate constant at 425 K]. At ~760 Torr total pressure of helium, argon or air diluent this OH radical addition reaction is close to the high-pressure limit, since the low pressure rate constant k_o is expected to be greater than that for propene.

As for ethene and propene, the thermal decomposition of the OH-1-butene adduct will become increasingly important above ~550–650 K, with the result that, unless this adduct radical rapidly rearranges and/or decomposes to products other than the original reactants, the addition process becomes of no importance above ~700 K. Above this temperature, H-atom abstraction, together with any "direct" reaction arising from the OH radical addition pathway (which cannot exceed the rate constant for OH addition to 1-butene to form the initially energy rich adduct), are the only reaction channels observed. That the major reaction pathway in this temperature region is H-atom abstraction is supported by the deuterium isotope effect observed by Tully[16] for the OH radical reactions with 1-butene and 1-butene-d_8, with[16]

$$k(\text{1-butene})/k(\text{1-butene-}d_8) = 1.05 \ e^{258/T}$$

Furthermore, the magnitude of this isotope effect is consistent with the H- or D-atom abstraction occurring mainly from the allylic C—H or C—D bonds

$$OH + CH_3CH_2CH{=}CH_2 \rightarrow H_2O + CH_3CHCH{=}CH_2$$

over the temperature range 650–830 K. As the temperature increases, H- or D-atom abstraction from the terminal —CH_3 or —CD_3 group will become increasingly important.[15]

(4) 2-Methylpropene

The available kinetic data are given in Table 5 and are plotted in Arrhenius form in Fig. 43. At temperatures ≲425 K the sole absolute study carried out is that of Atkinson and Pitts,[26] which is also the only temperature dependence study. Thus, this temperature dependence,[26] equivalent to $B = -504$ K, is used in combination with the best-fit 295 K rate constant derived from the relative rate constant data of Atkinson and Aschmann[12] of

$$k(\text{2-methylpropene}) = 5.23$$

$$\times \ 10^{-11} \ cm^3 \ molecule^{-1} \ s^{-1} \ \text{at 295 K},$$

to yield the recommended Arrhenius expression of

$$k(\text{2-methylpropene; } T \leqslant 425 \ \text{K}) = 9.47$$

$$\times \ 10^{-12} \ e^{504/T} \ cm^3 \ molecule^{-1} \ s^{-1}$$

over the temperature range 297–424 K, and

$$k(\text{2-methylpropene}) = 5.14 \times 10^{-11} \ cm^3 \ molecule^{-1} \ s^{-1}$$

at 298 K, with an estimated overall uncertainty at 298 K of ±20%.

The relative rate constants obtained by Wu et al.,[27] Barnes et al.[33] and Ohta[44] at room temperature are in good agreement with this recommendation.

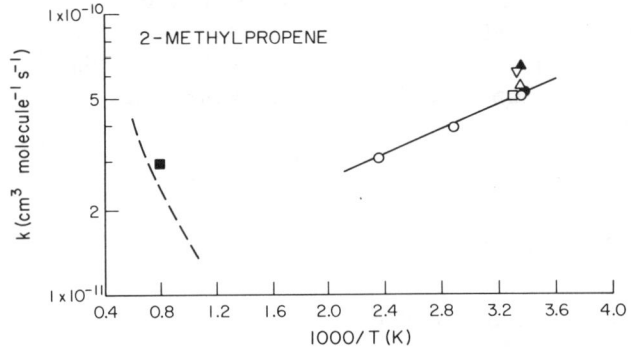

FIG. 43. Arrhenius plot of rate constants for the reaction of the OH radical with 2-methylpropene. (▲) Morris and Niki;[42] (◯) Atkinson and Pitts;[26] (☐) Wu et al.;[27] (▽) Barnes et al.;[33] (△) Ohta;[44] (●) Atkinson and Aschmann;[12] (■) Smith;[15] (———) recommendation for $T \lesssim 425$ K (see text); (— — —) tentatively recommended rate constant for H-atom abstraction (see text).

As for propene and 1-butene, at temperatures ≲425 K the OH radical reaction with 2-methylpropene proceeds predominantly by OH radical addition, with the rate constant being close to the high pressure limit at total pressures of ~50 Torr of helium, argon or air diluent. The OH-2-methylpropene adduct will thermally decompose at temperatures ≳550–750 K and, unless this adduct can rearrange and/or decompose more rapidly to products other than the initial reactants, at elevated temperatures only the H-atom abstraction reaction from the two —CH_3 groups will be important. The rate constant measured by Smith[15] of $2.96 \times 10^{-11} \ cm^3 \ molecule^{-1} \ s^{-1}$ at 1259 K is consistent with this expectation, since this measured rate constant for 2-methylpropene[15] is 2.5 k(propene).

Hence, it is tentatively recommended that the rate constant for H-atom abstraction from 2-methylpropene is given by

$$k(\text{2-methylpropene; } T \geqslant 700 \ \text{K}) \simeq 1.5$$

$$\times \ 10^{-17} \ T^2 \ e^{30/T} \ cm^3 \ molecule^{-1} \ s^{-1}.$$

Extrapolation of this H-atom abstraction rate constant to room temperature leads to the conclusion that at 298 K the H-atom abstraction channel accounts for ~3% of the overall high pressure rate constant, in agreement with the product study of Hoyermann and Sievert[79] (which yielded <5% abstraction at ~1 Torr total pressure).

(5) cis-2-Butene

The available rate constants are given in Table 5 and are plotted in Arrhenius form in Fig. 44. At room temperature the absolute rate constant determined by Atkinson and Pitts[26] is in good agreement with the relative rate constants derived by Lloyd et al.,[5] Ohta[44] and Atkinson and Aschmann.[12] As for the alkenes discussed above, the temperature dependence determined by Atkinson and Pitts[26] is used, together with the best-fit rate constant at 295 K derived from the relative rate constant data of Atkinson and Aschmann[12] and the available absolute rate data for a series of alkenes and dialkenes (see above), to recommend that

$$k(cis\text{-}2\text{-butene}) = 5.71$$
$$\times 10^{-11} \text{ cm}^3 \text{ molecule}^{-1} \text{ s}^{-1} \text{ at 295 K,}$$

$$k(cis\text{-}2\text{-butene}; T \lesssim 425 \text{ K}) = 1.10$$
$$\times 10^{-11} e^{487/T} \text{ cm}^3 \text{ molecule}^{-1} \text{ s}^{-1},$$

and

$$k(cis\text{-}2\text{-butene}) = 5.64$$
$$\times 10^{-11} \text{ cm}^3 \text{ molecule}^{-1} \text{ s}^{-1} \text{ at 298 K,}$$

with an estimated overall uncertainty at 298 K of ±20%.

The relative rate constant of Ohta[44] is in excellent agreement with this recommendation. However, as discussed previously, the absolute rate constant determined by Ravishankara et al.[30] at 298 K appears to be ~20% low, possibly because of wall losses of the cis-2-butene reactant in the static reaction system used.

(6) trans-2-Butene

The available kinetic data (apart from that of Cox,[61] as noted above) are given in Table 5 and those of Morris and Niki,[42] Atkinson and Pitts,[26] Wu et al.,[27] Ohta,[45] Atkinson and Aschmann,[12] Edney et al.[46] and Smith[15] are plotted in Arrhenius form in Fig. 45. The room temperature rate constants of Morris and Niki,[42] Atkinson and Pitts,[26] Wu et al.,[27] Ohta,[45] Atkinson and Aschmann,[12] Edney et al.[46] and Rogers[47] are in reasonable agreement. Consistent with the previous criteria, the temperature dependence determined by Atkinson and Pitts[26] of $B = -550$ K is used, together with the best-fit of the relative rate constants of Atkinson and Aschmann[12] to the abso-

lute rate constant data for a series of alkenes and dialkenes (see above), to yield

$$k(trans\text{-}2\text{-butene}) = 6.51$$
$$\times 10^{-11} \text{ cm}^3 \text{ molecule}^{-1} \text{ s}^{-1} \text{ at 295 K,}$$

$$k(trans\text{-}2\text{-butene}; T \lesssim 425 \text{ K}) = 1.01$$
$$\times 10^{-11} e^{550/T} \text{ cm}^3 \text{ molecule}^{-1} \text{ s}^{-1},$$

and

$$k(trans\text{-}2\text{-butene}) = 6.40$$
$$\times 10^{-11} \text{ cm}^3 \text{ molecule}^{-1} \text{ s}^{-1} \text{ at 298 K,}$$

with an estimated overall uncertainty at 298 K of ±20%.

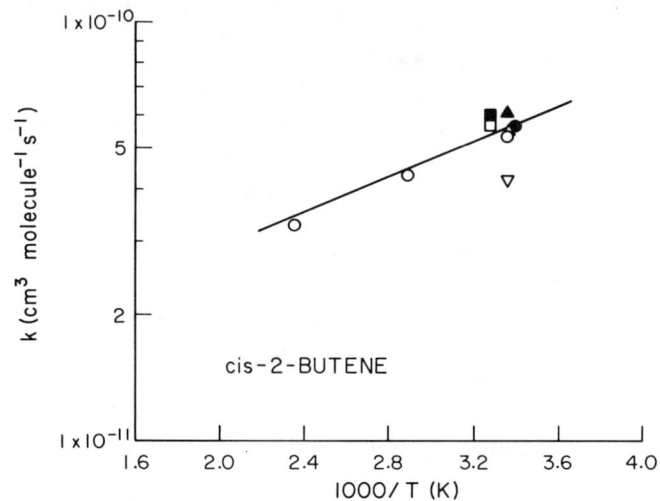

FIG. 44. Arrhenius plot of rate constants for the reaction of the OH radical with cis-2-butene. (▲) Morris and Niki;[42] (○) Atkinson and Pitts;[26] (□) Lloyd et al.;[5] (■) Winer et al.;[28] (▽) Ravishankara et al.;[30] (△) Ohta;[44] (●) Atkinson and Aschmann;[12] (———) recommendation (see text).

As for 2-methylpropene, 1-butene and propene, at temperatures $\lesssim 425$ K this OH radical reaction with trans-2-butene proceeds predominantly by OH radical addition, with the rate constant being close to the high pressure limit at total pressures of $\gtrsim 50$ Torr of helium, argon or air diluent. The OH-trans-2-butene adduct will thermally decompose at temperatures $\gtrsim 550$–750 K, and hence at higher temperatures only the rate constant for a "direct" reaction involving OH radical addition with rapid rearrangement and decomposition to products other than the initial reactants (which must have a rate constant $\leqslant k_\infty$) and/or H-atom abstraction is measured.

The rate constant determined by Smith[15] at 1275 K of $2.70 \times 10^{-11} \text{ cm}^3 \text{ molecule}^{-1} \text{ s}^{-1}$ is a factor of 2.25 higher than that calculated from the recommended expression for propene at this temperature, consistent with the number of substituent $-CH_3$ groups. Hence, it is tentatively recommended that the rate constant for H-atom abstraction from *trans*-2-butene is given by

$$k(\text{trans-2-butene}); \ T \gtrsim 700 \text{ K}) \simeq 1.5$$

$$\times 10^{-17} \ T^2 \ e^{30/T} \ cm^3 \ molecule^{-1} \ s^{-1},$$

identical to that for 2-methylpropene (and *cis*-2-butene). Extrapolation of this expression to room temperature leads to the conclusion that at 298 K the contribution of the H-atom abstraction pathway to the overall high-pressure rate constant is ~2%. This is consistent with the product study of Hoyermann and Sievert,[79] which showed that the abstraction channel accounts for <10% of the overall reaction at ~1 Torr total pressure.

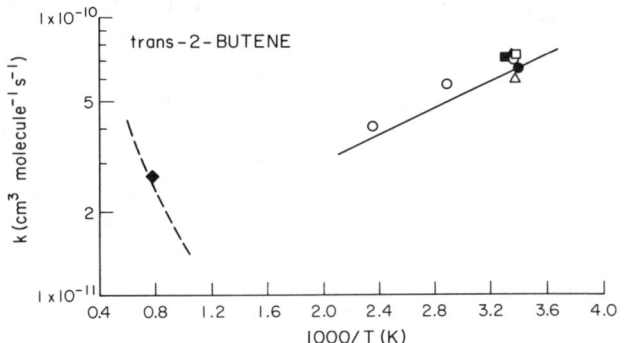

FIG. 45. Arrhenius plot of rate constants for the reaction of the OH radical with *trans*-2-butene. (▲) Morris and Niki;[42] (○) Atkinson and Pitts;[26] (□) Wu *et al.*;[27] (△) Ohta;[45] (●) Atkinson and Aschmann;[12] (⊡) Edney et al.;[46] (♦) Smith;[15] (———) recommendation for T ≲ 425 K (see text); (— — —) tentatively recommended rate constant for H-atom abstraction (see text).

(7) 1-Pentene

The available rate constants, all obtained at room temperature, are given in Table 5. These rate constants exhibit a significant amount of scatter, with those of Morris and Niki[42] and Nip and Paraskevopoulos[31] being ~30% higher than the remaining data. As for the C_2 through C_4 alkenes (see above), the rate constant derived from the relative rate study of Atkinson and Aschmann[12] is used to recommend that

$$k(\text{1-pentene}) = 3.19 \times 10^{-11} \ cm^3 \ molecule^{-1} \ s^{-1}$$

at 295 K. Combined with an estimated temperature dependence of $B = -500$ K (similar to the recommended temperature dependencies for propene, 1-butene, 2-

methylpropene and *cis*- and *trans*-2-butene for temperatures ≤425 K), this leads to

$$k(\text{1-pentene}) = 5.86 \times 10^{-12} \ e^{500/T} \ cm^3 \ molecule^{-1} \ s^{-1}$$

at around 300 K (this expression is definitely not applicable above ~425 K), and

$$k(\text{1-pentene}) = 3.14 \times 10^{-11} \ cm^3 \ molecule^{-1} \ s^{-1}$$

at 298 K, with an estimated overall uncertainty at 298 K of ±20%.

(8) 3-Methyl-1-butene

The only rate constants available for this alkene (Table 5) are those from the absolute rate constant study of Atkinson *et al.*[48] and the relative rate constant study of Atkinson and Aschmann,[12] and these data are plotted in Arrhenius form in Fig. 46.

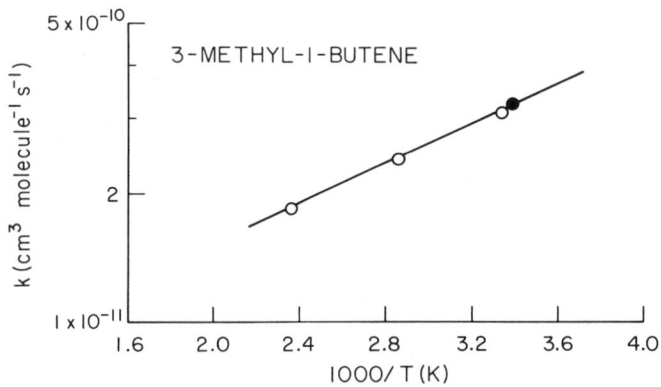

FIG. 46. Arrhenius plot of rate constants for the reaction of the OH radical with 3-methyl-1-butene. (○) Atkinson *et al.*;[48] (●) Atkinson and Aschmann;[12] (———) recommendation (see text).

These data are in excellent agreement. The best-fit rate constant from an analysis of the relative rate data of Atkinson and Aschmann[12] and the absolute rate constants for a series of alkenes and dialkenes (see above) leads to

$$k(\text{3-methyl-1-butene}) = 3.24$$

$$\times 10^{-11} \ cm^3 \ molecule^{-1} \ s^{-1}$$

at 295 K. Combined with the temperature dependence of Atkinson *et al.*,[48] this leads to the recommended Arrhenius expression of

$$k(\text{3-methyl-1-butene}) = 5.32$$

$$\times 10^{-12} \ e^{533/T} \ cm^3 \ molecule^{-1} \ s^{-1}$$

over the temperature range 299–423 K, and

$$k(\text{3-methyl-1-butene}) = 3.18 \times 10^{-11} \text{ at } 298 \text{ K,}$$

with an estimated overall uncertainty at 298 K of ±20%. This OH radical reaction proceeds mainly by OH radical addition at temperatures $\lesssim$425 K, with the rate constants being close to the high-pressure limit above ~20 Torr total pressure of diluent.

(9) 2-Methyl-2-butene

The available kinetic data are given in Table 5 and are plotted in Arrhenius form in Fig. 47. It can be seen that at room temperature the more recent absolute and relative rate constants of Atkinson and Pitts,[50] Atkinson et al.,[9,51] Ohta[45] and Atkinson and Aschmann[12] are in excellent agreement [the absolute rate constant study of Atkinson et al.[49] has been superseded by that of Atkinson and Pitts,[50] although it is in agreement with this later study[50] within the experimental error limits].

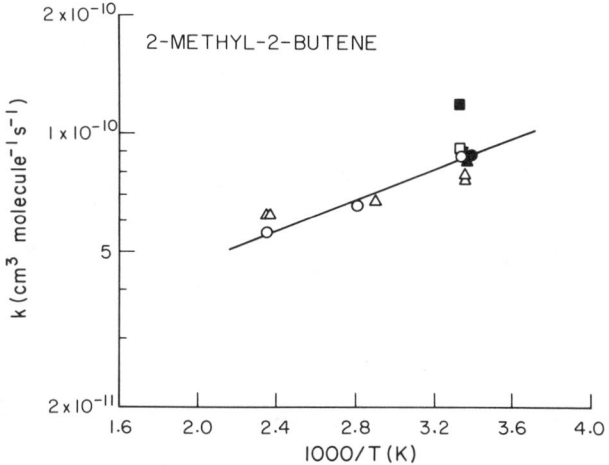

FIG. 47. Arrhenius plot of rate constants for the reaction of the OH radical with 2-methyl-2-butene. (■) Morris and Niki;[42] (△) Atkinson et al.;[49] (○) Atkinson and Pitts;[50] (□) Atkinson et al.;[51] (▼) Atkinson et al.;[9] (▲) Ohta;[45] (●) Atkinson and Aschmann;[12] (———) recommendation (see text).

Again, consistent with the above recommendations for the alkenes, the temperature dependence of Atkinson and Pitts,[50] of $B = -450$ K, is used, together with the best-fit rate constant obtained from the relative rate study of Atkinson and Aschmann[12] and the absolute rate constants for a series of alkenes and dialkenes of

$$k(\text{2-methyl-2-butene}) = 8.84$$

$$\times 10^{-11} \text{ cm}^3 \text{ molecule}^{-1} \text{ s}^{-1} \text{ at } 295 \text{ K,}$$

to recommend the Arrhenius expression of

$$k(\text{2-methyl-2-butene}) = 1.92$$

$$\times 10^{-11} \, e^{450/T} \text{ cm}^3 \text{ molecule}^{-1} \text{ s}^{-1}$$

over the temperature range 299–426 K, and

$$k(\text{2-methyl-2-butene}) = 8.69$$

$$\times 10^{-11} \text{ cm}^3 \text{ molecule}^{-1} \text{ s}^{-1}$$

at 298 K, with an estimated overall uncertainty at 298 K of ±20%. As for ethene, propene and 1-butene, at elevated temperatures ($\gtrsim$500 K) this OH radical addition reaction will changeover to an H-atom abstraction process (from the three $-CH_3$ groups).

(10) 2,3-Dimethyl-2-butene

The available rate constants (all but one obtained at around room temperature) are given in Table 5. At room temperature the most recent rate constants of Atkinson et al.,[9,53–56] Atkinson and Aschmann[12,57] and Ohta[44] are in good agreement. Using an assumed temperature dependence of $B = -500$ K for this reaction to extrapolate the measured rate constants to 298 K, a unit-weighted average of these data[9,12,44,53–57] yields the recommendation of

$$k(\text{2,3-dimethyl-2-butene}) = 1.10$$

$$\times 10^{-10} \text{ cm}^3 \text{ molecule}^{-1} \text{ s}^{-1} \text{ at } 298 \text{ K,}$$

with an estimated overall uncertainty of ±20%.

As discussed previously,[63,65] the room temperature rate constant obtained by Ravishankara et al.[30] is low, by a factor of ~2, presumably due to wall losses of the 2,3-dimethyl-2-butene reactant in the static reaction system used.

As for the alkenes discussed above, this rate constant primarily reflects OH radical addition. At elevated temperatures in the region of ~550–750 K the OH-2,3-dimethyl-2-butene adduct will thermally decompose and, unless this adduct more rapidly rearranges and/or decomposes to products other than the reactants, at temperatures $\gtrsim$700 K the observed reaction process will be H-atom abstraction from the four $-CH_3$ groups. Indeed, the rate constant of

$$k = 3.70 \times 10^{-11} \text{ cm}^3 \text{ molecule}^{-1} \text{ s}^{-1} \text{ at } 1237 \text{ K}$$

measured by Smith[15] is consistent with an H-atom abstraction reaction, being a factor of 3.3 times that of the calculated H-atom abstraction reaction rate constant for propene at 1237 K.

Thus, it is tentatively recommended that H-atom abstraction from 2,3-dimethyl-2-butene has a rate constant of

$$k(\text{2,3-dimethyl-2-butene}; T > 700 \text{ K}) = 3.0$$

$$\times 10^{-17} T^2 e^{30/T} \text{ cm}^3 \text{ molecule}^{-1} \text{ s}^{-1},$$

with this H-atom abstraction process being a minor contributor ($\sim 3\%$) to the overall reaction at room temperature and the high-pressure limit.

(11) Other Acyclic Monoalkenes

For the remaining acyclic monoalkenes for which data are available, only one or two studies have been carried out, and no specific recommendations are made. However, in general it is recommended that the room temperature rate constants derived from the relative rate studies of Ohta[44,45] and Atkinson and Aschmann[12] be used.

(12) Propadiene

The available rate constants of Bradley et al.,[22] Atkinson et al.,[48] Ohta,[45] Atkinson and Aschmann[12] and Liu et al.[59] are given in Table 6. At temperatures $\lesssim 425$ K the rate constants of Atkinson et al.,[48] Ohta,[45] Atkinson and Aschmann[12] and Liu et al.[59] are in good agreement and the data from these studies are plotted in Arrhenius form in Fig. 48. Atkinson et al.[48] showed that at room temperature the rate constant for this reaction exhibits fall-off behavior between second- and third-order kinetics below ~ 50 Torr total pressure of argon, with the bimolecular rate constant at 100 Torr total pressure being $\sim 10\%$ higher than that at 25 Torr total pressure.

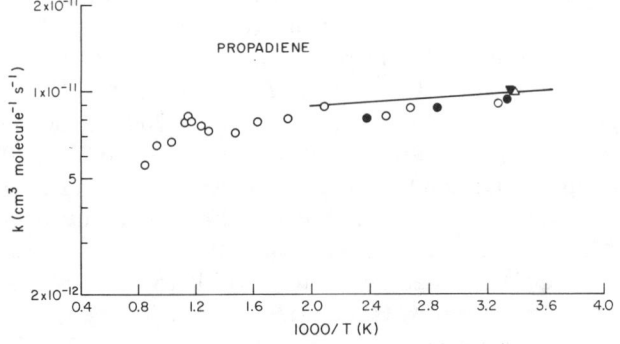

FIG. 48. Arrhenius plot of rate constants obtained at, or close to, the high-pressure limit for the reaction of the OH radical with propadiene. (●) Atkinson et al.;[48] (▼) Ohta;[45] (△) Atkinson and Aschmann;[12] (○) Liu et al.;[59] (——) recommendation (see text).

These observations indicate that the rate constants determined by Atkinson et al.[48] at 100 Torr total pressure of argon diluent were not totally at the high-pressure limit. At 298 K, a value of the low pressure third-order rate constant of

$$k_o^{\text{Ar}} \sim 4 \times 10^{-28} \text{ cm}^6 \text{ molecule}^{-2} \text{ s}^{-1},$$

combined with a limiting high-pressure second-order rate constant of

$$k_\infty = 1.0 \times 10^{-11} \text{ cm}^3 \text{ molecule}^{-1} \text{ s}^{-1},$$

accounts for this observed fall-off behavior (though obviously the value of k_o is only very approximate). The rate constants at 298 K and 100 Torr and 760 Torr total pressures of argon diluent are then calculated to be $\sim 10\%$ and $\sim 5\%$ below the high-pressure limit, respectively.

A unit-weighted least-squares analysis of the absolute rate constants obtained by Atkinson et al.[48] and Liu et al.[59] at temperatures $\leqslant 500$ K yields the Arrhenius expression of

$$k(\text{propadiene}; T \leqslant 500 \text{ K}) = (7.08^{+1.64}_{-1.33})$$

$$\times 10^{-12} e^{(74 \pm 75)/T} \text{ cm}^3 \text{ molecule}^{-1} \text{ s}^{-1}$$

over the temperature range 299–478 K and at a total pressure of argon diluent of 100–760 Torr, where the indicated error limits are two least-squares standard deviations. From a best-fit of the relative rate constants of Atkinson and Aschmann[12] and selected literature room temperature absolute rate constants for a series of alkenes and dialkenes (see above), a rate constant of

$$k(\text{propadiene}) = 9.84$$

$$\times 10^{-12} \text{ cm}^3 \text{ molecule}^{-1} \text{ s}^{-1} \text{ at 295 K}$$

is recommended, applicable to ~ 760 Torr total pressure of air. Combined with the temperature dependence derived above, this yields the recommendation of

$$k(\text{propadiene}; T < 500 \text{ K}) = (7.66^{+1.78}_{-1.44})$$

$$\times 10^{-12} e^{(74 \pm 75)/T} \text{ cm}^3 \text{ molecule}^{-1} \text{ s}^{-1}$$

over the temperature range 295–478 K, where the indicated errors are two least-squares standard deviations, and

$$k(\text{propadiene}) = 9.82 \times 10^{-12} \text{ cm}^3 \text{ molecule}^{-1} \text{ s}^{-1}$$

at 298 K, with an estimated overall uncertainty at 298 K of $\pm 20\%$. This recommendation is applicable to one atmosphere total pressure of air, and is expected to be slightly ($\lesssim 5\%$) into the fall-off region at 298 K (and will be further into the fall-off region at higher temperatures).

The rate constant reported by Bradley et al.,[22] obtained at a total pressure of ~ 1 Torr, is almost certainly well into the fall-off region. The data of Liu et al.[59] show no obvious deviation from a linear Arrhenius plot, with a slight negative temperature dependence, up to ~900 K. Above this temperature, a slightly enhanced decrease in the rate constants with increasing temperature is evident.[59] By analogy with the monoalkenes, this could be interpreted as the onset of thermal decomposition of the OH-propadiene adduct, although this phenomenon is observed to occur at ~500–600 K for the monoalkenes. It is also possible that this observed enhanced decrease in the rate constants with increasing temperatures above ~900 K is due to fall-off behavior. Clearly, further kinetic data are needed at elevated temperatures, $\gtrsim$ 500 K, to better define the onset of thermal decomposition of the OH-propadiene adduct and of the H-atom abstraction process.

(13) 1,3-Butadiene

The available kinetic data of Lloyd et al.,[5] Atkinson et al.,[48] Barnes et al.,[33] Ohta,[45] Atkinson and Aschmann[12] and Liu et al.[59] are given in Table 6 and are plotted in Arrhenius form in Fig. 49. It can be seen that the room temperature rate constants from the studies of Lloyd et al.,[5] Atkinson et al.,[48] Barnes et al.,[33] Ohta,[45] Atkinson et al.[12] and Liu et al.[59] are in very good agreement. Furthermore, the temperature-dependencies obtained by Atkinson et al.[48] and Liu et al.[59] are in good agreement.

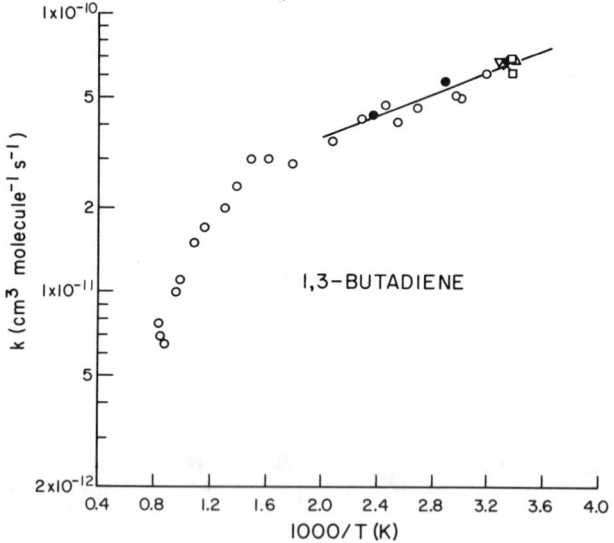

FIG. 49. Arrhenius plot of rate constants for the reaction of the OH radical with 1,3-butadiene. (∇) Lloyd et al.;[5] ($\bullet$) Atkinson et al.;[48] (x) Barnes et al.;[33] ($\square$) Ohta;[45] ($\triangle$) Atkinson and Aschmann;[12] ($\bigcirc$) Liu et al.;[59] (———) recommendation (see text).

A unit-weighted least-squares analysis of the data of Atkinson et al.[48] and Liu et al.[59] at temperatures <500 K yields the Arrhenius expression of

$$k(1,3\text{-butadiene}; T < 500 \text{ K}) = (1.44^{+0.54}_{-0.40})$$
$$\times 10^{-11} e^{(448 \pm 117)/T} \text{ cm}^3 \text{ molecule}^{-1} \text{ s}^{-1}$$

over the temperature range 299–483 K, where the indicated errors are two least-squares standard deviations. The relative rate study of Atkinson and Aschmann[12] for a series of alkenes and dialkenes, when combined with the literature absolute rate constants (see above), leads to the recommendation of

$$k(1,3\text{-butadiene}) = 6.78 \times 10^{-11} \text{ cm}^3 \text{ molecule}^{-1} \text{ s}^{-1}$$

at 295 K. Combined with the above temperature dependence, this yields the recommended Arrhenius expression of

$$k(1,3\text{-butadiene}; T < 500 \text{ K}) = (1.48^{+0.56}_{-0.42})$$
$$\times 10^{-11} e^{(448 \pm 117)/T} \text{ cm}^3 \text{ molecule}^{-1} \text{ s}^{-1}$$

over the temperature range 295–483 K, where the indicated errors are two least-squares standard deviations, and

$$k(1,3\text{-butadiene}) = 6.66 \times 10^{-11} \text{ cm}^3 \text{ molecule}^{-1} \text{ s}^{-1}$$

at 298 K, with an estimated overall uncertainty at 298 K of ±20%.

The relative rate constants of Lloyd et al.,[5] Barnes et al.,[33] and Ohta[45] are in very good agreement with this recommendation.

By analogy with the alkenes such as ethene, propene and the butenes, this rate constant is that for the OH radical addition pathway and will be very close to the high-pressure limit, at least up to ~500 K. Indeed, the above recommendation provides a good fit to the rate constants measured by Liu et al.[59] in one atmosphere of argon diluent up to ~700 K. Above ~700 K, the rate constants measured by Liu et al.[59] decrease more rapidly with increasing temperature than calculated from extrapolation of the above recommendation, and this is expected to be due to the onset of thermal decomposition of the OH-1,3-butadiene adduct back to reactants. At higher temperatures, H-atom abstraction will be the major process observed, and some evidence of the contribution of this reaction pathway is seen from the data at 1153–1203 K,[59] which may indicate an increasing rate constant with increasing temperature.

As for propadiene, more data are needed at temperatures >500 K (preferably as a function of pressure) to quantitatively define any fall-off behavior and the onset of thermal decomposition of the addition adduct and of the H-atom abstraction process.

(14) 2-Methyl-1,3-butadiene

The available rate constants of Cox et al.,[32] Atkinson et al.,[9] Kleindienst et al.,[60] Ohta,[45] Atkinson and Aschmann[12] and Edney *et al.*[46] are given in Table 6 and are plotted in Arrhenius form in Fig. 50.

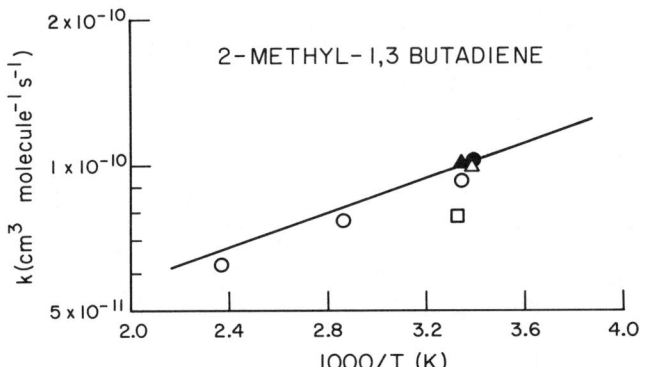

FIG. 50. Arrhenius plot of rate constants for the reaction of the OH radical with 2-methyl-1,3-butadiene (isoprene). ($\square$) Cox *et al.*;[32] ($\blacktriangle$) Atkinson *et al.*;[9] ($\bigcirc$) Kleindienst *et al.*;[60] ($\triangle$) Ohta;[45] ($\bullet$) Atkinson and Aschmann,[12] Edney *et al.*;[46] (———) recommendation (see text).

The most recent room temperature data of Atkinson *et al.*,[9] Atkinson and Aschmann,[12] Kleindienst *et al.*,[60] Ohta[45] and Edney *et al.*[46] are in good agreement. Consistent with the above recommendations for the alkenes, the temperature dependence observed by Kleindienst *et al.*[60] of $B = -410$ K is used, together with the rate constant resulting from a best fit of the relative rate constants of Atkinson and Aschmann[12] for a series of alkenes and dialkenes to the available absolute rate constant data of

$$k(\text{2-methyl-1,3-butadiene}) = 1.02$$

$$\times 10^{-10} \text{ cm}^3 \text{ molecule}^{-1} \text{ s}^{-1} \text{ at } 295 \text{ K},$$

to recommend

$$k(\text{2-methyl-1,3-butadiene}) = 2.54$$

$$\times 10^{-11} e^{410/T} \text{ cm}^3 \text{ molecule}^{-1} \text{ s}^{-1}$$

over the temperature range 295–422 K, and

$$k(\text{2-methyl-1,3-butadiene}) = 1.01$$

$$\times 10^{-10} \text{ cm}^3 \text{ molecule}^{-1} \text{ s}^{-1} \text{ at } 298 \text{ K},$$

with an estimated overall uncertainty at 298 K of ±25%.

This reaction proceeds by OH radical addition, and the above rate constant will be very close to the high-pressure limit over this temperature range. At elevated temperatures thermal decomposition of the OH-2-methyl-1,3-butadiene adduct will become important, and

the reaction pathways observed will then be OH radical addition followed by rapid rearrangement and/or decomposition of the adduct to products other than the reactants and/or H-atom abstraction from the $-\text{CH}_3$ group and vinyl C$-$H bonds.

(15) Remaining Acyclic Di- and Trialkenes

For the remaining acyclic di- and trialkenes, only single studies have been carried out, and no recommendations are made.

(16) Cyclohexene

The available rate constants of Wu *et al.*,[27] Darnall *et al.*,[58] Cox *et al.*,[32] Barnes *et al.*,[33] Ohta,[45] Atkinson *et al.*[53] and Rogers[47] are given in Table 7. While no temperature dependent data are available, the reported room temperature rate constants of Wu *et al.*,[27] Darnall *et al.*,[58] Cox *et al.*,[32] Barnes *et al.*,[33] Ohta[45] and Atkinson *et al.*[53] are in good agreement (though ~20% higher than that of Rogers[47]). Based upon the recent study of Atkinson *et al.*,[53] it is recommended that

$$k(\text{cyclohexene}) = 6.77$$

$$\times 10^{-11} \text{ cm}^3 \text{ molecule}^{-1} \text{ s}^{-1} \text{ at } 298 \text{ K},$$

with an estimated overall uncertainty of ±25%.

(17) α-Pinene

The available kinetic data of Winer *et al.*,[28] Kleindienst *et al.*[60] and Atkinson *et al.*[55] are given in Table 7 and are plotted in Arrhenius form in Fig. 51. The room temperature rate constants of Winer *et al.*,[28] Kleindienst *et al.*[60] and Atkinson *et al.*[55] are in reasonable agreement. The temperature dependence obtained from the data of Kleindienst *et al.*[60] of $B = -444$ K is used together with the 294 K rate constant of Atkinson *et al.*[55] to recommend

$$k(\alpha\text{-pinene}) = (1.21^{+0.54}_{-0.38})$$

$$\times 10^{-11} e^{(444 \pm 127)/T} \text{ cm}^3 \text{ molecule}^{-1} \text{ s}^{-1}$$

over the temperature range 294–422 K, where the indicated errors are two least-squares standard deviations, and

$$k(\alpha\text{-pinene}) = 5.37$$

$$\times 10^{-11} \text{ cm}^3 \text{ molecule}^{-1} \text{ s}^{-1} \text{ at } 298 \text{ K},$$

with an estimated overall uncertainty at 298 K of ±25%.

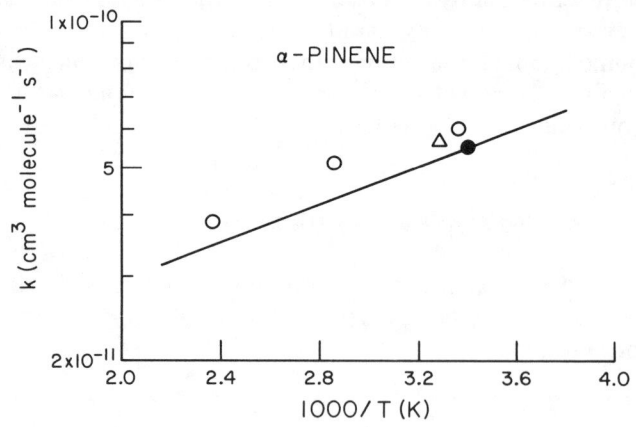

FIG. 51. Arrhenius plot of rate constants for the reaction of the OH radical with α-pinene. (Δ) Winer *et al.*;[28] (○) Kleindienst *et al.*;[60] (●) Atkinson *et al.*;[55] (———) recommendation (see text).

(18) β-Pinene

The available kinetic data of Winer *et al.*,[28] Kleindienst *et al.*[60] and Atkinson *et al.*[55] are given in Table 7 and are plotted in Arrhenius form in Fig. 52. The relative rate constant of Atkinson *et al.*[55] is in excellent agreement with the absolute room temperature rate constant of Kleindienst *et al.*[60] and in reasonable agreement with that of Winer *et al.*[28] As for α-pinene, the recommendation uses the temperature dependence determined by Kleindienst *et al.*,[60] of $B = -357$ K, in conjunction with the 294 K rate constant of Atkinson *et al.*[55] to derive

$$k(\beta\text{-pinene}) = (2.38^{+0.91}_{-0.67})$$
$$\times\ 10^{-11}\ e^{(357\pm111)/T}\ cm^3\ molecule^{-1}\ s^{-1}$$

over the temperature range 294–423 K, where the indicated errors are two least-squares standard deviations, and

$$k(\beta\text{-pinene}) = 7.89$$
$$\times\ 10^{-11}\ cm^3\ molecule^{-1}\ s^{-1}\ at\ 298\ K,$$

with an estimated overall uncertainty at 298 K of ±25%.

(19) Other Cycloalkenes

For the other cycloalkenes, cyclodialkenes and cyclotrialkenes listed in Table 7, no specific recommendations are made. However, it is recommended that the room temperature rate constants derived from the relative rate constant studies of Ohta[45] and Atkinson *et al.*[53-55] be used (see, for example, the excellent agreement between the

room temperature rate constants of Ohta[45] and Atkinson *et al.*[53] for 1,4-cyclohexadiene).

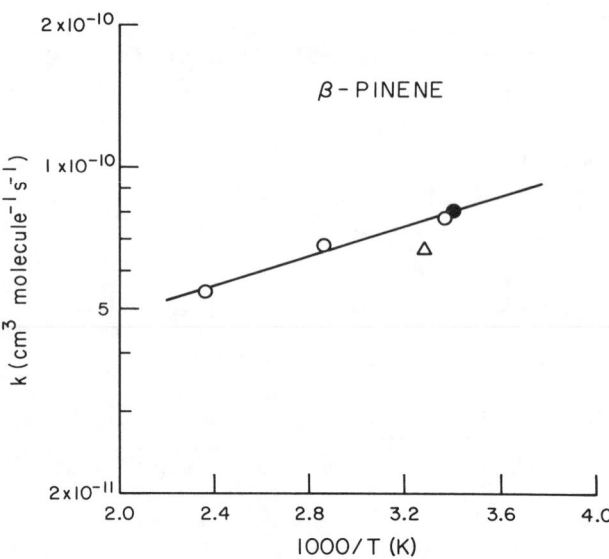

FIG. 52. Arrhenius plot of rate constants for the reaction of the OH radical with β-pinene. (Δ) Winer *et al.*;[28] (○) Kleindienst *et al.*;[60] (●) Atkinson *et al.*;[55] (———) recommendation (see text).

As discussed in detail previously,[63,81] the room temperature rate constants for the monoalkenes increase monotonically with the number of substituents around the double bond, and the rate constants for the acyclic and cyclic mono-alkenes and the non-conjugated di- and trialkenes can be estimated to a good degree of accuracy (±30%) from the number and position(s) of alkyl substituents around the double bond(s). Similarly, for alkenes containing conjugated double bond systems, reasonably accurate predictions of the room temperature rate constants can be made from the rate constants for $>C=C-C=C<$ systems with the varying numbers of substituents around this double bond system.[63,81]

b. Mechanisms

The kinetic data discussed above for the reactions of the OH (or OD) radical with alkenes show that three distinct temperature regimes exist, for temperatures ≲500 K, ~500–700 K, and ≳700 K, with the precise temperatures which define these regimes depending on the specific alkene, the total pressure and identity of the third body, and the experimental technique used. These temperature regimes, and the behaviors of the measured high pressure rate constants in the low and high temperature regimes (which approximate the recommendations for *trans*-2-butene) are shown in Fig. 53. To an approximation, these temperature regimes correspond to: (a)

$\lesssim 500$ K, OH radical addition to the $>C=C<$ bond(s) which may be in the fall-off regime between second- and third-order kinetics, with the rate constant at a given total pressure decreasing with increasing temperature, (b) ~ 500–700 K, the occurrence of thermal decomposition of the OH radical-alkene addition adduct and, for the smaller alkenes such as ethene, propene and possibly propadiene, increasing fall-off behavior with increasing temperature, and (c) $\gtrsim 700$ K, the occurrence of H-atom abstraction as the major or sole reaction pathway, with the rate constant increasing rapidly with increasing temperature. These processes are discussed briefly below.

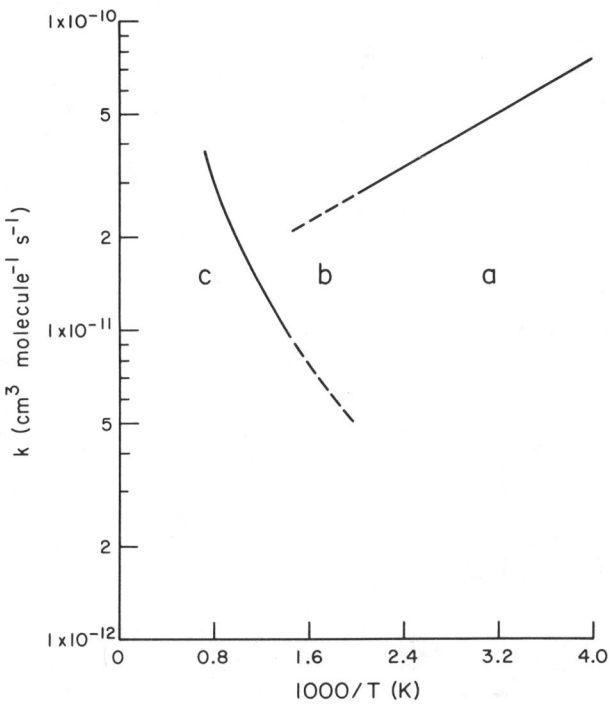

FIG. 53. Schematic Arrhenius plot (which approximates the OH radical rate constant behavior of *trans*-2-butene at the high-pressure limit) showing the three temperature regimes (a) $\lesssim 500$ K, (b) ~ 500–700 K and (c) $\gtrsim 700$ K (see text).

The available kinetic and mechanistic data show that at $\lesssim 500$–600 K the reaction of OH radicals to the alkenes proceeds predominantly via addition of the OH radical to the carbon-carbon double bond(s). Thus, in the discharge flow-mass spectrometric study of Morris *et al.*,[21] mass peaks corresponding to the OH-alkene adducts were observed for ethene and propene. These adduct peaks increased in intensity as the total pressure was increased from 1 to 4 Torr,[21] showing that OH radical addition was occurring and that these addition adducts were being collisionally stabilized.

As noted above, numerous kinetic studies have shown for ethene[7,8,10,11,13,67,72,73] and propene[11,13,37] that the rate

constants are in the fall-off region between second-order and third-order kinetics at total pressures below approximately one atmosphere for ethene and below ~ 30 Torr for propene. These observations show that these reactions proceed via initial addition of OH radicals to the alkene to form an initially energy-rich OH-alkene adduct, which can decompose back to the reactants or be collisionally stabilized. For example, for ethene

$$OH + C_2H_4 \rightleftharpoons HOC_2H_4*$$

$$HOC_2H_4* + M \rightarrow HOC_2H_4 + M$$

For ethene, Howard[67] has shown from a kinetic study over the total pressure range 0.7–7 Torr of helium that the rate constant extrapolates to essentially zero at zero pressure. Thus, as expected from the high C$-$H bond energy in ethene, H-atom abstraction from ethene is essentially negligible at room temperature. This prediction from kinetic studies is confirmed by the recent discharge flow-mass spectrometric study of Bartels *et al.*,[68] in which the abstraction channel was shown to account for $<2.5\%$ of the overall reaction channels at 295 K and at ~ 2 Torr total pressure. These investigations[67,68] thus show, in contradiction to the earlier product study of Meagher and Heicklen[82] (involving a difficult to interpret final product analysis which can be re-interpreted as indicating an $\sim 10\%$ H-atom abstraction route at the high-pressure limit),[65] that H-atom abstraction from ethene under atmospheric conditions is totally negligible.

For propene and the butenes, Hoyermann and Sievert[78,79] have shown from discharge flow-mass spectroscopic studies that H-atom abstraction from these alkenes is also insignificant, being $<5\%$ for propene and 2-methylpropene and $<10\%$ for 1-butene and *cis*- and *trans*-2-butene at room temperature. That H-atom abstraction from propene is negligible is totally consistent with the product study of Cvetanovic,[77] who, from a comprehensive investigation of the products formed and their formation reactions (mainly via radical-radical processes), concluded that the OH radical addition pathway was the major, if not exclusive, reaction pathway, and that addition to the terminal carbon atom

$$OH + CH_3CH{=}CH_2 \rightarrow CH_3\dot{C}HCH_2OH$$

occurs $\sim 65\%$ of the time at room temperature.[77] As shown above, extrapolation of elevated temperature (>650 K) kinetic data to 298 K indicates that H-atom abstraction from the vinyl C$-$H bonds and/or C$-$H bonds of substituent alkyl groups for ethene, propene, the butenes and 2,3-dimethyl-2-butene contributes $<5\%$ of the overall reaction rate at the high-pressure limit.

Hence, it appears that at $\lesssim 425$ K H-atom abstraction from acyclic alkenes containing $\leqslant C_2$ side chains is of minimal importance, and that at room temperature the reactions of OH radicals with these alkenes can be considered to proceed almost totally via OH radical addition to the $>C=C<$ double bonds. Of course, for the 1-alkenes and other alkenes with long side chains it must be expected that H-atom abstraction from the $>CH-$, $-CH_2-$, and $-CH_3$ groups will occur, with rate constants for H-abstraction from these groups being approximately similar to those for the corresponding alkane groups.[81] Moreover, Ohta[83] has shown that benzene is a minor, but significant, product formed during room temperature irradiations of $CH_3ONO-NO$-cyclohexadiene-air mixtures, accounting for 8.9% and 15.3% of the overall reaction pathways for 1,3-cyclohexadiene and 1,4-cyclohexadiene, respectively. These data[83] show that H-atom abstraction from the allylic $C-H$ bonds (of bond dissociation energy 73 ± 5 kcal mol^{-1}[84]) in these cyclohexadienes does occur, with a rate constant per allylic $C-H$ bond of $\sim 3.7 \times 10^{-12}$ cm^3 molecule^{-1} s^{-1} for both 1,3- and 1,4-cyclohexadiene.

The formation of the OH-ethene adduct is calculated to be ~ 32 kcal mol^{-1} exothermic[76] (formation of the other OH-alkene adducts have similar calculated exothermicities) and formation of an H-atom together with $HOCH=CH_2$ is endothermic from the OH radical and ethene reactants by ~ 7 kcal mol^{-1}.[76] Melius et al.[85] and Sosa and Schlegel[86] have calculated that the thermochemically most favorable decomposition pathway for the OH-ethene adduct involves redissociation back to the reactants. At elevated temperatures decomposition of the thermalized OH-ethene adduct, and the thermalized OH-alkene adducts in general, then occurs as experimentally observed by Tully[10] and Tully and Goldsmith.[37] For example, in the recent flash photolysis studies of Tully[10] and Tully and Goldsmith[37] for ethene[10] and propene,[37] non-exponential OH radical decays were observed to occur at temperatures of $\sim 500-700$ K, and the derived rate constants decreased rapidly with increasing temperature over this temperature range.

This is totally consistent with the increasing importance of thermal decomposition of the thermalized OH-alkene adducts at elevated temperature, with the adduct decomposing within the time-scale of these experimental observations[10,37] for temperatures $\sim 500-700$ K. At still higher temperatures the addition pathway, at least for ethene, becomes unimportant due to the extremely rapid decomposition rate of the OH-alkene adduct back to reactants, and for ethene the reaction is then expected to proceed via H-atom abstraction from the vinyl $C-H$ bonds[85] with a positive temperature dependence.

For the higher alkenes, as noted above, other decomposition pathways of the OH-alkene adducts (for example, CH_3 radical elimination and isomerization followed by decomposition reactions) may also become of importance in this temperature regime. Thus, the situation at elevated temperatures where thermal decomposition of the OH-alkene adduct becomes important may be more complex,[37,85] involving other reaction pathways as well as direct H-atom abstraction. However, for the methyl-substituted ethenes for which high temperature (>650 K) kinetic data are available (propene, 2-methylpropene, trans-2-butene and 2,3-dimethyl-2-butene), the magnitude of the rate constants depends almost linearly

on the number of substituent $-CH_3$ groups, as shown in Fig. 54. This indicates that for these alkenes the high temperature reaction pathway involves H-atom abstraction from the $-CH_3$ groups, with the rate constant per $-CH_3$ group being approximately constant at a given temperature.

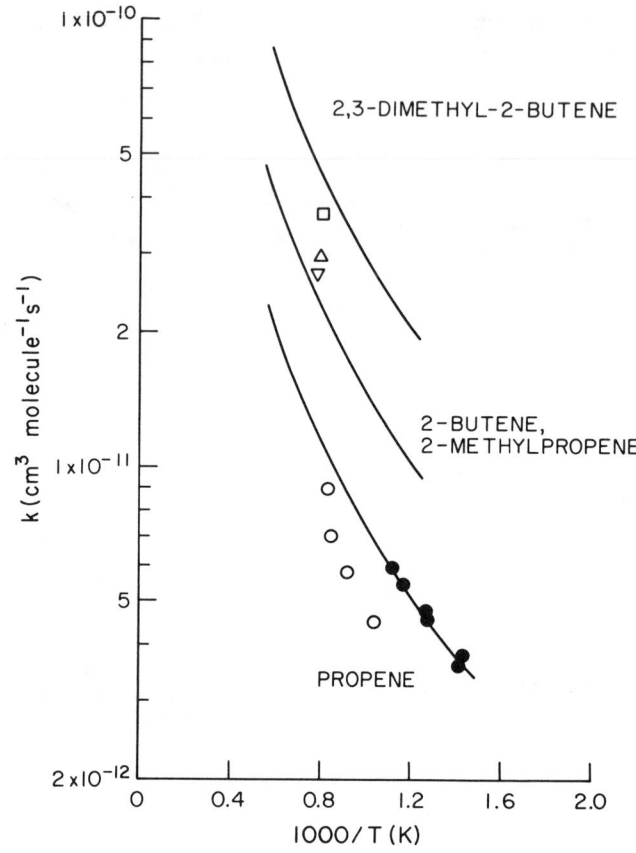

FIG. 54. Arrhenius plot of the available high-temperature ($\geqslant 650$ K) rate constants for the reactions of the OH radical with methyl-substituted ethenes. Propene: ($\bigcirc$) Smith et al.;[38] ($\bullet$) Tully.[16] 2-Methylpropene: ($\triangle$) Smith.[15] trans-2-Butene: ($\triangledown$) Smith.[15] 2,3-Dimethyl-2-butene: ($\square$) Smith.[15] (———) Calculated based upon the recommended H-atom abstraction rate constant for propene and the number of substituent $-CH_3$ groups (see text).

At low total pressures other reactions of the chemically activated OH-alkene adducts can occur.[86] Thus, for example, Bartels et al.[68] have observed the formation of $HCHO + CH_3$ and $CH_3CHO + H$ as decomposition products of the OH-ethene adduct at total pressures of ~ 2 Torr. Under these low pressure conditions the reaction sequence for ethene appears to be[68,86]

$$OH + C_2H_4 \rightleftharpoons [CH_2CH_2OH]^* \xrightarrow{M} HOCH_2\dot{C}H_2$$

$$\downarrow isom$$

$$[CH_3CH_2O]^* \rightarrow CH_3 + HCHO$$

$$\rightarrow CH_3CHO + H$$

$$\rightarrow HOCHCH_2 + H$$

Clearly, at low total pressures and/or high temperatures, the dynamics of these OH radical reactions, even for ethene, are complex,[85,86] and further experimental and theoretical studies are required.

References

[1] A. A. Westenberg and R. M. Fristrom, 10th International Symposium on Combustion, 1964; The Combustion Institute, Pittsburgh, PA, 1965, p. 473.

[2] R. R. Baldwin, R. F. Simmons, and R. W. Walker, Trans. Faraday Soc. **62**, 2486 (1966).

[3] D. E. Hoare and M. Patel, Trans. Faraday Soc. **65**, 1325 (1969).

[4] S. Gordon and W. A. Mulac, Int. J. Chem. Kinet., **Symp. 1**, 289 (1975).

[5] A. C. Lloyd, K. R. Darnall, A. M. Winer, and J. N. Pitts, Jr., J. Phys. Chem. **80**, 789 (1976).

[6] J. N. Bradley, W. D. Capey, R. W. Fair, and D. K. Pritchard, Int. J. Chem. Kinet. **8**, 549 (1976).

[7] R. Atkinson, R. A. Perry, and J. N. Pitts, Jr., J. Chem. Phys. **66**, 1197 (1977).

[8] R. Overend and G. Paraskevopoulos, J. Chem. Phys. **67**, 674 (1977).

[9] R. Atkinson, S. M. Aschmann, A. M. Winer, and J. N. Pitts, Jr., Int. J. Chem. Kinet. **14**, 507 (1982).

[10] F. P. Tully, Chem. Phys. Lett. **96**, 148 (1983).

[11] R. Zellner and K. Lorenz, J. Phys. Chem. **88**, 984 (1984).

[12] R. Atkinson and S. M. Aschmann, Int. J. Chem. Kinet. **16**, 1175 (1984).

[13] Th. Klein, I. Barnes, K. H. Becker, E. H. Fink, and F. Zabel, J. Phys. Chem. **88**, 5020 (1984).

[14] V. Schmidt, G. Y. Zhu, K. H. Becker, and E. H. Fink, Ber. Bunsenges Phys. Chem. **89**, 321 (1985).

[15] G. P. Smith, Int. J. Chem. Kinet. **19**, 269 (1987).

[16] F. P. Tully, Chem. Phys. Lett. **143**, 510 (1988).

[17] A.-D. Liu, W. A. Mulac, and C. D. Jonah, Int. J. Chem. Kinet. **19**, 25 (1987).

[18] A. Liu, W. A. Mulac, and C. D. Jonah, J. Phys. Chem., **92**, 3828 (1988).

[19] F. P. Tully, 10th International Symposium on Gas Kinetics, University College of Swansea, Swansea, UK, July 24–29, 1988; private communication (1988).

[20] H. Niki, P. D. Maker, C. M. Savage, and L. P. Breitenbach, J. Phys. Chem. **82**, 132 (1978).

[21] E. D. Morris, Jr., D. H. Stedman, and H. Niki, J. Amer. Chem. Soc. **93**, 3570 (1971).

[22] J. N. Bradley, W. Hack, K. Hoyermann, and H. Gg. Wagner, J. Chem. Soc. Faraday Trans. 1, **69**, 1889 (1973).

[23] F. Stuhl, Ber. Bunsenges Phys. Chem. **77**, 674 (1973).

[24] R. A. Gorse and D. H. Volman, J. Photochem. **3**, 115 (1974).

[25] A. V. Pastrana and R. W. Carr, Jr., J. Phys. Chem. **79**, 765 (1975).

[26] R. Atkinson and J. N. Pitts, Jr., J. Chem. Phys. **63**, 3591 (1975).

[27] C. H. Wu, S. M. Japar, and H. Niki, J. Environ. Sci. Health, **A11**, 191 (1976).

[28] A. M. Winer, A. C. Lloyd, K. R. Darnall, and J. N. Pitts, Jr., J. Phys. Chem. **80**, 1635 (1976).

[29] A. M. Winer, A. C. Lloyd, K. R. Darnall, R. Atkinson, and J. N. Pitts, Jr., Chem. Phys. Lett. **51**, 221 (1977).

[30] A. R. Ravishankara, S. Wagner, S. Fischer, G. Smith, R. Schiff, R. T. Watson, G. Tesi, and D. D. Davis, Int. J. Chem. Kinet. **10**, 783 (1978).

[31] W. S. Nip and G. Paraskevopoulos, J. Chem. Phys. **71**, 2170 (1979).

[32] R. A. Cox, R. G. Derwent, and M. R. Williams, Environ. Sci. Technol. **14**, 57 (1980).

[33] I. Barnes, V. Bastian, K. H. Becker, E. H. Fink, and F. Zabel, Atmos. Environ. **16**, 545 (1982).

[34] R. H. Smith, J. Phys. Chem. **87**, 1596 (1983).

[35] R. R. Baldwin, M. W. M. Hisham, and R. W. Walker, 20th International Symposium on Combustion, 1984; The Combustion Institute, Pittsburgh, PA, 1985, p. 743.

[36] G. P. Smith, 20th International Symposium on Combustion, 1984; The Combustion Institute, Pittsburgh, PA, 1985, p. 750.

[37] F. P. Tully and J. E. M. Goldsmith, Chem. Phys. Lett. **116**, 345 (1985).

[38] G. P. Smith, P. W. Fairchild, J. B. Jeffries, and D. R. Crosley, J. Phys. Chem. **89**, 1269 (1985).

[39] I. Barnes, V. Bastian, K. H. Becker, E. H. Fink, and W. Nelsen, J. Atmos. Chem. **4**, 445 (1986).

[40] T. J. Wallington, Int. J. Chem. Kinet. **18**, 487 (1986).

[41] R. Atkinson and S. M. Aschmann, Int. J. Chem. Kinet., **21**, 355 (1989).

[42] E. D. Morris, Jr. and H. Niki, J. Phys. Chem. **75**, 3640 (1971).

[43] H. W. Biermann, G. W. Harris, and J. N. Pitts, Jr., J. Phys. Chem. **86**, 2958 (1982).

[44] T. Ohta, Int. J. Chem. Kinet. **16**, 879 (1984).

[45] T. Ohta, J. Phys. Chem. **87**, 1209 (1983).

[46] E. O. Edney, T. E. Kleindienst, and E. W. Corse, Int. J. Chem. Kinet. **18**, 1355 (1986).

[47] J. D. Rogers, Environ. Sci. Technol. **23**, 177 (1989).

[48] R. Atkinson, R. A. Perry, and J. N. Pitts, Jr., J. Chem. Phys. **67**, 3170 (1977).

[49] R. Atkinson, R. A. Perry, and J. N. Pitts, Jr., Chem. Phys. Lett. **38**, 607 (1976).

[50] R. Atkinson and J. N. Pitts, Jr., J. Chem. Phys. **68**, 2992 (1978).

[51] R. Atkinson, K. R. Darnall, and J. N. Pitts, Jr., unpublished data (1978); cited in references 63 and 65.

[52] R. A. Perry, Ph.D. Thesis, 1977, University of California, Riverside, cited in references 63 and 65.

[53] R. Atkinson, S. M. Aschmann, and W. P. L. Carter, Int. J. Chem. Kinet. **15**, 1161 (1983).

[54] R. Atkinson, S. M. Aschmann, and W. P. L. Carter, Int. J. Chem. Kinet. **16**, 967 (1984).

[55] R. Atkinson, S. M. Aschmann, and J. N. Pitts, Jr., Int. J. Chem. Kinet. **18**, 287 (1986).

[56] R. Atkinson, S. M. Aschmann, M. A. Goodman, and A. M. Winer, Int. J. Chem. Kinet. **20**, 273 (1988).

[57] R. Atkinson and S. M. Aschmann, Int. J. Chem. Kinet., in press (1989).

[58] K. R. Darnall, A. M. Winer, A. C. Lloyd, and J. N. Pitts, Jr., Chem. Phys. Lett. **44**, 415 (1976).

[59] A. Liu, W. A. Mulac, and C. D. Jonah, J. Phys. Chem. **92**, 131 (1988).

[60] T. E. Kleindienst, G. W. Harris, and J. N. Pitts, Jr., Environ. Sci. Technol. **16**, 844 (1982).

[61] R. A. Cox, Int. J. Chem. Kinet., **Symp. 1**, 379 (1975).

[62] R. Atkinson, D. L. Baulch, R. A. Cox, R. F. Hampson, Jr., J. A. Kerr, and J. Troe, J. Phys. Chem. Ref. Data, **18**, 881 (1989).

[63] R. Atkinson, Chem. Rev. **86**, 69 (1986).

[64] R. Simonaitis and J. Heicklen, Int. J. Chem. Kinet. **5**, 231 (1973).

[65] R. Atkinson, K. R. Darnall, A. C. Lloyd, A. M. Winer, and J. N. Pitts, Jr., Adv. Photochem. **11**, 375 (1979).

[66] E. P. Grimsrud, H. H. Westberg, and R. A. Rasmussen, Int. J. Chem. Kinet., **Symp. 1**, 183 (1975).

[67] C. J. Howard, J. Chem. Phys. **65**, 4771 (1976).

[68] M. Bartels, K. Hoyermann, and R. Sievert, 19th International Symposium on Combustion, 1982; The Combustion Institute, Pittsburgh, PA, 1982, p. 61.

[69] W. E. Wilson and A. A. Westenberg, 11th International Symposium on Combustion, 1966; The Combustion Institute, Pittsburgh, PA, 1967, p. 1143.

[70]N. R. Greiner, J. Chem. Phys. **53**, 1284 (1970).

[71]I. W. M. Smith and R. Zellner, J. Chem. Soc. Faraday Trans. 2, **69**, 1617 (1973).

[72]D. D. Davis, S. Fischer, R. Schiff, R. T. Watson, and W. Bollinger, J. Chem. Phys. **63**, 1707 (1975).

[73]G. K. Farquharson and R. H. Smith, Aust. J. Chem. **33**, 1425 (1980).

[74]J. Troe, J. Phys. Chem. **83**, 114 (1979).

[75]R. G. Gilbert, K. Luther, and J. Troe, Ber. Bunsenges Phys. Chem. **87**, 169 (1983).

[76]S. W. Benson, *Thermochemical Kinetics*, 2nd Ed. (Wiley, New York, NY 1976).

[77]R. J. Cvetanovic, 12th International Symposium on Free Radicals, Laguna Beach, CA, January 4–9, 1976.

[78]K. Hoyermann and R. Sievert, Ber. Bunsenges Phys. Chem. **83**, 933 (1979).

[79]K. Hoyermann and R. Sievert, Ber. Bunsenges Phys. Chem. **87**, 1027 (1983).

[80]R. Atkinson, E. C. Tuazon, and W. P. L. Carter, Int. J. Chem. Kinet. **17**, 725 (1985).

[81]R. Atkinson, Int. J. Chem. Kinet. **19**, 799 (1987).

[82]J. F. Meagher and J. Heicklen, J. Phys. Chem. **80**, 1645 (1976).

[83]T. Ohta, Int. J. Chem. Kinet. **16**, 1495 (1984).

[84]D. F. McMillen and D. M. Golden, Ann. Rev. Phys. Chem. **33**, 493 (1982).

[85]C. F. Melius, J. S. Binkley, and M. L. Koszykowski, 8th International Symposium on Gas Kinetics, Univ. Nottingham, Nottingham, UK, July 15–20, 1984.

[86]C. Sosa and H. B. Schlegel, J. Am. Chem. Soc. **109**, 7007 (1987).

2.4. Haloalkenes

a. Kinetics

The available second-order rate constants obtained at, or close to, the high pressure limit are listed in Table 9. In addition, Howard[5] has determined, using a discharge flow-laser magnetic resonance (DF-LMR) technique, rate constants for the reactions of OH radicals with $CH_2=CHCl$, $CH_2=CF_2$ and $CF_2=CFCl$ at 296 K over the total pressure range of 0.7 to 7 Torr of helium. For these three haloalkenes the measured rate constants were in the fall-off region between second- and third-order kinetics,[5] with second-order rate constants at 296 K and 7 Torr total pressure of helium diluent of 2.1×10^{-12} cm^3 molecule^{-1} s^{-1} for $CH_2=CF_2$[5] and 7×10^{-12} cm^3 molecule^{-1} s^{-1} for $CF_2=CFCl$.[5] For $CHCl=CCl_2$, the rate constant at 296 K is in the fall-off region below ~2 Torr total pressure of helium.[5]

Kinetic data for the individual haloalkenes for which multiple studies have been carried out are discussed below.

(1) Trichloroethene

The available rate constants of Winer et al.,[4] Howard,[5] Davis et al.,[6] Chang and Kaufman,[7] Kirchner,[8] Klöpffer et al.[9] and Edney et al.[2] are given in Table 9, and those of Howard,[5] Davis et al.,[6] Chang and Kaufman,[7] Kirchner,[8] and Edney et al.,[2] which are in reasonably good agreement, are plotted in Arrhenius form in Fig. 55. The relative rate constant of Winer et al.[4] was at, or close to, the lower limit of values able to be derived by the experi-

mental technique used,[13] and is hence neglected in the evaluation. The rate constants reported by Kirchner[8] and Klöpffer et al.[9] at around room temperature have not been used in the evaluation because of a lack of details available.

Thus, a unit-weighted least-squares analysis of the rate constant data of Howard,[5] Davis et al.[6] and Chang and Kaufman[7] yields the recommended Arrhenius expression of

$$k(\text{trichloroethene}) = (5.63^{+1.54}_{-1.20})$$

$$\times 10^{-13} e^{(427 \pm 70)/T} \text{ cm}^3 \text{ molecule}^{-1} \text{ s}^{-1}$$

over the temperature range 234–420 K, where the indicated errors are two least-squares standard deviations, and

$$k(\text{trichloroethene}) = 2.36$$

$$\times 10^{-12} \text{ cm}^3 \text{ molecule}^{-1} \text{ s}^{-1} \text{ at 298 K,}$$

with an estimated overall uncertainty at 298 K of ±30%. This recommendation is identical to that of Atkinson.[13]

(2) Tetrachloroethene

The kinetic data of Winer et al.,[4] Howard,[5] Davis et al.,[6] Chang and Kaufman[7] and Kirchner[8] are given in Table 9, and those of Howard,[5] Davis et al.,[6] Chang and Kaufman[7] and Kirchner,[8] which are in good agreement, are plotted in Arrhenius form in Fig. 56. (Only the reported rate constants at 298 K and 305 K for the studies of Davis et al.[6] and Kirchner,[8] respectively, can be plotted, together with the reported Arrhenius expressions,[6,8] since the individual rate constants at the temperatures studied were not given.[6,8]). Analogous to the case for trichloroethene, the relative rate constant reported by Winer et al.[4] has been neglected and the rate constant of Kirchner[8] has not been used in the evaluation.

A unit-weighted least-squares analysis of the rate constants of Howard,[5] Chang and Kaufman[7] and the 298 K rate constant of Davis et al.[6] leads to the recommended Arrhenius expression of

$$k(\text{tetrachloroethene}) = (9.64^{+2.92}_{-2.24})$$

$$\times 10^{-12} e^{-(1209 \pm 90)/T} \text{ cm}^3 \text{ molecule}^{-1} \text{ s}^{-1}$$

over the temperature range 296–420 K, where the indicated errors are two least-squares standard deviations, and

$$k(\text{tetrachloroethene}) = 1.67$$

$$\times 10^{-13} \text{ cm}^3 \text{ molecule}^{-1} \text{ s}^{-1} \text{ at 298 K,}$$

with an estimated overall uncertainty at 298 K of ±30%. This recommendation is identical to that of Atkinson.[13]

TABLE 9. Rate constants k and temperature-dependent parameters for the gas-phase reactions of the OH radical with haloalkenes at, or close to, the high pressure limit

Haloalkene	$10^{12} \times A$ (cm^3 molecule^{-1} s^{-1})	B (K)	$10^{12} \times k$ (cm^3 molecule^{-1} s^{-1})	at T(K)	Technique	Reference	Temperature range covered (K)
CH$_2$=CHF			5.56 ± 0.56	299.2	FP-RF	Perry et al.[1]	299–426
			4.44 ± 0.45	346.8			
	1.48	−390 ± 151	3.76 ± 0.38	426.1			
CH$_2$=CHCl			6.60 ± 0.66	299.2	FP-RF	Perry et al.[1]	299–423
			5.01 ± 0.51	357.8			
	1.14	−526 ± 151	3.95 ± 0.40	422.5			
CH$_2$=CHBr			6.81 ± 0.69	298.6	FP-RF	Perry et al.[1]	299–424
			6.00 ± 0.60	350.0			
	1.79	−405 ± 151	4.56 ± 0.46	423.7			
CH$_2$=CCl$_2$			14.8 ± 2.1	296	RR [relative to k(n-butane) = 2.51 × 10^{-12}][a]	Edney et al.[2]	
			14.6	298	RR [relative to k(n-pentane) = 3.94 × 10^{-12}][a]	Edney et al.[2]	
			8.11 ± 0.24	298 ± 2	RR [relative to k(dimethyl ether) = 2.98 × 10^{-12}][a]	Tuazon et al.[3]	
cis-CHCl=CHCl			2.38 ± 0.14	298 ± 2	RR [relative to k(dimethyl ether) = 2.98 × 10^{-12}][a]	Tuazon et al.[3]	
trans-CHCl=CHCl			1.80 ± 0.03	298 ± 2	RR [relative to k(dimethyl ether) = 2.98 × 10^{-12}][a]	Tuazon et al.[3]	
CHCl=CCl$_2$			4.3 ± 1.3	305 ± 2	RR [relative to k(2-methylpropene) = 4.94 × 10^{-11}][a]	Winer et al.[4]	
			2.0 ± 0.4	296	DF-LMR	Howard[5]	
			2.35 ± 0.25	298	FP-RF	Davis et al.[6]	
			3.12 ± 0.24	234	DF-RF	Chang and Kaufman[7]	234–420
			3.65 ± 0.21	237			
			3.73 ± 0.18	243			
			3.14 ± 0.16	250			
			3.06 ± 0.07	260			
			2.78 ± 0.17	268			
			2.37 ± 0.10	296			
			1.74 ± 0.04	343			
			1.86 ± 0.13	357			
			1.67 ± 0.03	420			
			1.55 ± 0.06	420			
	0.532 ± 0.071	−445 ± 41	1.68 ± 0.04	420			
			2.11	305	DF-MS	Kirchner[8]	
			2.8	300	RR [relative to k(toluene) = 5.91 × 10^{-12}][a]	Klöpffer et al.[9]	
			2.85 ± 0.40	296	RR [relative to k(n-butane) = 2.51 × 10^{-12}][a]	Edney et al.[2]	

TABLE 9. Rate constants k and temperature-dependent parameters for the gas-phase reactions of the OH radical with haloalkenes at, or close to, the high pressure limit — Continued

Haloalkene	$10^{12} \times A$ (cm^3 molecule^{-1} s^{-1})	B (K)	$10^{12} \times k$ (cm^3 molecule^{-1} s^{-1})	at T(K)	Technique	Reference	Temperature range covered (K)
CCl$_2$=CCl$_2$			2.2 ± 0.7	305 ± 2	RR [relative to k(2-methylpropene) = 4.94 × 10^{-11}]^a	Winer et al.[4]	
			0.170 ± 0.034	296	DF-LMR	Howard[5]	
	10.6 ± 5	1295 ± 150	0.155 ± 0.015	298	FP-RF	Davis et al.[6]	250–375
			0.169 ± 0.007	297	DF-RF	Chang and Kaufman[7]	297–420
			0.270 ± 0.009	341			
			0.276 ± 0.010	341			
			0.303 ± 0.034	350			
			0.424 ± 0.016	378			
			0.477 ± 0.014	403			
	9.44 ± 1.34	1199 ± 55	0.526 ± 0.061	420			
	5.53	1034	0.179	305	DF-MS	Kirchner[8]	~305–430^b
CH$_2$ClCH=CH$_2$			17 ± 7	298	RR [relative to k(n-butane) = 2.54 × 10^{-12}]^a	Edney et al.[10]	
			19.5 ± 3.2	296	RR [relative to k(n-butane) = 2.51 × 10^{-12}]^a	Edney et al.[2]	
			16.9 ± 0.7	298 ± 2	RR [relative to k(propene) = 2.63 × 10^{-11}]^a	Tuazon et al.[11]	
cis-CH$_2$ClCH=CHCl			7.36 ± 0.12^c	295 ± 2	RR [relative to k(n-octane) = 8.57 × 10^{-12}]^a	Tuazon et al.[12]	
			8.41 ± 0.40	298 ± 2	RR [relative to k(n-octane) = 8.68 × 10^{-12}]^a	Tuazon et al.[3]	
trans-CH$_2$ClCH=CHCl			12.4 ± 0.4^c	295 ± 2	RR [relative to k(n-octane) = 8.57 × 10^{-12}]^a	Tuazon et al.[12]	
			14.3 ± 0.8	298 ± 2	RR [relative to k(n-octane) = 8.68 × 10^{-12}]^a	Tuazon et al.[3]	
(CH$_2$Cl)$_2$C=CH$_2$			40.2 ± 5.4	295 ± 2	RR [relative to k(2-methyl-1,3-butadiene) = 1.02 × 10^{-10}]^a	Tuazon et al.[12]	
			33.5 ± 3.0	298 ± 2	RR [relative to k(2-methyl-1,3-butadiene) = 1.01 × 10^{-10}]^a	Tuazon et al.[3]	

^{a}From the present recommendations (see text).
^{b}Temperature range covered estimated from the graphical presentation given.
^{c}No effort made to minimize possible effects of Cl atom reactions,[3] and these rate constants are superseded by those determined by Tuazon et al.[3]

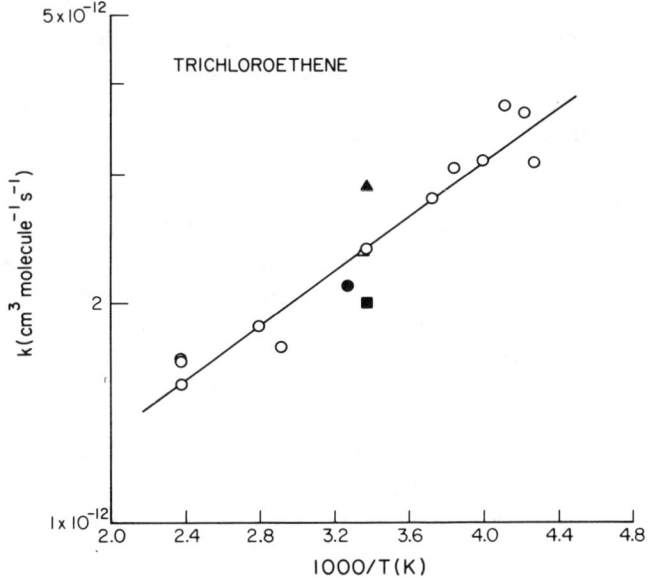

FIG. 55. Arrhenius plot of rate constants for the reaction of the OH radical with trichloroethene. (■) Howard;[5] (Δ) Davis et al.;[6] (O) Chang and Kaufman;[7] (●) Kirchner;[8] (▲) Edney et al.;[2] (———) recommendation (see text).

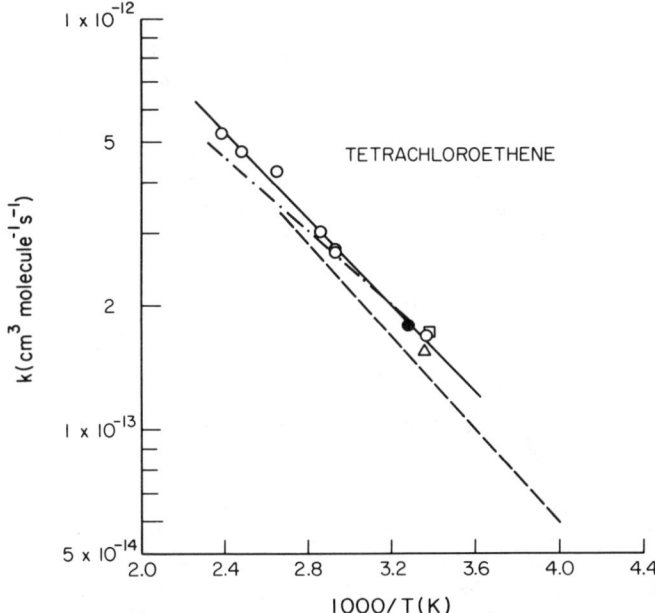

FIG. 56. Arrhenius plot of rate constants for the reaction of the OH radical with tetrachloroethene. (□) Howard;[5] (Δ, — — —) Davis et al.;[6] (O) Chang and Kaufman;[7] (●, – · –) Kirchner;[8] (———) recommendation (see text).

(3) 3-Chloropropene (allyl chloride)

Rate constants have been obtained at room temperature by Edney et al.[2,10] and Tuazon et al.[11] Relative rate techniques were used in all three of these studies,[2,10,11] and the involvement of Cl atom reactions (with the Cl

atoms being generated from the overall OH radical reaction with 3-chloropropene) was observed by Edney et al.,[2,10] but not by Tuazon et al.[11] The rate constants[2,10,11] given in Table 9 were obtained under experimental conditions designed to minimize the effects of any Cl atom reactions with the 3-chloropropene and the reference organic, and the rate constants obtained[2,10,11] are in agreement within the experimental error limits.

Based mainly upon the study of Tuazon et al.[11] in which any Cl atom reactions were suppressed by the addition of ethane to the reactant mixtures, a rate constant of

$$k(\text{3-chloropropene}) = 1.7 \times 10^{-11} \text{ cm}^3 \text{ molecule}^{-1} \text{ s}^{-1}$$

at 298 K is recommended, with an estimated overall uncertainty of $\pm 30\%$. This rate constant is expected to exhibit a small negative temperature dependence at around room temperature.

(4) Other Haloalkenes

For the remaining haloalkenes for which kinetic data are available at, or close to, the high pressure second-order limit, there are either significant discrepancies between the reported data (1,1-dichloroethene), or data have only been obtained by one research group. Accordingly, until further studies are carried out these data must be used with correspondingly large uncertainty limits. In particular, relative rate studies of the reaction of the OH radical with chloroalkenes can be complicated by the production of Cl atoms from reaction steps subsequent to the initial OH radical addition step, followed by reactions of these Cl atoms with the haloalkene and the reference organic(s). The most recent relative rate studies of Edney et al.[2,10] and Tuazon et al.[3,11] have attempted to avoid or take into account these complicating Cl atom reactions (although for 1,1-dichloroethene discrepancies of a factor of ~ 2 still remain between the room temperature rate constants obtained by these two groups[2,3]).

b. Mechanism

For the haloalkenes studied, the available kinetic data show that the reactions proceed via OH radical addition to the carbon-carbon double bond,[13] with H-atom abstraction from the vinyl C—H bonds or the alkyl-substituent C—H bonds being of essentially negligible significance. This situation is analogous to the OH radical reactions with the alkenes at temperatures $\lesssim 425$ K, and is consistent with the observed fall-off dependence of several of these rate constants over the range 0.7–7 Torr total pressure of helium diluent.[5] Hence, taking the vinyl halides as an example, these reactions proceed via addition of the OH radical to the haloalkenes to form an initially energy-rich OH-haloalkene adduct. This adduct can decompose back to the reactants or be collisionally stabilized, as shown, for example, for the vinyl halides (X = F, Cl, or Br).

$$OH + CH_2=CHX \rightleftharpoons HOC_2H_3X^*$$

$$HOC_2H_3X^* + M \rightarrow HOC_2H_3X + M$$

For $CH_2=CHF$, $CH_2=CHCl$ and $CH_2=CHBr$, the limiting high-pressure second-order kinetic regime is approached at total pressures of argon diluent of ~ 50 Torr at 298 K,[1] with the high pressure limit for $CHCl=CCl_2$ being approached at a total pressure of ~ 2 Torr of helium[5] at 296 K. The reaction to form the OH-haloalkene adduct is ~ 35 kcal mol^{-1} exothermic for all three halogen substituents[14] (similar to the situation for the alkenes). Elimination of an F-atom from the initially-formed OH-vinyl fluoride adduct is endothermic, with the overall reaction process

$$OH + CH_2=CHX \rightarrow HOC_2H_3X^*$$
$$\downarrow$$
$$CH_2=CHOH + X$$

being ~ 19 kcal mol^{-1} endothermic for X = F. Hence, for the reactions of the OH radical with vinyl fluoride, and with other haloalkenes containing no Cl or Br atoms attached to the double bond, the rate constants at temperatures $\lesssim 500-600$ K will exhibit similar behavior to that for ethene, showing fall-off behavior from second- to third-order kinetics as the total pressure decreases.

For the reaction of OH radicals with vinyl chloride and vinyl bromide (and by analogy, probably also for other haloalkenes with Cl or Br atoms attached to the $>C=C<$ double bond), the elimination of Cl or Br atoms from the OH-haloalkene adducts is themochemically favorable,[5] with the overall reactions being exothermic by ~ 11 kcal mol^{-1} and ~ 24 kcal mol^{-1} for X = Cl and Br, respectively.[1] In order for these Cl or Br atom eliminations to occur, the OH radical must add at the carbon atom to which the halogen substituent is located (the α-carbon atom) or, after OH radical addition to the β-carbon atom, a rapid 1,2-migration of OH must occur. If these elimination reactions occur, then the observed rate constants will exhibit second-order kinetics even at low total pressures where collisional stabilization of the OH-haloalkene adducts is not effective. At higher total pressures collisional stabilization of the adducts will become competitive with Cl or Br atom elimination, although the observed rate constant will remain pressure-independent and still be that for the initial reaction to form the adduct.

However, Howard[5] has shown that for the reaction of OH radicals with vinyl chloride at 296 K, the rate constant approaches a limiting low pressure value of $\lesssim 1 \times 10^{-12}$ cm^3 $molecule^{-1}$ s^{-1}, a factor of $\gtrsim 7$ lower than the limiting high pressure rate constant.[1] Thus, the elimination of a Cl atom is, at most, a relatively minor reaction pathway. This then implies that for the reaction of OH radicals with vinyl chloride (and presumably for other haloalkenes with Cl or Br atoms attached to the double bond) the two extreme reaction pathways involve either (a) OH radical addition only to the β-carbon atom and

that a 1,2-migration of OH has an activation energy of $\gtrsim 35$ kcal $mole^{-1}$, so that this 1,2-migration becomes rate determining, or (b) OH radical addition occurs at both the α- and β-positions, but mainly at the β position, the 1,2-migration of OH is negligible slow, and hence the elimination reaction occurs only after OH radical addition at the α position. While this latter situation is the most likely,[1] further work concerning both the pressure dependencies of the overall rate constants and the amount of reaction proceeding via halogen atom elimination is required for this class of organic compounds.

At elevated temperatures of $\gtrsim 600$ K, H-atom abstraction from the vinyl C—H bonds or alkyl-substituent C—H bonds is expected to become significant. Furthermore, at temperatures $\gtrsim 500-700$ K the thermal back decomposition of the thermalized OH-haloalkene adducts to reactants will become sufficiently rapid that, unless these adducts rearrange and/or decompose more rapidly by other channels, the OH radical addition pathway (with a rate constant which cannot exceed that for the addition of the OH radical to yield the energy-rich adduct) will become of no consequence. While this is expected to be the case for $CH_2=CHF$ and other fluoroalkenes, the rearrangement and/or decomposition pathways for other OH-haloalkene adducts are not experimentally known, and hence it cannot be predicted whether or not the observed rate constants will exhibit a discontinuity from the "high" ($\gtrsim 600$ K) to the "low" ($\lesssim 450$ K) temperature regimes (as for the alkenes; see Fig. 54) or exhibit a smooth, curved, Arrhenius plot with a minimum in the $\sim 500-700$ K temperature region. Clearly, further kinetic and mechanistic data are required for this class of organic compounds at temperatures > 450 K.

References

[1] R. A. Perry, R. Atkinson, and J. N. Pitts, Jr., J. Chem. Phys. **67**, 458 (1977).

[2] E. O. Edney, T. E. Kleindienst, and E. W. Corse, Int. J. Chem. Kinet. **18**, 1355 (1986).

[3] E. C. Tuazon, R. Atkinson, S. M. Aschmann, M. A. Goodman, and A. M. Winer, Int. J. Chem. Kinet. **20**, 241 (1988).

[4] A. M. Winer, A. C. Lloyd, K. R. Darnall, and J. N. Pitts, Jr., J. Phys. Chem. **80**, 1635 (1976).

[5] C. J. Howard, J. Chem. Phys. **65**, 4771 (1976).

[6] D. D. Davis, U. Machado, G. Smith, S. Wagner, and R. T. Watson, unpublished data (1977); cited in R. T. Watson, J. Phys. Chem. Ref. Data **6**, 871 (1977) and in references 5 and 7.

[7] J. S. Chang and F. Kaufman, J. Chem. Phys. **66**, 4989 (1977).

[8] K. Kirchner, Chimia **37**, 1 (1983).

[9] W. Klöpffer, R. Frank, E.-G. Kohl and F. Haag, Chemiker-Zeitung **110**, 57 (1986); "Methods of the Ecotoxicological Evaluation of Chemicals. Photochemical Degradation in the Gas Phase," Vol. 6 *OH Reaction Rate Constants and Tropospheric Lifetimes of Selected Environmental Chemicals*, Report 1980–1983, K. H. Becker, H. M. Biehl, P. Bruckmann, E. H. Fink, F. Führ, W. Klöpffer, R. Zellner and C. Zetzsch, Editors, Kernforschungsanlage Jülich GmbH, November 1984.

[10] E. O. Edney, P. B. Shepson, T. E. Kleindienst, and E. W. Corse, Int. J. Chem. Kinet. **18**, 597 (1986).

[11] E. C. Tuazon, R. Atkinson, and S. M. Aschmann, Int. J. Chem. Kinet., to be submitted for publication (1989).

[12]E. C. Tuazon, R. Atkinson, A. M. Winer, and J. N. Pitts, Jr., Arch. Environ. Contam. Toxicol. **13**, 691 (1984).

[13]R. Atkinson, Chem. Rev. **86**, 69 (1986).

[14]S. W. Benson, *Thermochemical Kinetics*, 2nd Ed. (Wiley, New York, NY, 1976).

2.5. Alkynes

a. Kinetics and Mechanisms

The available kinetic data reported to be at, or close to, the high-pressure limit, or obtained at one atmosphere total pressure of argon diluent,[16] are given in Table 10. In addition to these cited kinetic data, a number of rate constant studies have been carried out for acetylene which are now recognized to have been in the fall-off region between second- and third-order kinetics.[22-26] The data for the individual reactions are discussed below.

(1) Acetylene

Despite earlier evidence that at around room temperature the rate constant for the reaction of OH radicals with acetylene did not exhibit a pressure dependence,[4,23-25] the more recent flash or laser photolysis studies of Perry *et al.*,[5] Michael *et al.*,[7] Perry and Williamson,[8] Schmidt *et al.*[11] and Wahner and Zetzsch[12] show conclusively that at ~298 K the rate constant for this reaction exhibits fall-off behavior below ~1000 Torr total pressure of argon or nitrogen diluent.

Analogous to the situation for the reaction of OH radicals with ethene, these kinetic data[5,7,8,11,12] show that at temperatures $\lesssim$450 K the reaction of the OH radical with acetylene proceeds by initial addition, with the rate constant being in the fall-off region at total pressures of less than one atmosphere. At temperatures $\gtrsim$500–800 K the OH radical addition pathway becomes increasingly less important due to the increasingly important effects of fall-off with increasing temperature, and to the expectation that the thermal decomposition of the thermalized C_2H_2OH adduct will begin to become important at temperatures $\gtrsim$650 K.[13] Thus, at elevated temperatures representative of combustion conditions the observed reaction pathway involves H-atom abstraction[13]

$$OH + C_2H_2 \rightarrow H_2O + C_2H .$$

TABLE 10. Rate constants k and temperature-dependent parameters for the gas-phase reactions of the OH radical with alkynes at, or close to, the high pressure limit

Alkyne	$10^{12} \times A$ (cm³ molecule⁻¹ s⁻¹)	B (K)	$10^{12} \times k$ (cm³ molecule⁻¹ s⁻¹)	at T(K)	Technique	Reference	Temperature range covered (K)
Acetylene			3.3	1700–2000	Flame-equilibrium calculations	Fenimore and Jones[1]	1700–2000
			2.0	1600	Flame-RA	Porter *et al.*[2]	
	10 (C₂H + H₂O)ᵃ	3520		1000–1600	Flame-RA; product analysis	Browne *et al.*[3]	1000–1600
			0.165 ± 0.015	300	FP-RF	Davis *et al.*[4]	
			0.679 ± 0.070	298.1	FP-RF	Perry *et al.*[5]	298–422
			0.763 ± 0.100	350.2			
	1.91	312 ± 201	0.926 ± 0.120	422.4			
	0.53 (H + CH₂CO)ᵇ	101		570–850	Flame-MS	Vandooren and Van Tiggelen[6]	570–850
	91 (CH₃ + CO)ᵇ	6895		650–1100	Flame-MS	Vandooren and Van Tigglen[6]	650–1100
			0.384 ± 0.025	228	FP-RF	Michael *et al.*[7]	228–413
			0.597 ± 0.050	257			
			0.776 ± 0.073	298			
			1.056 ± 0.156	362			
	6.83 ± 1.19	646 ± 47	1.499 ± 0.163	413			
			0.675 ± 0.070	297	FP-RF	Perry and Williamson[8]	297–429
			0.798 ± 0.100	429			

TABLE 10. Rate constants k and temperature-dependent parameters for the gas-phase reactions of the OH radical with alkynes at, or close to, the high pressure limit — Continued

Alkyne	$10^{12} \times A$ (cm^3 molecule^{-1} s^{-1})	B (K)	$10^{12} \times k$ (cm^3 molecule^{-1} s^{-1})	at T(K)	Technique	Reference	Temperature range covered (K)
			2.2 ± 0.5 (H + CH$_2$CO)[b]	1700–1900	Flame-MS	Bittner and Howard[9]	1700–1900
			0.88 ± 0.11	298 ± 2	RR [relative to k(cyclohexane) = 7.49×10^{-12}][c]	Atkinson and Aschmann[10]	
			0.83 ± 0.08	295	LP-LIF	Schmidt et al.[11]	
			0.9[d]	298 ± 3	LP-RA	Wahner and Zetzsch[12]	
			0.09 ± 0.06[e] 0.27 ± 0.06 0.58 ± 0.08	880 ± 60 1140 ± 90 1330 ± 60	LH-LIF	Smith et al.[13]	880–1330
			0.81	300	RR [relative to k(propane) = 1.17×10^{-12}][c]	Klöpffer et al.[14]	
			0.87 ± 0.19	297 ± 2	RR [relative to k(cyclohexane) = 7.47×10^{-12}][c]	Hatakeyama et al.[15]	
			1.05 1.23 1.26 1.35 1.46 1.42 1.37 1.34 1.41 1.08 0.918 0.595 0.525 0.548 0.640 0.583 0.687 0.869 1.17	333 353 363 373 393 423 478 518 573 673 723 773 873 973 1073 1123 1173 1223 1273	PR-RA	Liu et al.[16]	333–1273
Acetylene-d_2			1.26 1.32 1.93 1.84 2.18 1.97 1.75 1.73 1.05 0.480 0.530	358 383 443 448 478 573 673 773 878 1073 1173	PR-RA	Liu et al.[16]	358–1173
Propyne			0.95 ± 0.17	300	DF-EPR	Bradley et al.[17]	
			6.15 ± 0.30	298 ± 2	RR [relative to k(cyclohexane) = 7.49×10^{-12}][c]	Atkinson and Aschmann[10]	

TABLE 10. Rate constants k and temperature-dependent parameters for the gas-phase reactions of the OH radical with alkynes at, or close to, the high pressure limit

Alkyne	$10^{12} \times A$ (cm³ molecule⁻¹ s⁻¹)	B (K)	$10^{12} \times k$ (cm³ molecule⁻¹ s⁻¹)	at T(K)	Technique	Reference	Temperature range covered (K)
			5.63 ± 0.15	297 ± 2	RR [relative to k(cyclohexane) = 7.47×10^{-12}][c]	Hatakeyama et al.[15]	
			3.73 ± 0.28	253	DF-RF	Boodaghians et al.[18]	253–343
			3.05 ± 0.14	298			
			2.39 ± 0.12	343			
1-Butyne			8.16 ± 0.23	298 ± 2	RR [relative to k(cyclohexane) = 7.49×10^{-12}][c]	Atkinson and Aschmann[10]	
			6.58 ± 1.24	253	DF-RF	Boodaghians et al.[18]	253–343
			4.95 ± 0.91	273			
			10.42 ± 1.38	300			
			8.81 ± 0.88	323			
			6.32 ± 0.93	343			
2-Butyne			29.7 ± 2.7	297 ± 2	RR [relative to k(cyclohexane) = 7.47×10^{-12}][c]	Hatakeyama et al.[15]	
			25.5 ± 1.8	253	DF-RF	Boodaghians et al.[18]	253–343
			24.6 ± 1.9	298			
			18.9 ± 1.5	343			
1-Pentyne			9.63 ± 0.81	253	DF-RF	Boodaghians et al.[18]	253–343
			11.17 ± 0.80	298			
			11.51 ± 0.61	343			
1-Hexyne			13.5 ± 1.1	253	DF-RF	Boodaghians et al.[18]	253–343
			12.6 ± 0.4	298			
			12.6 ± 0.7	343			
Butadiyne (Diacetylene)			83 ± 33	1700–1900	Flame-MS	Bittner and Howard[9]	1700–1900
			50	f	DF-RF	Homann et al.[19]	
			16.0 ± 0.6	297 ± 2	RR [relative to k(cyclohexane) = 7.47×10^{-12}][c]	Atkinson and Aschmann[20]	
			15.6 ± 0.2	297 ± 2	RR [relative to k(n-octane) = 8.65×10^{-12}][c]	Atkinson and Aschmann[20]	
			22.0 ± 1.1	296	FP-RF	Perry[21]	296–688
			19.5 ± 1.4	365			
			18.2 ± 2.0	475			
	11.1	-206 ± 151	14.5 ± 0.9	688			

[a]Assumed products.

[b]Products assumed; rate constants determined are dependent on the product species assumed.

[c]From the present recommendations (see text).

[d]Rate constants of $(8.3 \pm 0.6) \times 10^{-13}$ cm³ molecule⁻¹ s⁻¹ and $(8.1 \pm 0.7) \times 10^{-13}$ cm³ molecule⁻¹ s⁻¹ were determined at total pressures of N_2 diluent of 749 Torr and 771 Torr, respectively.[12]

[e]Rate constant extrapolated to zero pressure; rate constant of $(3.5 \pm 1.0) \times 10^{-13}$ cm³ molecule⁻¹ s⁻¹ determined at 100 Torr effective pressure of N_2.[13]

[f]Room temperature, not reported.

The rate constants obtained at temperatures $\gtrsim 500$ K and those obtained at $\lesssim 500$ K which were either reported to be at, or close to, the high-pressure limit or obtained at ~ 760 Torr total pressure of argon or air are given in Table 10. In the studies of Vandooren and Van Tiggelen[6] and Bittner and Howard,[9] rate constants were obtained by assuming specific reaction products, involving initial OH radical addition, to be formed. The study of Bittner and Howard[9] was carried out at sufficiently high temperatures (1700–1900 K) that the OH radical addition pathway would be of negligible importance, and hence the rate constant obtained[9] was not utilized in this evaluation. The study of Vandooren and Van Tiggelen[6] was carried out at a low total pressure (40 Torr) and at temperatures such that the addition channel would again be expected to be well into the fall-off region. Because of the uncertainties of the reaction pathways and the difficulties of extracting rate data from complex reaction systems, the data obtained by Vandooren and Van Tiggelen[6] were also not used in this evaluation.

The high-pressure rate constants of Perry et al.,[5] Michael et al.,[7] Perry and Williamson,[8] Atkinson and Aschmann,[10] Schmidt et al.,[11] Wahner and Zetzsch,[12] Hatakeyama et al.[15] and Liu et al.[16] and the elevated temperature data of Fenimore and Jones,[1] Porter et al.,[2] Browne et al.[3] and Smith et al.[13] (including the rate constants determined at 880 K at 100 Torr total pressure and extrapolated to zero pressure)[13] are plotted in Arrhenius form in Fig. 57.

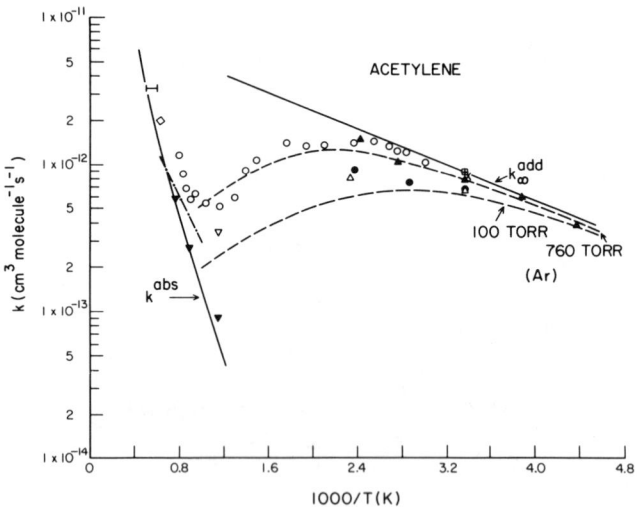

FIG. 57. Arrhenius plot of rate constants, reportedly at, or close to, the high pressure limit for argon or air diluent, for the reaction of the OH radical with acetylene. ($\vdash\dashv$) Fenimore and Jones;[1] ($\Diamond$) Porter et al.;[2] (–·–·–) Browne et al.;[3] (●) Perry et al.;[5] (▲) Michael et al.;[7] (△) Perry and Williamson;[8] (+) Atkinson and Aschmann,[10] Hatakeyama et al.;[15] (x) Schmidt et al.,[11] (☐) Wahner and Zetzsch;[12] (▼) Smith et al.,[13] pressure-independent (1140 K and 1330 K) or extrapolated to zero pressure (880 K); (▽) Smith et al.,[13] at 100 Torr effective pressure of N_2 diluent; (○) Liu et al.;[16] (——, – – –) recommendations (see text).

The laser heating-laser induced fluorescence study of Smith et al.[13] provided important information concerning the kinetics and reaction processes occurring. At 1140 and 1330 K, the rate constants obtained were observed to be independent of the total pressure,[13] while at 880 K the rate constant increased with the diluent pressure. The observation that the rate constants at 1140 and 1330 K and the extrapolated zero pressure rate constant at 880 K increase markedly with increasing temperature (Fig. 57), in conjunction with the lower temperature data showing the rate constant to be highly pressure dependent (at least up to several hundred Torr of argon diluent) at 295–429 K[5,7,8,11,12] and to decrease with increasing temperature over the temperature range ~ 500–900 K at a given pressure,[16] shows that the rate constants of Smith et al.[13] given in Table 10 and plotted as the filled invented triangles in Fig. 57 are those for the H-atom abstraction process. Furthermore, only at temperatures $\lesssim 1000$ K is the OH radical addition process of importance.[13]

The recent studies of Schmidt et al.[11] and Wahner and Zetzsch[12] show that the rate constants reported by Perry et al.[5] and Perry and Williamson[8] were still in the fall-off regime. The kinetic data of Schmidt et al.[11] and Wahner and Zetzsch,[12] obtained over wide pressure ranges utilizing argon and nitrogen as the diluent gases, allow the limiting low-pressure third-order rate constant k_o and high-pressure second-order rate constant k_∞ to be derived. Based upon the study of Wahner and Zetzsch[12] with the more efficient N_2 as the diluent gas, it is recommended that

$$k_\infty(\text{acetylene}) = 9.0$$

$$\times 10^{-13} \text{ cm}^3 \text{ molecule}^{-1} \text{ s}^{-1} \text{ at 298 K.}$$

From the data of Schmidt et al.[11] and Wahner and Zetzsch,[12] values of

$$k_o^{Ar}(\text{acetylene}) = 2.5 \times 10^{-30} \text{ cm}^6 \text{ molecule}^{-2} \text{ s}^{-1}$$

and

$$k_o^{N_2}(\text{acetylene}) = 5.0 \times 10^{-30} \text{ cm}^6 \text{ molecule}^{-2} \text{ s}^{-1},$$

both at 298 K, are recommended. These rate constants k_∞ and k_o^{Ar}, together with $F = 0.6$[11-13] and the Troe fall-off expression,

$$k = \left(\frac{k_o[M]}{1 + k_o[M]/k_\infty} \right) F^{\left\{ 1 + [\log(k_o[M]/k_\infty)]^2 \right\}^{-1}}$$

allow the experimental room temperature rate constants measured by Perry et al.,[5] Michael et al.[7] and Perry and Williamson[8] to be fit reasonably well.

The temperature dependence of $k_\infty(\text{acetylene})$ can be derived from the high pressure rate constants determined by Michael et al.[8] and the rate constants of Liu et al.[16] at 760 Torr total pressure of argon. With the above value of k_∞ at 298 K, this leads to the recommendation of

$$k_\infty(\text{acetylene}) = 9.4 \times 10^{-12} \, e^{-700/T} \, \text{cm}^3 \, \text{molecule}^{-1} \, \text{s}^{-1}$$

over the temperature range $\sim$230–500 K, and this expression is plotted in Fig. 57. While no definitive experimental data exist for the limiting low pressure rate constant k_o at other than room temperature, a $T^{-1.5}$ dependence of k_o allows the $\sim$350–360 K and $\sim$410–430 K rate constant data of Perry et al.,[5] Michael et al.[7] and Perry and Williamson[8] to be duplicated well, and this temperature dependence of k_o is very similar to that derived by Smith et al.[13] from transition state calculations.

Accordingly, rate constants k_o of

$$k_o^{\text{Ar}}(\text{acetylene}) = 2.5$$
$$\times 10^{-30} \, (T/298)^{-1.5} \, \text{cm}^6 \, \text{molecule}^{-2} \, \text{s}^{-1}$$

and

$$k_o^{\text{N}_2}(\text{acetylene}) = 5.0$$
$$\times 10^{-30} \, (T/298)^{-1.5} \, \text{cm}^6 \, \text{molecule}^{-2} \, \text{s}^{-1}$$

are recommended. With $F = 0.6$ at 298 K and $F = e^{-T/T^*} + e^{-4T^*/T}$, a value of $T^* = 580$ K is obtained.

The rate constants obtained by Schmidt et al.[11] and Wahner and Zetzsch[12] and, to a much lesser extent, by Perry et al.[5] and Perry and Williamson[8] show that any limiting low pressure bimolecular reaction is negligible at room temperature (with a rate constant of $\leqslant 8 \times 10^{-14}$ cm^3 molecule^{-1} s^{-1} [11]). This is in contrast to the data of Michael et al.[7] which suggested a limiting low pressure bimolecular reaction with a rate constant of $\sim 4 \times 10^{-13}$ cm^3 molecule^{-1} s^{-1}, independent of temperature over the range 228–413 K. The reasons for this observation[7] are not known, but the more recent data show that at temperatures $\lesssim$450 K the OH radical reaction with acetylene proceeds entirely by OH radical addition and that the rate constant exhibits the expected fall-off behavior with no observable limiting low-pressure bimolecular component. Also plotted in Fig. 57 as the dashed lines are the addition rate constants calculated from the above values of k_∞, k_o^{Ar} and F for total pressures of argon diluent of 100 and 760 Torr.

For the H-atom abstraction reaction pathway, the pressure-independent rate constants measured by Smith et al.[13] at 1140 and 1330 K are employed, using the expression $k = CT^2 e^{-D/T}$ with $C \sim 5 \times 10^{-18}$ cm^3 molecule^{-1} s^{-1} (similar to the values of C for methane and ethene), to yield the recommendation of

$$k^{\text{abs}}(\text{acetylene}) = 4.9$$
$$\times 10^{-18} \, T^2 \, e^{-3600/T} \, \text{cm}^3 \, \text{molecule}^{-1} \, \text{s}^{-1}$$

over the temperature range $\sim$1100–1350 K. This expression is also plotted in Fig. 57. It can be seen from Fig. 57 that these expressions k_∞^{add}, $k_{\text{Ar}}^{\text{add}}$ (760 Torr total pressure) and k^{abs} provide a reasonably good representation of the experimental kinetic data for acetylene plotted in Fig. 57. In particular, the calculated values of $k_{\text{Ar}}^{\text{add}}$ at 760 Torr total pressure agree reasonably well with the data obtained by Michael et al.[7] (at total pressures of argon diluent varying from 10–100 Torr at 228 K to 450–1100 Torr at 413 K) and with the rate constants of Liu et al.[16] determined at 760 Torr of argon diluent. Furthermore, the rate constant calculated for one atmosphere total pressure of air (making the reasonable assumption that O$_2$ and N$_2$ are equally efficient third-bodies) of $k = 8.15 \times 10^{-13}$ cm^3 molecule^{-1} s^{-1} at 298 K agrees well with the relative rate data of Atkinson and Aschmann,[10] Klöpffer et al.[14] and Hatakeyama et al.[15] Also relevant is the good agreement between the pressure-dependent portion of the 880 K rate constant determined by Smith et al.[13] at an effective pressure of 100 Torr of N$_2$ of $(2.6 \pm 0.4) \times 10^{-13}$ cm^3 molecule^{-1} s^{-1} and the calculated value of

$$k_{\text{N}_2}^{\text{add}} \, (100 \, \text{Torr}) = 3.5 \times 10^{-13} \, \text{cm}^3 \, \text{molecule}^{-1} \, \text{s}^{-1}.$$

Thus, at temperatures $<$650 K the OH radical reaction with acetylene is characterized by initial OH radical addition to form the initially energy-rich C$_2$H$_2$OH adduct, which can decompose back to reactants or be stabilized

$$\text{OH} + \text{C}_2\text{H}_2 \rightleftharpoons \text{CHCHOH*}$$

$$\text{CHCHOH*} + \text{M} \rightarrow \text{CHCHOH} + \text{M}$$

A further possible decomposition pathway for the adduct is via the elimination of an H atom[5]

$$[\text{HOCH} = \overset{\cdot}{\text{C}}\text{H}]^* \longrightarrow \text{H} + \text{C}_2\text{H}_2\text{O}$$

The overall reaction

$$\text{OH} + \text{C}_2\text{H}_2 \rightarrow \text{H} + \text{C}_2\text{H}_2\text{O}$$

is exothermic by $\sim$26 kcal mole^{-1} if the C$_2$H$_2$O product is ketene, but if the initial product formed after H-atom elimination is HOC$\equiv$CH, then the elimination reaction will be much less exothermic.[5]

The formation of C$_2$H$_2$O and C$_2$DHO from the reaction of OH radicals with C$_2$H$_2$ and C$_2$D$_2$, respectively, has been observed by Gutman and co-workers[27] using crossed molecular beams with photoionization mass spectrometric detection. These observations indicate that this elimination reaction does occur, with the H (or D) atom eliminated originating from the acetylene.[27] More recently, the C$_2$H$_2$O product has been identified as ketene by Hack et al.[26] from a discharge flow-mass spectrometry study of this reaction at a total pressure of $\sim$2 Torr. Under these low pressure conditions, the initially formed, energy-rich, OH—C$_2$H$_2$ adduct can thus either be stabilized or isomerize (presumably to the vinoxy radical) with subsequent decomposition:[26]

$$OH + C_2H_2 \rightleftharpoons [\dot{C}H=CHOH]^* \xrightarrow{M} \dot{C}H=CHOH$$
$$\downarrow \text{isom}$$
$$[CH_2\dot{C}HO]^* \rightarrow CH_2CO + H$$

This reaction sequence explains the observed formation of CHDCO from the reactions of OH radicals with C_2D_2[27] and of OD radicals with C_2H_2.[26] Recent room temperature product data at higher pressure indicates that the thermalized OH—C_2H_2 adduct can also isomerize to the vinoxy radical.[11]

As the temperature increases above ~650 K the thermal decomposition of the C_2H_2OH adduct is expected to become increasingly important and, unless this adduct can rapidly react with other species (such as O_2) or undergo isomerization and decomposition to products other than the reactants, the addition pathway will become of no significance. At these elevated temperatures, the H-atom abstraction pathway will then take over, with a rate constant which increases rapidly with increasing temperature.

These expectations are borne out by the experimental data and the recommended expressions shown in Fig. 57. In particular, the study of Liu et al.[16] may not have clearly differentiated the abstraction/addition regimes, with the OH radical addition channel contributing to the measure overall rate constants at temperatures up to $\gtrsim 1000$ K. The abstraction rate constant expression is consistent with a C—H bond strength in acetylene of 133 kcal mol^{-1},[28] although the parameters in the recommended expression for k^{abs} are not well determined. The same is true for the temperature dependencies of the rate constants k_∞ and k_o. However, for example, values of

$$k_\infty^{add} = 9.0 \times 10^{-13} (T/298)^{1.5} \text{ or } 7.0$$
$$\times 10^{-12} e^{-610/T} \text{ cm}^3 \text{ molecule}^{-1} \text{ s}^{-1}$$

do not change the calculated addition rate constants for 100 and 760 Torr total pressure of argon diluent shown in Fig. 57 to any significant extent.

(2) Acetylene-d_2

Rate constants for the reaction of the OH radical with acetylene-d_2 are available only from the pulsed radiolysis study of Liu et al.[16] carried out at 760 Torr total pressure of argon diluent. The reaction mechanism is expected to be totally analogous to that for the reaction of OH radicals with C_2H_2. No OD production from this reaction was observed,[16] showing that scrambling in the initially formed CDCDOH radical does not occur. These data of Liu et al.[16] are plotted in Arrhenius form in Fig. 58, together with the limiting high-pressure addition rate constant k_∞^{add} assuming (by analogy with the OH radical reactions with ethene and ethene-d_4) that

$$k_\infty^{add}(\text{acetylene-}d_2) = k_\infty^{add}(\text{acetylene}) = 9.4$$
$$\times 10^{-12} e^{-700/T} \text{ cm}^3 \text{ molecule}^{-1} \text{ s}^{-1}.$$

While no experimental data are available concerning the value of the low pressure third-order rate constant k_o, this rate constant for C_2D_2 is expected to be higher than that for C_2H_2 on account of the higher density of states in C_2D_2OH than in C_2H_2OH.[29,30] For the reactions of the OH radical with ethene and ethene-d_4, $k_o(\text{ethene-}d_4) \approx 3$ $k_o(\text{ethene})$. Since there are less C—D bonds in C_2D_2 than in C_2D_4, a ratio of $k_o(\text{acetylene-}d_2)/k_o(\text{acetylene}) \approx 2$, leading to

$$k_o^{Ar}(\text{acetylene-}d_2) = 5.0$$
$$\times 10^{-30} (T/298)^{-1.5} \text{ cm}^6 \text{ molecule}^{-2} \text{ s}^{-1},$$

has been used to calculate the addition rate constant at 760 Torr total pressure of argon diluent, and these calculated rate constants are also plotted in Fig. 58 [use of

$$k_o^{Ar}(\text{acetylene-}d_2) = 7.5$$
$$\times 10^{-30} (T/298)^{-1.5} \text{ cm}^6 \text{ molecule}^{-2} \text{ s}^{-1}$$

increases the addition rate constant at 760 Torr of argon by <10% at 500 K].

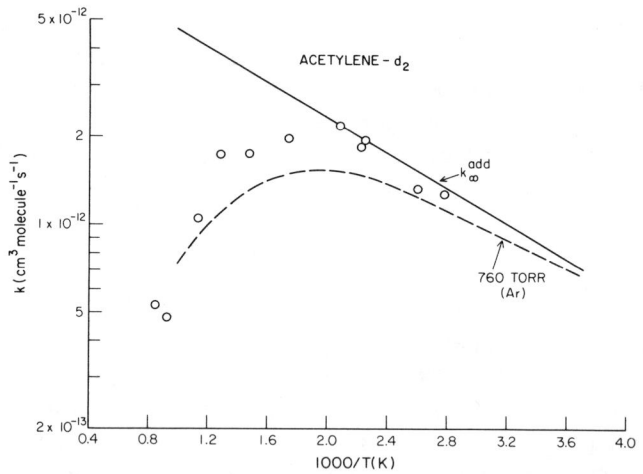

FIG. 58. Arrhenius plot of rate constants obtained at one atmosphere total pressure of argon diluent for the reaction of the OH radical with acetylene-d_2. (O) Liu et al.;[16] (——, – – –) recommendations (see text).

The agreement between the calculated and experimental data[16] is reasonable (Fig. 58), and it is clear that further rate data need to be obtained as a function of both pressure and temperature. The H-atom abstraction rate constant will be significantly lower than that for acetylene.

(3) Propyne

The rate constants obtained by Bradley et al.,[17] Atkinson and Aschmann,[10] Hatakeyama et al.[15] and Boodaghians et al.[18] are given in Table 10. These measured rate constants vary by a factor of ~6 at room temperature, with those studies conducted at lower total pressures yielding the lower rate constants. It is thus likely that this reaction, which is expected to proceed by OH radical addition,

$$OH + CH_3CH\equiv CH \rightleftharpoons CH_3\dot{C}H=CHOH^*$$

$$CH_3\dot{C}H=CHOH^* + M \rightarrow CH_3\dot{C}H=CHOH + M$$

is in the fall-off region at the total pressures characteristic of discharge flow system studies, although Boodaghians et al.[18] did not observe any effect of pressure on the room temperature rate constant over the total pressure range 1.7–6.4 Torr of helium diluent. That the data of Boodaghians et al.[18] for propyne are in the fall-off regime is supported by the good agreement of their data for 1- and 2-butyne, the more complex alkynes, with the relative rate data of Atkinson and Aschmann[10] and Hatakeyama et al.[15] obtained at ~750 Torr total pressure of air. Accordingly, a unit-weighted average of the atmospheric pressure rate constants of Atkinson and Aschmann[10] and Hatakeyama et al.[15] leads to the recommendation of

$$k(\text{propyne}) = 5.9 \times 10^{-12} \text{ cm}^3 \text{ molecule}^{-1} \text{ s}^{-1}$$

at 298 K, with an estimated overall uncertainty of ±40%. As for acetylene and acetylene-d_2, at elevated temperatures ($\gtrsim$800–1000 K) this reaction of the OH radical with propyne is expected to change over from OH radical addition to H-atom abstraction, mainly from the —CH$_3$ group:

$$OH + CH_3C\equiv CH \rightarrow H_2O + \dot{C}H_2C\equiv CH$$

(4) 1-Butyne

The rate constants obtained by Atkinson and Aschmann[10] and Boodaghians et al.[18] are given in Table 10 and are plotted in Arrhenius form in Fig. 59. While the data of Boodaghians et al.[18] as a function of temperature exhibit a significant degree of scatter, they are consistent with the room temperature rate constant of Atkinson and Aschmann.[10] This presumably indicates that for 1-butyne the OH radical addition reaction is at, or close to, the high pressure limit at total pressures of a few Torr at around room temperature. Based upon these rate constants of Atkinson and Aschmann[10] and Boodaghians et al.,[18] it is recommended that

$$k(\text{1-butyne}) = 8.0 \times 10^{-12} \text{ cm}^3 \text{ molecule}^{-1} \text{ s}^{-1},$$

independent of temperature over the range 253–343 K, with an estimated overall uncertainty of ±30% at 298 K.

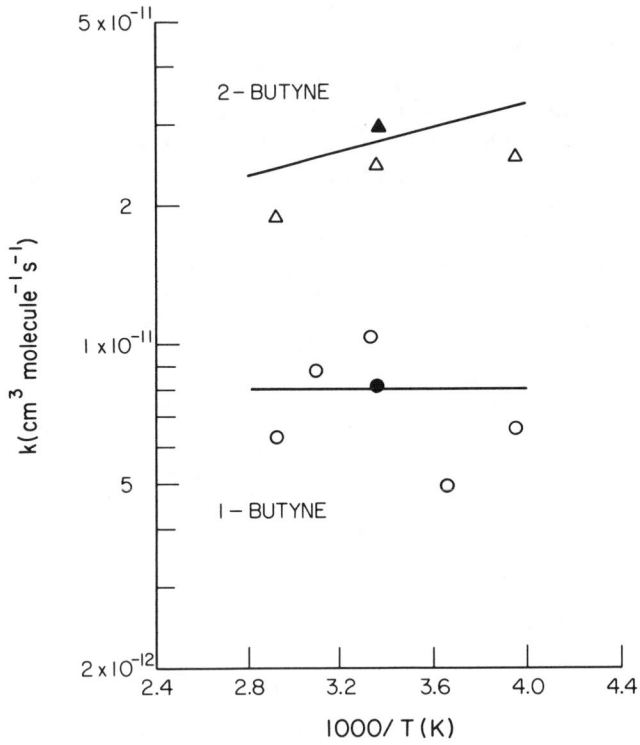

FIG. 59. Arrhenius plots of rate constants for the reactions of the OH radical with 1-butyne and 2-butyne. (●) Atkinson and Aschmann[10] (1-butyne); (▲) Hatakeyama et al.[15] (2-butyne); (○,△) Boodaghians et al.;[18] (——) recommendations (see text).

(5) 2-Butyne

The rate constants obtained by Hatakeyama et al.[15] and Boodaghians et al.[18] are given in Table 10 and are plotted in Arrhenius form in Fig. 59. The agreement at room temperature is reasonable, and a unit-weighted average of the room temperature rate constants of Hatakeyama et al.[15] and Boodaghians et al.,[18] combined with the temperature dependence of Boodaghians et al.,[18] leads to the tentative recommendation of

$$k(\text{2-butyne}) = 1.0 \times 10^{-11} \, e^{(300 \pm 300)/T} \text{ cm}^3 \text{ molecule}^{-1} \text{ s}^{-1}$$

over the temperature range 253–343 K, and

$$k(\text{2-butyne}) = 2.74 \times 10^{-11} \text{ cm}^3 \text{ molecule}^{-1} \text{ s}^{-1}$$

at 298 K, with an estimated overall uncertainty at 298 K of ±35%. As for 1-propyne and 1-butyne, at around room temperature this OH radical reaction with 2-butyne is expected to proceed by OH radical addition,

$$OH + CH_3C\equiv CCH_3 \rightarrow CH_3C(OH)=\dot{C}CH_3$$

with the reaction being at, or close to, the high pressure limit at total pressures of a few Torr.

(6) Butadiyne

The rate constants of Bittner and Howard,[9] Homann et al.,[19] Atkinson and Aschmann[20] and Perry[21] are given in Table 10 and are plotted in Arrhenius form in Fig. 60.

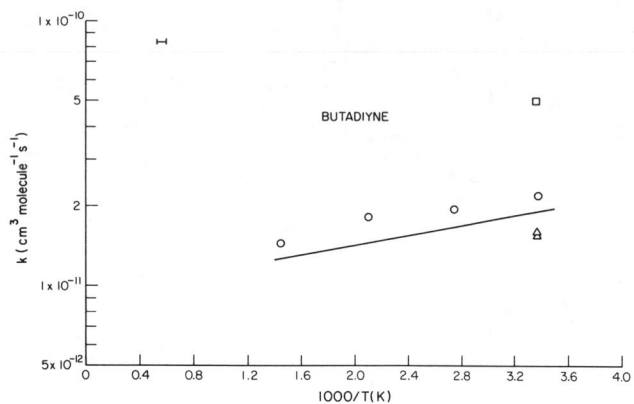

FIG. 60. Arrhenius plot of rate constants for the reaction of the OH radical with butadiyne. ($\vdash$) Bittner and Howard;[9] ($\square$) Homann et al.,[19] assuming a temperature of 298 K; ($\triangle$) Atkinson and Aschmann;[20] ($\bigcirc$) Perry;[21] (———) recommendation (see text).

At room temperature the reported rate constants of Homann et al.,[19] Atkinson and Aschmann[20] and Perry[21] disagree by a factor of ~3. This reaction will proceed by OH radical addition at temperatures $\lesssim 700$ K, and the temperature dependence measured by Perry[21] is consistent with such an OH radical addition process, which is expected to be at the high pressure limit at total pressures of a few Torr. The data obtained by Perry[21] and Atkinson and Aschmann[20] at room temperature disagree by ~35%, for reasons which are not presently understood, although the rate constants obtained by Atkinson and Aschmann[20] using two different reference organics are in excellent agreement. The recommended room temperature rate constant of

$$k(\text{butadiyne}) = 1.89 \times 10^{-11} \text{ cm}^3 \text{ molecule}^{-1} \text{ s}^{-1}$$

at 298 K, with an estimated overall uncertainty of ±40%, is a unit-weighted average of the room temperature rate constants from these studies,[20,21] using an average rate constant of 1.58×10^{-11} cm^3 molecule^{-1} s^{-1} for the Atkinson and Aschmann[20] study. The temperature dependence obtained by Perry[21] has been combined with

this 298 K value to yield the recommendation of

$$k(\text{butadiyne}) = 9.45 \times 10^{-12} \text{ e}^{206/T} \text{ cm}^3 \text{ molecule}^{-1} \text{ s}^{-1}$$

over the temperature range 296–688 K.

No recommendation is made for temperatures >700 K.

(7) Other Alkynes

Boodaghians et al.[18] have obtained rate constant data for 1-pentyne and 1-hexyne. Since these are the only available data, no recommendations are made.

References

[1]C. P. Fenimore and G. W. Jones, J. Chem. Phys. 41, 1887 (1964).

[2]R. P. Porter, A. H. Clark, W. E. Kaskan, and W. E. Browne, 11th International Symposium on Combustion, 1966; The Combustion Institute, Pittsburgh, PA, 1967, p. 907.

[3]W. G. Browne, R. P. Porter, J. D. Verlin, and A. H. Clark, 12th International Symposium on Combustion, 1968; The Combustion Institute, Pittsburgh, PA, 1969, p. 1035.

[4]D. D. Davis, S. Fischer, R. Schiff, R. T. Watson, and W. Bollinger, J. Chem. Phys. 63, 1707 (1975).

[5]R. A. Perry, R. Atkinson, and J. N. Pitts, Jr., J. Chem. Phys. 67, 5577 (1977).

[6]J. Vandooren and P. J. Van Tiggelen, 16th International Symposium on Combustion, 1976; The Combustion Institute, Pittsburgh, PA, 1977, p. 1133.

[7]J. V. Michael, D. F. Nava, R. P. Borkowski, W. A. Payne, and L. J. Stief, J. Chem. Phys. 73, 6108 (1980).

[8]R. A. Perry and D. Williamson, Chem. Phys. Lett. 93, 331 (1982).

[9]J. D. Bittner and J. B. Howard, 19th International Symposium on Combustion, 1982; The Combustion Institute, Pittsburgh, PA, 1982, p. 211.

[10]R. Atkinson and S. M. Aschmann, Int. J. Chem. Kinet. 16, 259 (1984).

[11]V. Schmidt, G. Y. Zhu, K. H. Becker, and E. H. Fink, Ber. Bunsenges Phys. Chem. 89, 321 (1985).

[12]A. Wahner and C. Zetzsch, Ber. Bunsenges Phys. Chem. 89, 323 (1985).

[13]G. P. Smith, P. W. Fairchild, and D. R. Crosley, J. Chem. Phys. 81, 2667 (1984).

[14]W. Klöpffer, R. Frank, E.-G. Kohl, and F. Haag, Chemiker-Zeitung 110, 57 (1986); "Methods of the Ecotoxicological Evaluation of Chemicals, Photochemical Degradation in the Gas Phase," Vol. 6, OH Reaction Rate Constants and Tropospheric Lifetimes of Selected Environmental Chemicals. Report 1980–1983, K. H. Becker, H. M. Biehl, P. Bruckmann, E. H. Fink, F. Führ, W. Klöpffer, R. Zellner, and C. Zetzsch, Editors, Kernforschungsanlage Jülich GmbH, November 1984.

[15]S. Hatakeyama, N. Washida, and H. Akimoto, J. Phys. Chem. 90, 173 (1986).

[16]A. Liu, W. A. Mulac, and C. D. Jonah, J. Phys. Chem., 92, 5942 (1988).

[17]J. N. Bradley, W. Hack, K. Hoyermann, and H. Gg. Wagner, J. Chem. Soc. Faraday Trans 1, 69, 1889 (1973).

[18]R. B. Boodaghians, I. W. Hall, F. S. Toby, and R. P. Wayne, J. Chem. Soc. Faraday Trans. 2, 83, 2073 (1987).

[19]K. H. Homann, M. Schottler, and J. Warnatz, unpublished data, cited in J. Warnatz, H. Bockhorn, A. Möser, and H. W. Wenz, 19th International Symposium on Combustion, 1982; The Combustion Institute, Pittsburgh, PA, 1982, p. 197.

[20]R. Atkinson and S. M. Aschmann, Combust. Flame **58**, 217 (1984).

[21]R. A. Perry, Combust. Flame **58**, 221 (1984).

[22]W. E. Wilson and A. A. Westenberg, 11th International Symposium on Combustion, 1966; The Combustion Institute, Pittsburgh, PA, 1967, p. 1143.

[23]J. E. Breen and G. P. Glass, Int. J. Chem. Kinet. **3**, 145 (1971).

[24]I. W. M. Smith and R. Zellner, J. Chem. Soc. Faraday Trans. 2, **69**, 1617 (1973).

[25]A. Pastrana and R. W. Carr, Jr., Int. J. Chem. Kinet. **6**, 587 (1974).

[26]W. Hack, K. Hoyermann, R. Sievert, and H. Gg. Wagner, Oxid. Commun. **5**, 101 (1983).

[27]J. R. Kanofsky, D. Lucas, F. Pruss, and D. Gutman, J. Phys. Chem. **78**, 311 (1974).

[28]R. Atkinson, D. L. Baulch, R. A. Cox, R. F. Hampson, Jr., J. A. Kerr, and J. Troe, J. Phys. Chem. Ref. Data, **18**, 881 (1989).

[29]J. Troe, J. Chem. Phys. **66**, 4758 (1977).

[30]J. Troe, J. Phys. Chem. **83**, 114 (1979).

2.6. Oxygen-Containing Organics

a. Kinetics and Mechanisms

The available kinetic data are given in Tables 11 (OH radical reactions) and 12 (OD radical reactions), and are discussed below by class of oxygenate. The experimental data concerning the mechanisms of these reactions are discussed for the individual oxygenates.

(1) Aldehydes

(a) Formaldehyde, Formaldehyde-^{13}C and Formaldehyde-d_1

The available kinetic data are given in Table 11. The rate constants obtained by Hoare,[1,5] Baldwin and Cowe,[2] Blundell et al.,[3] Westenberg and Fristrom,[4] Hoare and Peacock,[6] Morris and Niki,[8,9] Peeters and Mahnen,[10]

Vandooren and Van Tiggelen,[11] Niki et al.,[12] Atkinson and Pitts,[13] Stief et al.,[15] Temps and Wagner[16] and Zabarnick et al.[17] for $^{12}CH_2O$ and of Niki et al.[18] for $^{13}CH_2O$ are plotted in Arrhenius form in Fig. 61. A significant amount of scatter in these data is evident. Since the rate constant for the self reaction of OH radicals is subject to significant uncertainties,[105] the rate constants derived from the study of Smith[14] are not plotted in Fig. 61 and are not used in the evaluation of the rate constant for this reaction.

It can be seen from Fig. 61 that the rate constant for this reaction appears to be approximately independent of temperature over the range ~230–500 K, but that at temperatures >500 K the rate constant increases with increasing temperature. At around room temperature, absolute rate constants have been determined by Morris and Niki,[8] Atkinson and Pitts,[13] Stief et al.,[15] Temps and Wagner[16] and Zabarnick et al.[17] Again, a significant amount of scatter is observed, with Morris and Niki[8] and Zabarnick et al.[17] obtaining rate constants of $(1.2–1.4) \times 10^{-11}$ cm^3 molecule^{-1} s^{-1}, Atkinson and Pitts[13] and Stief et al.[15] rate constants of $(9.4–9.9) \times 10^{-12}$ cm^3 molecule^{-1} s^{-1}, and Temps and Wagner[16] a rate constant of 8.1×10^{-12} cm^3 molecule^{-1} s^{-1}. At elevated temperatures, there are again significant discrepancies, especially between the various studies of Hoare and co-workers[1,3,5,6] and that of Baldwin and Cowe,[2] and between those of Zabarnick et al.[17] and Vandooren and Van Tiggelen[11] (the rate constants from this study[11] being dependent upon those for the self reaction of OH radicals and the reaction of OH radicals with H_2, and cannot be reevaluated). Interestingly, the rate constants obtained from the flame studies of Westenberg and Fristrom[4] and Peeters and Mahnen[10] are in good agreement as to the magnitude of the rate constant at ~1400 K.

TABLE 11. Rate constants k and temperature-dependent parameters for the gas-phase reactions of the OH radical with oxygen-containing organics

Oxygenate	$10^{12} \times A$ (cm^3 mole-cule^{-1} s^{-1})	n	B (K)	$10^{12} \times k$ (cm^3 molecule^{-1} s^{-1})	at T (K)	Technique	Reference	Temperature range covered (K)
Aldehydes								
Formaldehyde				28 ± 8	773	RR [relative to $k(CH_4) = 6.95 \times 10^{-18}$ $T^2 e^{-1282/T}$]^a	Hoare[1]	773–923
				29	798			
				33	873			
				33	923			
				47	813	RR [relative to $k(H_2)$ $= 1.12 \times 10^{-12}$]^b	Baldwin and Cowe[2]	
				26 ± 3	773	RR [relative to $k(CH_4)$ $= 7.91 \times 10^{-13}$]^a	Blundell et al.[3]	
	~880		~4265		1250–1400	RR [relative to $k(CO)$ $= 1.12 \times 10^{-13} e^{0.000907T}$]^b	Westenberg and Fristrom[4]	1250–1400

TABLE 11. Rate constants k and temperature-dependent parameters for the gas-phase reactions of the OH radical with oxygen-containing organics — Continued

Oxygenate	$10^{12} \times A$ (cm^3 molecule^{-1} s^{-1})	n	B (K)	$10^{12} \times k$ (cm^3 molecule^{-1} s^{-1})	at T (K)	Technique	Reference	Temperature range covered (K)
				25	723	RR [relative to $k(CH_4)$ = 6.95×10^{-18} $T^2 e^{-1282/T}$]^a	Hoare[5]	723–923
				29	798			
				33	923			
				25	798	RR [relative to $k(CH_4)$ = 8.88×10^{-13}]^a	Hoare and Peacock[6]	
				14	798	RR [relative to $k(C_2H_6)$ = 5.07×10^{-12}]^a	Hoare and Peacock[6]	
				$\geqslant 6.6$	300	DF-MS	Herron and Penzhorn[7]	
				14 ± 3.5	298	DF-MS	Morris and Niki[8]	
				15.3	298	DF-MS	Morris and Niki[9]	
			~ 500	42	1600	Flame-MS	Peeters and Mahnen[10]	1400–1800
				17	485	Flame-MS	Vandooren and Van Tiggelen[11]	485–570
				22	570			
				15.8 ± 0.9	298 ± 2	RR [relative to k(ethene-d_4) = 8.78×10^{-12}]^c	Niki et al.[12]	
				9.4 ± 1.0	299.3	FP-RF	Atkinson and Pitts[13]	299–426
				9.4 ± 1.0	356.5			
	12.5		88 ± 151	10.3 ± 1.1	426.4			
				5.5 ± 0.7	268	DF-MS [relative to k(OH + OH) = 4.2×10^{-12} $e^{-240/T}$]^d	Smith[14]	268–334
				5.5 ± 0.7	298			
				7.4 ± 1.0	334			
				11.22 ± 0.98	228	FP-RF	Stief et al.[15]	228–362
				10.28 ± 0.90	257			
				9.86 ± 1.13	298			
	10.5 ± 1.1		0	10.46 ± 1.50	362			
				8.1 ± 1.7	296	DF-LMR	Temps and Wagner[16]	
				10.7 ± 1.3	296	LP-LIF	Zabarnick et al.[17]	296–576
				12.5 ± 0.5	297			
				11.4 ± 0.6	297			
				13.6 ± 1.0	297			
				13.2 ± 0.4	298			
				13.1 ± 0.4	298			
				13.7 ± 0.5	299			
				12.0 ± 0.3	301			
				13.9 ± 0.4	378			
				13.3 ± 0.3	473			
				11.4 ± 0.5	567			
				15.8 ± 1.2	572			
				14.1 ± 0.5	574			
	16.6 ± 2.0		86 ± 40	16.7 ± 0.8	576			
Formaldehyde-^{13}C				8.40 ± 0.51	299 ± 2	RR [relative to k(ethene) = 8.48×10^{-12}]^a	Niki et al.[18]	
Formaldehyde-d_1				~ 14	298	DF-MS	Morris and Niki[8]	
Acetaldehyde				15 ± 3.8	300	DF-MS	Morris et al.[19]	

TABLE 11. Rate constants k and temperature-dependent parameters for the gas-phase reactions of the OH radical with oxygen-containing organics — Continued

Oxygenate	$10^{12} \times A$ (cm³ molecule⁻¹ s⁻¹)	n	B (K)	$10^{12} \times k$ (cm³ molecule⁻¹ s⁻¹)	at T (K)	Technique	Reference	Temperature range covered (K)
				15.3	298	DF-MS	Morris and Niki[9]	
				≤14	295 ± 2	RR [relative to k(HONO) = 4.80 × 10⁻¹²][b]	Cox et al.[20]	
				≥3.0	1100	RR [relative to k(CO) = 3.04 × 10⁻¹³][b]	Colket et al.[21]	
				16.2 ± 1.8	298 ± 2	RR [relative to k(ethene) = 8.52 × 10⁻¹²][a]	Niki et al.[12]	
				16.0 ± 1.6	299.4	FP-RF	Atkinson and Pitts[13]	299–426
				14.4 ± 1.5	355.0			
	6.87		−257 ± 151	12.4 ± 1.3	426.1			
				12.8 ± 4.3	298 ± 4	RR [relative to k(ethene) = 8.52 × 10⁻¹²][a]	Kerr and Sheppard[22]	
				14.0 ± 3.1	253	FP-RF	Semmes et al.[23]	253–424
				12.2 ± 2.7	298			
				10.7 ± 2.3	356			
	7.1 ± 0.2		−165 ± 91	11.0 ± 2.3	424			
				21.0 ± 1.4	244	DF-RF	Michael et al.[24]	244–528
				19.2 ± 0.6	244			
				18.9 ± 1.4	259			
				17.9 ± 1.2	259			
				15.6 ± 0.8	273			
				16.3 ± 1.2	273			
				17.8 ± 0.6	273			
				19.6 ± 1.2	273			
				14.2 ± 1.0	298			
				14.7 ± 2.8	298			
				13.0 ± 1.2	333			
				14.0 ± 0.8	355			
				14.0 ± 0.4	367			
				14.3 ± 1.0	367			
				15.0 ± 1.0	373			
				11.6 ± 0.6	393			
				11.7 ± 0.8	420			
				10.6 ± 0.6	424			
				11.0 ± 0.6	433			
				11.5 ± 0.8	466			
				10.4 ± 0.4	468			
				10.4 ± 0.4	492			
				9.2 ± 1.4	499			
	5.52 ± 0.80		−307 ± 52	9.9 ± 0.4	528			
Glycolaldehyde [HOCH₂CHO]				9.9 ± 1.0	298 ± 2	RR [relative to k(acetaldehyde) = 1.58 × 10⁻¹¹][a]	Niki et al.[25]	
1-Propanal				5.2	713	RR [relative to k(CO) = 2.14 × 10⁻¹³][b]	Baldwin et al.[26]	
				30.6	298	DF-MS	Morris and Niki[9]	
				22.2 ± 0.9	298 ± 2	RR [relative to k(ethene) = 8.52 × 10⁻¹²][a]	Niki et al.[12]	

TABLE 11. Rate constants k and temperature-dependent parameters for the gas-phase reactions of the OH radical with oxygen-containing organics — Continued

Oxygenate	$10^{12} \times A$ (cm³ molecule⁻¹ s⁻¹)	n	B (K)	$10^{12} \times k$ (cm³ molecule⁻¹ s⁻¹)	at T (K)	Technique	Reference	Temperature range covered (K)
				19.4 ± 1.5	298 ± 4	RR [relative to k(ethene) = 8.52×10^{-12}]ᵃ	Kerr and Sheppard[22]	
				18.0 ± 2.1	298	RR [relative to k(acetaldehyde) = 1.58×10^{-11}]ᵃ	Audley et al.[27]	
				$\leqslant 30$	553	RR [relative to k(trans-2-butene) = 2.73×10^{-11}]ᵃ	Kaiser[28]	
				17.1 ± 2.4	298	FP-RF	Semmes et al.[23]	
				$\leqslant 28$	296	RR [relative to k(HONO) = 4.82×10^{-12}]ᵇ	Kerr and Stocker[29]	
1-Butanal [CH₃(CH₂)₂CHO]				25.2 ± 0.6	298 ± 4	RR [relative to k(ethene) = 8.52×10^{-12}]ᵃ	Kerr and Sheppard[22]	
				25.6 ± 3.2	298	RR [relative to k(acetaldehyde) = 1.58×10^{-11}]ᵃ	Audley et al.[27]	
				30.8 ± 4.2	258	FP-RF	Semmes et al.[23]	258–422
				20.6 ± 3.0	298			
				18.2 ± 2.6	361			
	5.7 ± 0.3		-411 ± 164	15.4 ± 2.3	422			
2-Methyl-1-propanal [(CH₃)₂CHCHO]				29.0 ± 5.7	298 ± 4	RR [relative to k(ethene) = 8.52×10^{-12}]ᵃ	Kerr and Sheppard[22]	
				17.7 ± 2.1	298	RR [relative to k(acetaldehyde) = 1.58×10^{-11}]ᵃ	Audley et al.[27]	
				33.4 ± 4.5	255	FP-RF	Semmes et al.[23]	255–423
				24.2 ± 3.3	298			
				19.7 ± 2.7	354			
	6.8 ± 0.3		-393 ± 125	18.2 ± 2.7	423			
1-Pentanal [CH₃(CH₂)₃CHO]				27.6 ± 4.2	298 ± 4	RR [relative to k(ethene) = 8.52×10^{-12}]ᵃ	Kerr and Sheppard[22]	
				13.9 ± 1.8	298	RR [relative to k(acetaldehyde) = 1.58×10^{-11}]ᵃ	Audley et al.[27]	
				38.9 ± 5.7	253	FP-RF	Semmes et al.[23]	253–410
				26.9 ± 3.9	298			
				23.3 ± 3.4	355			
	6.3 ± 0.2		-451 ± 108	19.0 ± 2.8	410			
3-Methyl-1-butanal [(CH₃)₂CHCH₂CHO]				28.9 ± 0.9	298 ± 4	RR [relative to k(ethene) = 8.52×10^{-12}]ᵃ	Kerr and Sheppard[22]	
				18.6 ± 2.1	298	RR [relative to k(acetaldehyde) = 1.58×10^{-11}]ᵃ	Audley et al.[27]	

TABLE 11. Rate constants k and temperature-dependent parameters for the gas-phase reactions of the OH radical with oxygen-containing organics — Continued

Oxygenate	$10^{12} \times A$ (cm^3 molecule^{-1} s^{-1})	n	B (K)	$10^{12} \times k$ (cm^3 molecule^{-1} s^{-1})	at T (K)	Technique	Reference	Temperature range covered (K)
				25.8 ± 4.0	298	FP-RF	Semmes et al.[23]	
2,2-Dimethyl-1-propanal [(CH$_3$)$_3$CCHO]				22.4 ± 6.3	298 ± 4	RR [relative to k(ethene) = 8.52×10^{-12}][a]	Kerr and Sheppard[22]	
				8.53 ± 0.95	298	RR [relative to k(acetaldehyde) = 1.58×10^{-11}][a]	Audley et al.[27]	
				33.9 ± 6.4	254	FP-RF	Semmes et al.[23]	254–425
				30.6 ± 4.4	298			
				21.8 ± 3.1	354			
	6.7 ± 0.3		-423 ± 154	17.6 ± 2.9	425			
Benzaldehyde [C$_6$H$_5$CHO]				14.0 ± 0.9	298 ± 2	RR [relative to k(ethene-d_4) = 8.78×10^{-12}][c]	Niki et al.[12]	
				11.8 ± 2.3	298 ± 4	RR [relative to k(ethene) = 8.52×10^{-12}][a]	Kerr and Sheppard[22]	
CCl$_3$CHO				1.73	298 ± 3	RR [relative to k(ethyl acetate) = 1.6×10^{-12}][a]	Nelson et al.[30]	
CH$_3$CClO				0.068	298 ± 3	RR [relative to k(trichloromethane) = 1.03×10^{-13}][a]	Nelson et al.[30]	
Ketones								
Acetone				$\leqslant 0.53$	300	RR [relative to k(ethene) = 8.44×10^{-12}][a]	Cox et al.[31]	
				0.23 ± 0.03	300	FP-RF	Zetzsch[32]	
				0.63 ± 0.09	298	RR [relative to k(n-hexane) = 5.61×10^{-12}][a]	Chiorboli et al.[33]	
				0.27 ± 0.01	303 ± 2	RR [relative to k(ethene) = 8.32×10^{-12}][a]	Kerr and Stocker[34]	
				0.145 ± 0.015	240	FP-RF	Wallington and Kurylo[35]	240–440
				0.216 ± 0.016	296			
				0.292 ± 0.023	350			
				0.407 ± 0.030	400			
	1.7 ± 0.4		600 ± 75	0.436 ± 0.050	440			
2-Butanone				3.5 ± 1.0	305 ± 2	RR [relative to k(2-methylpropene) = 4.94×10^{-11}][a]	Winer et al.[36]	
				2.74	300	RR [relative to k(ethene) = 8.44×10^{-12}][a]	Cox et al.[31]	
				0.95 ± 0.09	295 ± 2	RR [relative to k(ethene) = 8.65×10^{-12}][a]	Cox et al.[37]	

TABLE 11. Rate constants k and temperature-dependent parameters for the gas-phase reactions of the OH radical with oxygen-containing organics — Continued

Oxygenate	$10^{12} \times A$ (cm^3 molecule^{-1} s^{-1})	n	B (K)	$10^{12} \times k$ (cm^3 molecule^{-1} s^{-1})	at T (K)	Technique	Reference	Temperature range covered (K)
				1.2 ± 0.2	300	FP-RF	Zetzsch[32]	
				0.97 ± 0.17	297	RR [relative to k(propane) = 1.14×10^{-12}][a]	Edney et al.[38]	
				1.23 ± 0.10	240	FP-RF	Wallington and Kurylo[35]	240–440
				1.15 ± 0.10	296			
				1.41 ± 0.09	350			
				1.55 ± 0.07	400			
	2.3 ± 1.1		170 ± 120	1.65 ± 0.09	440			
2-Pentanone				4.70 ± 0.14	299 ± 2	RR [relative to k(cyclohexane) = 7.51×10^{-12}][a]	Atkinson et al.[39]	
				4.00 ± 0.29	296	FP-RF	Wallington and Kurylo[35]	
				5.07 ± 0.26	296 ± 2	RR [relative to k(cyclohexane) = 7.45×10^{-12}][a]	Atkinson and Aschmann[40]	
3-Pentanone				1.84 ± 0.34	299 ± 2	RR [relative to k(cyclohexane) = 7.51×10^{-12}][a]	Atkinson et al.[39]	
				2.85 ± 0.17	240	FP-RF	Wallington and Kurylo[35]	240–440
				2.74 ± 0.13	296			
				2.91 ± 0.17	350			
				2.79 ± 0.32	400			
	2.8 ± 0.3		-10 ± 35	2.78 ± 0.40	440			
				2.09 ± 0.15	296 ± 2	RR [relative to k(cyclohexane) = 7.45×10^{-12}][a]	Atkinson and Aschmann[40]	
2-Hexanone				9.09 ± 0.61	299 ± 2	RR [relative to k(cyclohexane) = 7.51×10^{-12}][a]	Atkinson et al.[39]	
				6.64 ± 0.56	296	FP-RF	Wallington and Kurylo[35]	
				9.09 ± 0.45	296 ± 2	RR [relative to k(cyclohexane) = 7.45×10^{-12}][a]	Atkinson and Aschmann[40]	
3-Hexanone				6.90 ± 0.29	299 ± 2	RR [relative to k(cyclohexane) = 7.51×10^{-12}][a]	Atkinson et al.[39]	
4-Methyl-2-pentanone				15 ± 5	305 ± 2	RR [relative to k(2-methylpropene) = 4.94×10^{-11}][a]	Winer et al.[36]	
				13.1	300	RR [relative to k(ethene) = 8.44×10^{-12}][a]	Cox et al.[31]	
				13.9 ± 0.4	295 ± 2	RR [relative to k(ethene) = 8.65×10^{-12}][a]	Cox et al.[37]	

TABLE 11. Rate constants k and temperature-dependent parameters for the gas-phase reactions of the OH radical with oxygen-containing organics — Continued

Oxygenate	$10^{12} \times A$ (cm³ molecule⁻¹ s⁻¹)	n	B (K)	$10^{12} \times k$ (cm³ molecule⁻¹ s⁻¹)	at T (K)	Technique	Reference	Temperature range covered (K)
				14.3 ± 0.7	299 ± 2	RR [relative to k(cyclohexane) $=7.51 \times 10^{-12}$][a]	Atkinson et al.[39]	
3,3-Dimethyl-2-butanone				1.21 ± 0.05	296	FP-RF	Wallington and Kurylo[35]	
2-Heptanone				8.67 ± 0.84	296	FP-RF	Wallington and Kurylo[35]	
2,4-Dimethyl-3-pentanone				5.38 ± 0.41	299 ± 2	RR [relative to k(cyclohexane) $=7.51 \times 10^{-12}$][a]	Atkinson et al.[39]	
2-Octanone				11.0 ± 0.9	296	FP-RF	Wallington and Kurylo[35]	
2-Nonanone				12.2 ± 1.3	296	FP-RF	Wallington and Kurylo[35]	
2,6-Dimethyl-4-heptanone				25 ± 8	305 ± 2	RR [relative to k(2-methylpropene) $= 4.94 \times 10^{-11}$][a]	Winer et al.[36]	
				27.5 ± 1.5	299 ± 2	RR [relative to k(cyclohexane) $=7.51 \times 10^{-12}$][a]	Atkinson et al.[39]	
2-Decanone				13.2 ± 1.2	296	FP-RF	Wallington and Kurylo[35]	
Cyclobutanone				0.87 ± 0.06	298	FP-RF	Dagaut et al.[41]	
Cyclopentanone				2.94 ± 0.18	298	FP-RF	Dagaut et al.[41]	
Cyclohexanone				6.39 ± 0.51	298	FP-RF	Dagaut et al.[41]	

α, β-Unsaturated Carbonyls

Oxygenate	$10^{12} \times A$ (cm³ molecule⁻¹ s⁻¹)	n	B (K)	$10^{12} \times k$ (cm³ molecule⁻¹ s⁻¹)	at T (K)	Technique	Reference	Temperature range covered (K)
Acrolein [$CH_2=CHCHO$]				25.4 ± 3.2	298 ± 2	RR [relative to k(n-butane) $= 2.54 \times 10^{-12}$][a]	Maldotti et al.[42]	
				20.3 ± 2.4	298 ± 4	RR [relative to k(ethene) $= 8.52 \times 10^{-12}$][a]	Kerr and Sheppard[22]	
				19.0 ± 1.4	299 ± 2	RR [relative to k(propene) $= 2.62 \times 10^{-11}$][a]	Atkinson et al.[43]	
				20.4 ± 0.1	297	RR [relative to k(propene) $= 2.65 \times 10^{-11}$][a]	Edney et al.[38]	
Crotonaldehyde [*trans*-$CH_3CH=CHCHO$]				35.1 ± 6.9	298 ± 4	RR [relative to k(ethene) $= 8.52 \times 10^{-12}$][a]	Kerr and Sheppard[22]	
				36.4 ± 4.2	299 ± 2	RR [relative to k(propene) $= 2.62 \times 10^{-11}$][a]	Atkinson et al.[43]	
Methacrolein [$CH_2=C(CH_3)CHO$]				31.4 ± 4.9	300	FP-RF	Kleindienst et al.[44]	300–423
				29.9 ± 4.8	350			
	17.7		−175 ± 52	26.5 ± 3.9	423			

TABLE 11. Rate constants k and temperature-dependent parameters for the gas-phase reactions of the OH radical with oxygen-containing organics — Continued

Oxygenate	$10^{12} \times A$ (cm³ mole-cule⁻¹ s⁻¹)	n	B (K)	$10^{12} \times k$ (cm³ molecule⁻¹ s⁻¹)	at T (K)	Technique	Reference	Temperature range covered (K)
				29.6 ± 2.4	299 ± 2	RR [relative to k(propene) = 2.62 × 10⁻¹¹]ᵃ	Atkinson et al.[43]	
				39.2 ± 3.1	298	RR [relative to k(propene) = 2.63 × 10⁻¹¹]ᵃ	Edney et al.[38]	
Methyl vinyl ketone [CH₂=CHCOCH₃]				14.8	300	RR [relative to k(ethene) = 8.44 × 10⁻¹²]ᵃ	Cox et al.[31]	
				17.9 ± 2.8	298	FP-RF	Kleindienst et al.[44]	298–424
				13.5 ± 2.4	350			
	3.85		−456 ± 73	11.4 ± 2.1	424			
				19.6 ± 1.5	299 ± 2	RR [relative to k(propene) = 2.62 × 10⁻¹¹]ᵃ	Atkinson et al.[43]	
Ketenes								
Ketene [CH₂=CO]		0		46.5	480–1000	Flame-MS	Vandooren and Van Tiggelen[11]	480–1000
				>1.7	295	RR [relative to k(C₃O₂) = 1.4 × 10⁻¹²]ᵉ	Faubel et al.[45]	
				17.3 ± 2.3	299 ± 2	RR [relative to k(cyclohexane) = 7.51 × 10⁻¹²]ᵃ	Hatakeyama et al.[46]	
Methylketene [CH₃CH=CO]				60 ± 13	299 ± 2	RR [relative to k(cyclohexane) = 7.51 × 10⁻¹²]ᵃ	Hatakeyama et al.[46]	
				79 ± 14	299 ± 2	RR [relative to k(propene) = 2.62 × 10⁻¹¹]ᵃ	Hatakeyama et al.[46]	
Ethylketene [C₂H₅CH=CO]				118 ± 29	299 ± 2	RR [relative to k(propene) = 2.62 × 10⁻¹¹]ᵃ	Hatakeyama et al.[46]	
Dimethylketene [(CH₃)₂C=CO]				107 ± 29	299 ± 2	RR [relative to k(propene) = 2.62 × 10⁻¹¹]ᵃ	Hatakeyama et al.[46]	
Dicarbonyls								
Glyoxal [(CHO)₂]				11.4 + 0.4	298 ± 2	RR [relative to k(cyclohexane) = 7.49 × 10⁻¹²]ᵃ	Plum et al.[47]	
				14.2 ± 2.1	298 ± 2	RR [relative to k(cyclohexane) = 7.49 × 10⁻¹²]ᵃ	Becker and Klein[48]	
Methylglyoxal [CH₃COCHO]				7.1 ± 1.6	297	FP-RF	Kleindienst et al.[44]	
				17.2 ± 1.2	298 ± 2	RR [relative to k(cyclohexane) = 7.49 × 10⁻¹²]ᵃ	Plum et al.[47]	

TABLE 11. Rate constants k and temperature-dependent parameters for the gas-phase reactions of the OH radical with oxygen-containing organics — Continued

Oxygenate	$10^{12} \times A$ (cm^3 molecule^{-1} s^{-1})	n	B (K)	$10^{12} \times k$ (cm^3 molecule^{-1} s^{-1})	at T (K)	Technique	Reference	Temperature range covered (K)
2,3-Butanedione				$0.24^{+0.08}_{-0.06}$	298	FP-RF	Darnall et al.[49]	
				0.19 ± 0.02	240	FP-RF	Dagaut et al.[41]	240–440
				0.23 ± 0.02	298			
				0.26 ± 0.02	350			
				0.39 ± 0.02	400			
	1.12 ± 0.65		450 ± 90	0.44 ± 0.03	440			
Pentane-1,5-dial [CHO(CH$_2$)$_3$CHO]				25.2 ± 1.1	298 ± 3	RR [relative to k(propene) = 2.63×10^{-11}]^a	Rogers[50]	
				22.4 ± 1.1	298 ± 3	RR [relative to k(trans-2-butene) = 6.40×10^{-11}]^a	Rogers[50]	
2,4-Pentanedione				1.15 ± 0.15	298	FP-RF	Dagaut et al.[41]	
2,5-Hexanedione				9.4 ± 1.2	240	FP-RF	Dagaut et al.[41]	240–440
				7.13 ± 0.34	298			
				5.07 ± 0.47	350			
				4.29 ± 0.38	400			
	1.49 ± 0.43		-450 ± 90	4.35 ± 0.53	440			
Unsaturated 1,4-Dicarbonyls								
cis-3-Hexene-2,5-dione				63.1 ± 6.1	298 ± 2	RR [relative to k(propene) = 2.63×10^{-11}]^a	Tuazon et al.[51]	
trans-3-Hexene-2,5-dione				53.1 ± 2.4	298 ± 2	RR [relative to k(propene) = 2.63×10^{-11}]^a	Tuazon et al.[51]	
Alcohols								
Methanol				1.01 ± 0.11	292	RR [relative to k(n-butane) = 2.47×10^{-12}]^a	Campbell et al.[52]	
				1.06 ± 0.10	296 ± 2	FP-RA	Overend and Paraskevopoulos[53]	
				1.00 ± 0.10	298	FP-RF	Ravishankara and Davis[54]	
	80		2265		1000–2000	Flame-MS	Vandooren and Van Tiggelen[55]	1000–2000
				1.10	300	RR [relative to k(ethene) = 8.44×10^{-12}]^a	Barnes et al.[56]	
				0.75 ± 0.15	293	LP-RF	Hägele et al.[57]	293–420
				0.94 ± 0.19	294			
				0.71 ± 0.15	295			
				0.97 ± 0.20	324			
				1.33 ± 0.27	372			
	12 ± 3		810 ± 50	1.74 ± 0.35	420			
				0.945 ± 0.073	300 ± 3	RR [relative to k(dimethyl ether) = 3.01×10^{-12}]^a	Tuazon et al.[58]	

TABLE 11. Rate constants k and temperature-dependent parameters for the gas-phase reactions of the OH radical with oxygen-containing organics — Continued

Oxygenate	$10^{12} \times A$ (cm³ mole-cule^{-1} s^{-1})	n	B (K)	$10^{12} \times k$ (cm³ molecule^{-1} s^{-1})	at T (K)	Technique	Reference	Temperature range covered (K)
	11 ± 3		798 ± 45	0.77·	300	DF-LIF	Meier et al.[59,60]	300–1010
				0.71 ± 0.08	296	FP-RF	Zetzsch[61]	
				0.54 ± 0.04	260	FP-RA	Greenhill and O'Grady[62]	260–803
				0.76 ± 0.04	292			
				0.75 ± 0.08	300			
				1.13 ± 0.05	331			
				1.44 ± 0.06	362			
				1.44 ± 0.09	453			
				1.35 ± 0.08	465			
				2.06 ± 0.17	570			
				2.67 ± 0.24	597.5			
				2.79 ± 0.25	669			
	8.0 ± 1.9		664 ± 88	5.76 ± 0.59	803			
				0.657 ± 0.046	240	FP-RF	Wallington and Kurylo[63]	240–440
				0.861 ± 0.047	296			
				1.25 ± 0.080	350			
				1.41 ± 0.12	400			
	4.8 ± 1.2		480 ± 70	1.62 ± 0.14	440			
				0.934 ± 0.041	294	LP-LIF	Hess and Tully[64]	294–866
				1.09 ± 0.05	332			
				1.33 ± 0.06	380			
				1.69 ± 0.07	441			
				2.10 ± 0.09	505			
				2.31 ± 0.10	527.5			
				3.01 ± 0.13	626			
				3.96 ± 0.20	709			
				5.05 ± 0.24	786.5			
	5.89×10^{-8}	2.65	-444	6.18 ± 0.32	865.5			
				1.01 ± 0.10	298 ± 2	DF-LIF	McCaulley et al.[65]	
Methanol-d_3 [CD₃OH]				0.50 ± 0.02	293	FP-RA	Greenhill and O'Grady[62]	
				0.435 ± 0.019	293	LP-LIF	Hess and Tully[64]	293–862
				0.529 ± 0.022	331			
				0.682 ± 0.030	384			
				0.920 ± 0.041	438.5			
				1.27 ± 0.06	530			
				1.88 ± 0.09	634			
				2.81 ± 0.12	730			
	1.28×10^{-10}	3.48	-642	4.85 ± 0.25	861.5			
				0.335 ± 0.072	298 ± 2	DF-LIF	McCaulley et al.[65]	
Methanol-d_4 [CD₃OD]				0.193 ± 0.045	298 ± 2	DF-LIF	McCaulley et al.[65]	
Ethanol				3.2 ± 0.4	292	RR [relative to k(n-butane) = 2.47×10^{-12}][a]	Campbell et al.[52]	
				3.74 ± 0.37	296 ± 2	FP-RA	Overend and Paraskevopoulos[53]	
				2.62 ± 0.36	298	FP-RF	Ravishankara and Davis[54]	
				3.5 ± 0.6	295 ± 2	RR [relative to k(propene = 2.68×10^{-11}][a]	Cox and Goldstone[66]	

TABLE 11. Rate constants k and temperature-dependent parameters for the gas-phase reactions of the OH radical with oxygen-containing organics — Continued

Oxygenate	$10^{12} \times A$ (cm^3 mole-cule^{-1} s^{-1})	n	B (K)	$10^{12} \times k$ (cm^3 molecule^{-1} s^{-1})	at T (K)	Technique	Reference	Temperature range covered (K)
	5.16 ± 1		274 ± 90	2.07	300	DF-LIF	Meier et al.[60,67,68]	300–1000
				3.0 ± 0.6	296	LP-RF	Lorenz et al.[69]	296–609
				2.9 ± 0.6	296			
				3.0 ± 0.6	298			
				2.5 ± 0.5	339			
				3.3 ± 0.6	386			
				3.1 ± 0.6	386			
				3.6 ± 0.7	452			
				4.3 ± 0.8	524			
				4.0 ± 0.8	525			
	5.6 ± 0.6		200 ± 50	3.7 ± 0.8	609			
				3.66 ± 0.42	303 ± 2	RR [relative to k(ethene) = 8.32×10^{-12}]^a	Kerr and Stocker[34]	
				2.84 ± 0.15	255	FP-RA	Greenhill and O'Grady[62]	255–459
				3.40 ± 0.14	273			
				3.80 ± 0.24	289			
				3.40 ± 0.17	293			
				4.26 ± 0.19	331			
				4.26 ± 0.18	360			
				5.21 ± 0.36	369			
	12.5 ± 2.4		360 ± 52	5.63 ± 0.48	459			
				2.75 ± 0.14	240	FP-RF	Wallington and Kurylo[63]	240–440
				3.33 ± 0.23	296			
				3.25 ± 0.39	350			
				4.07 ± 0.40	400			
	7.4 ± 3.2		240 ± 110	4.58 ± 0.29	440			
				3.26 ± 0.14	293	LP-LIF	Hess and Tully[70]	293–750
				3.32 ± 0.16^f	295			
				3.33 ± 0.14	326.5			
				3.63 ± 0.15	380			
				3.94 ± 0.16	441			
				4.65 ± 0.19	520.5			
				4.78 ± 0.23	544			
				4.74 ± 0.22	561			
				4.74 ± 0.22	582			
				4.65 ± 0.21	598			
				5.47 ± 0.34^f	599			
				4.79 ± 0.22	620.5			
				5.06 ± 0.23	645			
				5.66 ± 0.30	677			
				6.12 ± 0.35	706			
				6.62 ± 0.37	749.5			
1-Propanol				4.1 ± 0.4	292	RR [relative to k(n-butane) = 2.47×10^{-12}]^a	Campbell et al.[52]	
				5.33 ± 0.54	296 ± 2	FP-RA	Overend and Paraskevopoulos[53]	
				5.34 ± 0.29	296	FP-RF	Wallington and Kurylo[63]	
2-Propanol				6.9 ± 2.1	305 ± 2	RR [relative to k(2-methylpropene) = 4.94×10^{-11}]^a	Lloyd et al.[71]	
				5.48 ± 0.55	296 ± 2	FP-RA	Overend and Paraskevopoulos[53]	

TABLE 11. Rate constants k and temperature-dependent parameters for the gas-phase reactions of the OH radical with oxygen-containing organics — Continued

Oxygenate	$10^{12} \times A$ (cm³ mole-cule⁻¹ s⁻¹)	n	B (K)	$10^{12} \times k$ (cm³ molecule⁻¹ s⁻¹)	at T (K)	Technique	Reference	Temperature range covered (K)
				4.8	300	RR [relative to k(propane) = 1.17×10^{-12}][a]	Klöpffer et al.[72]	
				5.12 ± 0.31	240	FP-RF	Wallington and Kurylo[63]	240–440
				5.81 ± 0.34	296			
				5.27 ± 0.38	350			
				5.16 ± 0.44	400			
	5.8 ± 1.9		30 ± 90	5.75 ± 0.55	440			
1-Butanol				7.2 ± 1.1	292	RR [relative to k(n-butane) = 2.47×10^{-12}][a]	Campbell et al.[52]	
				8.31 ± 0.63	296	FP-RF	Wallington and Kurylo[63]	
2-Methyl-2-propanol [(CH₃)₃COH]				1.08 ± 0.13	295 ± 2	RR [relative to k(ethene) = 8.65×10^{-12}][a]	Cox and Goldstone[66]	
				1.00 ± 0.06	240	FP-RF	Wallington et al.[73]	240–440
				1.07 ± 0.08	298			
				1.23 ± 0.08	350			
				1.63 ± 0.07	400			
	3.3 ± 1.6		310 ± 150	1.77 ± 0.17	440			
1-Pentanol				10.8 ± 1.1	296	FP-RF	Wallington and Kurylo[63]	
2-Pentanol				11.8 ± 0.8	298	FP-RF	Wallington et al.[74]	
3-Pentanol				12.2 ± 0.7	298	FP-RF	Wallington et al.[74]	
Cyclopentanol				10.7 ± 0.7	298	FP-RF	Wallington et al.[74]	
3-Methyl-2-butanol				12.4 ± 0.7	298	FP-RF	Wallington et al.[74]	
1-Hexanol				12.4 ± 0.7	298	FP-RF	Wallington et al.[74]	
2-Hexanol				12.1 ± 0.7	298	FP-RF	Wallington et al.[74]	
1-Heptanol				13.6 ± 1.3	298	FP-RF	Wallington et al.[74]	
Allyl alcohol [CH₂=CHCH₂OH]				25.9 ± 3.4	440	PR-RA	Gordon and Mulac[75]	
2-Chloroethanol				1.4 ± 0.1	295	FP-RF	Wiedelmann and Zetzsch[76]	
Glycols, Hydroxyethers and Ketoethers								
1,2-Ethanediol [HOCH₂CH₂OH]				7.7 ± 1.1	295	FP-RF	Wiedelmann and Zetzsch[76]	
Hydroxyacetone [CH₃COCH₂OH]				3.02 ± 0.30	298	FP-RF	Dagaut et al.[77]	
1,2-Propanediol [HOCH₂CHOHCH₃]				12 ± 1	295	FP-RF	Wiedelmann and Zetzsch[76]	
2-Methoxyethanol [CH₃OCH₂CH₂OH]				18.8 ± 1.3	240	FP-RF	Dagaut et al.[77]	240–440
				12.5 ± 0.7	298			
				11.0 ± 0.6	350			

TABLE 11. Rate constants k and temperature-dependent parameters for the gas-phase reactions of the OH radical with oxygen-containing organics — Continued

Oxygenate	$10^{12} \times A$ (cm³ molecule⁻¹ s⁻¹)	n	B (K)	$10^{12} \times k$ (cm³ molecule⁻¹ s⁻¹)	at T (K)	Technique	Reference	Temperature range covered (K)
				10.4 ± 0.8	400			
	4.5 ± 1.4		-325 ± 100	10.1 ± 0.6	440			
Methoxyacetone [$CH_3OCH_2COCH_3$]				6.77 ± 0.61	298	FP-RF	Dagaut et al.[77]	
2-Hydroxyethyl ether [$HOCH_2CH_2OCH_2CH_2OH$]				30 ± 2	295	FP-RF	Wiedelmann and Zetzsch[76]	
2-Ethoxyethanol [$CH_3CH_2OCH_2CH_2OH$]				12 ± 3	298	LP-RF	Hartmann et al.[78]	298–485
	18 ± 4		120 ± 30	14 ± 3	485			
				18.7 ± 2.0	298	FP-RF	Dagaut et al.[77]	
3-Ethoxy-1-propanol [$CH_3CH_2OCH_2CH_2CH_2OH$]				22.0 ± 1.3	298	FP-RF	Dagaut et al.[77]	
3-Methoxy-1-butanol [$CH_3OCH(CH_3)CH_2CH_2OH$]				23.6 ± 1.6	298	FP-RF	Dagaut et al.[77]	
2-Butoxyethanol [$CH_3CH_2CH_2CH_2OCH_2CH_2OH$]				14 ± 3	298	LP-RF	Hartmann et al.[78]	298–505
	14 ± 3		0	14 ± 3	505			
				23.1 ± 0.9	298	FP-RF	Dagaut et al.[77]	
Ethers and Cycloethers								
Dimethyl ether				3.50 ± 0.35	298.9	FP-RF	Perry et al.[79]	299–424
				4.31 ± 0.43	350.5			
	12.9		388 ± 151	5.13 ± 0.51	423.9			
				2.95 ± 0.12	295	LP-LIF	Tully and Droege[80]	295–442
				3.40 ± 0.14	332			
				3.81 ± 0.16	377.5			
	10.4 ± 1.0		372 ± 34	4.52 ± 0.19	442			
				1.92 ± 0.22	240	FP-RF	Wallington et al.[81]	240–440
				2.49 ± 0.22	296			
				2.87 ± 0.40	350			
				3.02 ± 0.22	400			
	6.7 ± 1.5		300 ± 70	3.69 ± 0.35	440			
Diethyl ether				9.1 ± 1.8	305 ± 2	RR [relative to k(2-methylpropene) $= 4.94 \times 10^{-11}$][a]	Lloyd et al.[71]	
				13.4 ± 0.6	295	LP-LIF	Tully and Droege[80]	295–442
				12.9 ± 0.5	332			
				12.4 ± 0.5	377.5			
	9.13 ± 0.35		-115 ± 14	11.8 ± 0.5	442			
				17.7 ± 1.5	240	FP-RF	Wallington et al.[81]	240–440
				13.6 ± 0.9	296			
				11.4 ± 1.2	350			
				11.5 ± 1.2	400			
	5.6 ± 1.7		-270 ± 100	11.4 ± 1.7	440			
				12.0 ± 1.0	294 ± 2	RR [relative to k(2-methylpropene) $= 5.26 \times 10^{-11}$][a]	Bennett and Kerr[82]	

TABLE 11. Rate constants k and temperature-dependent parameters for the gas-phase reactions of the OH radical with oxygen-containing organics — Continued

Oxygenate	$10^{12} \times A$ (cm^3 mole-cule^{-1} s^{-1})	n	B (K)	$10^{12} \times k$ (cm^3 molecule^{-1} s^{-1})	at T (K)	Technique	Reference	Temperature range covered (K)
Diethyl ether-d_{10} [C$_2$D$_5$OC$_2$D$_5$]				6.70 ± 0.40	296	LP-LIF	Tully[83]	296–441
				6.54 ± 0.39	333			
				6.61 ± 0.40	375			
				6.84 ± 0.41	441			
Di-n-propyl ether				16.8 ± 3.4	305 ± 2	RR [relative to k(2-methylpropene) = 4.94 × 10^{-11}]^a	Lloyd et al.[71]	
				21.8 ± 2.4	240	FP-RF	Wallington et al.[81]	240–440
				18.0 ± 2.2	296			
				16.3 ± 1.8	350			
				15.9 ± 1.0	400			
	11 ± 3		−150 ± 80	16.4 ± 2.0	440			
				15.3 ± 1.7	294 ± 2	RR [relative to k(2-methylpropene) = 5.26 × 10^{-11}]^a	Bennett and Kerr[82]	
Methyl n-butyl ether				16.4 ± 0.6	298	FP-RF	Wallington et al.[74]	
Methyl t-butyl ether [CH$_3$OC(CH$_3$)$_3$]				2.85 ± 0.52	295 ± 2	RR [relative to k(ethene) = 8.65 × 10^{-12}]^a	Cox and Goldstone[66]	
				2.44 ± 0.39	295 ± 2	RR [relative to k(n-hexane) = 5.55 × 10^{-12}]^a	Cox and Goldstone[66]	
				2.74 ± 0.19	240	FP-RF	Wallington et al.[73]	240–440
				3.09 ± 0.15	298			
				3.20 ± 0.26	350			
				3.21 ± 0.25	400			
	5.1 ± 1.6		155 ± 100	3.97 ± 0.36	440			
Ethyl n-butyl ether [C$_2$H$_5$OCH$_2$CH$_2$CH$_2$CH$_3$]				22.8 ± 0.9	298	FP-RF	Wallington et al.[74]	
				13.4 ± 0.6	294 ± 2	RR [relative to k(2-methylpropene) = 5.26 × 10^{-11}]^a	Bennett and Kerr[82]	
Ethyl $tert$-butyl ether [C$_2$H$_5$OC(CH$_3$)$_3$]				8.12 ± 0.32	298	FP-RF	Wallington et al.[74]	
				5.63 ± 0.58	294 ± 2	RR [relative to k(2-methylpropene) = 5.26 × 10^{-11}]^a	Bennett and Kerr[82]	
Methyl $tert$-amyl ether [CH$_3$OC(CH$_3$)$_2$CH$_2$CH$_3$]				7.91 ± 0.42	298	FP-RF	Wallington et al.[74]	
Di-n-butyl ether				27.8 ± 3.6	296	FP-RF	Wallington et al.[81]	
				17.0 ± 0.9	294 ± 2	RR [relative to k(2-methylpropene) = 5.26 × 10^{-11}]^a	Bennett and Kerr[82]	
Di-isobutyl ether				26.0 ± 1.6	294 ± 2	RR [relative to k(2-methylpropene) = 5.26 × 10^{-11}]^a	Bennett and Kerr[82]	

TABLE 11. Rate constants k and temperature-dependent parameters for the gas-phase reactions of the OH radical with oxygen-containing organics — Continued

Oxygenate	$10^{12} \times A$ (cm^3 molecule^{-1} s^{-1})	n	B (K)	$10^{12} \times k$ (cm^3 molecule^{-1} s^{-1})	at T (K)	Technique	Reference	Temperature range covered (K)
Di-n-pentyl ether				34.7 ± 2.0	296	FP-RF	Wallington et al.[81]	
Trimethylene oxide[g]				10.3 ± 0.6	298	FP-RF	Dagaut et al.[77]	
Tetrahydrofuran[g]				14.3 ± 2.9	305 ± 2	RR [relative to k(2-methylpropene) $= 4.94 \times 10^{-11}$][a]	Winer et al.[84]	
				16.2 ± 2.3	298	FP-RF	Ravishankara and Davis[54]	
				17.8 ± 1.6	296	FP-RF	Wallington et al.[81]	
Tetrahydropyran[g]				13.8 ± 0.7	298	FP-RF	Dagaut et al.[77]	
Oxepane[g]				15.4 ± 1.3	298	FP-RF	Dagaut et al.[77]	
1,1-Dimethoxyethane [(CH$_3$O)$_2$CHCH$_3$]				8.89 ± 0.95	298	FP-RF	Dagaut et al.[77]	
Diethoxymethane [CH$_3$CH$_2$OCH$_2$OCH$_2$CH$_3$]				16.8 ± 1.6	298	FP-RF	Dagaut et al.[77]	
2,2-Dimethoxypropane [CH$_3$OC(CH$_3$)$_2$OCH$_3$]				4.09 ± 0.89	240	FP-RF	Dagaut et al.[77]	240–440
				3.92 ± 0.22	298			
				3.75 ± 0.18	350			
				3.80 ± 0.51	400			
	3.55 ± 0.39		−30 ± 35	3.93 ± 0.35	440			
1,2-Dimethoxypropane [CH$_3$OCH$_2$CH(CH$_3$)OCH$_3$]				14.3 ± 1.5	298	FP-RF	Dagaut et al.[77]	
2,2-Diethoxypropane [CH$_3$CH$_2$OC(CH$_3$)$_2$OCH$_2$CH$_3$]				11.1 ± 1.7	240	FP-RF	Dagaut et al.[77]	240–440
				11.7 ± 1.3	298			
				10.5 ± 1.0	350			
				11.7 ± 1.0	400			
	1.06 ± 0.25		−15 ± 15	10.6 ± 0.7	440			
2-Methoxyethyl ether [CH$_3$OCH$_2$CH$_2$OCH$_2$CH$_2$OCH$_3$]				17.5 ± 1.1	298	FP-RF	Dagaut et al.[77]	
1,1,3-Trimethoxypropane [(CH$_3$O)$_2$CHCH$_2$CH$_2$OCH$_3$]				19.2 ± 1.0	298	FP-RF	Dagaut et al.[77]	
2-Ethoxyethyl ether [CH$_3$CH$_2$OCH$_2$CH$_2$OCH$_2$CH$_2$OCH$_2$CH$_3$]				26.8 ± 2.4	298	FP-RF	Dagaut et al.[77]	
1,3-Dioxane[g]				10.0 ± 0.5	240	FP-RF	Dagaut et al.[77]	240–440
				9.15 ± 0.43	298			
				10.6 ± 0.2	350			
				9.65 ± 0.31	400			
	9.4 ± 0.2		−10 ± 60	9.72 ± 1.18	440			
1,4-Dioxane[g]				11.8 ± 0.8	240	FP-RF	Dagaut et al.[77]	240–440
				10.9 ± 0.5	298			
				9.55 ± 0.58	350			
				9.68 ± 0.83	400			
	8.3 ± 2.2		−80 ± 90	10.4 ± 0.9	440			
4-Methyl-1,3-dioxane[g]				11.3 ± 0.6	298	FP-RF	Dagaut et al.[77]	

TABLE 11. Rate constants k and temperature-dependent parameters for the gas-phase reactions of the OH radical with oxygen-containing organics — Continued

Oxygenate	$10^{12} \times A$ (cm^3 mole-cule^{-1} s^{-1})	n	B (K)	$10^{12} \times k$ (cm^3 molecule^{-1} s^{-1})	at T (K)	Technique	Reference	Temperature range covered (K)
1,3,5-Trioxane[g]				6.71 ± 0.21	292	LP-LIF	Zabarnick et al.[17]	292–597
				6.85 ± 0.18	293			
				5.85 ± 0.15	294			
				6.35 ± 0.17	373			
				7.64 ± 0.17	434			
				7.94 ± 0.37	487			
				9.82 ± 0.14	542			
	13.6 ± 2.0		232 ± 50	9.86 ± 0.15	597			
Vinyl methyl ether [CH$_2$=CHOCH$_3$]				33.5 ± 3.4	299.1	FP-RF	Perry et al.[79]	299–427
				26.0 ± 2.6	352.4			
	6.10		−511 ± 151	20.1 ± 2.0	427.0			
Furan[g]				105 ± 8	295 ± 1	DF-RF	Lee and Tang[85]	
				39.5 ± 2.9	298 ± 2	RR [relative to k(n-hexane) = 5.61 × 10^{-12}][a]	Atkinson et al.[86]	
				49.6 ± 3.3	254	FP-RF	Wine and Thompson[87]	254–424
				40.8 ± 1.8	297			
				43.1 ± 1.2	297			
				38.7 ± 2.2	299			
				41.6 ± 3.5	299			
				38.3 ± 4.5	300			
				31.9 ± 1.6	365			
	13.2 ± 2.9		−333 ± 67	29.9 ± 2.0	424			
				42.3 ± 3.2	295 ± 2	RR [relative to k(2-methyl-1,3-butadiene) = 1.02 × 10^{-10}][a]	Tuazon et al.[88]	
				46.6 ± 3.9	298	FP-RF	Witte and Zetzsch[89]	298–440
				46.6 ± 9.0	299			
				49.1 ± 2.7	299			
				43.6 ± 2.3	323			
				46.4 ± 2.9	349			
				38.9 ± 2.8	350			
				34.4 ± 5.4	372			
				38.5 ± 2.5	373			
				35.3 ± 1.9	399			
				34.0 ± 2.2	422			
				35.1 ± 2.2	422			
				32.4 ± 1.4	424			
				35.7 ± 3.1	424			
	12 ± 3		−430 ± 100	29.4 ± 1.7	440			
3-Methylfuran[g]				93.5 ± 2.4	296 ± 2	RR [relative to k(2,3-dimethyl-2-butene) = 1.11 × 10^{-10}][a]	Atkinson et al.[90]	
Oxazole[g]				10.11 ± 0.24	299	FP-RF	Witte and Zetzsch[89]	299–468
				9.83 ± 0.26	299			
				8.96 ± 0.38	299			
				9.17 ± 0.55	299			
				8.29 ± 0.19	324			
				7.68 ± 0.38	348			
				7.68 ± 0.24	349			
				7.25 ± 0.23	373			
				6.84 ± 0.10	398			
				6.67 ± 0.18	398			
				6.44 ± 0.18	423			

TABLE 11. Rate constants k and temperature-dependent parameters for the gas-phase reactions of the OH radical with oxygen-containing organics — Continued

Oxygenate	$10^{12} \times A$ (cm^3 molecule^{-1} s^{-1})	n	B (K)	$10^{12} \times k$ (cm^3 molecule^{-1} s^{-1})	at T (K)	Technique	Reference	Temperature range covered (K)
				6.04 ± 0.17	448			
				6.18 ± 0.19	449			
	2.8 ± 0.1		-350 ± 30	5.99 ± 0.22	468			
Esters								
Methyl formate [HC(O)OCH$_3$]				0.227 ± 0.034	296	FP-RF	Wallington et al.[91]	
Ethyl formate [HC(O)OCH$_2$CH$_3$]				1.02 ± 0.14	296	FP-RF	Wallington et al.[91]	
n-Propyl formate [HC(O)OCH$_2$CH$_2$CH$_3$]				2.38 ± 0.27	296	FP-RF	Wallington et al.[91]	
n-Butyl formate [HC(O)OCH$_2$CH$_2$CH$_2$CH$_3$]				3.12 ± 0.33	296	FP-RF	Wallington et al.[91]	
Methyl acetate [CH$_3$C(O)OCH$_3$]				0.17 ± 0.06	292	RR [relative to k(n-butane) = 2.47×10^{-12}][a]	Campbell and Parkinson[92]	
				0.486 ± 0.037	240	FP-RF	Wallington et al.[91]	240–440
				0.419 ± 0.032	263			
				0.341 ± 0.029	296			
				0.414 ± 0.030	350			
				0.395 ± 0.038	400			
	0.83 ± 0.35		260 ± 150 (296–440 K)	0.474 ± 0.066	440			
Methyl trifluoroacetate [CF$_3$C(O)OCH$_3$]				0.037 ± 0.003	240	FP-RF	Wallington et al.[91]	240–440
				0.052 ± 0.008	296			
				0.064 ± 0.005	350			
				0.083 ± 0.011	400			
	0.30 ± 0.07		512 ± 78	0.099 ± 0.008	440			
Ethyl acetate [CH$_3$C(O)OCH$_2$CH$_3$]				1.84 ± 0.37	292	RR [relative to k(n-butane) = 2.47×10^{-12}][a]	Campbell and Parkinson[92]	
				1.7 ± 0.2	296	FP-RF	Zetzsch[61]	
				3.26 ± 0.21	240	FP-RF	Wallington et al.[91]	240–440
				2.40 ± 0.15	263			
				1.51 ± 0.14	296			
				1.57 ± 0.10	350			
				1.72 ± 0.13	400			
	2.3 ± 0.2		131 ± 28 (296–440 K)	1.73 ± 0.13	440			
n-Propyl acetate [CH$_3$C(O)OCH$_2$CH$_2$CH$_3$]				4.2 ± 0.9	305 ± 2	RR [relative to k(2-methylpropene) = 4.94×10^{-11}][a]	Winer et al.[84]	
				2.50 ± 0.25	303 ± 2	RR [relative to k(ethene) = 8.32×10^{-12}][a]	Kerr and Stocker[34]	
				3.45 ± 0.34	296	FP-RF	Wallington et al.[91]	
Isopropyl acetate [CH$_3$C(O)OCH(CH$_3$)$_2$]				3.08 ± 0.84	303 ± 2	RR [relative to k(ethene) = 8.32×10^{-12}][a]	Kerr and Stocker[34]	

TABLE 11. Rate constants k and temperature-dependent parameters for the gas-phase reactions of the OH radical with oxygen-containing organics — Continued

Oxygenate	$10^{12} \times A$ (cm^3 mole-cule^{-1} s^{-1})	n	B (K)	$10^{12} \times k$ (cm^3 molecule^{-1} s^{-1})	at T (K)	Technique	Reference	Temperature range covered (K)
				3.72 ± 0.29	296	FP-RF	Wallington et al.[91]	
n-Butyl acetate [CH$_3$C(O)OCH$_2$-CH$_2$CH$_2$CH$_3$]	31 ± 7		594 ± 126	4.3 ± 0.8 6.8 ± 1.3 10 ± 2	298 400 516	LP-RF	Hartmann et al.[78]	298–516
				4.15 ± 0.30	296	FP-RF	Wallington et al.[91]	
sec-Butyl acetate [CH$_3$C(O)OCH(CH$_3$)-CH$_2$CH$_3$]				5.4 ± 1.1	305 ± 2	RR [relative to k(2-methylpropene) $= 4.94 \times 10^{-11}$][a]	Winer et al.[84]	
				5.65 ± 0.59	296	FP-RF	Wallington et al.[91]	
Methyl propionate [CH$_3$CH$_2$C(O)OCH$_3$]				0.27 ± 0.11	292	RR [relative to k(n-butane) $= 2.47 \times 10^{-12}$][a]	Campbell and Parkinson[92]	
				1.03 ± 0.04	296	FP-RF	Wallington et al.[91]	
Ethyl propionate [CH$_3$CH$_2$C(O)OCH$_2$CH$_3$]				1.68 ± 0.36	292	RR [relative to k(n-butane) $= 2.47 \times 10^{-12}$][a]	Campbell and Parkinson[92]	
				2.14 ± 0.30	296	FP-RF	Wallington et al.[91]	
n-Propyl propionate [CH$_3$CH$_2$C(O)OCH$_2$-CH$_2$CH$_3$]				4.02 ± 0.32	296	FP-RF	Wallington et al.[91]	
Methyl butyrate [CH$_3$CH$_2$CH$_2$C(O)-OCH$_3$]				3.04 ± 0.33	296	FP-RF	Wallington et al.[91]	
Ethyl butyrate [CH$_3$CH$_2$CH$_2$C(O)-OCH$_2$CH$_3$]				4.94 ± 0.38	296	FP-RF	Wallington et al.[91]	
n-Propyl butyrate [CH$_3$CH$_2$CH$_2$C(O)-OCH$_2$CH$_2$CH$_3$]				7.41 ± 0.32	296	FP-RF	Wallington et al.[91]	
n-Butyl butyrate [CH$_3$CH$_2$CH$_2$C(O)-OCH$_2$CH$_2$CH$_2$CH$_3$]				10.6 ± 1.3	296	FP-RF	Wallington et al.[91]	
1-Acetoxy-2-ethoxyethane [CH$_3$C(O)OCH$_2$CH$_2$OCH$_2$CH$_3$]	3.6 ± 0.8		−383 ± 80	13 ± 2 9 ± 2 8 ± 2	298 401 506	LP-RF	Hartmann et al.[78]	298–506
Carboxylic Acids								
Formic acid				0.32 ± 0.10	298	FP-RF	Zetzsch and Stuhl[93]	
				0.461 ± 0.051 0.405 ± 0.047 0.545 ± 0.012 0.448 ± 0.032 0.432 ± 0.065 0.446 ± 0.011 0.449 ± 0.026 0.428 ± 0.049	298 298 298 298 298 298 298 298	FP-RF	Wine et al.[94]	298–430

TABLE 11. Rate constants k and temperature-dependent parameters for the gas-phase reactions of the OH radical with oxygen-containing organics — Continued

Oxygenate	$10^{12} \times A$ (cm³ molecule⁻¹ s⁻¹)	n	B (K)	$10^{12} \times k$ (cm³ molecule⁻¹ s⁻¹)	at T (K)	Technique	Reference	Temperature range covered (K)
				0.481 ± 0.059	298			
				0.482 ± 0.042	298			
				0.523 ± 0.030	299			
				0.466 ± 0.007	299			
				0.480 ± 0.075	299			
				0.464 ± 0.037	299			
				0.495 ± 0.081	299			
				0.490 ± 0.094	300			
				0.539 ± 0.076	300			
				0.446 ± 0.033	300			
				0.443 ± 0.053	300			
				0.495 ± 0.050	320			
				0.406 ± 0.024	337			
				0.433 ± 0.037	374			
				0.505 ± 0.002	378			
				0.479 ± 0.068	402			
				0.407 ± 0.034	428			
				0.409 ± 0.051	428			
				0.439 ± 0.072	430			
	0.363 ± 0.089 [0.462 ± 0.078	-77 ± 75 0]		0.434 ± 0.053	430			
				0.490 ± 0.012	296	FP-RA	Jolly et al.[95]	
				0.37 ± 0.04	298	FP-RF	Dagaut et al.[96]	
				0.447 ± 0.028	296.9	LP-RA	Singleton et al.[97]	297–445
				0.365 ± 0.030	326.3			
				0.369 ± 0.032	356.2			
				0.367 ± 0.012	396.2			
	0.291 ± 0.159 [0.365 ± 0.033	-102 ± 194 0]		0.390 ± 0.028	445.2			
Formic acid dimer				$\leqslant 0.025$	296	FP-RA	Jolly et al.[95,98]	
				0.0802 ± 0.0206	296.9	LP-RA	Singleton et al.[97]	297–326
				0.223 ± 0.103	326.3			
Formic acid-d_1 [DCOOH]				0.435 ± 0.038	298	FP-RF	Wine et al.[94]	
				0.498 ± 0.099	298			
				0.456 ± 0.028	298			
				0.400 ± 0.033	296.0	LP-RA	Singleton et al.[97]	
Formic acid dimer-d_2 [(DCOOH)$_2$]				-0.0203 ± 0.0113	296.0	LP-RA	Singleton et al.[97]	
Acetic acid				0.6 ± 0.2	298	FP-RF	Zetzsch and Stuhl[93]	
				0.74 ± 0.06	298	FP-RF	Dagaut et al.[96]	298–440
				0.81 ± 0.09	350			
				0.87 ± 0.14	400			
	1.3 ± 0.1		170 ± 20	0.88 ± 0.08	440			
Propionic acid				1.6 ± 0.5	298	FP-RF	Zetzsch and Stuhl[93]	
				1.22 ± 0.12	298	FP-RF	Dagaut et al.[96]	298–440
				1.28 ± 0.13	350			
				1.36 ± 0.13	400			
	1.8 ± 0.2		120 ± 30	1.37 ± 0.10	440			

TABLE 11. Rate constants k and temperature-dependent parameters for the gas-phase reactions of the OH radical with oxygen-containing organics — Continued

Oxygenate	$10^{12} \times A$ (cm^3 molecule^{-1} s^{-1})	n	B (K)	$10^{12} \times k$ (cm^3 molecule^{-1} s^{-1})	at T (K)	Technique	Reference	Temperature range covered (K)
Butyric acid				2.4 ± 0.7	298	FP-RF	Zetzsch and Stuhl[93]	
Isobutyric acid [(CH$_3$)$_2$CHCOOH]				2.00 ± 0.20	298	FP-RF	Dagaut et al.[96]	298–440
				2.09 ± 0.18	350			
				2.12 ± 0.13	400			
	2.6 ± 0.2		70 ± 25	2.17 ± 0.17	440			
Oxides								
Epoxyethane				0.080 ± 0.016	297	LP-RF	Lorenz and Zellner[99]	297–515
				0.18 ± 0.04	377			
	11 ± 4		1460 ± 150 (297–435 K)	0.40 ± 0.08	435			
				1.6 ± 0.1	501			
				2.7 ± 0.5	515			
				0.053 ± 0.01	295	FP-RF	Zetzsch[100]	
				<0.10	300	RR [relative to k(propane) = 1.17×10^{-12}][a]	Klöpffer et al.[72]	
				0.095 ± 0.005	296	FP-RF	Wallington et al.[81]	
1,2-Epoxypropane				1.2 ± 0.7	300 ± 1	RR [relative to k(n-butane)-k(neopentane) = 1.70×10^{-12}][a]	Winer et al.[101]	
				0.52 ± 0.1	295	FP-RF	Zetzsch[100]	
				1.11 ± 0.75	296	RR [relative to k(n-butane) = 2.51×10^{-12}][a]	Edney et al.[38]	
				0.495 ± 0.052	296	FP-RF	Wallington et al.[81]	
1,2-Epoxybutane				2.1 ± 0.7	300 ± 1	RR [relative to k(n-butane)-k(neopentane) = 1.70×10^{-12}][a]	Winer et al.[101]	
1-Chloro-2,3-epoxypropane				0.44 ± 0.05	295	FP-RF	Zetzsch[100]	
				$\geqslant 0.55$	297	RR [relative to k(n-butane) = 2.53×10^{-12}][a]	Edney et al.[38]	
Hydroperoxides								
Methyl hydroperoxide				10.2 ± 0.8	h	RR [relative to k(ethene) = 8.52×10^{-12}][a]	Niki et al.[102]	
				10.7 ± 1.2	h	RR [relative to k(acetaldehyde) = 1.58×10^{-11}][a]	Niki et al.[102]	
				5.13 ± 0.19	203	FP/LP-LIF	Vaghjiani and Ravishankara[103]	203–348
				5.00 ± 0.29	223			
				4.33 ± 0.54	244			
				3.85 ± 0.23	298			
	1.78 ± 0.25		-220 ± 21	3.29 ± 0.32	348			
				$6.93 \pm 0.26^{\text{f}}$	223	FP/LP-LIF	Vaghjiani and Ravishankara[103]	223–373
				$6.45 \pm 0.20^{\text{f}}$	244			

TABLE 11. Rate constants k and temperature-dependent parameters for the gas-phase reactions of the OH radical with oxygen-containing organics — Continued

Oxygenate	$10^{12} \times A$ (cm³ molecule⁻¹ s⁻¹)	n	B (K)	$10^{12} \times k$ (cm³ molecule⁻¹ s⁻¹)	at T (K)	Technique	Reference	Temperature range covered (K)
				5.48 ± 0.20^f	298			
	3.08 ± 0.36		-179 ± 18	5.06 ± 0.14^f	373			
t-Butyl hydroperoxide [(CH₃)₃COOH]				3.0 ± 0.8	298	FP-RA	Anastasi et al.[104]	

[a]From the present recommendations (see text).
[b]See Introduction.
[c]From the rate constant ratio k(OH + ethene-d_4)/k(OH + ethene)[12] and the present recommendation for ethene (see text).
[d]From DeMore et al.[105]
[e]From Faubel et al.[45]
[f]Rate constants for reactions of the ¹⁸OH radical.

[g]Structures:

Trimethylene oxide, ; Tetrahydrofuran, ; Tetrahydropyran, ; Oxepane, ; 1,3-Dioxane, ;

1,4-Dioxane, ; 4-Methyl-1,3-dioxane, ; 1,3,5-Trioxane, ; Furan, ; 3-Methylfuran, ;

Oxazole, .

[h]Room temperature not reported. 298 K has been assumed, based on previous studies carried out by Niki and co-workers.[12,25]

TABLE 12. Rate constants k and temperature-dependent parameters for the gas-phase reactions of the OD radical with oxygen-containing organics

Oxygenate	$10^{12} \times A$ (cm³ molecule⁻¹ s⁻¹)	B (K)	$10^{12} \times k$ (cm³ molecule⁻¹ s⁻¹)	at T (K)	Technique	Reference	Temperature range covered (K)
Methanol			0.95 ± 0.12	298 ± 2	DF-LIF	McCaulley et al.[65]	
Methanol-d_1 (CH₃OD)			0.93 ± 0.11	298 ± 2	DF-LIF	McCaulley et al.[65]	
Methanol-d_3 (CD₃OH)			0.286 ± 0.037	298 ± 2	DF-LIF	McCaulley et al.[65]	
Methanol-d_4 (CD₃OD)			0.167 ± 0.016	298 ± 2	DF-LIF	McCaulley et al.[65]	
Formic acid-d_1 (HCOOD)			$\lesssim 0.15$	a	LP-RA	Singleton et al.[97]	
Formic acid-d_2 (DCOOD)			0.0636 ± 0.0130	298.0	LP-RA	Singleton et al.[97]	298–445
			0.0674 ± 0.0052	324.0			
			0.0864 ± 0.0039	355.9			
			0.0943 ± 0.0045	396.1			
	0.447 ± 0.169	594 ± 134	0.123 ± 0.0055	445.3			
Formic acid dimer-d_4 [(DCOOD)₂]			0.0137 ± 0.0135	298.0	LP-RA	Singleton et al.[97]	298–324
			0.0181 ± 0.0135	324.0			
Methyl hydroperoxide			6.29 ± 0.23	249	LP-LIF	Vaghjiani and Ravishankara[103]	249–423
			5.27 ± 0.13	298			
			4.97 ± 0.48	348			
	2.94 ± 0.38	-185 ± 24	4.61 ± 0.41	423			
Methyl hydroperoxide-d_1 (CH₃OOD)			1.94 ± 0.09	298	LP-LIF	Vaghjiani and Ravishankara[103]	

[a]Room temperature, not reported.

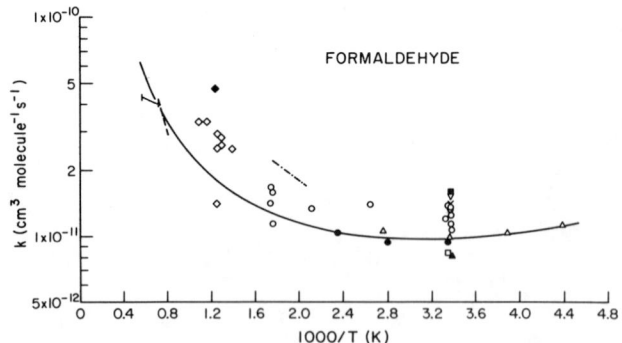

FIG. 61. Arrhenius plot of rate constants for the reaction of the OH radical with formaldehyde. ($\Diamond$) Hoare,[1,5] Blundell et al.,[3] Hoare and Peacock;[6] ($\blacklozenge$) Baldwin and Cowe;[2] (— — —) Westenberg and Fristrom;[4] (x) Morris and Niki;[8] (∇) Morris and Niki;[9] ($\vdash$) Peeters and Mahnen;[10] (–·–·–) Vandooren and Van Tiggelen;[11] ($\blacksquare$) Niki et al.;[12] ($\bullet$) Atkinson and Pitts;[13] ($\triangle$) Stief et al.;[15] ($\blacktriangle$) Temps and Wagner;[16] ($\bigcirc$) Zabarnick et al.;[17] ($\square$) Niki et al.[18] (for reaction with formaldehyde-¹³C); (———) recommendation (see text).

Based upon the data shown in Fig. 61, the Arrhenius plot exhibits significant curvature. A unit-weighted least-squares analysis of the absolute rate constants determined by Atkinson and Pitts[13] and Stief et al.[15] (which are in excellent agreement), using the expression $k = CT^2 e^{-D/T}$, yields the recommendation of

$$k\text{(formaldehyde)} = (1.25^{+0.20}_{-0.18})$$

$$\times 10^{-17} T^2 e^{(648 \pm 45)/T} \text{ cm}^3 \text{ molecule}^{-1} \text{ s}^{-1}$$

over the temperature range 228–426 K, where the indicated errors are two least-squares standard deviations, and

$$k\text{(formaldehyde)} = 9.77 \times 10^{-12} \text{ cm}^3 \text{ molecule}^{-1} \text{ s}^{-1}$$

at 298 K, with an estimated overall uncertainty at 298 K of $\pm 30\%$. This recommendation is $\sim 10\%$ higher than that recommended by Atkinson[106] of

$$k\text{(formaldehyde)} = 9.0 \times 10^{-12} \text{ cm}^3 \text{ molecule}^{-1} \text{ s}^{-1},$$

independent of temperature over the same range of 228–426 K. At elevated temperatures, the recommended expression yields calculated rate constants in good agreement with those obtained from the flame studies of Westenberg and Fristrom[4] and Peeters and Mahnen[10] (Fig. 61).

As expected, the rate constant for the reaction of OH radicals with formaldehyde-^{13}C is, within the likely experimental errors, essentially identical to that for formaldehyde-[12]C.[18] Similarly, Morris and Niki[8] determined that the room temperature rate constant for the reaction of formaldehyde-d_1 with the OH radical is essentially identical with that for the reaction of the OH radical with formaldehyde, showing that any deuterium isotope effect is small. This is consistent with the essential lack of a temperature dependence for the reaction of OH radicals with formaldehyde at around room temperature and indicates that this reaction of the OH radical with formaldehyde proceeds by an initial addition pathway (although the overall reaction involves H-atom abstraction) [compare with the kinetics of the reactions of the OH radical with diethyl ether and diethyl ether-d_{10}, see below].

This OH radical reaction with formaldehyde can proceed by the pathways

$$OH + HCHO \rightarrow HCO + H_2O \qquad (a)$$

$$\rightarrow HCOOH + H \qquad (b)$$

$$\rightarrow H + CO + H_2O \qquad (c)$$

Morrison and Heicklen,[107] Temps and Wagner[16] and Niki et al.[18] have shown from product studies that reaction pathway (b) is negligible, accounting for $\lesssim 2\%$ of the overall reaction.[18] Morrison and Heicklen[107] did not observe any formation ($<10\%$) of HCOOH, and concluded that reaction pathways (a) and (c) occur with approximately equal probability. More recently, Temps and Wagner,[16] using a discharge flow technique with LMR detection to monitor both OH and HCO radicals, have shown that reaction pathway (a) accounts for $100\pm5\%$ of the overall reaction. Thus, at room temperature the OH radical reaction with formaldehyde proceeds essentially entirely by the H-atom abstraction process.

$$OH + HCHO \rightarrow H_2O + HCO$$

(b) Acetaldehyde

The available kinetic data are given in Table 11, and those of Morris et al.,[19] Morris and Niki,[9] Niki et al.,[12] Atkinson and Pitts,[13] Kerr and Sheppard,[22] Semmes et al.[23] and Michael et al.[24] are plotted in Arrhenius form in Fig. 62. Within the cited experimental errors, the room temperature rate constants of Morris et al.,[19] Morris and Niki,[9] Niki et al.,[12] Atkinson and Pitts,[13] Kerr and Sheppard,[22] Semmes et al.[23] and Michael et al.[24] are in reason-

ably good agreement, although those of Kerr and Sheppard[22] and Semmes et al.[23] are somewhat lower than the remaining data. The Arrhenius plot (Fig. 62) does not show any definitive evidence of curvature, and hence the experimental data have been fitted to the Arrhenius expression $k = A e^{-B/T}$.

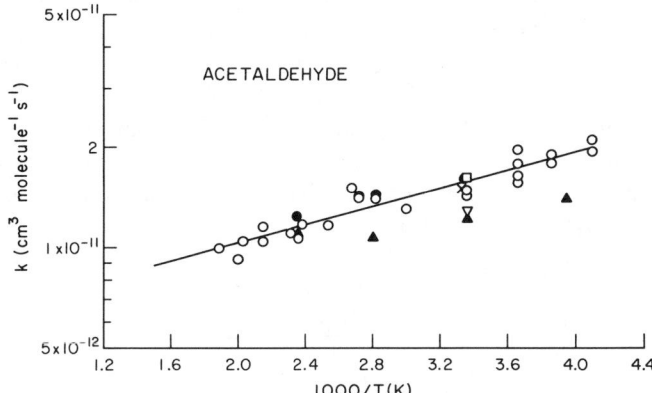

FIG. 62. Arrhenius plot of rate constants for the reaction of the OH radical with acetaldehyde. (x) Morris et al.,[19] Morris and Niki;[9] (□) Niki et al.;[12] (●) Atkinson and Pitts;[13] (▽) Kerr and Sheppard;[22] (▲) Semmes et al.;[23] (○) Michael et al.;[24] (———) recommendation (see text).

A unit-weighted least-squares analysis of the rate constants of Niki et al.,[12] Atkinson and Pitts[13] and Michael et al.[24] leads to the recommendation of

$$k(\text{acetaldehyde}) = (5.55^{+0.76}_{-0.66})$$
$$\times 10^{-12} e^{(311 \pm 42)/T} \text{ cm}^3 \text{ molecule}^{-1} \text{ s}^{-1}$$

over the temperature range 244–528 K, where the indicated errors are two least-squares standard deviations, and

$$k(\text{acetaldehyde}) = 1.58 \times 10^{-11} \text{ cm}^3 \text{ molecule}^{-1} \text{ s}^{-1}$$

at 298 K, with an estimated overall uncertainty at 298 K of $\pm20\%$.

Using the three-parameter expression $k = CT^2 e^{-D/T}$, a unit-weighted least-squares analysis of these same kinetic data[12,13,24] yields

$$k(\text{acetaldehyde}) = (6.03^{+1.07}_{-0.92})$$
$$\times 10^{-18} T^2 e^{(999 \pm 54)/T} \text{ cm}^3 \text{ molecule}^{-1} \text{ s}^{-1}$$

over the temperature range 244–528 K, where the indicated errors are again the two least-squares standard deviations, and

$$k(\text{acetaldehyde}) = 1.53 \times 10^{-11} \text{ cm}^3 \text{ molecule}^{-1} \text{ s}^{-1}$$

at 298 K. Over a wider temperature range extending to temperatures $\gtrsim 600$ K, curvature in the Arrhenius plot is expected, and the above three-parameter expression should probably then be used. Clearly, rate constants are needed for this reaction at temperatures $\gtrsim 600$ K.

The recent rate constants of Semmes et al.[23] were not included in the evaluation of this rate constant since they reported difficulties in adequately determining the acetaldehyde concentrations in their reactant mixtures.

While definite product and mechanistic data are not available for the OH radical reaction with acetaldehyde, the observation of peroxyacetyl nitrate (PAN) from the reaction of the OH radical with CH_3CHO in air in the presence of NO_x[108] shows that at room temperature this reaction must also proceed via overall H-atom abstraction from the —CHO group.

$$OH + CH_3CHO \rightarrow H_2O + CH_3\dot{C}O$$

$$CH_3\dot{C}O + O_2 \rightarrow CH_3C(O)O\dot{O}$$

$$CH_3C(O)O\dot{O} + NO_2 \rightleftharpoons CH_3C(O)OONO_2$$
$$(PAN)$$

This is consistent with the observation that the room temperature rate constants for the $\geqslant C_2$ aldehydes are reasonably similar, increasing only slightly with the length of the alkyl side chain (Table 11) and showing that the alkyl substituent group has only a minimal effect on the OH radical rate constant. As for formaldehyde, the observed negative temperature dependence suggests that, although the reaction proceeds by overall H-atom abstraction, the reaction involves initial OH radical addition followed by rapid decomposition of the adduct to the observed products. While H-atom abstraction from athe —CH_3 group is expected to be of minimal importance at room temperature,[109] this process will become of more significance at higher temperatures.[109]

(c) Glycolaldehyde

The sole kinetic study conducted to date is that of Niki et al.[25] (Table 11). From the associated product study, Niki et al.[25] determined the branching ratio for the two reaction pathways,

$$OH + HOCH_2CHO \begin{cases} \rightarrow H_2O + HOCH_2\dot{C}O & (a) \\ \rightarrow H_2O + HO\dot{C}HCHO & (b) \end{cases}$$

and rate constant ratios of $k_a/(k_a + k_b) = 0.80$ and $k_b/(k_a + k_b) = 0.20$ were obtained from the $(CHO)_2$, HCHO and CO_2 products observed in the presence of NO and air at atmospheric pressure.

(d) 1-Propanal

The available rate constants, or upper limits to the rate constants, for 1-propanal are given in Table 11. The room temperature rate constants of Morris and Niki,[9] Niki et al.,[12] Kerr and Sheppard,[22] Audley et al.[27] and Semmes et al.[23] and the upper limit to the rate constant of Kerr and Stocker[29] are in reasonable agreement. While the rate constant of Audley et al.[27] for 1-propanal agrees well with those of Kerr and Sheppard[22] and Semmes et al.[23] (and with the rate constant of Niki et al.[12]), significant discrepancies exist between the data of Audley et al.[27] and those of Kerr and Sheppard[22] and Semmes et al.[23] for the other aldehydes studied.[23,106] Accordingly, the rate constants of Audley et al.[27] were not used in the rate constant evaluations for any of the aldehydes. Hence, a unit-weighted average of the room temperature rate constants of Niki et al.,[12] Kerr and Sheppard[22] and Semmes et al.[23] yields the recommendation of

$$k(\text{1-propanal}) = 1.96 \times 10^{-11} \text{ cm}^3 \text{ molecule}^{-1} \text{ s}^{-1}$$

at 298 K, with an estimated overall uncertainty of $\pm 25\%$.

Rate constants have been derived from relative rate studies carried out at elevated temperatures by Baldwin et al.[26] and Kaiser.[28] The rate constant cited in Table 11 from the study of Kaiser[28] is an upper limit, since under the conditions employed (~ 50 Torr total pressure, mainly of O_2) the rate constant for the reaction of the OH radical with *trans*-2-butene may have been somewhat into the fall-off regime, and a fraction (measured to be $\sim 0.17 \pm 0.10$) of the 1-propanal decay rate was possibly due to loss processes other than reaction with the OH radical.[28] Based upon the discussion in Sec. 2.3 for ethene and propene, the rate constant for the reaction of the OH radical with *trans*-2-butene at 553 K and 50 Torr total pressure of O_2 is estimated to be $\sim 2.1 \times 10^{-11}$ cm^3 molecule^{-1} s^{-1} ($\sim 20\%$ below the high-pressure limit), and this yields a value of

$$k(\text{1-propanal}) \sim 1.7 \times 10^{-11} \text{ cm}^3 \text{ molecule}^{-1} \text{ s}^{-1}$$

at 553 K. The observation that the relative decay rates of propene and *trans*-2-butene were 0.45:1[28] (close to the calculated relative high-pressure limit values) further indicates that these OH radical reactions with the alkenes were not far into the fall-off region.

These data[28] suggest that the rate constant for the overall OH radical reaction with 1-propanal at ~ 550 K is similar to that at 298 K, and that the rate constant derived from the experimental study of Baldwin et al.[26] at 713 K is erroneously low. At room temperature, the major reaction pathway is by H-atom abstraction from the —CHO group[23,29,106,109]

$$OH + CH_3CH_2CHO \rightarrow H_2O + CH_3CH_2\dot{C}O$$

At elevated temperatures, H-atom abstraction from the .
—CH_2— and, at still higher temperatures, the —CH_3
groups will become significant.[109]

$$OH + CH_3CH_2CHO \rightarrow H_2O + CH_3\dot{C}HCHO$$

$$\rightarrow H_2O + \dot{C}H_2CH_2CHO$$

(e) 1-Butanal, 2-Methyl-1-propanal, 1-Pentanal and 2,2-Dimethyl-1-propanal

The available rate constants of Kerr and Sheppard,[22]
Audley *et al.*[27] and Semmes *et al.*[23] are given in Table 11,
and those of Kerr and Sheppard[22] and Semmes *et al.*[23] are
plotted in Arrhenius form in Figs. 63 to 66. At 298 K the
rate constants obtained by Kerr and Sheppard[22] and
Semmes *et al.*[23] are in reasonable agreement, especially
when the rate constants of Semmes *et al.*[23] as calculated
from their Arrhenius expressions are used.

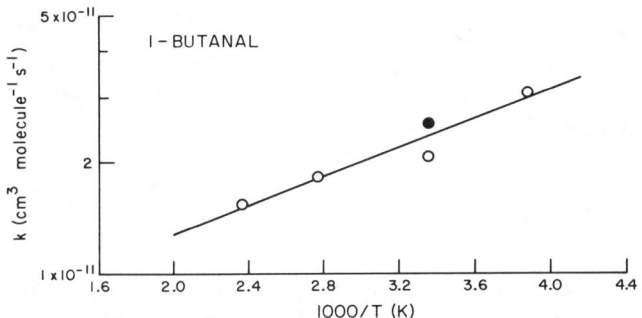

FIG. 63.　Arrhenius plot of rate constants for the reaction of the OH
radical with 1-butanal. (●) Kerr and Sheppard;[22] (○)
Semmes *et al.*;[23] (———) recommendation (see text).

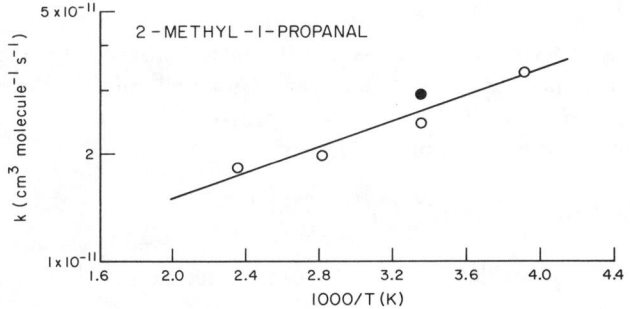

FIG. 64.　Arrhenius plot of rate constants for the reaction of the OH
radical with 2-methyl-1-propanal. (●) Kerr and Sheppard;[22]
(○) Semmes *et al.*;[23] (———) recommendation (see text).

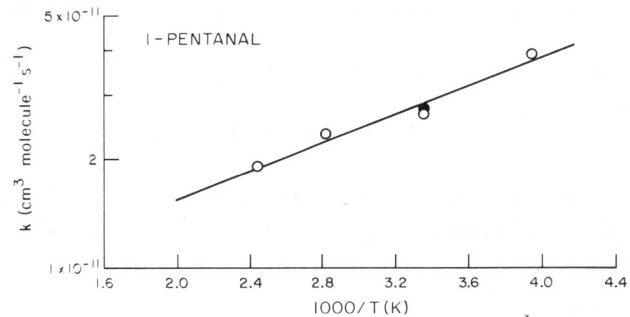

FIG. 65.　Arrhenius plot of rate constants for the reaction of the OH
radical with 1-pentanal. (●) Kerr and Sheppard;[22] (○)
Semmes *et al.*;[23] (———) recommendation (see text).

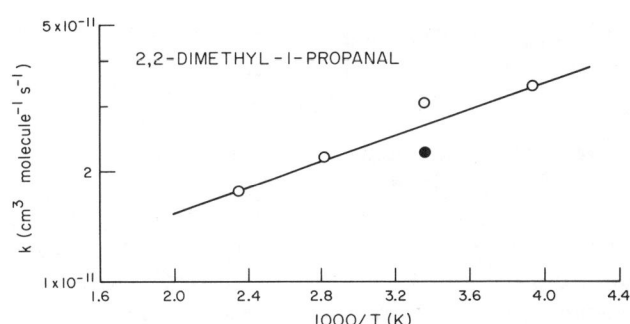

FIG. 66.　Arrhenius plot of rate constants for the reaction of the OH
radical with 2,2-dimethyl-1-propanal. (●) Kerr and Sheppard;[22] (○) Semmes *et al.*;[23] (———) recommendation (see
text).

However, for 2-methyl-1-propanal, 1-pentanal and, especially, 2,2-dimethyl-1-propanal, the data of Audley *et
al.*[27] are significantly lower, and are hence not used in
the rate constant evaluations.

Unit-weighted least-squares analyses of the rate constant data of Kerr and Sheppard[22] and Semmes *et al.*,[23]
using the Arrhenius expression (since no clear evidence
of curvature in the Arrhenius plots is evident), leads to
the recommendations of:

$$k(\text{1-butanal}) = (5.26^{+3.33}_{-2.04})$$

$$\times 10^{-12} e^{(446 \pm 154)/T} \text{ cm}^3 \text{ molecule}^{-1} \text{ s}^{-1}$$

over the temperature range 258–422 K,

$$k(\text{2-methyl-1-propanal}) = (6.61^{+4.02}_{-2.51})$$

$$\times 10^{-12} e^{(411 \pm 149)/T} \text{ cm}^3 \text{ molecule}^{-1} \text{ s}^{-1}$$

over the temperature range 255–423 K,

$$k(\text{1-pentanal}) = (6.34^{+2.12}_{-1.59})$$

$$\times 10^{-12} e^{(448 \pm 90)/T} \text{ cm}^3 \text{ molecule}^{-1} \text{ s}^{-1}$$

over the temperature range 253–410 K, and

$$k\text{(2,2-dimethyl-1-propanal)} = (6.82^{+6.72}_{-3.39})$$

$$\times\ 10^{-12}\ e^{(405\ \pm\ 214)/T}\ cm^3\ molecule^{-1}\ s^{-1}$$

over the temperature range 254–425 K, where in all cases the indicated error limits are two least-squares standard deviations, and

$$k\text{(1-butanal)} = 2.35 \times 10^{-11}\ cm^3\ molecule^{-1}\ s^{-1},$$

$$k\text{(2-methyl-1-propanal)} = 2.63$$

$$\times\ 10^{-11}\ cm^3\ molecule^{-1}\ s^{-1},$$

$$k\text{(1-pentanal)} = 2.85 \times 10^{-11}\ cm^3\ molecule^{-1}\ s^{-1}, \text{ and}$$

$$k\text{(2,2-dimethyl-1-propanal)} = 2.65$$

$$\times\ 10^{-11}\ cm^3\ molecule^{-1}\ s^{-1}$$

at 298 K, with estimated overall uncertainties at 298 K of ±30% for all four of these reactions.

At room temperature and below, these reactions are expected to proceed almost entirely by H-atom abstraction from the —CHO group,[106,109] consistent with the independence of the rate constants at 298 K on the substituent alkyl group.[23] At elevated temperatures, however, H-atom abstraction from the alkyl substituent groups will become increasingly important, and hence the above recommended Arrhenius expressions should not be used outside of the temperature ranges from which they were derived.

(f) 3-Methyl-1-butanal

Rate constants have been determined at 298 K by Kerr and Sheppard,[22] Audley et al.[27] and Semmes et al.[23] (Table 11) and, consistent with the above recommendations for the aldehydes, a unit-weighted average of the rate constants of Kerr and Sheppard[22] and Semmes et al.[23] leads to the recommendation of

$$k\text{(3-methyl-1-butanal)} = 2.74$$

$$\times\ 10^{-11}\ cm^3\ molecule^{-1}\ s^{-1}$$

at 298 K, with an estimated overall uncertainty of ±30%.

This OH radical reaction at room temperature and below is again expected to proceed mainly by H-atom abstraction from the —CHO group.

(g) Benzaldehyde

Room temperature rate constants have been determined by Niki et al.[12] and Kerr and Sheppard[22] (Table

11). These rate constants are in good agreement, and it is recommended from a unit-weighted average of these data[12,22] that

$$k\text{(benzaldehyde)} = 1.29$$

$$\times\ 10^{-11}\ cm^3\ molecule^{-1}\ s^{-1}$$

at 298 K, with an estimated overall uncertainty of ±25%.

Benzaldehyde is included in the aldehydes rather than with the aromatic compounds since it is apparent[12,106,109] that at room temperature the reaction proceeds essentially totally ($\gtrsim$90%) by overall H-atom abstraction from the —CHO group,

$$OH + C_6H_5CHO \rightarrow H_2O + C_6H_5\overset{\cdot}{C}O$$

and not by OH radical addition to the aromatic ring. This H-atom abstraction process is expected to be by far the dominant reaction pathway up to at least 1000 K.

(2) Ketones

The available kinetic data are given in Table 11. Only for acetone, 2-butanone, 2- and 3-pentanone, 2-hexanone, 4-methyl-2-pentanone and 2,6-dimethyl-4-heptanone have more than one study been carried out.

(a) Acetone

The available rate constant data of Cox et al.,[31] Zetzsch,[32] Chiorboli et al.,[33] Kerr and Stocker[34] and Wallington and Kurylo[35] are given in Table 11, and those of Zetzsch,[32] Kerr and Stocker[34] and Wallington and Kurylo[35] are plotted in Arrhenius form in Fig. 67. The rate constant reported by Chiorboli et al.[33] was obtained from irradiations of NO$_x$-organic-air mixtures, and it is possible that photolysis of acetone contributed to its removal. The upper limit to the rate constant of Cox et al.[31] is consistent with the remaining data,[32,34,35] which are in good agreement. The only temperature-dependent study is that of Wallington and Kurylo.[35]

A unit-weighted least-squares analysis of the rate constant data of Zetzsch,[32] Kerr and Stocker[34] and Wallington and Kurylo,[35] using the expression $k = CT^2e^{-D/T}$, leads to the recommendation of

$$k\text{(acetone)} = (2.13^{+0.85}_{-0.61})$$

$$\times\ 10^{-18}\ T^2\ e^{(53\ \pm\ 106)/T}\ cm^3\ molecule^{-1}\ s^{-1}$$

over the temperature range 240–440 K, where the indicated error limits are two least-squares standard deviations, and

$$k\text{(acetone)} = 2.26 \times 10^{-13}\ cm^3\ molecule^{-1}\ s^{-1}$$

at 298 K, with an estimated overall uncertainty at 298 K of $\pm 35\%$.

This reaction proceeds by H-atom abstraction from the —CH_3 groups.

$$OH + CH_3COCH_3 \rightarrow H_2O + CH_3CO\dot{C}H_2$$

The magnitude of the temperature dependence is somewhat less than may be expected by analogy with the alkanes[109] (for example, ethane, which has a similar room temperature rate constant).

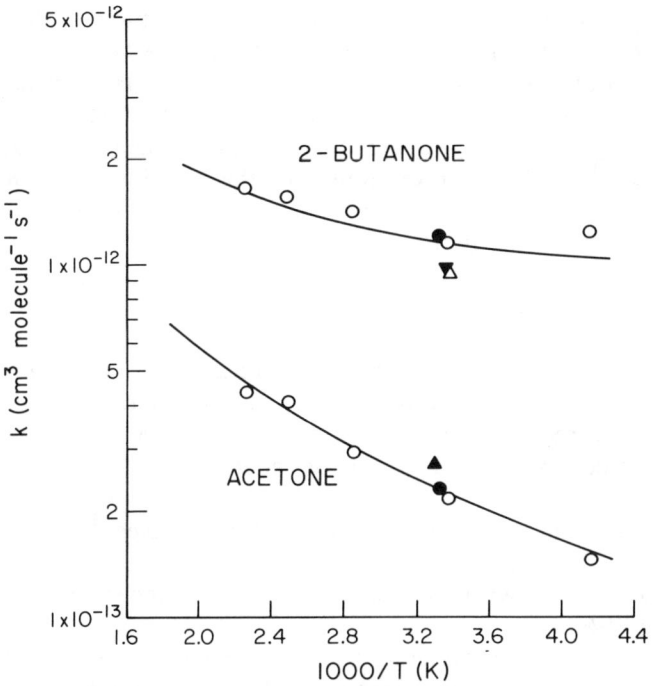

FIG. 67. Arrhenius plot of rate constants for the reactions of the OH radical with acetone and 2-butanone. (Δ) Cox *et al.*;[37] ($\bullet$) Zetzsch;[32] ($\blacktriangle$) Kerr and Stocker;[34] ($\blacktriangledown$) Edney *et al.*;[38] ($\bigcirc$) Wallington and Kurylo;[35] (———) recommendations (see text).

(b) 2-Butanone

The available rate constants of Winer *et al.*,[36] Cox *et al.*,[31,37] Zetzsch,[32] Edney *et al.*[38] and Wallington and Kurylo[35] are given in Table 11, and those of Cox *et al.*,[37] Zetzsch,[32] Edney *et al.*[38] and Wallington and Kurylo[35] are plotted in Arrhenius form in Fig. 67. The rate constants obtained from the relative rate studies of Winer *et al.*[36] and Cox *et al.*[31] are a factor of ~2–3 higher than the more recent absolute[32,35] and relative[37,38] rate data, and are not consistent with the kinetic data of Atkinson *et al.*[39] for the higher ketones.

Accordingly, a unit-weighted least-squares analysis of the rate constants of Cox *et al.*,[37] Zetzsch,[32] Edney *et al.*[38] and Wallington and Kurylo[35] (the only temperature-dependent study), using the expression $k = CT^2e^{-D/T}$, leads to the recommendation of

$$k(\text{2-butanone}) = (3.24^{+2.29}_{-1.34})$$
$$\times 10^{-18}\ T^2\ e^{(414 \pm 167)/T}\ \text{cm}^3\ \text{molecule}^{-1}\ \text{s}^{-1}$$

over the temperature range 240–440 K, where the indicated errors are two least-squares standard deviations, and

$$k(\text{2-butanone}) = 1.15 \times 10^{-12}\ \text{cm}^3\ \text{molecule}^{-1}\ \text{s}^{-1}$$

at 298 K, with an estimated overall uncertainty at 298 K of $\pm 25\%$. As for acetone, the magnitude of the temperature dependence is less than expected by analogy with the reactions of the OH radical with alkanes[109] (for example, propane, which has a similar room temperature rate constant).

The OH radical reaction with 2-butanone, as for the other ketones, occurs by H-atom abstraction from the C—H bonds:

$$OH + CH_3COCH_2CH_3 \longrightarrow$$
$$H_2O + \dot{C}H_2COCH_2CH_3 \quad (a)$$
$$H_2O + CH_3CO\dot{C}HCH_3 \quad (b)$$
$$H_2O + CH_3COCH_2\dot{C}H_2 \quad (c)$$

From a product study, Cox *et al.*[37] determined that at 295 ± 2 K the rate constant ratio $k_b/(k_a + k_b + k_c) = 0.62 \pm 0.02$, and approximate values of the fractions of the overall reaction proceeding by the three pathways (a), (b) and (c) are available from estimation methods.[109]

(c) 2-Pentanone, 3-Pentanone and 2-Hexanone

For these three ketones, rate constants have been determined by Atkinson *et al.*,[39] Wallington and Kurylo[35] and Atkinson and Aschmann[40] (Table 11). As discussed by Atkinson and Aschmann,[40] the rate constants obtained from the relative rate studies[39,40] (which are in good agreement) exhibit significant discrepancies with the room temperature absolute rate constants of Wallington and Kurylo[35] which are independent of the uncertainties associated with the rate constant for the reference organic (cyclohexane) used in the relative rate studies. Furthermore, these discrepancies are of a random nature, ranging from -25% for 2-pentanone to $+25$–35% for 3-pentanone and 2-hexanone. These data suggest that the absolute rate constants obtained by Wallington and Kurylo[35] for these, and possibly the higher, ketones were subject to significant systematic uncertainties which are not reflected in the cited error limits. From the relative rate studies of Atkinson *et al.*[39] and Atkinson and Aschmann,[40] the following 298 K rate constants are recommended,

$$k(\text{2-pentanone}) = 4.9 \times 10^{-12}\ \text{cm}^3\ \text{molecule}^{-1}\ \text{s}^{-1},$$

k(3-pentanone) = 2.0×10^{-12} cm^3 molecule^{-1} s^{-1}, and

k(2-hexanone) = 9.1×10^{-12} cm^3 molecule^{-1} s^{-1},

all with estimated overall uncertainties of $\pm30\%$.

(d) 4-Methyl-2-pentanone

The four reported room temperature rate constants of Winer et al.,[36] Cox et al.[31,37] and Atkinson et al.[39] are in good agreement (Table 11), and a unit-weighted average of the two most recent (and supposedly accurate) determinations of Cox et al.[37] and Atkinson et al.[39] yields the recommendation of

k(4-methyl-2-pentanone) = 1.41

$\times 10^{-11}$ cm^3 molecule^{-1} s^{-1}

at 298 K, with an estimated overall uncertainty of $\pm30\%$.

(e) 2,6-Dimethyl-4-heptanone

The two reported room temperature rate constants of Winer et al.[36] and Atkinson et al.[39] are in good agreement (Table 11), and the most recent and precise rate constant of Atkinson et al.[39] of

k(2,6-dimethyl-4-heptanone) = 2.75

$\times 10^{-11}$ cm^3 molecule^{-1} s^{-1}

is recommended at 298 K, with an estimated overall uncertainty of $\pm30\%$.

(f) Other Ketones

For the remaining ketones, only single studies have been carried out to date and no recommendations are made. All of these OH radical reactions with the ketones proceed by H-atom abstraction from the C—H bonds.[31,37,106] The kinetic study of Atkinson et al.[39] shows that at room temperature the carbonyl >C=O group decreases the reactivity of the C—H bonds on the α-carbon atom towards attack by the OH radical, relative to the C—H bonds in the analogous alkane, but increases the reactivity of the C—H bonds on the β-carbon atom. Those effects have been incorporated into an estimation method[109] for the calculation of the overall OH radical reaction rate constants and the contributions of the differing C—H bond abstraction pathways to the overall reaction rate constant over the temperature range ~250–1000 K.

(3) α,β-Unsaturated Carbonyls

The available rate constant data are given in Table 11. For all four of the α,β-unsaturated carbonyls investigated, two or more studies have been carried out.

(a) Acrolein

The four room temperature rate constants obtained by Maldotti et al.,[42] Kerr and Sheppard,[22] Atkinson et al.[43] and Edney et al.[38] are in reasonable agreement. A unit-weighted average of the rate constants from the three most recent studies of Kerr and Sheppard,[22] Atkinson et al.[43] and Edney et al.[38] leads to the recommendation of

k(acrolein) = 1.99×10^{-11} cm^3 molecule^{-1} s^{-1}

at 298 K, with an estimated overall uncertainty of $\pm30\%$.

This recommended room temperature rate constant is of a similar magnitude to those for the saturated aldehydes. By analogy, it is expected that the OH radical reaction with acrolein proceeds mainly by H-atom abstraction from the —CHO group, with the OH radical addition pathway being of minor importance at 298 K.[106,109,110] Thus, it is expected that this reaction will have a negative temperature dependence of $B \sim -250$ K at around room temperature.

(b) Crotonaldehyde

The room temperature rate constants of Kerr and Sheppard[22] and Atkinson et al.[43] (Table 11) are in excellent agreement, and a unit-weighted average of these rate constants yields the recommendation of

k(crotonaldehyde) = 3.6×10^{-11} cm^3 molecule^{-1} s^{-1}

at 298 K, with an estimated overall uncertainty of $\pm30\%$.

At around room temperature this reaction proceeds by H-atom abstraction from the —CHO group and OH radical addition to the >C=C< bond,

OH + CH$_3$CH=CHCHO

$\longrightarrow$ H$_2$O + CH$_3$CH=CH$\overset{\cdot}{C}$O (a)

$\longrightarrow$ CH$_3$$\overset{\cdot}{C}$HCHOHCHO and (b)
CH$_3$CHOH$\overset{\cdot}{C}$HCHO

with $k_a/(k_a + k_b)$ being estimated to be ~0.5 at 298 K.[109] By analogy with methacrolein (see below), a negative temperature dependence equivalent to $B \sim -150$ K is expected at around room temperature. At elevated temperatures $\gtrsim 500$–600 K, only the H-atom abstraction route is expected to be of importance due to thermal decomposition of the addition radicals formed in reaction pathway (b).

(c) Methacrolein

The available rate constants of Kleindienst et al.,[44] Atkinson et al.[43] and Edney et al.[38] are given in Table 11 and are plotted in Arrhenius form in Fig. 68.

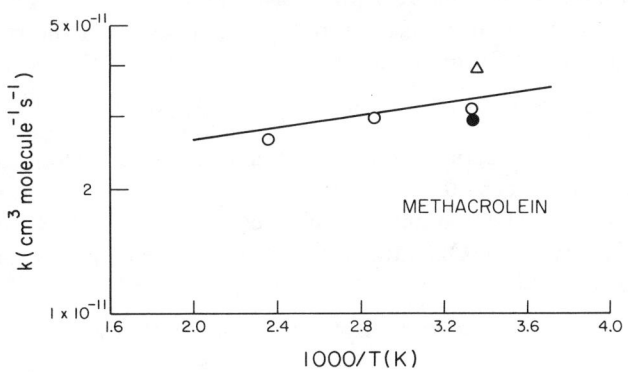

FIG. 68. Arrhenius plot of rate constants for the reaction of the OH radical with methacrolein. (○) Kleindienst et al.;[44] (●) Atkinson et al.;[43] (△) Edney et al.;[38] (———) recommendation (see text).

At room temperature the rate constants of Kleindienst et al.[44] and Atkinson et al.[43] are in excellent agreement, with that of Edney et al.[38] being ~30% higher. Since the sole temperature-dependent study is that of Kleindienst et al.,[44] a unit-weighted average of the three room temperature rate constants[38,43,44] is used in conjunction with the temperature dependence obtained from the study of Kleindienst et al.[44] to recommend that

$$k(\text{methacrolein}) = (1.86^{+0.51}_{-0.41})$$

$$\times 10^{-11} e^{(175 \pm 83)/T} \text{ cm}^3 \text{ molecule}^{-1} \text{ s}^{-1}$$

over the temperature range 298–423 K, where the indicated errors are two least-squares standard deviations, and

$$k(\text{methacrolein}) = 3.35 \times 10^{-11} \text{ cm}^3 \text{ molecule}^{-1} \text{ s}^{-1}$$

at 298 K, with an estimated overall uncertainty at 298 K of ±30%.

As for crotonaldehyde, at room temperature this OH radical reaction proceeds by H-atom abstraction from the —CHO group and OH radical addition to the >C=C< bond,

$$OH + CH_2=C(CH_3)CHO \longrightarrow$$

$$\longrightarrow H_2O + CH_2=C(CH_3)\dot{C}O \quad (a)$$

$$\longrightarrow HOCH_2\dot{C}(CH_3)CHO \text{ and } \dot{C}H_2C(OH)CHO \quad (b)$$
$$\quad\quad\quad\quad\quad CH_3$$

with $k_a/(k_a + k_b)$ being estimated to be ~0.5 at 298 K.[106,109]

(d) Methyl vinyl ketone

The available rate constants of Cox et al.,[31] Kleindienst et al.[44] and Atkinson et al.[43] are given in Table 11 and are plotted in Arrhenius form in Fig. 69.

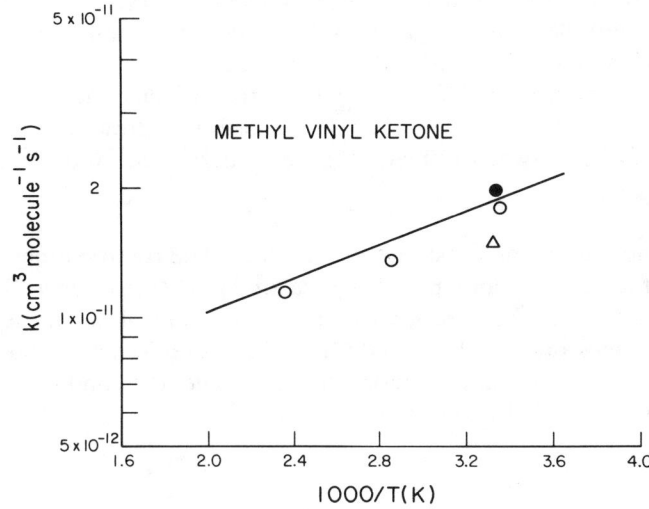

FIG. 69. Arrhenius plot of rate constants for the reaction of the OH radical with methyl vinyl ketone. (△) Cox et al.;[31] (○) Kleindienst et al.;[44] (●) Atkinson et al.;[43] (———) recommendation (see text).

Again, the room temperature rate constants of Kleindienst et al.[44] and Atkinson et al.[43] are in good agreement, with that of Cox et al.[31] being ~20% lower. From a unit-weighted average of the room temperature rate constants of Kleindienst et al.[44] and Atkinson et al.[43] and the temperature dependence of Kleindienst et al.,[44] it is recommended that

$$k(\text{methyl vinyl ketone}) = (4.13^{+1.88}_{-1.30})$$

$$\times 10^{-12} e^{(452 \pm 130)/T} \text{ cm}^3 \text{ molecule}^{-1} \text{ s}^{-1}$$

over the temperature range 298–424 K, where the indicated errors are two least-squares standard deviations, and

$$k(\text{methyl vinyl ketone}) = 1.88$$

$$\times 10^{-11} \text{ cm}^3 \text{ molecule}^{-1} \text{ s}^{-1}$$

at 298 K, with an estimated overall uncertainty at 298 K of ±30%.

This OH radical reaction will proceed essentially totally by OH radical addition to the >C=C< bond at temperatures ≲500 K, with the rate constant for this reaction being at, or close to, the high-pressure limit at total pressures ≳50 Torr. At 298 ± 2 K, the $CH_3COCHCH_2OH$ and $CH_3COCHOHCH_2$ radicals are formed in an ~70%:30% distribution.[111]

$$OH + CH_3COCH=CH_2 \rightarrow CH_3COCHCH_2OH \text{ and}$$

$$CH_3COC\overset{\cdot}{H}OHCH_2$$

(4) Ketenes

The available kinetic data for the reactions of the OH radical with ketene, methylketene, ethylketene and dimethylketene are given in Table 11. No recommendations are made. The rate constant for ketene derived by Vandooren and Van Tiggelen[11] from flame measurements was based upon the measured formation of HCHO. It was assumed that the reaction occurred by

$$OH + CH_2CO \rightarrow HCHO + HCO$$

and was hence not a direct rate study. The room temperature kinetic and product data of Hatakeyama et al.[46] show that these reactions of the OH radical with the ketenes proceed by initial OH radical addition, with the ultimate formation (under the experimental conditions employed) of carbonyl compounds.[46]

$$OH + R_1R_2CCO \longrightarrow \begin{bmatrix} OH \\ | \quad \cdot \\ R_1R_2CCO \end{bmatrix}$$

followed by either

$$\begin{bmatrix} OH \\ | \\ R_1R_2C\overset{\cdot}{C}O \end{bmatrix} \longrightarrow R_1R_2\overset{\cdot}{C}OH + CO$$

$$\downarrow O_2$$

$$R_1R_2CO + HO_2$$

or

$$\begin{bmatrix} OH \\ | \\ R_1R_2C\overset{\cdot}{C}O \end{bmatrix} \longrightarrow R_1R_2CO + HCO$$

in a direct reaction. In the presence of O_2 these reactions yield identical products, since HCO reacts rapidly with O_2 to form HO_2 and CO.[105]

(5) Dicarbonyls

Rate constants have been measured for the α-dicarbonyls glyoxal,[47,48] methylglyoxal,[44,47] and 2,3-butanedione (biacetyl)[41,49] and for pentane-1,5-dial,[50] 2,4-pentanedione[41] and 2,5-hexanedione.[41]

(a) Glyoxal

The available rate constant data of Plum et al.[47] and Becker and Klein[48] are given in Table 11. Both of these studies were relative rate measurements carried out at room temperature. The agreement is good, and the rate constant of Plum et al.[47] of

$$k(\text{glyoxal}) = 1.14 \times 10^{-11} \text{ cm}^3 \text{ molecule}^{-1} \text{ s}^{-1}$$

at 298 K is recommended, with an estimated overall uncertainty of $\pm 40\%$.

As for the aliphatic aldehydes such as HCHO and CH_3CHO, the OH radical reaction proceeds by H-atom abstraction,

$$OH + (CHO)_2 \rightarrow H_2O + HCO\overset{\cdot}{C}O$$

and the subsequent reactions of the HCOCO radical (involving decomposition and reaction with O_2) have been investigated at room temperature by Niki et al.[112] in the presence of 700 Torr of $O_2 + N_2$ diluent.

(b) Methylglyoxal

The available rate constants of Kleindienst et al.[44] and Plum et al.[47] are given in Table 11. The rate constant derived from the relative rate study of Plum et al.[47] is a factor of 2.4 higher than the absolute value of Kleindienst et al.[44] It is possible that the methylglyoxal sample prepared and used by Kleindienst et al.[44] contained a significant amount of non-reactive impurities (such as CO and CO_2), and the rate constant of Plum et al.[47] of

$$k(\text{methylglyoxal}) = 1.72 \times 10^{-11} \text{ cm}^3 \text{ molecule}^{-1} \text{ s}^{-1}$$

at 298 K is recommended, with an estimated overall uncertainty of $\pm 40\%$.

At around room temperature this reaction will proceed by H-atom abstraction from the —CHO group

$$OH + CH_3COCHO \rightarrow H_2O + CH_3C\overset{\cdot}{O}O$$

By analogy with the $HCO\overset{\cdot}{C}O$ radical formed from glyoxal, the resulting CH_3COCO radical is expected to decompose or react with O_2, with decomposition (to CH_3CO and CO) being expected to dominate at ~ 298 K and atmospheric pressure of air. At elevated temperatures H-atom abstraction from the —CH_3 group will become significant.[109]

(c) 2,3-Butanedione

The rate constants of Darnall et al.[49] and Dagaut et al.[41] are given in Table 11 and are plotted in Arrhenius form in Fig. 70. At room temperature the measured rate constants are in excellent agreement. A unit-weighted least-squares analysis of the kinetic data of Darnall et al.[49] and Dagaut et al.,[41] using the expression $k = CT^2 e^{-D/T}$, leads to the recommendation of

$$k(\text{2,3-butanedione}) = (1.40^{+0.59}_{-0.42})$$

$$\times 10^{-18} T^2 e^{(194 \pm 112)/T} \text{ cm}^3 \text{ molecule}^{-1} \text{ s}^{-1}$$

over the temperature range 240–440 K, where the indicated errors are two least-squares standard deviations, and

$$k(\text{2,3-butanedione}) = 2.38 \times 10^{-13} \text{ cm}^3 \text{ molecule}^{-1} \text{ s}^{-1}$$

at 298 K, with an estimated overall uncertainty at 298 K of ±35%.

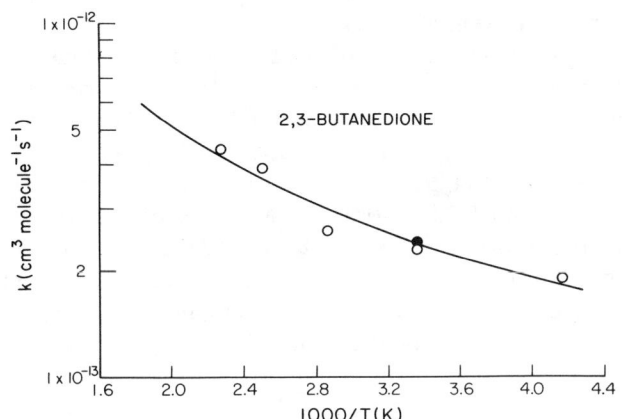

FIG. 70. Arrhenius plot of rate constants for the reaction of the OH radical with 2,3-butanedione (biacetyl). (●) Darnall et al.;[49] (○) Dagaut et al.;[41] (———) recommendation (see text).

This OH radical reaction proceeds by H-atom abstraction from the —CH₃ groups

$$\text{OH} + \text{CH}_3\text{COCOCH}_3 \rightarrow \text{H}_2\text{O} + \text{CH}_3\text{COCOCH}_2$$

At room temperature this rate constant is similar to those for the reactions of the OH radical with ethane and acetone and, as for acetone, the magnitude of the temperature dependence appears somewhat low when compared to the alkanes (for example, ethane for which $C = 1.42 \times 10^{-17}$ cm³ molecule⁻¹ s⁻¹ and $D = 462$ K). Accordingly, until further temperature-dependent studies are carried out, the above recommended expression should be used with caution outside of the temperature range ~240–440 K.

(6) Unsaturated 1,4-Dicarbonyls

The sole kinetic data for this class of organic compounds arise from the room temperature relative rate study of Tuazon et al.[51] for cis- and trans-3-hexene-2,5-dione. At around room temperature these reactions are expected to proceed essentially entirely by OH radical addition to the >C=C< bond,

$$\text{OH} + \text{CH}_3\text{COCH}=\text{CHCOCH}_3 \rightarrow$$

$$\text{CH}_3\text{COCHOHCHCOCH}_3$$

with the rate constant being at, or very close to, the high-pressure limit under atmospheric conditions.

(7) Alcohols and Glycols

(a) Methanol and Methanol-d₃(CD₃OH)

The available rate constants of Campbell et al.,[52] Overend and Paraskevopoulos,[53] Ravishankara and Davis,[54] Vandooren and Van Tiggelen,[55] Barnes et al.,[56] Hägele et al.,[57] Tuazon et al.,[58] Meier et al.,[59,60] Zetzsch,[61] Greenhill and O'Grady,[62] Wallington and Kurylo,[63] Hess and Tully[64] and McCaulley et al.[65] are given in Table 11 and those of Campbell et al.,[52] Overend and Paraskevopoulos,[53] Ravishankara and Davis,[54] Vandooren and Van Tiggelen,[55] Barnes et al.,[56] Hägele et al.,[57] Tuazon et al.,[58] Meier et al.,[59,60] Greenhill and O'Grady,[62] Wallington and Kurylo[63] and Hess and Tully[64] for methanol are plotted in Arrhenius form in Fig. 71.

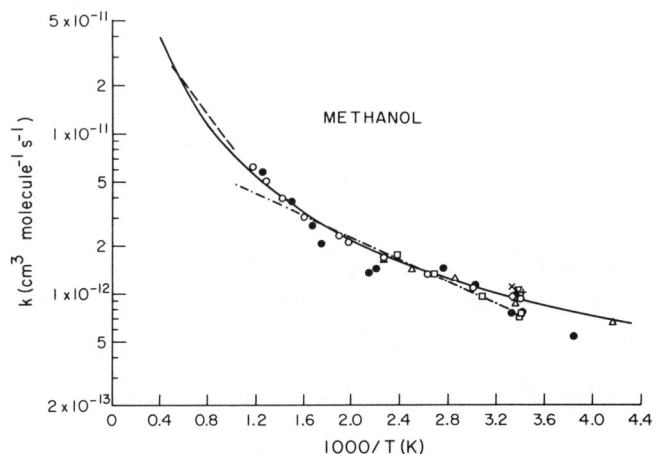

FIG. 71. Arrhenius plot of rate constants for the reaction of the OH radical with methanol. (+) Campbell et al.;[52] (▽) Overend and Paraskevopoulos;[53] (▲) Ravishankara and Davis;[54] (———) Vandooren and Van Tiggelen;[55] (x) Barnes et al.;[56] (□) Hägele et al.;[57] (◇) Tuazon et al.;[58] (–··–) Meier et al.;[59,60] (●) Greenhill and O'Grady;[62] (△) Wallington and Kurylo;[63] (○) Hess and Tully;[64] (———) recommendation (see text).

In addition to the rate constants given in Table 11, Osif et al.[113] also determined rate constants for methanol at 298 and 345 K, relative to those for the reaction of OH radicals with CO, at total pressures of 28–203 Torr of CH₃OH + N₂O + CO. While no quantitative estimate of this reference reaction rate constant can be made, a lower limit of 1.5×10^{-13} cm³ molecule⁻¹ s⁻¹ at 298 K is applicable, and the data of Osif et al.[113] yield a rate constant of $\geq (9.5 \pm 1.5) \times 10^{-14}$ cm³ molecule⁻¹ s⁻¹ at 298 K, with a likely upper limit at this temperature of

$\lesssim 2 \times 10^{-13}$ cm^3 molecule^{-1} s^{-1}. Since this rate constant is lower by a factor of ~ 4–10 than the room temperature rate constants listed in Table 11, the data of Osif et al.[113] are neglected in the evaluation of the rate constant.

In general, the agreement between these studies is reasonably good, although the rate constants determined by Greenhill and O'Grady[62] exhibit a significant degree of scatter and, together with the rate data of Hägele et al.[57] and Zetzsch,[61] are substantially lower at ~ 290–300 K than the rate constants from the remaining studies. Of particular concern for atmospheric purposes is the disagreement at temperatures < 290 K between the measurement of Greenhill and O'Grady[62] and that of Wallington and Kurylo.[63]

The data from the studies of Vandooren and Van Tiggelen[55] and Meier et al.[59,60] cannot be used in the evaluation because the rate constants at the specific temperatures studied were not tabulated. The absolute rate constants of Overend and Paraskevopoulos,[53] Ravishankara and Davis,[54] Wallington and Kurylo[63] and Hess and Tully[64] and the rate constant from the relative rate study of Tuazon et al.[58] have been used to evaluate the rate constant for the methanol reaction. A unit-weighted least-squares analysis of these data,[53,54,58,63,64] using the expression $k = CT^2 e^{-D/T}$, yields the recommendation of

$$k(\text{methanol}) = (6.39^{+0.60}_{-0.54})$$

$$\times 10^{-18} T^2 e^{(148 \pm 33)/T} \text{ cm}^3 \text{ molecule}^{-1} \text{ s}^{-1}$$

over the temperature range 240–866 K, where the indicated errors are two least-squares standard deviations, and

$$k(\text{methanol}) = 9.32 \times 10^{-13} \text{ cm}^3 \text{ molecule}^{-1} \text{ s}^{-1}$$

at 298 K, with an estimated overall uncertainty at 298 K of $\pm 25\%$.

The data of Vandooren and Van Tiggelen[55] at 1000–2000 K are in excellent agreement with this recommendation (Fig. 71), suggesting that the recommended expression can be used with some confidence up to ~ 2000 K. However, it is clear that further rate constant data are needed at temperatures $\lesssim 290$ K.

The reaction of the OH radical with methanol proceeds by H-atom abstraction, from either the —CH$_3$ group or the —OH group.

$$\text{OH} + \text{CH}_3\text{OH} \longrightarrow \begin{cases} \text{H}_2\text{O} + \text{CH}_3\dot{\text{O}} \quad \text{(a)} \\ \\ \text{H}_2\text{O} + \dot{\text{C}}\text{H}_2\text{OH} \quad \text{(b)} \end{cases}$$

Based upon the C—H and O—H bond dissociation energies of 94.1 kcal mol^{-1} and 104.5 kcal mol^{-1}, respectively,[114] reaction pathway (b) would be expected to

totally dominate at room temperature and below. The room temperature rate constants of Greenhill and O'Grady,[62] Hess and Tully[64] and McCaulley et al.[65] for CD$_3$OH are in reasonable agreement, and the deuterium isotope effect observed by Hess and Tully[64] of

$$k(\text{CH}_3\text{OH})/k(\text{CD}_3\text{OH}) = k^{\text{H}}/k^{\text{D}} = 460 \, T^{-0.83} e^{-198/T}$$

(equivalent to $k^{\text{H}}/k^{\text{D}} = 1.15 \, e^{217/T}$ centered at 500 K) is consistent (Section 2.1) with H- (or D-) atom abstraction from the —CH$_3$ (or —CD$_3$) group dominating over the temperature range studied.

However, Hägele et al.[57] and Meier et al.[59,60] have experimentally investigated the relative importance of pathways (a) and (b), and derived, from LIF measurements of the CH$_3$O radical, ratios of $k_a/(k_a + k_b)$ at ~ 298 K of 0.11 ± 0.03[57] (increasing to 0.22 ± 0.07 at 393 K[57]) and 0.25 ± 0.08.[59,60] Meier et al.[59] also determined a ratio of $k_b/(k_a + k_b)$ of 0.83 ± 0.13 at room temperature using mass spectrometry. From their kinetic measurements on the various methanol isotopes, McCaulley et al.[65] derived a ratio of

$$k_a/(k_a + k_b) = 0.15 \pm 0.08 \text{ at } 298 \pm 2 \text{ K.}$$

Thus, for CH$_3$OH H-atom abstraction occurs to a significant extent from both the C—H and O—H bonds, with abstraction from the stronger O—H bonds increasing in importance with increasing temperature.

(b) Ethanol

The rate constant data of Campbell et al.,[52] Overend and Paraskevopoulos,[53] Ravishankara and Davis,[54] Cox and Goldstone,[66] Meier et al.,[68] Lorenz et al.,[69] Kerr and Stocker,[34] Greenhill and O'Grady,[62] Wallington and Kurylo[63] and Hess and Tully[70] are given in Table 11 and are plotted in Arrhenius form in Fig. 72. Clearly, there is a large amount of scatter between the various studies, with the rate constants of Ravishankara and Davis[54] and Meier et al.[68] being lower than the data from the remaining studies.

As recognized and experimentally demonstrated by Hess and Tully,[70] kinetic studies which involve monitoring the decay rates of the OH radical in the presence of C$_2$H$_5$OH are subject to regeneration of the OH radical at elevated temperatures from the reaction pathway involving H-atom abstraction from the —CH$_3$ group, leading to erroneously low measured rate constants. The three possible reaction pathways are,

$$\text{OH} + \text{CH}_3\text{CH}_2\text{OH} \longrightarrow \begin{cases} \text{H}_2\text{O} + \dot{\text{C}}\text{H}_2\text{CH}_2\text{OH} \quad \text{(a)} \\ \text{H}_2\text{O} + \text{CH}_3\dot{\text{C}}\text{HOH} \quad \text{(b)} \\ \text{H}_2\text{O} + \text{CH}_3\text{CH}_2\dot{\text{O}} \quad \text{(c)} \end{cases}$$

and the CH_2CH_2OH radical formed in pathway (a) is identical to that formed from the addition reaction of the OH radical to ethene. As discussed in Section 2.3 above, the thermalized CH_2CH_2OH radical thermally decomposes to the OH radical and ethene at a significant rate at temperatures above ~ 450–500 K. Thus, in absolute studies employing ^{16}OH and ^{16}O-ethanol, at temperatures $\lesssim 450$–500 K all three channels are observed and the measured rate constant is $k_{obs} = k_a + k_b + k_c$. At temperatures $\gtrsim 500$ K where thermal decomposition of the CH_2CH_2OH radical is sufficiently rapid, only channels (b) and (c) are observed, with $k_{obs} = k_b + k_c$. At intermediate temperatures (~ 450–700 K, depending on the experimental conditions employed), bi-exponential OH radical decays should be observed.

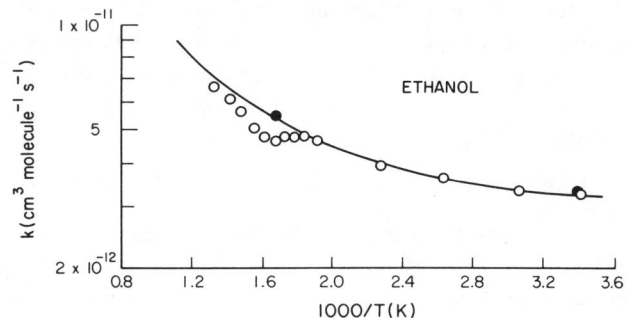

FIG. 73. Arrhenius plot of rate constants obtained by Hess and Tully[70] for the reactions of ^{16}OH and ^{18}OH radicals with ethanol. (○) ^{16}OH radical reaction; (●) ^{18}OH radical reaction; (——) recommendation for the overall reaction rate constant, $k = k_a + k_b + k_c$ (see text).

Accordingly, the rate constant data of Hess and Tully[70] have been used in the recommendation of the rate constant for this reaction, using only the ^{18}OH radical rate constant data above 500 K. A unit-weighted least-squares analysis of these data,[70] using the expression $k = CT^2 e^{-D/T}$, yields the recommendation of

$$k(\text{ethanol}) = (6.18^{+0.36}_{-0.34})$$
$$\times 10^{-18} T^2 e^{(532 \pm 21)/T} \text{ cm}^3 \text{ molecule}^{-1} \text{ s}^{-1}$$

over the temperature range 293–599 K, where the indicated errors are two least-squares standard deviations, and

$$k(\text{ethanol}) = 3.27 \times 10^{-12} \text{ cm}^3 \text{ molecule}^{-1} \text{ s}^{-1}$$

at 298 K, with an estimated overall uncertainty at 298 K of $\pm 20\%$.

The rate constants of Greenhill and O'Grady[62] are somewhat higher than the recommendation, while those of Lorenz et al.[69] (obtained at $\leqslant 500$ K) and Wallington and Kurylo[63] are in reasonable agreement. It is possible that the decrease in the measured rate constants of Lorenz et al.[69] as the temperature was raised above 525 K was due to the changeover from observing all of the reaction channels to observing only channels (b) and (c).

The sole direct product study carried out to date is that of Meier et al.,[60,67] in which mass spectrometry was used to show that reaction channel (b) accounted for 75 $\pm$ 15% of the overall reaction at 300 K. As noted above, the kinetic data of Hess and Tully[70] indicate that channel (a) accounts for $\sim 15\%$ of the overall reaction at 600 K. At combustion temperatures (indeed, above ~ 500 K), channel (a) in effect proceeds by

$$OH + CH_3CH_2OH \rightarrow H_2O + OH + C_2H_4$$

(c) 1-Propanol

Rate constants have been determined at room temperature by Campbell et al.,[52] Overend and Paraskevopou-

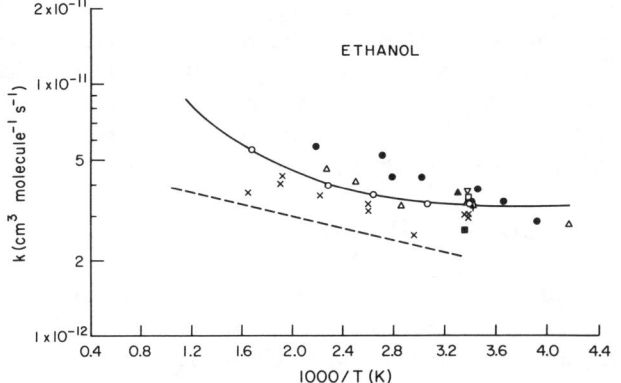

FIG. 72. Arrhenius plot of rate constants for the reaction of the OH radical with ethanol. (+) Campbell et al.;[52] (∇) Overend and Paraskevopoulos;[53] (■) Ravishankara and Davis;[54] (□) Cox and Goldstone;[66] (— — —) Meier et al.;[68] (x) Lorenz et al.;[69] (▲) Kerr and Stocker;[34] (●) Greenhill and O'Grady;[62] (△) Wallington and Kurylo;[63] (○) Hess and Tully,[70] rate constant for ^{18}OH reaction only plotted for temperatures >441 K; (——) recommendation for overall reaction rate constant (see text).

The recent absolute study of Hess and Tully[70] confirms these expectations.[70] As shown by the Arrhenius plot in Fig. 73, the measured rate constants for the reaction of the ^{16}OH radical with ethanol exhibit a plateau region at ~ 520–600 K. Moreover, the measured rate constant for the reaction of the ^{18}OH radical with ethanol [which is not subject to regeneration of the ^{18}OH radical from the thermal decomposition of the $CH_2CH_2^{16}OH$ radical formed in pathway (a)] at 599 K is $\sim 15\%$ higher than that measured for the ^{16}OH reaction. These data indicate that pathway (a) accounts for $\sim 15\%$ of the overall reaction at ~ 600 K (which agrees well with the calculated value of 20% from the estimation procedure of Atkinson[109]), and show that the measured rate constants for the reaction of the ^{16}OH radical with ^{16}O-ethanol at temperatures $\gtrsim 500$ K cannot be used to derive the overall rate constant for this reaction.

los[53] and Wallington and Kurylo[63] (Table 11). The absolute rate constants of Overend and Paraskevopoulos[53] and Wallington and Kurylo[63] are in excellent agreement, and it is recommended from a unit-weighted average of these data[53,63] that

$$k(1\text{-propanol}) = 5.34 \times 10^{-12} \text{ cm}^3 \text{ molecule}^{-1} \text{ s}^{-1}$$

at 298 K, with an estimated overall uncertainty of ±40%. (The rate constant of Campbell et al.[52] was not used in the evaluation because of questions concerning the validity of the experimental technique used[106]). Consistent with the discussion above for ethanol, H-atom abstraction from the β —CH$_2$— group will lead to the formation of the radical CH$_3$CHCH$_2$OH, identical to that formed by terminal OH radical addition to propene. This radical will thermally decompose to regenerate the OH radical above ~500 K and hence this reaction channel will not be observed at temperatures $\gtrsim$500 K in absolute techniques monitoring the decay rates of the ^{16}OH radical.

(d) 2-Propanol

The available rate constant data of Lloyd et al.,[71] Overend and Paraskevopoulos,[53] Klöpffer et al.[72] and Wallington and Kurylo[63] are given in Table 11 and are plotted in Arrhenius form in Fig. 74.

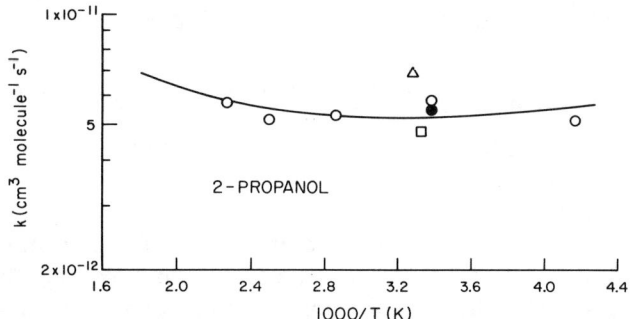

FIG. 74. Arrhenius plot of rate constants for the reaction of the OH radical with 2-propanol. (△) Lloyd et al.;[71] (●) Overend and Paraskevopoulos;[53] (□) Klöpffer et al.;[72] (○) Wallington and Kurylo;[63] (———) recommendation (see text).

At 296 K, the absolute rate constants of Overend and Paraskevopoulos[53] and Wallington and Kurylo[63] are in good agreement. A unit-weighted least-squares analysis of the data of Overend and Paraskevopoulos[53] and Wallington and Kurylo,[63] using the expression $k = CT^2e^{-D/T}$, yields the recommendation of

$$k(2\text{-propanol}) = (7.32^{+2.89}_{-2.07})$$

$$\times 10^{-18} T^2 e^{(620 \pm 106)/T} \text{ cm}^3 \text{ molecule}^{-1} \text{ s}^{-1}$$

over the temperature range 240–440 K, where the indi-

cated errors are two least-squares standard deviations, and

$$k(2\text{-propanol}) = 5.21 \times 10^{-12} \text{ cm}^3 \text{ molecule}^{-1} \text{ s}^{-1}$$

at 298 K, with an estimated overall uncertainty at 298 K of ±40%.

At around room temperature and below, the dominant reaction pathway will be H-atom abstraction from the tertiary C—H bond:

$$OH + (CH_3)_2CHOH \rightarrow H_2O + CH_3\overset{\cdot}{C}(OH)CH_3$$

At elevated temperatures H-atom abstraction from the —CH$_3$ groups will become significant,

$$OH + (CH_3)_2CHOH \rightarrow H_2O + CH_3CHOH\overset{\cdot}{C}H_2$$

leading to the radical also formed by OH radical addition to propene. At temperatures $\gtrsim$500 K this radical will thermally decompose to regenerate the OH radical together with propene,[70] and hence under these conditions this reaction channel becomes an OH radical catalyzed conversion of the alcohol to the alkene.[70]

$$OH + (CH_3)_2CHOH \rightarrow H_2O + OH + CH_3CH=CH_2$$

Hess and Tully[70] have confirmed the occurrence of this reaction process from kinetic studies of the reactions of ^{16}OH and ^{18}OH radicals with (CH$_3$)$_2$CH^{16}OH.

(e) 2-Methyl-2-propanol (tert-butyl alcohol)

The rate constants of Cox and Goldstone[66] and Wallington et al.[73] are given in Table 11 and are plotted in Arrhenius form in Fig. 75.

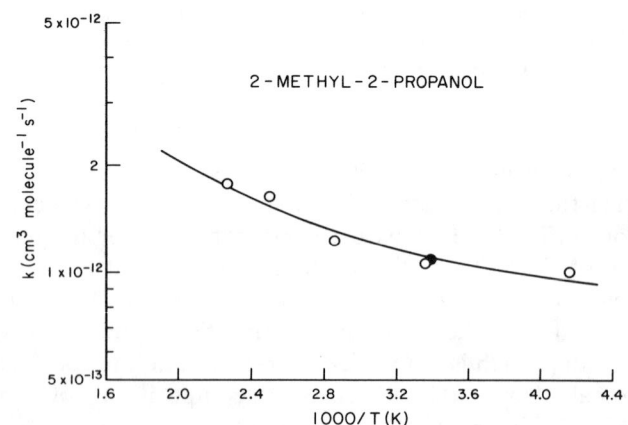

FIG. 75. Arrhenius plot of rate constants for the reaction of the OH radical with 2-methyl-2-propanol. (●) Cox and Goldstone;[66] (○) Wallington et al.;[73] (———) recommendation (see text).

The room temperature rate constants from these studies are in excellent agreement, and a unit-weighted least-squares analysis of these data,[66,73] using the expression $k = CT^2 e^{-D/T}$, leads to the recommendation of

$$k(\text{2-methyl-2-propanol}) = (4.29^{+1.24}_{-0.97})$$
$$\times 10^{-18} \, T^2 \, e^{(322 \pm 81)/T} \, \text{cm}^3 \, \text{molecule}^{-1} \, \text{s}^{-1}$$

over the temperature range 240–440 K, where the indicated errors are two least-squares standard deviations, and

$$k(\text{2-methyl-2-propanol}) = 1.12$$
$$\times 10^{-12} \, \text{cm}^3 \, \text{molecule}^{-1} \, \text{s}^{-1}$$

at 298 K, with an estimated overall uncertainty at 298 K of $\pm 40\%$.

The two reaction channels involve H-atom abstraction

$$\text{OH} + (\text{CH}_3)_3\text{COH} \longrightarrow \begin{cases} \text{H}_2\text{O} + (\text{CH}_3)_3\text{C\overset{\bullet}{O}} & \text{(a)} \\[2ex] \text{H}_2\text{O} + \overset{\bullet}{\text{C}}\text{H}_2\overset{\overset{\text{OH}}{|}}{\text{C}}(\text{CH}_3)_2 & \text{(b)} \end{cases}$$

It is expected that the major reaction pathway will be (b).[109] The radical formed from this reaction channel is that also formed from the addition reaction of the OH radical with 2-methylpropene and, as discussed above, this radical will thermally decompose to the OH radical and 2-methylpropene at temperatures $\gtrsim 500$ K. Hence, at these temperatures, reaction channel (b) will proceed by the overall reaction

$$\text{OH} + (\text{CH}_3)_3\text{COH} \rightarrow \text{H}_2\text{O} + \text{OH} + (\text{CH}_3)_2\text{C}=\text{CH}_2$$

to regenerate the OH radical.

(f) Other Alcohols and Glycols, Hydroxyethers and Ketoethers

The available kinetic data for a number of alcohols and glycols not dealt with above are given in Table 11. Apart from 1-butanol, 2-ethoxyethanol and 2-butoxyethanol these data were obtained from single studies (for 1-butanol one of the two studies was that of Campbell et al.,[52] which is not used in the evaluations). No recommendations are made for these alcohols and glycols.

(8) Ethers and Cycloethers

The available kinetic data are given in Table 11, and it can be seen that studies have been carried out by more than one research group for dimethyl ether, diethyl ether, di-n-propyl ether, methyl t-butyl ether, ethyl n-butyl ether, ethyl t-butyl ether, di-n-butyl ether, tetrahydrofuran and furan. The data for these compounds are discussed as follows.

(a) Dimethyl ether

The available rate constant data of Perry et al.,[79] Tully and Droege[80] and Wallington et al.,[81] all obtained using

flash or laser photolysis techniques, are given in Table 11 and are plotted in Arrhenius form in Fig. 76.

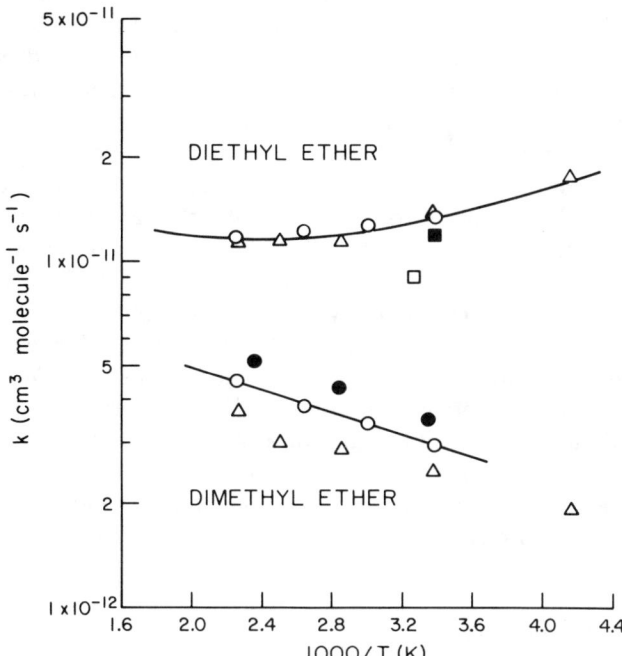

FIG. 76. Arrhenius plots of rate constants for the reactions of the OH radical with dimethyl ether and diethyl ether. ($\square$) Lloyd et al.;[71] ($\bullet$) Perry et al.;[79] ($\bigcirc$) Tully and Droege;[80] ($\triangle$) Wallington et al.;[81] ($\blacksquare$) Bennett and Kerr;[82] (———) recommendations (see text).

The rate constants determined by Perry et al.[79] are uniformly $\sim 15\%$ higher than those of Tully and Droege[80] over the entire temperature range studied, while those of Wallington et al.[81] are $\sim 20\%$ lower over the temperature range common to both studies. This implies the existence of systematic errors in at least two of these studies. The rate constants of Tully and Droege,[80] which lie in between those of the other studies,[79,81] are used to recommend, from a unit-weighted least-squares analysis, that

$$k(\text{dimethyl ether}) = (1.04^{+0.13}_{-0.11})$$
$$\times 10^{-11} \, e^{-(372 \pm 39)/T} \, \text{cm}^3 \, \text{molecule}^{-1} \, \text{s}^{-1}$$

over the temperature range 295–442 K, where the indicated errors are two least-squares standard deviations, and

$$k(\text{dimethyl ether}) = 2.98 \times 10^{-12} \, \text{cm}^3 \, \text{molecule}^{-1} \, \text{s}^{-1}$$

at 298 K, with an estimated overall uncertainty at 298 K of $\pm 25\%$. This recommended Arrhenius expression is applicable only over the temperature range cited, since it is expected that non-Arrhenius behavior will be observed over a wider temperature range.[109] This reaction proceeds by H-atom abstraction from the —CH$_3$ groups.

$$\text{OH} + \text{CH}_3\text{OCH}_3 \rightarrow \text{H}_2\text{O} + \text{CH}_3\overset{\bullet}{\text{O}}\text{CH}_2$$

(b) Diethyl ether and Diethyl ether-d₁₀

The rate constants for diethyl ether obtained by Lloyd et al.,[71] Tully and Droege,[80] Wallington et al.[81] and Bennett and Kerr[82] are given in Table 11 and are plotted in Arrhenius form in Fig. 76. In this case, the absolute rate constants of Tully and Droege[80] and Wallington et al.[81] are in good agreement and agree reasonably well with the room temperature rate constant of Bennett and Kerr.[82] However, at room temperature the rate constants from these studies[80-82] are ~30–50% higher than the rate constant derived from the relative rate study of Lloyd et al.[71] While the data of Tully and Droege,[80] obtained over the temperature range 295–442 K, show no evidence of non-Arrhenius behavior, the combined data set of Tully and Droege[80] and Wallington et al.[81] do suggest that the Arrhenius plot exhibits curvature. A unit-weighted least-squares analysis of the rate constants of Tully and Droege,[80] Wallington et al.[81] and Bennett and Kerr,[82] using the expression $k = CT^2 e^{-D/T}$, yields the recommendation of

$$k(\text{diethyl ether}) = (8.80^{+1.73}_{-1.46})$$

$$\times 10^{-18} \ T^2 \ e^{(844 \pm 60)/T} \ \text{cm}^3 \ \text{molecule}^{-1} \ \text{s}^{-1}$$

over the temperature range 240–442 K, where the indicated errors are two least-squares standard deviations, and

$$k(\text{diethyl ether}) = 1.33 \times 10^{-11} \ \text{cm}^3 \ \text{molecule}^{-1} \ \text{s}^{-1}$$

at 298 K, with an estimated overall uncertainty at 298 K of ±25%.

The kinetic data of Tully[83] for diethyl ether-d₁₀ (Table 11) show a significant deuterium isotope effect of k(diethyl ether)/k(diethyl ether-d₁₀) = 1.7–2.0 over the temperature range 295–440 K, showing that these OH radical reactions with diethyl ether and diethyl ether-d₁₀ proceed by H (or D) atom abstraction from the C—H (or C—D) bonds.

$$\text{OH} + \text{CH}_3\text{CH}_2\text{OCH}_2\text{CH}_3 \longrightarrow \begin{cases} \text{H}_2\text{O} + \text{CH}_3\dot{\text{C}}\text{HOCH}_2\text{CH}_3 & \text{(a)} \\ \text{H}_2\text{O} + \dot{\text{C}}\text{H}_2\text{CH}_2\text{OCH}_2\text{CH}_3 & \text{(b)} \end{cases}$$

Pathway (a) is expected to dominate at essentially all temperatures.[109]

(c) Di-n-propyl ether

The available rate constants of Lloyd et al.,[71] Wallington et al.[81] and Bennett and Kerr[82] are given in Table 11 and are plotted in Arrhenius form in Fig. 77. The room temperature rate constants from these studies[71,81,82] are in good agreement, and a unit-weighted least-squares analysis of the data of Wallington et al.[81] and Bennett and Kerr,[82] using the expression $k = CT^2 e^{-D/T}$, yields the recommendation of

$$k(\text{di-}n\text{-propyl ether}) = (1.42^{+0.47}_{-0.36})$$

$$\times 10^{-17} \ T^2 \ e^{(778 \pm 90)/T} \ \text{cm}^3 \ \text{molecule}^{-1} \ \text{s}^{-1}$$

over the temperature range 240–440 K, where the indicated errors are two least-squares standard deviations, and

$$k(\text{di-}n\text{-propyl ether}) = 1.72 \times 10^{-11} \ \text{cm}^3 \ \text{molecule}^{-1} \ \text{s}^{-1}$$

at 298 K, with an estimated overall uncertainty at 298 K of ±35%. As for dimethyl ether and diethyl ether, this OH radical reaction will proceed by H-atom abstraction from the various C—H bonds.

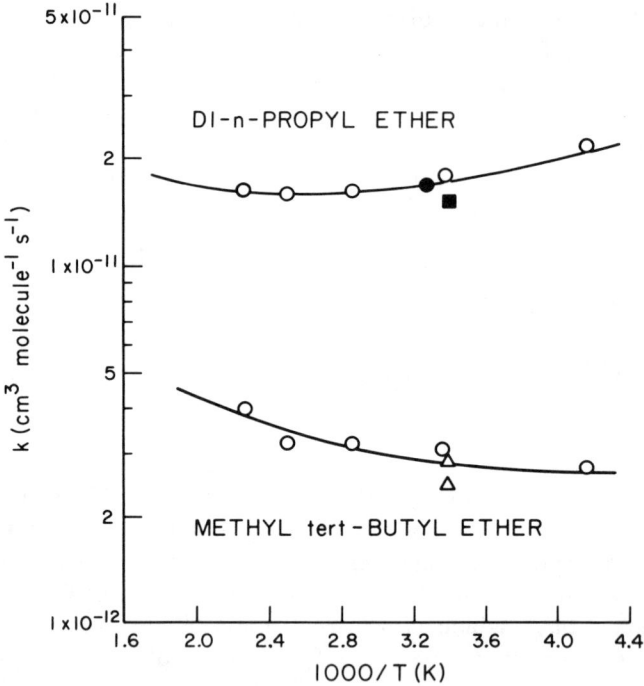

FIG. 77. Arrhenius plots of rate constants for the reactions of the OH radical with di-n-propyl ether and methyl t-butyl ether. (●) Lloyd et al.;[71] (△) Cox and Goldstone;[66] (○) Wallington et al.[81] (di-n-propyl ether) and Wallington et al.[73] (methyl t-butyl ether); (■) Bennett and Kerr;[82] (——) recommendations (see text).

(d) Methyl tert-butyl ether

The rate constants of Cox and Goldstone[66] and Wallington et al.[73] are given in Table 11 and are plotted in Arrhenius form in Fig. 77. Within the combined overall experimental error limits, the room temperature rate constants from these studies[66,73] are in agreement. The rate constants obtained by Cox and Goldstone[66] at 295 ± 2 K using both n-hexane and ethene as the reference organics are in good agreement, showing a self-consistency of the data and suggesting the absence of significant systematic errors.

Although the Arrhenius plot (Fig. 77) does not show clear evidence for curvature, a unit-weighted least-squares analysis of the rate constant data of Cox and Goldstone[66] and Wallington et al.,[73] using the equation $k = CT^2 e^{-D/T}$, has been carried out to yield the recommendation of

$$k(\text{methyl } t\text{-butyl ether}) = (6.81^{+2.91}_{-2.04})$$
$$\times 10^{-18} \, T^2 \, e^{(460 \pm 112)/T} \, \text{cm}^3 \, \text{molecule}^{-1} \, \text{s}^{-1}$$

over the temperature range 240–440 K, where the indicated errors are two least-squares standard deviations, and

$$k(\text{methyl } t\text{-butyl ether}) = 2.83$$
$$\times 10^{-12} \, \text{cm}^3 \, \text{molecule}^{-1} \, \text{s}^{-1}$$

at 298 K, with an estimated overall uncertainty at 298 K of $\pm 35\%$.

This OH radical reaction will proceed by H-atom abstraction from all of the C—H bonds[109]

$$\text{OH} + \text{CH}_3\text{OC(CH}_3)_3 \longrightarrow
\begin{cases}
\text{H}_2\text{O} + \dot{\text{C}}\text{H}_2\text{OC(CH}_3)_3 & \text{(a)} \\
\\
\text{H}_2\text{O} + \text{CH}_3\text{OC(CH}_3)_2\overset{\dot{\text{C}}\text{H}_2}{|} & \text{(b)}
\end{cases}$$

with pathways (a) and (b) being calculated[109] to be of approximately comparable importance at around room temperature, with channel (b) becoming increasingly important with increasing temperature.

(e) Tetrahydrofuran

The available rate constants of Winer et al.,[84] Ravishankara and Davis[54] and Wallington et al.,[81] all obtained at room temperature, are given in Table 11. These room temperature rate constants[54,81,84] are in reasonably good agreement. A unit-weighted average of these rate constants[54,81,84] leads to the recommendation of

$$k(\text{tetrahydrofuran}) = 1.61 \times 10^{-11} \, \text{cm}^3 \, \text{molecule}^{-1} \, \text{s}^{-1}$$

at ~ 298 K, with an estimated uncertainty of $\pm 30\%$. The temperature dependence of the rate constant for this reaction is expected to be essentially zero at around room temperature.

(f) Furan

The available rate constants of Lee and Tang,[85] Atkinson et al.,[86] Wine and Thompson,[87] Tuazon et al.[88] and Witte and Zetzsch[89] are given in Table 11 and are plotted in Arrhenius form in Fig. 78.

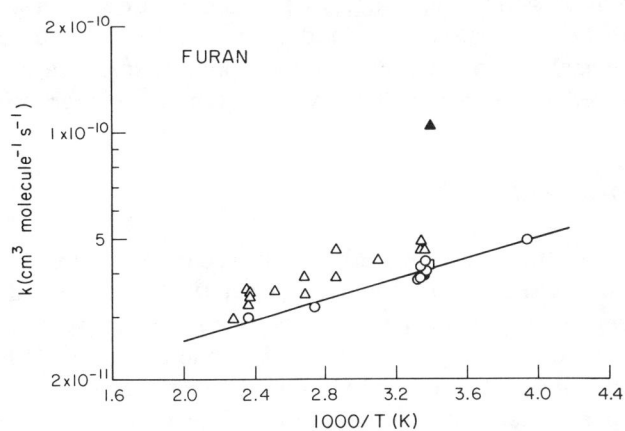

FIG. 78. Arrhenius plot of rate constants for the reaction of the OH radical with furan. (▲) Lee and Tang;[85] (●) Atkinson et al.;[86] (○) Wine and Thompson;[87] (□) Tuazon et al.;[88] (△) Witte and Zetzsch;[89] (——) recommendation (see text).

The room temperature rate constants from the studies of Atkinson et al.,[86] Wine and Thompson,[87] Tuazon et al.[88] and Witte and Zetzsch[89] are a factor of ~ 2.0–2.5 lower than that of Lee and Tang.[85] It should be noted that a similar discrepancy occurs for the analogous reaction of OH radicals with thiophene (see Sec. 2.7 below), suggesting the occurrence of a systematic error in the discharge flow study of Lee and Tang,[85] at least for these two heterocycles [their room temperature rate constant for ethane[85] is consistent with other literature data (Sec. 2.1)].

The rate constants of Witte and Zetzsch[89] are uniformly $\sim 20\%$ higher than those of Atkinson et al.,[86] Wine and Thompson[87] and Tuazon et al.,[88] which are in excellent agreement at room temperature. A unit-weighted least-squares analysis of the data of Atkinson et al.,[86] Wine and Thompson[87] and Tuazon et al.[88] yields the recommended Arrhenius expression of

$$k(\text{furan}) = (1.32^{+0.30}_{-0.24})$$
$$\times 10^{-11} \, e^{(334 \pm 61)/T} \, \text{cm}^3 \, \text{molecule}^{-1} \, \text{s}^{-1}$$

over the temperature range 254–424 K, where the indicated errors are two least-squares standard deviations, and

$$k(\text{furan}) = 4.05 \times 10^{-11} \, \text{cm}^3 \, \text{molecule}^{-1} \, \text{s}^{-1} \text{ at 298 K,}$$

with an estimated uncertainty at 298 K of $\pm 25\%$ (which encompasses the rate constants of Witte and Zetzsch[89]).

Consistent with the magnitude of the room temperature rate constant and the negative temperature dependency, the OH radical reaction with furan (and 3-methylfuran) almost certainly proceeds via initial OH radical addition to the $>\text{C}=\text{C}<$ double bond, with the measured rate constants[86–89] being at, or close to, the high-pressure limit. At elevated temperatures the resulting OH radical addition adduct is expected to undergo thermal decomposition, and hence kinetic studies are required to determine the rate constant at temperatures ≳500 K.

(g) Other Ethers

For the remaining ethers, no recommendations are made, although it should be noted that significant discrepancies exist between the room temperature rate constants of Wallington et al.[74] and Bennett and Kerr[82] for ethyl n-butyl ether, ethyl t-butyl ether and di-n-butyl ether. These discrepancies, of up to a factor of 1.7, indicate the presence of systematic errors in at least one of these two studies.[74,82] The OH radical reactions with the saturated ethers will proceed by H-atom abstraction from the C—H bonds, as discussed above for dimethyl, diethyl, di-n-propyl and methyl t-butyl ether.

For methyl vinyl ether, at temperatures ≲400–500 K the OH radical reaction will proceed mainly by OH radical addition to the $>C=C<$ bond.

$$OH + CH_3OCH=CH_2 \rightarrow CH_3OCHOHCH_2$$

$$\text{and (mainly) } CH_3OCHCH_2OH$$

Analogous to the reactions of the OH radical with the alkenes, at elevated temperatures (≳500–700 K) this OH radical adduct will undergo rapid thermal decomposition and the major reaction pathway then observed will be H-atom abstraction from the —CH₃ group:

$$OH + CH_3OCH=CH_2 \rightarrow H_2O + \dot{C}H_2OCH=CH_2$$

(9) Esters

The available rate constant data are given in Table 11. For methyl acetate, ethyl acetate, n-propyl acetate, isopropyl acetate, n-butyl acetate, sec-butyl acetate, methyl propionate and ethyl propionate, studies have been carried out by two or three groups. However, as discussed previously[106] (and above), the rate constants obtained by Campbell and Parkinson[92] are suspect due to questions concerning the validity of the experimental technique used (and discrepancies between the data of Campbell and Parkinson[92] and Wallington et al.[91] are evident for methyl acetate and methyl propionate, but not for ethyl acetate or ethyl propionate). For ethyl acetate, n-propyl acetate, isopropyl acetate, n-butyl acetate and sec-butyl acetate, the room temperature rate constants measured by various combinations of the studies of Winer et al.,[84] Zetzsch,[61] Kerr and Stocker,[34] Hartmann et al.[78] and Wallington et al.[91] are in reasonable agreement.

However, the temperature dependences (for $T \geq 296$ K) obtained by Wallington et al.[91] for methyl acetate and ethyl acetate (and methyl trifluoroacetate) are surprisingly low considering the low magnitude of the room temperature rate constants for these acetates (≲1.5 ×

10^{-12} cm³ molecule⁻¹ s⁻¹), especially when compared with the significant temperature dependence observed by Hartmann et al.[78] for n-butyl acetate ($B = 594 \pm 126$ K for n-butyl acetate;[78] $B = 131 \pm 28$ K for ethyl acetate[91]). It would be expected that the absolute magnitude of the temperature dependences would decrease with the increasing room temperature rate constants associated with the increasing —OR chain lengths.

Accordingly, no recommendations are made concerning the temperature dependencies of the OH radical reactions with the esters. However, based upon the reasonable agreement of the studies of Winer et al.,[84] Zetzsch,[61] Kerr and Stocker,[34] Hartmann et al.[78] and Wallington et al.[91] for the room temperature rate constants for the esters studied by two or more of these groups, the following recommendations are made at 298 K

$$k(\text{ethyl acetate}) = 1.6 \times 10^{-12} \text{ cm}^3 \text{ molecule}^{-1} \text{ s}^{-1},$$

based upon the data of Zetzsch[61] and Wallington et al.;[91]

$$k(n\text{-propyl acetate}) = 3.4 \times 10^{-12} \text{ cm}^3 \text{ molecule}^{-1} \text{ s}^{-1},$$

based upon the data of Winer et al.,[84] Kerr and Stocker[34] and Wallington et al.[91] (though significant discrepancies exist between the relative rate constants obtained by Winer et al.[84] and Kerr and Stocker[34]);

$$k(\text{isopropyl acetate}) = 3.4 \times 10^{-12} \text{ cm}^3 \text{ molecule}^{-1} \text{ s}^{-1},$$

based upon the data of Kerr and Stocker[34] and Wallington et al.,[91]

$$k(n\text{-butyl acetate}) = 4.2 \times 10^{-12} \text{ cm}^3 \text{ molecule}^{-1} \text{ s}^{-1}$$

based upon the data of Hartmann et al.[78] and Wallington et al.,[91] and

$$k(\text{sec-butyl acetate}) = 5.5 \times 10^{-12} \text{ cm}^3 \text{ molecule}^{-1} \text{ s}^{-1}$$

based upon the data of Winer et al.[84] and Wallington et al.[91]

For these and the other esters for which no recommendations are made (including $CH_3C(O)OCH_2CH_2OCH_2CH_3$), the rate constant data given in Table 11 indicate that the OH radical reaction with $R_1C(O)OR_2$ occurs mainly at the —OR_2 entity rather than at the R_1CO-entity (thus the room temperature rate constants for ethyl formate, ethyl acetate and ethyl propionate increase only slowly with increasing length of the $R_1C(O)O$- chain, while there is a marked increase in the room temperature rate constant for the formates, acetates, propionates and butyrates as the —OR_2 chain length increases[91]). The magnitude of the rate constants for the acetate series further shows that they increase with the number of secondary and tertiary C—H bonds, as expected for H-atom abstraction reactions.[106,109] Furthermore, analogous to the ethers, the rate constants per C—H bond for the —OR$_2$ entities are higher than those for the corresponding alkanes.[106,109]

(10) Carboxylic Acids

The available kinetic data for the carboxylic acids are given in Table 11.

(a) Formic Acid, Formic Acid-d$_1$ (DCOOH and HCOOD) and Formic Acid-d$_2$ (DCOOD)

The available rate constant data of Zetzsch and Stuhl,[93] Wine et al.,[94] Jolly et al.,[95] Dagaut et al.[96] and Singleton et al.[97] for the reactions of the OH radical with HCOOH and DCOOH are given in Table 11, and rate data for the reactions of the OD radical with HCOOD[97] and DCOOD[97] are given in Table 12. The data for HCOOH[93-97] are plotted in Arrhenius form in Fig. 79.

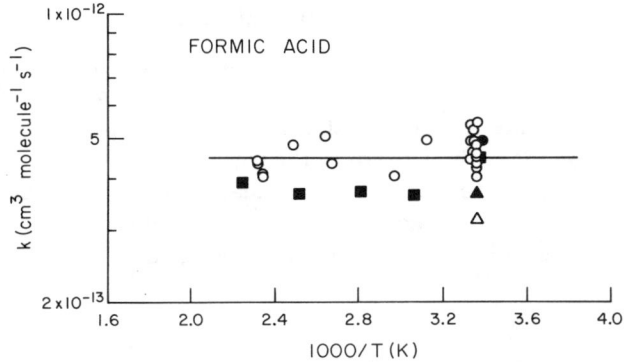

FIG. 79. Arrhenius plot of rate constants for the reaction of the OH radical with formic acid. (△) Zetzsch and Stuhl;[93] (○) Wine et al.;[94] (●) Jolly et al.;[95] (▲) Dagaut et al.;[96] (■) Singleton et al.;[97] (———) recommendation (see text).

At room temperature the rate constants determined by Wine et al.,[94] Jolly et al.[95] and Singleton et al.[97] are in good agreement, with those of Zetzsch and Stuhl[93] and Dagaut et al.[96] being somewhat lower (but in agreement within the combined overall error limits). The rate constants obtained by Wine et al.[94] and Singleton et al.[97] over the temperature range 297–445 K show no temperature dependence within the experimental uncertainties.

Problems associated with dimerization of HCOOH occur during the measurement of the rate constant for the reaction of the OH radical with the formic acid monomer. In the studies of Wine et al.,[94] Jolly et al.[95] and Singleton et al.,[97] the HCOOH concentrations were monitored by ultraviolet absorption, and hence the data determined from these studies[94,95,97] are used in the evaluation of the rate constant for this reaction. From a unit-weighted average of the rate constants reported by Wine et al.,[94] Jolly et al.[95] and Singleton et al.,[97] a rate constant of

$$k(\text{formic acid}) = 4.5 \times 10^{-13} \text{ cm}^3 \text{ molecule}^{-1} \text{ s}^{-1}$$

is recommended, independent of temperature over the range 296–445 K, with an estimated overall uncertainty of ±35% over this temperature range.

The data of Wine et al.[94] and Singleton et al.[97] show that the rate constant for the reaction of the OH radical with DCOOH is identical to that for HCOOH at 298 K, within the experimental uncertainties. Using resonance fluorescence detection of H-atoms, Wine et al.[94] estimated that the H-atom production yield from the OH radical reaction with HCOOH was 0.75 ± 0.25, indicating that the major reaction pathway proceeds via,

$$\text{OH} + \text{HCOOH} \rightarrow \text{H}_2\text{O} + \text{CO}_2 + \text{H}$$

and this is consistent with the observations of Jolly et al.[95]

Singleton et al.[97] have determined that the room temperature rate constants for the reactions of the OD radical with HCOOD and DCOOD (Table 12) are significantly lower than those for the reactions of the OH radical with HCOOH and DCOOH (Table 12). These observations, together with the similarity of the room temperature rate constants for the reactions of the OH radical with HCOOH and DCOOH,[94,97] indicate that abstraction of the H (or D) atom from the —OH (or —OD) group is the major reaction pathway at around room temperature

$$\text{OH} + \text{DCOOH} \rightarrow \text{H}_2\text{O} + \text{DCO}_2$$
$$\downarrow$$
$$\text{D} + \text{CO}_2$$

This conclusion[97] is consistent with the low reactivity of the various formic acid dimers studied.[95,97,98] The detailed reaction dynamics are not known, i.e., initial formation of a HO—HCOOH adduct followed by decomposition to H$_2$O + (mainly) HCOO, or direct H-atom abstraction to yield H$_2$O and (mainly) HCOO. The observation of an essentially zero temperature dependence of the rate constant for HCOOH suggests the initial formation of an HO—HCOOH adduct.[94,95,97,98]

(b) Other Carboxylic Acids

Rate constants have been measured for acetic, propionic and butyric acids at room temperature by Zetzsch and Stuhl,[93] and for acetic, propionic and isobutyric acids over the temperature range 298–440 K by Dagaut et al.[96] For acetic and priopionic acid, the agreements between the room temperature rate constants determined from these studies are within the combined experimental error limits. The rate constants obtained by Zetzsch and Stuhl[93] required a knowledge of the vapor pressures of the carboxylic acids studied at 298 K, and are hence subject to significant uncertainties.[93] Consequently, no firm recommendations are made for the reactions of the OH radical with these carboxylic acids. However, until further experimental data become available, the rate constants of Dagaut et al.[96] should be used over the temperature range of 298–440 K studied, with accordingly large overall uncertainties.

These reactions are expected to proceed by an overall H-atom abstraction process; for example,

although the observed near-zero temperature dependences measured by Dagaut et al.[96] would indicate that these reactions proceed by initial OH radical addition to form a complex which then, presumably, rapidly decomposes to products.

(11) Epoxides

(a) Kinetics

The available kinetic data are given in Table 11. Only for epoxyethane, 1,2-epoxypropane and 1-chloro-2,3-epoxypropane have multiple studies been carried out, and for 1-chloro-2,3-epoxypropane the relative rate study of Edney et al.[38] leads only to a lower limit to the rate constant.

(i) Epoxyethane

At room temperature, the absolute rate constants determined by Lorenz and Zellner,[99] Zetzsch[100] and Wallington et al.[81] exhibit a spread of a factor of 1.8, but are consistent with the upper limit to the rate constant derived by Klöpffer et al.[72] The only temperature-dependent study is that of Lorenz and Zellner,[99] who observed a rapid increase in the rate constant above 435 K, leading to marked non-Arrhenius behavior. A unit-weighted average of the room temperature rate constants of Lorenz and Zellner,[99] Zetzsch[100] and Wallington et al.[81] yields the recommendation of

$$k(\text{epoxyethane}) = 7.6 \times 10^{-14} \text{ cm}^3 \text{ molecule}^{-1} \text{ s}^{-1}$$

at 298 K, with an estimated overall uncertainty of ±50%.

Because of the high magnitude of the temperature dependence measured by Lorenz and Zellner,[99] leading to an Arrhenius pre-exponential factor of 1.1×10^{-11} cm^3 molecule^{-1} s^{-1} for rate data obtained over the range 297–435 K and a markedly higher value for data obtained over the temperature range 435–515 K, no temperature dependence is recommended.

(ii) 1,2-Epoxypropane

The two absolute measurements of the room temperature rate constant by Zetzsch[100] and Wallington et al.[81] are in good agreement, but are a factor of ~2 lower than the relative rate data of Winer et al.[101] and Edney et al.[38] The absolute flash photolysis-resonance fluorescence

data are preferred, and a unit-weighted average of the room temperature rate constants of Zetzsch[100] and Wallington et al.[81] yields the recommendation of

$$k(\text{1,2-epoxypropane}) = 5.2 \times 10^{-13} \text{ cm}^3 \text{ molecule}^{-1} \text{ s}^{-1}$$

at 298 K (assuming a temperature dependence of $B \sim 1000$ K to extrapolate the measured data from 295–296 K to 298 K), with an estimated overall uncertainty of ±40%.

(b) Mechanism

The OH radical reactions with the epoxides proceed by initial H-atom abstraction from the C—H bonds, for example, for 1,2-epoxypropane

At around room temperature, reaction pathway (a) is estimated to dominate,[106,109] since the three-membered ring structure decreases the reactivity of the C—H bonds on the carbon atoms in the ring[106,109] (of course, for epoxyethane a reaction pathway analogous to (a) cannot occur). It is also expected that reaction pathways (b) and (c) will be followed by rapid ring cleavage:

with the resulting species being initially energy-rich.

For epoxyethane, the initially formed OH radical is

and it is expected that this radical will rapidly undergo ring cleavage.

Using LIF detection to monitor the vinoxy (CH$_2$CHO) radical, Lorenz and Zellner[99] have measured CH$_2$CHO yields at 298 K of 0.08 ± 0.03 and 0.23 ± 0.08 at 10 and 60 Torr total pressure of helium, respectively.

(12) Hydroperoxides

(a) Methylhydroperoxide and Methylhydroperoxide-d_1 (CH$_3$OOD)

The available rate constant data are given in Tables 11 (^{16}OH and ^{18}OH radical reactions with CH$_3$OOH) and 12 (^{16}OD radical reactions with CH$_3$OOH and CH$_3$OOD).

The most comprehensive of the two studies[102,103] is that of Vaghjiani and Ravishankara,[103] who studied the kinetics and mechanisms of the reactions of ^{16}OH, ^{18}OH and ^{16}OD radicals with CH_3OOH and of ^{16}OD radicals with CH_3OOD. The reaction of the OH radical with CH_3OOH proceeds by the two pathways

$$OH + CH_3OOH \longrightarrow \begin{cases} H_2O + CH_3OO \cdot & (a) \\ H_2O + \dot{C}H_2OOH & (b) \end{cases}$$

with the CH_2OOH radical rapidly decomposing to yield an OH radical and HCHO.[103] Thus, relative rate studies yield the overall rate constant $k = k_a + k_b$, while flash or laser photolysis studies monitoring the disappearance of the OH radical measure only the rate constant k_a if the OH radical regenerated contains the same oxygen isotope as the reactant OH radical.

Thus, reaction of the ^{16}OH radical with $CH_3O^{16}OH$ (or of the ^{16}OD radical with $CH_3O^{16}OD$) yields the rate constant k_a, while reaction of the ^{18}OH radical with $CH_3O^{16}OH$ yields $k = (k_a + k_b)$. Similarly, reaction of the OD radical with CH_3OOH yields the overall reaction rate constant $(k_a + k_b)$ for the OD radical reaction.

The overall room temperature rate constant $k = (k_a + k_b)$ derived from the data of Niki et al.[102] is a factor of ~2 higher than those obtained by Vaghjiani and Ravishankara[103] from the reactions of ^{18}OH and OD radicals with CH_3OOH, for unknown reasons. The rate constants for the reaction of the OD radical with CH_3OOH (Table 12) are essentially identical to those for the reaction of the ^{18}OH radical with CH_3OOH (Table 11), as expected, while a significant isotope effect on reaction channel (a) is shown by the lower room temperature rate constant for the OD radical reaction with CH_3OOD compared to the ^{16}OH radical reaction with CH_3OOH. These observations confirm the occurrence of the two H-atom abstraction channels (a) and (b).

Unit-weighted least-squares analyses of the data of Vaghjiani and Ravishankara[103] lead to:

from the data for the ^{16}OH radical reaction with CH_3OOH;

$$k_a = (1.79^{+0.36}_{-0.30}) \times 10^{-12}\, e^{(219 \pm 46)/T}\ cm^3\ molecule^{-1}\ s^{-1}$$

over the temperature range 203–348 K, where the indicated errors are two least-squares standard deviations, and

$$k_a = 3.73 \times 10^{-12}\ cm^3\ molecule^{-1}\ s^{-1}\ at\ 298\ K,$$

and from the data for the ^{18}OH and OD radical reactions with CH_3OOH;

$$(k_a + k_b) = (2.93^{+0.30}_{-0.28})$$
$$\times 10^{-12}\, e^{(190 \pm 28)/T}\ cm^3\ molecule^{-1}\ s^{-1}$$

over the temperature range 223–423 K, where the indicated errors are two least-squares standard deviations, and

$$(k_a + k_b) = 5.54 \times 10^{-12}\ cm^3\ molecule^{-1}\ s^{-1}\ at\ 298\ K.$$

These rate constant expressions yield the rate constant ratio $k_a/(k_a + k_b) = 0.611\, e^{29/T} = 0.67$ at 298 K. Despite the disagreement concerning the overall rate constant, this rate constant ratio derived from the data of Vaghjiani and Ravishankara[103] is in agreement with that of $k_a/(k_a + k_b) = 0.58 \pm 0.09$ obtained from the product study of Niki et al.[102]

(b) t-Butylhydroperoxide

For $(CH_3)_3COOH$, because of the stronger C—H bonds than the O—H bond, the reaction is expected to proceed mainly via H-atom abstraction from the weak O—H bond,

$$OH + (CH_3)_3COOH \rightarrow H_2O + (CH_3)_3COO\cdot$$

and this is consistent with the magnitude of the rate constant measured by Anastasi et al.[104]

References

[1] D. E. Hoare, Nature **194**, 283 (1962).

[2] R. R. Baldwin and D. W. Cowe, Trans. Faraday Soc. **58**, 1768 (1962).

[3] R. V. Blundell, W. G. A. Cook, D. E. Hoare, and G. S. Milne, 10th International Symposium on Combustion, 1964; The Combustion Institute, Pittsburgh, PA, 1965, p. 445.

[4] A. A. Westenberg and R. M. Fristrom, 10th International Symposium on Combustion, 1964; The Combustion Institute, Pittsburgh, PA, 1965, p. 473.

[5] D. E. Hoare, Proc. Roy. Soc. (London) **A291**, 73 (1966).

[6] D. E. Hoare and G. B. Peacock, Proc. Roy. Soc. (London) **A291**, 85 (1966).

[7] J. T. Herron and R. D. Penzhorn, J. Phys. Chem. **73**, 191 (1969).

[8] E. D. Morris, Jr. and H. Niki, J. Chem. Phys. **55**, 1991 (1971).

[9] E. D. Morris, Jr. and H. Niki, J. Phys. Chem. **75**, 3640 (1971).

[10] J. Peeters and G. Mahnen, 14th International Symposium on Combustion, 1972; The Combustion Institute, Pittsburgh, PA, 1973, p. 133.

[11] J. Vandooren and P. J. Van Tiggelen, 16th International Symposium on Combustion, 1976; The Combustion Institute, Pittsburgh, PA, 1977, p. 1133.

[12] H. Niki, P. D. Maker, C. M. Savage, and L. P. Breitenbach, J. Phys. Chem. **82**, 132 (1978).

[13] R. Atkinson and J. N. Pitts, Jr., J. Chem. Phys. **68**, 3581 (1978).

[14] R. H. Smith, Int. J. Chem. Kinet. **10**, 519 (1978).

[15] L. J. Stief, D. F. Nava, W. A. Payne, and J. V. Michael, J. Chem. Phys. **73**, 2254 (1980).

[16] F. Temps and H. Gg. Wagner, Ber. Bunsenges Phys. Chem. **88**, 415 (1984).

[17] S. Zabarnick, J. W. Fleming, and M. C. Lin, Int. J. Chem. Kinet. **20**, 117 (1988).

[18] H. Niki, P. D. Maker, C. M. Savage, and L. P. Breitenbach, J. Phys. Chem. **88**, 5342 (1984).

[19] E. D. Morris, Jr., D. H. Stedman, and H. Niki, J. Amer. Chem. Soc. **93**, 3570 (1971).

[20] R. A. Cox, R. G. Derwent, P. M. Holt, and J. A. Kerr, J. Chem. Soc. Faraday Trans. 1, **72**, 2061 (1976).

[21] M. B. Colket, III, D. W. Naegeli, and I. Glassman, 16th International Symposium on Combustion, 1976; The Combustion Institute, Pittsburgh, PA, 1977, p. 1023.

[22] J. A. Kerr and D. W. Sheppard, Environ. Sci. Technol., **15**, 960 (1981).

[23]D. H. Semmes, A. R. Ravishankara, C. A. Gump-Perkins, and P. H. Wine, Int. J. Chem. Kinet. **17**, 303 (1985).

[24]J. V. Michael, D. G. Keil, and R. B. Klemm, J. Chem. Phys. **83**, 1630 (1985).

[25]H. Niki, P. D. Maker, C. M. Savage, and M. D. Hurley, J. Phys. Chem. **91**, 2174 (1987).

[26]R. R. Baldwin, R. W. Walker, and D. H. Langford, Trans. Faraday Soc. **65**, 806 (1969).

[27]G. J. Audley, D. L. Baulch, and I. M. Campbell, J. Chem. Soc. Faraday Trans. 1, **77**, 2541 (1981).

[28]E. W. Kaiser, Int. J. Chem. Kinet. **15**, 997 (1983).

[29]J. A. Kerr and D. W. Stocker, J. Photochem. **28**, 475 (1985).

[30]L. Nelson, J. J. Treacy, and H. W. Sidebottom, Proceedings, 3rd European Symposium on the Physico-Chemical Behavior of Atmospheric Pollutants, 1984; D. Riedel Publishing Co., Dordrecht, Holland, 1984, p. 258.

[31]R. A. Cox, R. G. Derwent, and M. R. Williams, Environ. Sci. Technol. **14**, 57 (1980).

[32]C. Zetzsch, 7th International Symposium on Gas Kinetics, Univ. Göttingen, Göttingen, W. Germany, August 23–28, 1982.

[33]C. Chiorboli, C. A. Bignozzi, A. Maldotti, P. F. Giardini, A. Rossi, and V. Carassiti, Int. J. Chem. Kinet. **15**, 579 (1983).

[34]J. A. Kerr and D. W. Stocker, J. Atmos. Chem. **4**, 253 (1986).

[35]T. J. Wallington and M. J. Kurylo, J. Phys. Chem. **91**, 5050 (1987).

[36]A. M. Winer, A. C. Lloyd, K. R. Darnall, and J. N. Pitts, Jr., J. Phys. Chem. **80**, 1635 (1976).

[37]R. A. Cox, K. F. Patrick, and S. A. Chant, Environ. Sci. Technol. **15**, 587 (1981).

[38]E. O. Edney, T. E. Kleindienst, and E. W. Corse, Int. J. Chem. Kinet. **18**, 1355 (1986).

[39]R. Atkinson, S. M. Aschmann, W. P. L. Carter, and J. N. Pitts, Jr., Int. J. Chem. Kinet. **14**, 839 (1982).

[40]R. Atkinson and S. M. Aschmann, J. Phys. Chem., **92**, 4008 (1988).

[41]P. Dagaut, T. J. Wallington, R. Liu, and M. J. Kurylo, J. Phys. Chem. **92**, 4375 (1988).

[42]A. Maldotti, C. Chiorboli, C. A. Bignozzi, C. Bartocci, and V. Carassiti, Int. J. Chem. Kinet. **12**, 905 (1980).

[43]R. Atkinson, S. M. Aschmann, and J. N. Pitts, Jr., Int. J. Chem. Kinet. **15**, 75 (1983).

[44]T. E. Kleindienst, G. W. Harris, and J. N. Pitts, Jr., Environ. Sci. Technol. **16**, 844 (1982).

[45]C. Faubel, H. Gg. Wagner, and W. Hack, Ber. Bunsenges Phys. Chem. **81**, 689 (1977).

[46]S. Hatakeyama, S. Honda, N. Washida, and H. Akimoto, Bull. Chem. Soc. Jpn. **58**, 2157 (1985).

[47]C. N. Plum, E. Sanhueza, R. Atkinson, W. P. L. Carter, and J. N. Pitts, Jr., Environ. Sci. Technol. **17**, 479 (1983).

[48]K. H. Becker and Th. Klein, Proceedings, 4th European Symposium on the Physico-Chemical Behavior of Atmospheric Pollutants, 1986; D. Riedel Publishing Co., Dordrecht, Holland, 1987, p. 320.

[49]K. R. Darnall, R. Atkinson, and J. N. Pitts, Jr., J. Phys. Chem. **83**, 1943 (1979).

[50]J. D. Rogers, Environ. Sci. Technol. **23**, 177 (1989).

[51]E. C. Tuazon, R. Atkinson, and W. P. L. Carter, Environ. Sci. Technol. **19**, 265 (1985).

[52]I. M. Campbell, D. F. McLaughlin, and B. J. Handy, Chem. Phys. Lett. **38**, 362 (1976).

[53]R. Overend and G. Paraskevopoulos, J. Phys. Chem. **82**, 1329 (1978).

[54]A. R. Ravishankara and D. D. Davis, J. Phys. Chem. **82**, 2852 (1978).

[55]J. Vandooren and P. J. Van Tiggelen, 18th International Symposium on Combustion, 1980; The Combustion Institute, Pittsburgh, PA, 1981, p. 473.

[56]I. Barnes, V. Bastian, K. H. Becker, E. H. Fink, and F. Zabel, Atmos. Environ. **16**, 545 (1982).

[57]J. Hägele, K. Lorenz, D. Rhäsa, and R. Zellner, Ber. Bunsenges Phys. Chem. **87**, 1023 (1983); private communication (1984).

[58]E. C. Tuazon, W. P. L. Carter, R. Atkinson, and J. N. Pitts, Jr., Int. J. Chem. Kinet. **15**, 619 (1983).

[59]U. Meier, H. H. Grotheer, and Th. Just, Chem. Phys. Lett. **106**, 97 (1984).

[60]U. Meier, H. H. Grotheer, G. Riekert, and Th. Just, Ber. Bunsenges Phys. Chem. **89**, 325 (1985).

[61]C. Zetzsch, report to Bundeminister für Forschung und Technologie, Projektträger für Unweltchemikalien (1982); private communication (1985).

[62]P. G. Greenhill and B. V. O'Grady, Aust. J. Chem. **39**, 1775 (1986).

[63]T. J. Wallington and M. J. Kurylo, Int. J. Chem. Kinet. **19**, 1015 (1987).

[64]W. P. Hess and F. P. Tully, J. Phys. Chem. **93**, 1944 (1989).

[65]J. A. McCaulley, N. Kelly, M. F. Golde, and F. Kaufman, J. Phys. Chem. **93**, 1014 (1989).

[66]R. A. Cox and A. Goldstone, Proceedings, 2nd European Symposium on the Physico-Chemical Behavior of Atmospheric Pollutants; D. Riedel Publishing Co., Dordrecht, Holland, 1982, p. 112.

[67]U. Meier, H. H. Grotheer, G. Riekert, and Th. Just, Chem. Phys. Lett. **115**, 221 (1985).

[68]U. Meier, H.-H. Grotheer, G. Riekert, and Th. Just, Chem. Phys. Lett. **133**, 162 (1987).

[69]K. Lorenz, D. Rhäsa, and R. Zellner, private communication (1984).

[70]W. P. Hess and F. P. Tully, Chem. Phys. Lett. **152**, 183 (1988).

[71]A. C. Lloyd, K. R. Darnall, A. M. Winer, and J. N. Pitts, Jr., Chem. Phys. Lett. **42**, 205 (1976).

[72]W. Klöpffer, R. Frank, E.-G. Kohl, and F. Haag, Chemiker-Zeitung **110**, 57 (1986); "Methods of the Ecotoxicological Evaluation of Chemicals, Photochemical Degradation in the Gas Phase," Vol. 6, *OH Reaction Rate Constants and Tropospheric Lifetimes of Selected Environmental Chemicals*. Report 1980–1983, K. H. Becker, H. M. Biehl, P. Bruckmann, E. H. Fink, F. Führ, W. Klöpffer, R. Zellner, and C. Zetzsch, Editors, Kernforschungsanlage Jülich GmbH, November 1984.

[73]T. J. Wallington, P. Dagaut, R. Liu, and M. J. Kurylo, Environ. Sci. Technol. **22**, 842 (1988).

[74]T. J. Wallington, P. Dagaut, R. Liu, and M. J. Kurylo, Int. J. Chem. Kinet. **20**, 541 (1988).

[75]S. Gordon and W. A. Mulac, Int. J. Chem. Kinet. **Symp. 1**, 289 (1975).

[76]A. Weidelmann and C. Zetzsch, presented at Bunsentagung, Ulm und Neu-Ulm, May 20–22, 1982.

[77]P. Dagaut, T. J. Wallington, R. Liu, and M. J. Kurylo, 22nd International Symposium on Combustion, Seattle, August 14–19, 1988.

[78]D. Hartmann, A. Gedra, D. Rhäsa, and R. Zellner, Proceedings, 4th European Symposium on the Physico-Chemical Behavior of Atmospheric Pollutants, 1986; D. Riedel Publishing Co., Dordrecht, Holland, 1987, p. 225.

[79]R. A. Perry, R. Atkinson, and J. N. Pitts, Jr., J. Chem. Phys. **67**, 611 (1977).

[80]F. P. Tully and A. T. Droege, Int. J. Chem. Kinet. **19**, 251 (1987).

[81]T. J. Wallington, R. Liu, P. Dagaut, and M. J. Kurylo, Int. J. Chem. Kinet. **20**, 41 (1988).

[82]P. J. Bennett and J. A. Kerr, J. Atmos. Chem. **8**, 87 (1989).

[83]F. P. Tully, unpublished data, cited in reference 81.

[84]A. M. Winer, A. C. Lloyd, K. R. Darnall, R. Atkinson, and J. N. Pitts, Jr., Chem. Phys. Lett. **51**, 221 (1977).

[85]J. H. Lee and I. N. Tang, J. Chem. Phys. **77**, 4459 (1982).

[86]R. Atkinson, S. M. Aschmann, and W. P. L. Carter, Int. J. Chem. Kinet. **15**, 51 (1983).

[87]P. H. Wine and R. J. Thompson, Int. J. Chem. Kinet. **16**, 867 (1984).

[88]E. C. Tuazon, R. Atkinson, A. M. Winer, and J. N. Pitts, Jr., Arch. Environ. Contamin. Toxicol. **13**, 691 (1984).

[89]F. Witte and C. Zetzsch, 9th International Symposium on Gas Kinetics, University of Bordeaux, Bordeaux, France, July 20–25, 1986; private communication, 1988.

[90]R. Atkinson, S. M. Aschmann, E. C. Tuazon, J. Arey and B. Zielinska, Int. J. Chem. Kinet. **21** 593 (1989).

[91]T. J. Wallington, P. Dagaut, R. Liu, and M. J. Kurylo, Int. J. Chem. Kinet. **20**, 177 (1988).

[92]I. M. Campbell and P. E. Parkinson, Chem. Phys. Lett. **53**, 385 (1978).

[93]C. Zetzsch and F. Stuhl, Proceedings, 2nd European Symposium on the Physico-Chemical Behavior of Atmospheric Pollutants, 1981; D.

Riedel Publishing Co., Dordrecht, Holland, 1982, p. 129; private communication, 1985.

94P. H. Wine, R. J. Astalos, and R. L. Mauldin, III, J. Phys. Chem. **89**, 2620 (1985).

95G. S. Jolly, D. J. McKenney, D. L. Singleton, G. Paraskevopoulos, and A. R. Bossard, J. Phys. Chem. **90**, 6557 (1986).

96P. Dagaut, T. J. Wallington, R. Liu, and M. J. Kurylo, Int. J. Chem. Kinet. **20**, 331 (1988).

97D. L. Singleton, G. Paraskevopoulos, R. S. Irwin, G. S. Jolly and D. J. McKenney, J. Am. Chem. Soc. **110**, 7786 (1988).

98G. S. Jolly, D. J. McKenney, D. L. Singleton, G. Paraskevopoulos, and A. R. Bossard, 9th International Symposium on Gas Kinetics, University of Bordeaux, Bordeaux, France, July 20–25, 1986.

99K. Lorenz and R. Zellner, Ber. Bunsenges Phys. Chem. **88**, 1228 (1984); "Methods of the Ecotoxicological Evaluation of Chemicals, Photochemical Degradation in the Gas Phase," Vol. 6, *OH Reaction Rate Constants and Tropospheric Lifetimes of Selected Environmental Chemicals*. Report 1980–1983; K. H. Becker, H. M. Biehl, P. Bruckmann, E. H. Fink, F. Führ, W. Klöpffer, R. Zellner, and C. Zetzsch, Editors, Kernforschungsanlage Jülich GmbH, November 1984.

100C. Zetzsch, presented at Bunsen Colloquium, Göttingen, W. Germany, October 9, 1980; private communication, 1985.

101A. M. Winer, K. R. Darnall, R. Atkinson, and J. N. Pitts, Jr., unpublished data, 1978, cited in R. Atkinson, K. R. Darnall, A. C. Lloyd, A. M. Winer, and J. N. Pitts, Jr., Adv. Photochem. **11**, 375 (1979).

102H. Niki, P. D. Maker, C. M. Savage, and L. P. Breitenbach, J. Phys. Chem. **87**, 2190 (1983).

103G. L. Vaghjiani and A. R. Ravishankara, J. Phys. Chem. **93**, 1948 (1989).

104C. Anastasi, I. W. M. Smith, and D. A. Parkes, J. Chem. Soc. Faraday Trans. 1, **74**, 1693 (1978).

105W. B. DeMore, M. J. Molina, S. P. Sander, D. M. Golden, R. F. Hampson, M. J. Kurylo, C. J. Howard, and A. R. Ravishankara, "Chemical Kinetics and Photochemical Data for Use in Stratospheric Modeling," Evaluation No. 8, NASA Panel for Data Evaluation, JPL Publication 87–41, September 15, 1987.

106R. Atkinson, Chem. Rev. **86**, 69 (1986).

107B. M. Morrison, Jr. and J. Heicklen, J. Photochem. **13**, 189 (1980).

108R. Atkinson and A. C. Lloyd, J. Phys. Chem. Ref. Data **13**, 315 (1984).

109R. Atkinson, Int. J. Chem. Kinet. **19**, 799 (1987).

110R. Atkinson, S. M. Aschmann, A. M. Winer, and J. N. Pitts, Jr., Int. J. Chem. Kinet. **13**, 1133 (1981).

111E. C. Tuazon and R. Atkinson, Int. J. Chem. Kinet., in press (1989).

112H. Niki, P. D. Maker, C. M. Savage, and L. P. Breitenbach, Int. J. Chem. Kinet. **17**, 547 (1985).

113T. L. Osif, R. Simonaitis, and J. Heicklen, J. Photochem. **4**, 233 (1975).

114R. Atkinson, D. L. Baulch, R. A. Cox, R. F. Hampson, Jr., J. A. Kerr, and J. Troe, J. Phys. Chem. Ref. Data **18**, 881 (1989).

2.7. Sulfur-Containing Organics

The available kinetic data are given in Tables 13 and 14. Table 13 gives the rate constants obtained in the absence of O$_2$ and from relative rate studies carried out in the presence of one atmosphere total pressure of air, while Table 14 gives the available kinetic data obtained from studies designed to investigate the effect of the O$_2$ concentration on the measured rate constants. The rate constants and mechanisms of the reactions of the OH radical with the inorganic reduced sulfur compounds H$_2$S, COS and CS$_2$ are not included in this article; these reactions are dealt with in the NASA[28] and IUPAC[29] evaluations.

a. Thiols

(1) Kinetics

(a) *Methanethiol, Methanethiol-d$_1$ (CH$_3$SD) and Methanethiol-d$_3$ (CD$_3$SH)*

The available rate constants obtained by Atkinson *et al.*,[1] Cox and Sheppard,[2] Wine *et al.*,[3,7] Mac Leod *et al.*,[4,5] Lee and Tang,[6] Barnes *et al.*[8] and Hynes and Wine[9] are given in Tables 13 and 14. While the rate constant obtained by Cox and Sheppard[2] from a relative rate study involving the photolysis of HONO-NO-ethene-methanethiol-air mixtures is a factor of ~3 higher than the flash photolysis data,[1,3,7,9] the more recent data of Barnes *et al.*[8,21,26] show that secondary reactions, possibly involving CH$_3$SO,[18] occur in reaction systems which include NO. The relative rate study of Barnes *et al.*[8] utilized the photolysis of H$_2$O$_2$ to generate OH radicals in the absence of oxides of nitrogen, and the rate constants obtained for CH$_3$SH[8] are in reasonably good agreement with the absolute rate data of Atkinson *et al.*,[1] Wine *et al.*,[3,7] and Hynes and Wine.[9]

For both CH$_3$SH and CD$_3$SH, Hynes and Wine[9] have shown that the rate constant is independent of the O$_2$ concentration (Table 14). Furthermore, the rate constants for CH$_3$SH[1,3,7,9] and CD$_3$SH[9] are independent of the total pressure of the diluent gas.

The rate constants for the reaction of the OH radical with CH$_3$SH obtained by Atkinson *et al.*,[1] Wine *et al.*,[3,7] Mac Leod *et al.*,[4,5] Lee and Tang[6] and Barnes *et al.*[8] and the unit-weighted averages of the rate constants determined by Hynes and Wine[9] in the presence and absence of O$_2$ at 270 and 300 K are plotted in Arrhenius form in Fig. 80.

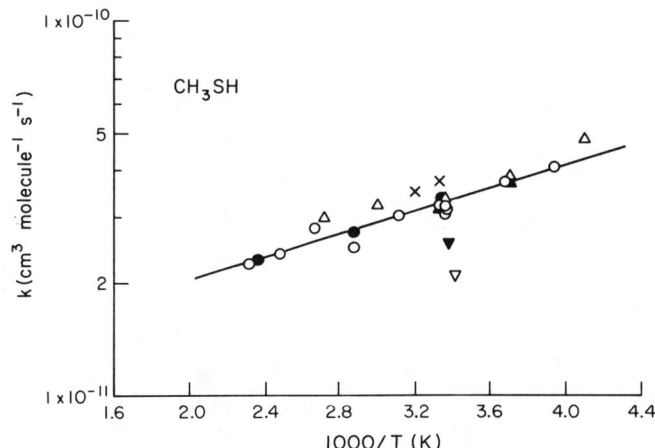

FIG. 80. Arrhenius plot of rate constants for the reaction of the OH radical with methanethiol. (●) Atkinson *et al.*;[1] (△) Wine *et al.*;[3] (▽) Mac Leod *et al.*;[4,5] (▼) Lee and Tang;[6] (○) Wine *et al.*;[7] (x) Barnes *et al.*;[8] (▲) Hynes and Wine;[9] (——) recommendation (see text).

The flash and laser photolysis studies of Atkinson et al.,[1] Wine et al.,[3,7] and Hynes and Wine[9] are in excellent agreement and agree within the experimental error limits with the rate constants obtained from the relative rate study of Barnes et al.[8] However, the room temperature rate constants of Atkinson et al.,[1] Wine et al.[3,7] and Hynes and Wine[9] are somewhat higher, by up to 50%, than the room temperature values of Mac Leod et al.[4,5] and Lee and Tang.[6] Since similar discrepancies occur for ethanethiol, the data of Mac Leod et al.[4,5] and Lee and Tang[6] are not used in the evaluation of the rate constant for CH_3SH.

A unit-weighted least-squares analysis of the rate constant data of Atkinson et al.,[1] Wine et al.,[3,7] Barnes et al.[8] and Hynes and Wine[9] leads to the recommended Arrhenius expression of

$$k\text{(methanethiol)} = (9.97^{+2.15}_{-1.77})$$
$$\times 10^{-12} e^{(356 \pm 60)/T} \text{ cm}^3 \text{ molecule}^{-1} \text{ s}^{-1}$$

over the temperature range 244–430 K, where the indicated errors are two least-squares standard deviations, and

$$k\text{(methanethiol)} = 3.29 \times 10^{-11} \text{ cm}^3 \text{ molecule}^{-1} \text{ s}^{-1}$$

at 298 K, with an estimated overall uncertainty at 298 K of ±25%. This recommendation is essentially identical to that of

$$k\text{(methanethiol)} = 9.70$$
$$\times 10^{-12} e^{366/T} \text{ cm}^3 \text{ molecule}^{-1} \text{ s}^{-1}$$

recommended by Atkinson.[30]

The rate constants determined by Wine et al.[7] for CH_3SD are essentially identical to those for CH_3SH, while the rate constants for CD_3SH[9] are ~15% lower than those for CH_3SH (Tables 13 and 14).

TABLE 13. Rate constants k and temperature-dependent parameters for the gas-phase reactions of the OH radical with sulfur-containing organics in the absence of O_2 (unless indicated)

Organic	$10^{12} \times A$ (cm³ molecule⁻¹ s⁻¹)	B (K)	$10^{12} \times k$ (cm³ molecule⁻¹ s⁻¹)	at T (K)	Technique	Reference	Temperature range covered (K)
Thiols							
Methanethiol [CH₃SH]			33.9 ± 3.4	299.8	FP-RF	Atkinson et al.[1]	300–423
			27.3 ± 2.8	347.2			
	8.89	−398 ± 151	23.0 ± 2.3	423.1			
			96.8 ± 9.5[a]	297 ± 2	RR [relative to k(ethene) = 8.57×10^{-12}][b]	Cox and Sheppard[2]	
			48.3 ± 9.8	244	FP-RF	Wine et al.[3]	244–366
			38.4 ± 5.8	270			
			33.7 ± 4.1	298			
			32.2 ± 6.2	333			
	11.5 ± 3.9	−338 ± 100	29.7 ± 4.7	366			
			21 ± 2	293	DF-EPR	Mac Leod et al.[4,5]	
			25.6 ± 4.4	296	DF-RF	Lee and Tang[6]	
			40.8 ± 4.2	254	FP-RF	Wine et al.[7]	254–430
			37.3 ± 4.3	272			
			32.2 ± 3.2	298			
			31.6 ± 4.3	298			
			30.4 ± 1.9	298			
			32.5 ± 0.9	298			
			30.9 ± 1.0	298			
			32.5 ± 2.8	299			
			32.3 ± 3.7	300			
			30.3 ± 2.6	322			
			24.9 ± 3.1	347			
			28.0 ± 2.8	375			
			23.9 ± 1.3	403			
	10.1 ± 1.9	−347 ± 59	22.5 ± 1.4	430			
			37.2 ± 3.7	300	RR [relative to k(propene) = $4.85 \times 10^{-12} e^{504/T}$][b]	Barnes et al.[8]	300–313
			35.0 ± 4.9	313			

TABLE 13. Rate constants k and temperature-dependent parameters for the gas-phase reactions of the OH radical with sulfur-containing organics in the absence of O_2 (unless indicated) — Continued

Organic	$10^{12} \times A$ (cm³ molecule⁻¹ s⁻¹)	B (K)	$10^{12} \times k$ (cm³ molecule⁻¹ s⁻¹)	at T (K)	Technique	Reference	Temperature range covered (K)
			36.6 ± 2.1	270	LP-LIF	Hynes and Wine[9]	270–300
			36.9^c	270			
			33.0	300			
			31.7^c	300			
Methanethiol-d_1 [CH₃SD]			40.4 ± 2.2	253	FP-RF	Wine et al.[7]	253–429
			34.3 ± 3.9	268			
			34.1 ± 4.1	276			
			31.9 ± 2.8	295			
			30.8 ± 2.1	297			
			28.7 ± 1.8	346			
			24.3 ± 1.0	384			
			24.0 ± 1.8	412			
	11.2 ± 1.5	-310 ± 43	23.4 ± 0.7	429			
Methanethiol-d_3 [CD₃SH]			27.4	273	LP-LIF	Hynes and Wine[9]	273–300
			27.9	300			
			27.6^c	300			
Ethanethiol [CH₃CH₂SH]			27 ± 2	293	DF-EPR	Mac Leod et al.[4,5]	
			36.7 ± 1.8	296	DF-RF	Lee and Tang[6]	
			65.5 ± 5.1	252	FP-RF	Wine et al.[7]	252–425
			51.5 ± 3.5	278			
			45.2 ± 6.2	298			
			43.1 ± 6.1	298			
			42.1 ± 3.2	298			
			46.5 ± 2.9	300			
			40.2 ± 1.4	343			
			33.2 ± 3.6	381			
			34.1 ± 3.1	397			
	12.3 ± 3.3	-396 ± 84	33.2 ± 2.7	425			
			46.5 ± 6.0	300	RR [relative to k(propene) = $4.85 \times 10^{-12}e^{504/T}$][b]	Barnes et al.[8]	300–313
			46.9 ± 4.7	313			
1-Propanethiol [CH₃CH₂CH₂SH]			63.1 ± 2.0	257	FP-RF	Wine et al.[7]	257–419
			45.6 ± 1.8	298			
			41.8 ± 5.7	298			
			45.5 ± 2.5	298			
			36.3 ± 1.6	353			
	8.89 ± 2.80	-489 ± 98	29.1 ± 0.9	419			
			55.4 ± 6.5	300	RR [relative to k(propene) = $4.85 \times 10^{-12}e^{504/T}$][b]	Barnes et al.[8]	300–313
			52.7 ± 7.3	313			
2-Propanethiol [(CH₃)₂CHSH]			56.9 ± 9.0	256	FP-RF	Wine et al.[7]	256–429
			40.7 ± 3.7	297			
			42.2 ± 7.1	299			
			39.5 ± 4.4	300			
			31.2 ± 0.9	358			
			33.0 ± 3.4	380			
			35.5 ± 2.6	423			
	11.6 ± 5.5	-386 ± 155	25.1 ± 2.3	429			
			40.8 ± 3.9	300	RR [relative to k(propene) = $4.85 \times 10^{-12}e^{504/T}$][b]	Barnes et al.[8]	300–313
			38.6 ± 2.7	313			

TABLE 13. Rate constants k and temperature-dependent parameters for the gas-phase reactions of the OH radical with sulfur-containing organics in the absence of O_2 (unless indicated) — Continued

Organic	$10^{12} \times A$ (cm³ molecule⁻¹ s⁻¹)	B (K)	$10^{12} \times k$ (cm³ molecule⁻¹ s⁻¹)	at T (K)	Technique	Reference	Temperature range covered (K)
1-Butanethiol [CH₃CH₂CH₂CH₂SH]			43.8 ± 6.6	298	FP-RF	Wine et al.[7]	
			58.2 ± 4.5	300	RR [relative to	Barnes et al.[8]	300–313
			57.3 ± 5.4	313	k(propene) = $4.85 \times 10^{-12}e^{504/T}$][b]		
2-Methyl-1-propanethiol [(CH₃)₂CHCH₂SH]			41.8 ± 6.3	298	FP-RF	Wine et al.[7]	
			47.8 ± 5.5	300	RR [relative to	Barnes et al.[8]	300–313
			39.1 ± 5.9	313	k(propene) = $4.85 \times 10^{-12}e^{504/T}$][b]		
2-Butanethiol [CH₃CH₂CH(CH₃)SH]			39.8 ± 5.9	298	FP-RF	Wine et al.[7]	
			39.5 ± 7.1	300	RR [relative to	Barnes et al.[8]	300–313
			32.8 ± 2.7	313	k(propene) = $4.85 \times 10^{-12}e^{504/T}$][b]		
2-Methyl-2-propanethiol [(CH₃)₃CSH]			47.2 ± 2.3	257	FP-RF	Wine et al.[7]	257–409
			34.2 ± 1.3	298			
			35.7 ± 0.8	298			
			26.3 ± 4.2	348			
	6.22 ± 1.35	−516 ± 67	22.7 ± 0.5	409			
			30.7 ± 3.9	300	RR [relative to	Barnes et al.[8]	300–313
			24.1 ± 3.7	313	k(propene) = $4.85 \times 10^{-12}e^{504/T}$][b]		
2-Methyl-1-butanethiol [CH₃CH₂CH(CH₃)CH₂SH]			54.3 ± 2.9	300	RR [relative to	Barnes et al.[8]	300–313
			44.0 ± 5.9	313	k(propene) = $4.85 \times 10^{-12}e^{504/T}$][b]		

Sulfides

Organic	$10^{12} \times A$ (cm³ molecule⁻¹ s⁻¹)	B (K)	$10^{12} \times k$ (cm³ molecule⁻¹ s⁻¹)	at T (K)	Technique	Reference	Temperature range covered (K)
Dimethyl sulfide [CH₃SCH₃]	5.47	−179 ± 151	9.8 ± 1.2	299.9	FP-RF	Atkinson et al.[10]	300–427
			9.3 ± 1.2	355.3			
			8.2 ± 1.2	426.5			
			10.98 ± 3.37	273	FP-RF	Kurylo[11]	273–400
			8.28 ± 0.87	296			
			10.75 ± 2.85	323			
			7.99 ± 1.37	362			
	6.25 ± 4.19	−131 ± 215	9.28 ± 2.01	400			
			9.77 ± 1.55[a]	297 ± 2	RR [relative to k(ethene) = 8.57×10^{-12}][b]	Cox and Sheppard[2]	
			3.89 ± 0.38	248	FP-RF	Wine et al.[3]	248–363
			4.15 ± 0.55	271			
			4.26 ± 0.56	298			
			4.50 ± 0.68	334			
	6.8 ± 1.1	138 ± 46	4.67 ± 0.51	363			
			9.2 ± 0.6	373	DF-EPR	Mac Leod et al.[4,5]	373–573
			7.8 ± 1	573			
			10.0 ± 0.5[a]	296 ± 2	RR [relative to k(n-hexane) = 5.57×10^{-12}][b]	Atkinson et al.[12]	

TABLE 13. Rate constants k and temperature-dependent parameters for the gas-phase reactions of the OH radical with sulfur-containing organics in the absence of O_2 (unless indicated) — Continued

Organic	$10^{12} \times A$ (cm^3 molecule^{-1} s^{-1})	B (K)	$10^{12} \times k$ (cm^3 molecule^{-1} s^{-1})	at T (K)	Technique	Reference	Temperature range covered (K)
			3.80 ± 0.30	273	DF-EPR	Martin et al.[13]	273–318
			3.22 ± 0.16	293			
			3.66 ± 0.19	318			
			3.6 ± 0.2	297	FP-RF	Wallington et al.[14]	297–400
			3.8 ± 0.7	320			
			3.7 ± 0.9	332			
			3.7 ± 0.4	359			
			3.4 ± 0.4	369			
			3.4 ± 0.4	377			
	$2.5^{+0.9}_{-0.6}$	-130 ± 102	3.3 ± 0.3	400			
			5.36 ± 0.44	296 ± 2	RR [relative to k(cyclohexane) = 7.45×10^{-12}][b]	Wallington et al.[14]	
			9.36 ± 0.67[a]	296 ± 2	RR [relative to k(n-hexane) = 5.57×10^{-12}][b]	Wallington et al.[14]	
			4.17 ± 0.87	276	FP-RF	Hynes et al.[15]	276–397
			4.09 ± 1.16	298			
			4.44 ± 0.23	298			
			4.75 ± 0.71	300			
			5.45 ± 0.89	359			
			5.97 ± 0.07	374			
			5.46 ± 0.52	374			
	13.6 ± 4.0	332 ± 96	5.69 ± 0.46	397			
			4.29 ± 0.48	261	LP-LIF	Hynes et al.[15]	261–321
			4.80 ± 0.11	298			
			4.75 ± 0.15	298			
			3.5 ± 0.4	d	PR-RA	Nielsen et al.[16]	
			4.94 ± 0.15	260	DF-RF	Hsu et al.[17]	260–393
			4.51 ± 0.15	265			
			5.09 ± 0.11	278			
			5.54 ± 0.15	298			
			5.92 ± 0.27	333			
			6.00 ± 0.13	363			
	11.8 ± 2.2	236 ± 150	6.44 ± 0.26	393			
			4.69 ± 0.43	298 ± 3	RR [relative to k(ethene) = 8.52×10^{-12}]	Barnes et al.[18]	
			4.85 ± 0.14	299	FP-RF	Witte and Zetzsch[19]	299–469
			4.76 ± 0.32	299			
			5.20 ± 0.16	323			
			5.43 ± 0.27	348			
			5.36 ± 0.30	348			
			5.79 ± 0.19	373			
			6.09 ± 0.20	398			
			6.13 ± 0.28	398			
			6.22 ± 0.38	423			
			6.63 ± 0.36	442			
			6.13 ± 0.28	447			
			6.29 ± 0.26	448			
			6.57 ± 0.38	453			
			6.28 ± 0.31	459			
			6.67 ± 0.24	463			

TABLE 13. Rate constants k and temperature-dependent parameters for the gas-phase reactions of the OH radical with sulfur-containing organics in the absence of O_2 (unless indicated) — Continued

Organic	$10^{12} \times A$ (cm³ molecule⁻¹ s⁻¹)	B (K)	$10^{12} \times k$ (cm³ molecule⁻¹ s⁻¹)	at T (K)	Technique	Reference	Temperature range covered (K)
			6.59 ± 0.47	468			
	11 ± 1	250 ± 30	6.43 ± 0.58	469			
Dimethyl sulfide-d_6 [CD₃SCD₃]			1.46 ± 0.14	253	FP-RF	Hynes et al.[15]	253–418
			1.95 ± 0.13	299			
			1.87 ± 0.16	299			
			1.98 ± 0.18	299			
			2.53 ± 0.19	360			
			2.72 ± 0.21	360			
	10.3 ± 1.7	498 ± 51	3.11 ± 0.18	418			
			1.82 ± 0.11	298	LP-LIF	Hynes et al.[15]	261–361
Methyl ethyl sulfide			8.50	299	FP-RF	Hynes et al.[15]	
Diethyl sulfide			12 ± 1.4	293	DF-EPR	Martin et al.[13]	
			11.6 ± 2.2	300	RR [relative to k(ethene) = 8.44×10^{-12}][b]	Barnes et al.[20]	
			14.2 ± 1.8	255	FP-RF	Hynes et al.[15]	255–370
			17.6 ± 2.5	255			
			15.4 ± 1.6	269			
			14.5 ± 1.2	299			
			16.1 ± 2.1	299			
			15.4 ± 2.3	338			
	13.9 ± 6.3 [15.5 ± 2.2	−31 ± 132 0]	15.1 ± 2.2	370			
			4.5 ± 0.5	d	PR-RA	Nielsen et al.[16]	
Ethyl propyl sulfide			4.9 ± 0.5	d	PR-RA	Nielsen et al.[16]	
Di-n-propyl sulfide			20.0 ± 2.2	300	RR [relative to k(ethene) = 8.44×10^{-12}][b]	Barnes et al.[20]	
			5.2 ± 0.5	d	PR-RA	Nielsen et al.[16]	
Dimethyl disulfide [CH₃SSCH₃]			240 ± 86[a]	297 ± 2	RR [relative to k(ethene) = 8.57×10^{-12}][b]	Cox and Sheppard[2]	
			280 ± 18	249	FP-RF	Wine et al.[3]	249–367
			198 ± 18	298			
	59 ± 33	−380 ± 160	171 ± 25	367			
			192 ± 24	300	RR [relative to k(trans-2-butene) = 6.32×10^{-11}][b]	Barnes et al.[20,21]	
			300 ± 30	d	PR-RA	Nielsen et al.[16]	
Di-$tert$-butyl disulfide [(CH₃)₃CSSC(CH₃)₃]			41 ± 4	d	PR-RA	Nielsen et al.[16]	

TABLE 13. Rate constants k and temperature-dependent parameters for the gas-phase reactions of the OH radical with sulfur-containing organics in the absence of O_2 (unless indicated) — Continued

Organic	$10^{12} \times A$ (cm^3 molecule^{-1} s^{-1})	B (K)	$10^{12} \times k$ (cm^3 molecule^{-1} s^{-1})	at T (K)	Technique	Reference	Temperature range covered (K)
Thioethers							
Tetrahydrothiophene[e]			23.2 ± 1.3	255	FP-RF	Wine and Thompson[22]	255–377
			20.9 ± 1.9	255			
			19.8 ± 3.4	298			
			18.4 ± 1.0	298			
			18.8 ± 1.8	338			
			19.5 ± 0.6	377			
			16.2 ± 1.4	377			
	11.3 ± 3.5	-166 ± 97	17.4 ± 1.3	377			
			21.2 ± 1.6	293	DF-EPR	Martin *et al.*[13]	
	9.7 ± 0.9	-240 ± 40	21.8^f	297	FP-RF	Witte and Zetzsch[19]	297–399
Thiophene[e]			47.7 ± 6.3	295 ± 1	DF-RF	Lee and Tang[23]	
			9.42 ± 0.34^a	298 ± 2	RR [relative to $k(n\text{-hexane}) = 5.61 \times 10^{-12}$]^b	Atkinson *et al.*[24]	
			50 ± 4	293	DF-EPR	Mac Leod *et al.*[5,25]	293–473
			22 ± 2	333			
			12 ± 2	373			
	0.13 ± 0.08	-1750 ± 200	5.2 ± 0.5	473			
			11.4 ± 0.6	255	FP-RF	Wine and Thompson[22]	255–425
			11.5 ± 0.9	255			
			9.57 ± 1.15	298			
			9.37 ± 0.66	298			
			8.20 ± 0.68	353			
			7.28 ± 0.41	419			
			6.06 ± 0.37	425			
	3.20 ± 0.70	-325 ± 71	7.37 ± 0.41	425			
			9.6 ± 1.5^g	300	RR [relative to $k(\text{propene}) = 2.60 \times 10^{-11}$]^b	Barnes *et al.*[26]	
			12 ± 1	293	DF-EPR	Martin *et al.*[13]	
			10.1 ± 0.5	274 ± 2	FP-RF	Wallington[27]	274–382
			8.9 ± 0.7	298 ± 2			
			6.1 ± 1.2	325			
			5.5 ± 0.3	349			
			6.3 ± 0.6	365			
			5.3 ± 0.5	379			
	$1.2^{+1.0}_{-0.6}$	-584 ± 217	5.8 ± 0.5	382			
			10.6 ± 0.5	298	FP-RF	Witte and Zetzsch[19]	298–471
			10.9 ± 0.7	298			
			13.1 ± 2.3	299			
			10.9 ± 0.5	300			
			11.7 ± 0.9	300			
			10.4 ± 0.5	312			
			10.5 ± 0.7	322			
			10.1 ± 0.5	335			
			9.4 ± 0.3	349			
			9.1 ± 0.5	360			
			8.7 ± 0.5	373			
			7.8 ± 0.3	380			
			8.3 ± 0.4	380			

TABLE 13. Rate constants k and temperature-dependent parameters for the gas-phase reactions of the OH radical with sulfur-containing organics in the absence of O_2 (unless indicated) — Continued

Organic	$10^{12} \times A$ (cm³ molecule⁻¹ s⁻¹)	B (K)	$10^{12} \times k$ (cm³ molecule⁻¹ s⁻¹)	at T (K)	Technique	Reference	Temperature range covered (K)
			7.7 ± 0.4	400			
			6.6 ± 0.3	414			
			7.5 ± 0.3	422			
			7.4 ± 0.2	434			
			7.5 ± 0.2	434			
			6.3 ± 0.2	438			
			7.5 ± 0.3	441			
			6.3 ± 0.3	442			
			7.2 ± 0.2	442			
			7.0 ± 0.2	442			
			6.0 ± 0.2	448			
			6.1 ± 0.2	452			
			5.6 ± 0.2	457			
			5.9 ± 0.4	463			
			5.4 ± 0.1	465			
			5.6 ± 0.3	468			
	1.9 ± 0.5	−540 ± 110	5.3 ± 0.5	471			
Thiazole[e]	0.94 ± 0.07	−120 ± 30	1.41[f]	297	FP-RF	Witte and Zetzsch[19]	297–423
Miscellaneous							
Dimethyl sulfoxide [(CH₃)₂SO]			62 ± 25[a]	300	RR [relative to k(cis-2-butene) = 5.58×10^{-11}][b]	Barnes et al.[21]	

[a]At atmospheric pressure of air.
[b]From the present recommendations (see text).
[c]In the presence and absence of O_2.
[d]Room temperature, not specified.

[e]Structures: tetrahydrothiophene, 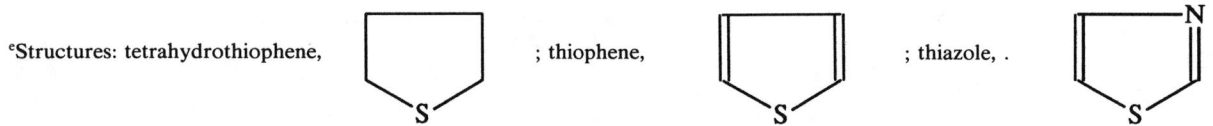 ; thiophene, ; thiazole, .

[f]Calculated from cited Arrhenius expression.
[g]Independent of O_2 pressure over the range 10–300 Torr.

TABLE 14. Rate constants k and temperature-dependent parameters for the gas-phase reactions of the OH radical with sulfur-containing organics as a function of the O_2 concentration

Organic	O_2 Pressure (Torr)	$10^{12} \times k$ (cm^3 molecule^{-1} s^{-1})	at T (K)	Technique	Reference	Temperature range covered (K)
Thiols						
Methanethiol	0	36.6 ± 2.1	270	LP-LIF	Hynes and Wine[9]	270–300
(CH$_3$SH)	69.3	35.9 ± 3.6	270			
	147	38.2 ± 3.6	270			
	0	32.8 ± 4.0	300			
	0	32.8 ± 1.8	300			
	0	33.8 ± 1.5	300			
	0	32.7 ± 3.6	300			
	31.5	29.3 ± 1.8	300			
	84.0	28.7 ± 1.4	300			
	94.5	33.0 ± 3.3	300			
	147	29.3 ± 1.6	300			
	147	31.5 ± 4.8	300			
	200	28.0 ± 8.0	300			
	520	35.0 ± 5.0	300			
	700	33.1 ± 2.2	300			
Methanethiol-d_3	0, 230	27.4 ± 1.3	273	LP-LIF	Hynes and Wine[9]	273–300
(CD$_3$SH)	0	27.6 ± 6.0	300			
	0	28.2 ± 3.1	300			
	21.0	26.4 ± 1.0	300			
	94.5	28.1 ± 3.3	300			
Sulfides						
Dimethyl sulfide	51	9.25 ± 1.06	296 ± 2	RR [relative to	Wallington et al.[14]	
(CH$_3$SCH$_3$)	110	9.13	296 ± 2	k(n-hexane)		
	154	9.36 ± 0.67	296 ± 2	$= 5.57 \times 10^{-12}$]^{a,b}		
	368	10.4 ± 0.8	296 ± 2			
	740	14.0 ± 0.7	296 ± 2			
	0	5.36 ± 0.44	296 ± 2	RR [relative to	Wallington et al.[14]	
	50	7.08 ± 0.10	296 ± 2	k(cyclohexane)		
	160	8.64 ± 0.15	296 ± 2	$= 7.45 \times 10^{-12}$]^{a,c}		
	740	10.7 ± 0.3	296 ± 2			
	0	4.29 ± 0.48	261	LP-LIF	Hynes et al.[15]	261–321
	147	12.5 ± 1.7	262			
	147	9.53 ± 0.28	279			
	0	4.80 ± 0.11	298			
	0	4.75 ± 0.15	298			
	10.5	4.68 ± 0.08	298			
	27.3	5.04 ± 0.14	298			
	71.4	5.18 ± 0.34	298			
	124	5.80 ± 0.16	298			
	158	6.28 ± 0.10	298			
	147	5.43 ± 0.30	321			
	0	4.69 ± 0.43	298 ± 3	RR [relative to	Barnes et al.[18]	
	50	5.62 ± 0.77	298 ± 3	k(ethene)		
	100	7.16 ± 0.94	298 ± 3	$= 8.52 \times 10^{-12}$]^{a,d}		
	155	8.52 ± 0.52	298 ± 3			
	760	12.4 ± 1.3	298 ± 3			
Dimethyl sulfide-d_6	147	11.6 ± 1.1	261	LP-LIF	Hynes et al.[15]	261–361
(CD$_3$SCD$_3$)	700	13.5 ± 1.2	266			
	700	11.9 ± 2.0	275			
	147	9.63 ± 0.63	276			
	147	5.29 ± 0.44	287			
	700	6.99 ± 0.53	287			
	0	1.82 ± 0.11	298			

TABLE 14. Rate constants k and temperature-dependent parameters for the gas-phase reactions of the OH radical with sulfur-containing organics as a function of the O_2 concentration — Continued

Organic	O_2 Pressure (Torr)	$10^{12} \times k$ (cm^3 molecule^{-1} s^{-1})	at T (K)	Technique	Reference	Temperature range covered (K)
	21.0	2.10 ± 0.15	298			
	63.0	2.68 ± 0.09	298			
	105	2.97 ± 0.13	298			
	147	3.40 ± 0.13	298			
	700	6.50 ± 0.72	298			
	147	3.02 ± 0.18	317			
	700	3.72 ± 0.27	321			
	147	2.32 ± 0.11	340			
	700	2.30 ± 0.28	340			
	147	2.66 ± 0.11	361			

[a]From present recommendations (see text).
[b]OH radicals generated from photolysis of CH_3ONO-NO-O_2-N_2 mixtures.
[c]OH radicals generated from the dark $N_2H_4 + O_3$ reaction.
[d]OH radicals generated from photolysis of H_2O_2 in N_2-O_2 mixtures.

(b) Ethanethiol

The available rate constants of Mac Leod et al.,[4,5] Lee and Tang,[6] Wine et al.,[7] and Barnes et al.[8] are given in Table 13 and are plotted in Arrhenius form in Fig. 81.

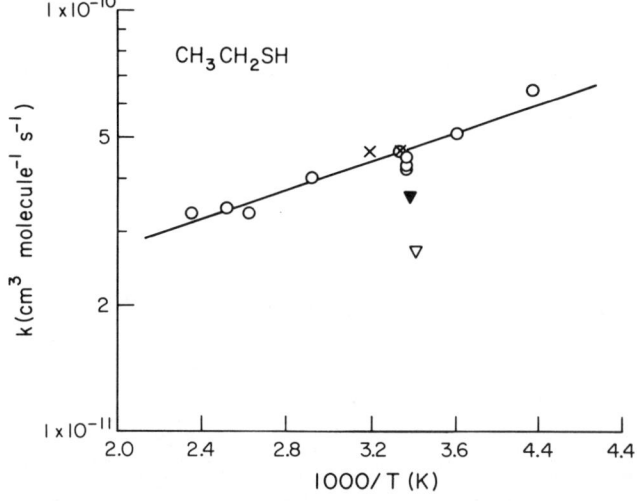

FIG. 81. Arrhenius plot of rate constants for the reaction of the OH radical with ethanethiol. ($\triangledown$) Mac Leod et al.;[4,5] ($\blacktriangledown$) Lee and Tang;[6] ($\bigcirc$) Wine et al.;[7] (x) Barnes et al.;[8] (———) recommendation (see text).

As for methanethiol, the two discharge flow studies of Mac Leod et al.[4,5] and Lee and Tang[6] yield somewhat lower room temperature rate constants than do the flash photolysis study of Wine et al.[7] and the relative rate study of Barnes et al.,[8] which are in excellent agreement. A unit-weighted least-squares analysis of the rate constant data of Wine et al.[7] and Barnes et al.[8] leads to the recommended Arrhenius expression of

$$k(\text{ethanethiol}) = (1.23^{+0.36}_{-0.28})$$

$$\times 10^{-11} e^{(398 \pm 80)/T} \text{ cm}^3 \text{ molecule}^{-1} \text{ s}^{-1}$$

over the temperature range 252–425 K, where the indicated error limits are two least-squares standard deviations, and

$$k(\text{ethanethiol}) = 4.68 \times 10^{-11} \text{ cm}^3 \text{ molecule}^{-1} \text{ s}^{-1}$$

at 298 K, with an estimated overall uncertainty at 298 K of ±25%. This recommendation is essentially identical to that of

$$k(\text{ethanethiol}) = 1.23 \times 10^{-11} e^{396/T} \text{ cm}^3 \text{ molecule}^{-1} \text{ s}^{-1}$$

recommended by Atkinson.[30]

(c) 1-Propanethiol, 2-Propanethiol, 1-Butanethiol, 2-Methyl-1-propanethiol, 2-Butanethiol and 2-Methyl-2-propanethiol

The available rate constants determined by Wine et al.[7] and Barnes et al.[8] are given in Table 13, and the data for 1-propanethiol, 2-propanethiol and 2-methyl-2-propanethiol are plotted in Arrhenius form in Figs. 82 through 84, respectively. (For 1-butanethiol, 2-methyl-1-propanethiol and 2-butanethiol rate constants are available only at 298, 300 and 313 K). The room temperature rate constants from these studies of Wine et al.[7] and Barnes et al.[8] are in agreement within ≲30% for these thiols.

Unit-weighted least-squares analyses of these rate constants of Wine et al.[7] and Barnes et al.[8] lead to the recommended Arrhenius expressions of

$k(\text{1-propanethiol}) = (8.93^{+7.95}_{-4.21})$

$\times\ 10^{-12}\ e^{(503\ \pm\ 197)/T}\ \text{cm}^3\ \text{molecule}^{-1}\ \text{s}^{-1}$

over the temperature range 257–419 K,

$k(\text{2-propanethiol}) = (1.17^{+0.60}_{-0.40})$

$\times\ 10^{-11}\ e^{(381\ \pm\ 133)/T}\ \text{cm}^3\ \text{molecule}^{-1}\ \text{s}^{-1}$

over the temperature range 256–429 K, and

$k(\text{2-methyl-2-propanethiol}) = (6.05^{+6.69}_{-3.18})$

$\times\ 10^{-12}\ e^{(506\ \pm\ 231)/T}\ \text{cm}^3\ \text{molecule}^{-1}\ \text{s}^{-1}$

over the temperature range 257–409 K, where the indicated errors are two least-squares standard deviations.

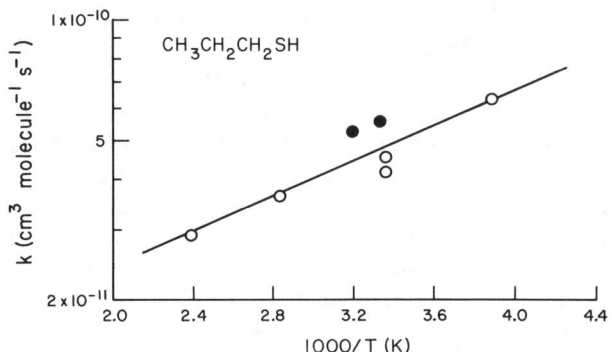

FIG. 82.   Arrhenius plot of rate constants for the reaction of the OH radical with 1-propanethiol. (○) Wine et al.;[7] (●) Barnes et al.;[8] (——) recommendation (see text).

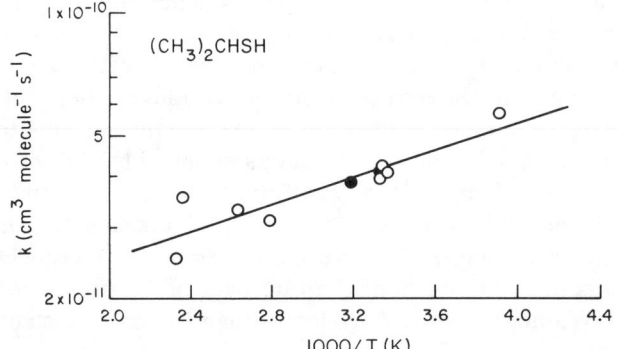

FIG. 83.   Arrhenius plot of rate constants for the reaction of the OH radical with 2-propanethiol. (○) Wine et al.;[7] (●) Barnes et al.;[8] (——) recommendation (see text).

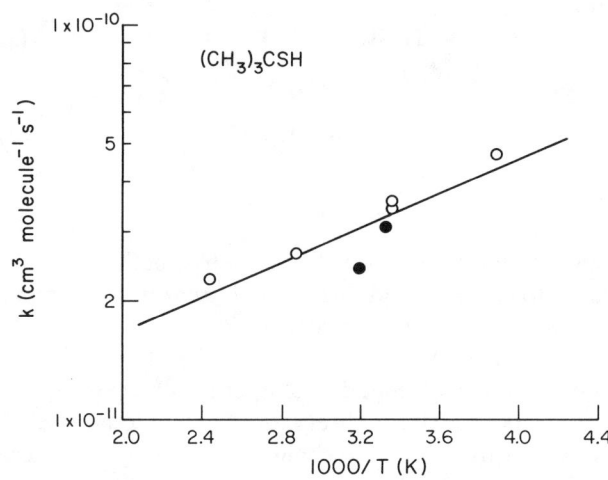

FIG. 84.   Arrhenius plot of rate constants for the reaction of the OH radical with 2-methyl-2-propanethiol. (○) Wine et al.;[7] (●) Barnes et al.;[8] (——) recommendation (see text).

At 298 K,

$k(\text{1-propanethiol}) = 4.83 \times 10^{-11}\ \text{cm}^3\ \text{molecule}^{-1}\ \text{s}^{-1}$,

$k(\text{2-propanethiol}) = 4.20 \times 10^{-11}\ \text{cm}^3\ \text{molecule}^{-1}\ \text{s}^{-1}$,

and

$k(\text{2-methyl-2-propanethiol}) = 3.31$

$\times\ 10^{-11}\ \text{cm}^3\ \text{molecule}^{-1}\ \text{s}^{-1}$,

all with estimated overall uncertainties of ±30% at 298 K. Unit-weighted averages of the 298–300 K rate constants of Wine et al.[7] and Barnes et al.[8] for 1- and 2-butanethiol and 2-methyl-1-propanethiol lead to the recommended rate constants at 298 K of

$k(\text{1-butanethiol}) = 5.1 \times 10^{-11}\ \text{cm}^3\ \text{molecule}^{-1}\ \text{s}^{-1}$,

$k(\text{2-methyl-1-propanethiol}) = 4.5$

$\times\ 10^{-11}\ \text{cm}^3\ \text{molecule}^{-1}\ \text{s}^{-1}$,

and

$k(\text{2-butanethiol}) = 4.0 \times 10^{-11}\ \text{cm}^3\ \text{molecule}^{-1}\ \text{s}^{-1}$,

with estimated overall uncertainties of ±35% at 298 K.

No recommendation is made for 2-methyl-1-butanethiol, since only a single kinetic study has been carried out.

(2) Mechanism

There are three possible pathways for the reaction of OH radicals with the thiols, taking methanethiol as an example:

$$OH + CH_3SH \rightarrow H_2O + \dot{C}H_2SH \qquad (a)$$

$$\rightarrow H_2O + CH_3S \qquad (b)$$

$$\rightarrow CH_3SH \qquad (c)$$
$$|$$
$$OH$$

The non-deuterated thiols for which kinetic data are available (Tables 13 and 14) all have reasonably similar room temperature rate constants, ranging from 3.3×10^{-11} cm^3 molecule^{-1} s^{-1} to $\sim 5.1 \times 10^{-11}$ cm^3 molecule^{-1} s^{-1}, and negative temperature dependencies of $B \approx -400$ K. These observations indicate no significant effect of the alkyl side chain on the kinetics of these reactions. This is further confirmed by the small isotope effect observed for CD$_3$SH compared to CH$_3$SH,[9] which suggests that H-atom abstraction from the —CH$_3$ group is a minor, but not totally negligible, reaction process for the OH radical reaction with CH$_3$SH. Furthermore, the rate constants for CH$_3$SD are virtually identical to those for CH$_3$SH,[7] indicating no deuterium isotope effect within the experimental error limits.

These kinetic observations show that over the temperature range ~ 250–400 K the H-atom abstraction channels (a) and (b) are of minor importance, and that the major reaction pathway involves OH radical addition to the S atom [reaction pathway (c)]. While this is in agreement with the conclusions of Hatakeyama and Akimoto[31] obtained from a product study carried out in air at atmospheric pressure in the presence of NO, the now recognized occurrence of secondary reactions removing CH$_3$SH in these chemical systems[26,30] may lead to added complexities in the analysis of the experimental data of Hatakeyama and Akimoto.[31] The fate of the RS(OH)H adduct requires further study.

b. Sulfides

The available kinetic data are given in Tables 13 and 14. The majority of these data deal with the reaction of the OH radical with dimethyl sulfide and, since these kinetic studies provide the most definitive data concerning the reaction mechanisms, the kinetics and mechanisms of these OH radical reactions are discussed together in the remainder of this section.

(1) Dimethyl Sulfide and Dimethyl Sulfide-d$_6$ (CD$_3$SCD$_3$)

The available kinetic data of Atkinson et al.,[10,12] Kurylo,[11] Cox and Sheppard,[2] Wine et al.,[3] Mac Leod et al.,[4,5] Martin et al.,[13] Wallington et al.,[14] Hynes et al.,[15] Nielsen et al.,[16] Hsu et al.,[17] Barnes et al.,[18] and Witte and Zetzsch[19] are given in Tables 13 and 14. In addition, preliminary data were reported for the reaction of CH$_3$SCH$_3$ at room temperature from the relative rate studies of Barnes et al.[26] and Nielsen et al.[16]

The relative rate studies of Barnes et al.,[26] Wallington et al.[14] and Nielsen et al.[16] show that erroneously high rate constants are obtained for the reaction of the OH radical with dimethyl sulfide (and for CH$_3$SH[26] and diethyl sulfide[16]) when these studies are conducted in the presence of oxides of nitrogen, due to secondary reactions removing CH$_3$SCH$_3$ (possibly involving the CH$_3$SO radical). Thus, the data obtained from the relative rate studies of Cox and Sheppard,[2] Atkinson et al.,[12] Barnes et al.[26] (not cited in Table 13 or Table 14) and Wallington et al.[14] using irradiated HONO—NO—air[2] and CH$_3$ONO—NO—air[12,14,26] mixtures to generate the OH radical are in error and are not discussed further here. It is also possible that the relative rate data obtained by Wallington et al.,[14] using the dark N$_2$H$_4$—O$_3$ reaction to generate OH radicals, were also subject to the occurrence of secondary reactions removing dimethyl sulfide, and these data must be judged to be of a qualitative nature only.[14]

Furthermore, the absolute rate constant data of Mac Leod et al.[4,5] have been shown by a subsequent study of Martin et al.[13] to be in error due to the occurrence of heterogeneous reactions on the flow tube walls. The absolute rate data of Atkinson et al.,[10] Kurylo,[11] Wine et al.,[3] Martin et al.,[13] Wallington et al.,[14] Hynes et al.,[15] Nielsen et al.,[16] Hsu et al.[17] and Witte and Zetzsch[19] and the relative rate constant of Barnes et al.[18] then remain to be considered. The laser photolysis-laser induced fluorescence study of Hynes et al.[15] showed that the measured rate constants are dependent on the O$_2$ concentration for both CH$_3$SCH$_3$ and CD$_3$SCD$_3$, increasing with increasing O$_2$ concentration, and this observation has been confirmed by the relative rate study of Barnes et al.,[18] using the photolysis of H$_2$O$_2$ as an OH radical source. Thus, the evaluation of the rate constants for these reactions can be best carried out by first considering the data obtained in the absence of O$_2$, and then dealing with the O$_2$ dependence of the rate constant.

The rate constants obtained in the absence of O$_2$ by Atkinson et al.,[10] Kurylo,[11] Wine et al.,[3] Martin et al.,[13] Wallington et al.,[14] Hynes et al.,[15] Hsu et al.,[17] Barnes et al.[18] and Witte and Zetzsch[19] are plotted in Arrhenius form in Fig. 85 (the absolute room temperature rate constant of Nielsen et al.[16] is not plotted since the temperature was not specified).

Clearly, there is a significant degree of scatter in the rate constants determined from the various studies, with the reported room temperature rate constants varying by a factor of ~ 5. The rate constants obtained by Atkinson et al.[10] and Kurylo[11] are significantly higher than those from the other studies plotted,[3,13–15,17–19] presumably due to the presence of reactive impurities in the CH$_3$SCH$_3$ reactant.[3,14] In addition, the rate data of Martin et al.[13] and Wallington et al.[14] are lower than the rate constants of Wine et al.,[3] Hynes et al.,[15] Hsu et al.,[17] Barnes et al.[18] and Witte and Zetzsch,[19] and exhibit essentially zero or slightly negative temperature dependencies, in contrast to the positive temperature dependencies observed by the absolute studies of Wine et al.,[3] Hynes et al.,[15] Hsu et al.[17] and Witte and Zetzsch.[19]

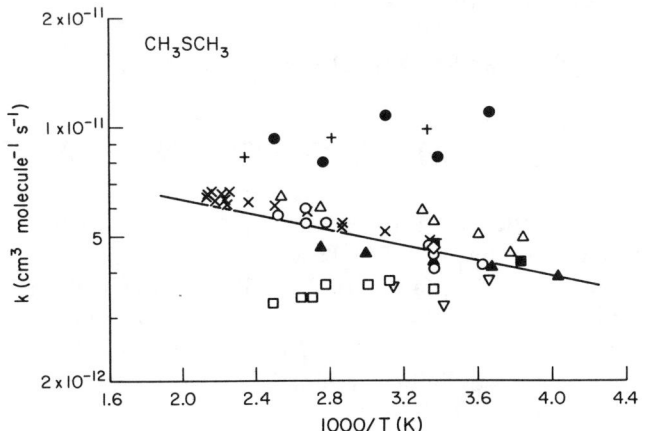

J. Phys. Chem. Ref. Data, Monograph 1 (1989)

FIG. 85. Arrhenius plot of rate constants, obtained in the absence of O_2, for the reaction of the OH radical with dimethyl sulfide. (+) Atkinson et al.;[10] (●) Kurylo;[11] (▲) Wine et al.;[3] (▽) Martin et al.;[13] (☐) Wallington et al.,[14] absolute rate data; (○) Hynes et al.,[15] FP-RF data; (■) Hynes et al.,[15] LP-LIF data; (△) Hsu et al.;[17] (◇) Barnes et al.;[18] (x) Witte and Zetzsch;[19] (———) recommendation (see text).

Since the rate constants determined from the discharge flow study of Hsu et al.[17] are ~20% higher than the data of Wine et al.,[3] Hynes et al.,[15] Barnes et al.[18] and Witte and Zetzsch,[19] and few details are available concerning the study of Witte and Zetzsch,[19] a unit-weighted least-squares analysis of the rate constant data of Wine et al.,[3] Hynes et al.[15] and Barnes et al.,[18] using the Arrhenius expression $k = A e^{-B/T}$, yields the recommendation of

$$k(\text{dimethyl sulfide}) = (1.03^{+0.28}_{-0.23})$$

$$\times \ 10^{-11} \ e^{-(243 \pm 76)/T} \ \text{cm}^3 \ \text{molecule}^{-1} \ \text{s}^{-1}$$

in the absence of O_2 over the temperature range 248–397 K, where the indicated error limits are two least-squares standard deviations, and

$$k(\text{dimethyl sulfide}) = 4.56 \times 10^{-12} \ \text{cm}^3 \ \text{molecule}^{-1} \ \text{s}^{-1}$$

in the absence of O_2 at 298 K, with an estimated overall uncertainty at 298 K of ±30%. The data of Witte and Zetzsch[19] are in excellent agreement (within ~10%) with this recommendation.

For the reaction of the OH radical with CD_3SCD_3 in the absence of O_2, the flash and laser photolysis data of Hynes et al.[15] lead to

$$k(\text{dimethyl sulfide-}d_6) = (1.05^{+0.22}_{-0.19})$$

$$\times \ 10^{-11} \ e^{-(505 \pm 60)/T} \ \text{cm}^3 \ \text{molecule}^{-1} \ \text{s}^{-1}$$

over the temperature range 253–418 K, where the indicated error limits are two least-squares standard deviations, and

$$k(\text{dimethyl sulfide-}d_6) = 1.93$$

$$\times \ 10^{-12} \ \text{cm}^3 \ \text{molecule}^{-1} \ \text{s}^{-1}$$

at 298 K, in the absence of O_2. These rate constant expressions for CH_3SCH_3 and CD_3SCD_3, in the absence of O_2, yield a deuterium isotope effect of

$$\frac{k(\text{CH}_3\text{SCH}_3)}{k(\text{CD}_3\text{SCD}_3)} = \frac{k^H}{k^D} = 0.98 \ e^{262/T}$$

This deuterium isotope effect is similar in magnitude to that observed for H-atom abstraction from secondary C—H bonds in the alkanes (Sec. 2.1). Since the C—H bonds in CH_3SCH_3 have a similar bond dissociation energy (96.6 kcal mol^{-1} [29]) to the alkane secondary C—H bonds (96.1 kcal mol^{-1} for propane[29]), this deuterium isotope effect indicates that in the absence of O_2 the OH radical reaction proceeds by H (or D) atom abstraction from the —CH_3 (or —CD_3) groups.

$$\text{OH} + \text{CH}_3\text{SCH}_3 \rightarrow \text{H}_2\text{O} + \text{CH}_3\text{S}\dot{\text{C}}\text{H}_2 \qquad \text{(a)}$$

Thus, the recommended rate expression given above for CH_3SCH_3 (and the analogous rate expression obtained from the data of Hynes et al.[15] for CD_3SCD_3) is for this reaction channel (a):

$$k_a(\text{CH}_3\text{SCH}_3) = (1.03^{+0.28}_{-0.23})$$

$$\times \ 10^{-11} \ e^{-(243 \pm 76)/T} \ \text{cm}^3 \ \text{molecule}^{-1} \ \text{s}^{-1}$$

over the temperature range 248–397 K, and

$$k_a(\text{CH}_3\text{SCH}_3) = 4.56 \times 10^{-12} \ \text{cm}^3 \ \text{molecule}^{-1} \ \text{s}^{-1}$$

at 298 K.

The observations of Hynes et al.[15] and Barnes et al.[18] that at around room temperature and below the measured rate constant k_{obs} increases with increasing O_2 concentration are interpreted as showing that reaction of the OH-dimethyl sulfide addition adduct with O_2 occurs in competition with rapid back-decomposition to reactants. Thus, the product of the OH radical addition channel (b),

$$\text{OH} + \text{CH}_3\text{SCH}_3 \rightarrow \text{CH}_3\text{S(OH)CH}_3 \qquad \text{(b)}$$

must react with O_2, in competition with dissociation back to the reactants:

$$\text{CH}_3\text{S(OH)CH}_3 + \text{O}_2 \rightarrow \text{products} \qquad \text{(d)}$$

$$\text{CH}_3\text{S(OH)CH}_3 \rightarrow \text{OH} + \text{CH}_3\text{SCH}_3 \qquad \text{(-b)}$$

Hence the measured rate constant k_{obs} is given by[15]

$$k_{obs} = k_a + \frac{k_b k_d \ [\text{O}_2]}{k_{-b} + k_d \ [\text{O}_2]} = k_{abstr} + k_{add}$$

The data of Hynes et al.[15] for CH_3SCH_3 and CD_3CD_3 and of Barnes et al.[18] for CH_3SCH_3 at 298 K exhibit this behavior, as shown by the plot of $k_{add} = (k_{obs}-k_a)$ in Fig. 86.

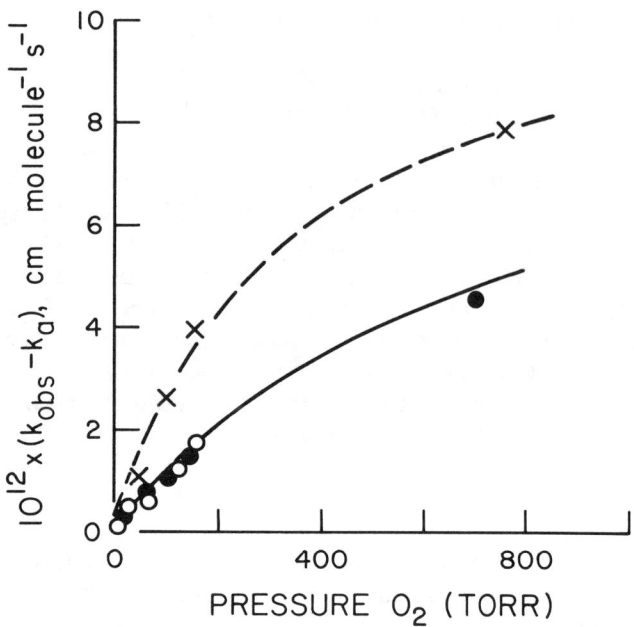

FIG. 86. Plot of the measured 298 K rate constants for OH radical addition to dimethyl sulfide and dimethyl sulfide-d_6 ($k_{obs}-k_a$, see text) as a function of the O_2 pressure. CH_3SCH_3: (O) Hynes et al.;[15] (x, — — —) Barnes et al.;[18] CD_3SCD_3: (●) Hynes et al.;[15] (———) recommendation (see text).

The effect of O_2 on the rate constant k_{add} is much more pronounced in the data obtained by Barnes et al.[18] from a relative rate study. While the reasons for this discrepancy are not clear, the absolute rate data of Hynes et al.[15] in the presence of O_2 (which are also available over the temperature ranges 262–321 K for CH_3SCH_3 and 261–361 K for CD_3SCD_3) are preferred. Hynes et al.[15] derived the expression

$$k_{add} = \frac{1.68 \times 10^{-42} [O_2] e^{7812/T}}{(1 + 5.53 \times 10^{-31} [O_2] e^{7460/T})} \text{ cm}^3 \text{ molecule}^{-1} \text{ s}^{-1}$$

from a best fit to their data for CD_3SCD_3 in the presence of 700 Torr total pressure of air or O_2 (assuming that the temperature dependence for channel (b) was $B = -350$ K, consistent with the temperature dependencies for OH radical addition to the thiols). This expression also fits the data of Hynes et al.[15] for CH_3SCH_3 and CD_3SCD_3 as a function of the O_2 concentration at 298 K (Fig. 86), as well as the rate constants for CH_3SCH_3 at 700 Torr total pressure of air at 262, 279 and 321 K, indicating no deuterium isotope effect on k_{add}.

Accordingly, the expression

$k_{add}(CH_3SCH_3$ and $CD_3SCD_3) =$

$$\frac{1.68 \times 10^{-42} [O_2] e^{7812/T}}{(1 + 5.53 \times 10^{-31} [O_2] e^{7460/T})} \text{ cm}^3 \text{ molecule}^{-1} \text{ s}^{-1}$$

is recommended for pressures of $O_2 \leqslant 700$ Torr over the temperature range 261–361 K, with an estimated overall uncertainty of ± a factor of 2 at all pressures and temperatures within the temperature and pressure limits of this recommendation. This OH radical addition channel becomes of negligible importance at temperatures $\gtrsim 350$ K.

Thus,

$k(CH_3SCH_3) = 1.03 \times 10^{-11} e^{-243/T} +$

$$\frac{1.68 \times 10^{-42} [O_2] e^{7812/T}}{(1 + 5.53 \times 10^{-31} [O_2] e^{7460/T})} \text{ cm}^3 \text{ molecule}^{-1} \text{ s}^{-1}$$

over the temperature range ~260–400 K for O_2 pressures $\leqslant 700$ Torr. At 760 Torr total pressure of air and 298 K the overall rate constant for CH_3SCH_3 calculated from the recommendation is

$$k(CH_3SCH_3) = 6.30 \times 10^{-12} \text{ cm}^3 \text{ molecule}^{-1} \text{ s}^{-1},$$

and ~70% of the OH radical reaction proceeds by H-atom abstraction under these conditions.

(2) Diethyl Sulfide

The available rate constant data of Martin et al.,[13] Barnes et al.,[20] Hynes et al.[15] and Nielsen et al.,[16] all obtained in the absence of O_2, are given in Table 13, and those of Martin et al.,[13] Barnes et al.[20] and Hynes et al.[15] are plotted in Arrhenius form in Fig. 87 (the temperature of the room temperature study of Nielsen et al.[16] was not specified).

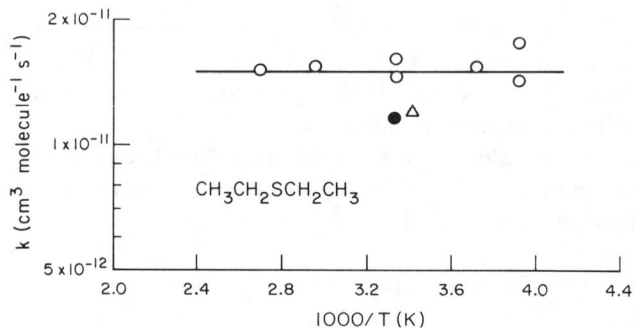

FIG. 87. Arrhenius plot of rate constants for the reaction of the OH radical with diethyl sulfide. (Δ) Martin et al.;[13] (●) Barnes et al.;[20] (O) Hynes et al.;[15] (———) recommendation (see text).

At room temperature, the rate constants of Martin *et al.*,[13] Barnes *et al.*[20] and Hynes *et al.*[15] are in reasonable agreement, but (Table 13) are a factor of ~ 3 higher than the pulsed radiolysis value of Nielsen *et al.*[16] Due to a lack of details concerning the relative rate study of Barnes *et al.*[20] and the significantly low rate constants obtained by Martin *et al.*[13] for dimethyl sulfide, these two studies are given a lower weight in the evaluation of the rate constant for this reaction. A rate constant of

$$k(\text{diethyl sulfide}) = 1.5 \times 10^{-11} \text{ cm}^3 \text{ molecule}^{-1} \text{ s}^{-1},$$

independent of temperature over the range 255–370 K, is recommended, with an estimated overall uncertainty of $\pm 35\%$ over this temperature range. By analogy with dimethyl sulfide, this recommended rate constant refers to the H-atom abstraction reactions,

$$\text{OH} + \text{CH}_3\text{CH}_2\text{SCH}_2\text{CH}_3 \begin{cases} \rightarrow \text{H}_2\text{O} + \text{CH}_3\text{CH}_2\text{S}\dot{\text{C}}\text{HCH}_3 & \text{(a)} \\ \rightarrow \text{H}_2\text{O} + \text{CH}_3\text{CH}_2\text{SCH}_2\dot{\text{C}}\text{H}_2 & \text{(b)} \end{cases}$$

with reaction channel (a) being estimated to dominate.[32] The addition pathway is expected to be operative in the presence of O_2, although on a relative basis its magnitude may be small compared to the H-atom abstraction process.[32]

(3) Other Sulfides

The only rate constants available for other sulfides are those determined by Nielsen *et al.*[16] for ethyl propyl sulfide and by Nielsen *et al.*[16] and Barnes *et al.*[20] for di-*n*-propyl sulfide, all being obtained at room temperature. The data of Nielsen *et al.*[16] and Barnes *et al.*[20] for di-*n*-propyl sulfide disagree by a factor of 4, and it appears that the rate constants measured by Nielsen *et al.*[16] for a series of sulfides are systematically low, exhibiting essentially no variation with the increasing complexity of the sulfide, contrary to other data.[15,18,20]

c. Disulfides

(1) Dimethyl Disulfide

The available kinetic data of Cox and Sheppard,[2] Wine *et al.*,[3] Barnes *et al.*[20,21] and Nielsen *et al.*[16] are given in Table 13 and (apart from the rate constant of Nielsen *et al.*,[16] for which the temperature was not specified) are plotted in Arrhenius form in Fig. 88. The room temperature rate constants of Cox and Sheppard,[2] Wine *et al.*[3] and Barnes *et al.*[20,21] are in good agreement, but are $\sim 50\%$ lower than the room temperature rate constant determined by Nielsen *et al.*[16] Because of the lack of details concerning the Barnes *et al.*[20,21] study and the fact that the relative rate study of Cox and Sheppard[2] was carried out in the presence of oxides of nitrogen (al-

though there is no evidence that secondary consumption of CH_3SSCH_3 was occurring), the recommended Arrhenius expression is derived from a unit-weighted least-squares analysis of the absolute rate data of Wine *et al.*,[3] with

$$k(\text{dimethyl disulfide}) = (5.83^{+4.53}_{-2.55})$$
$$\times 10^{-11} e^{(383 \pm 169)/T} \text{ cm}^3 \text{ molecule}^{-1} \text{ s}^{-1}$$

over the temperature range 249–367 K, where the indicated errors are two least-squares standard deviations, and

$$k(\text{dimethyl disulfide}) = 2.11 \times 10^{-10} \text{ cm}^3 \text{ molecule}^{-1} \text{ s}^{-1}$$

at 298 K, with an estimated overall uncertainty at 298 K of $\pm 35\%$. Note that due to a calculational error[30] this recommendation is somewhat different to that of Atkinson[30] based upon the same data set. Over the temperature range studied, this reaction appears to proceed by initial OH radical addition[3,31]

$$\text{OH} + \text{CH}_3\text{SSCH}_3 \rightarrow \text{CH}_3\text{SS(OH)CH}_3$$

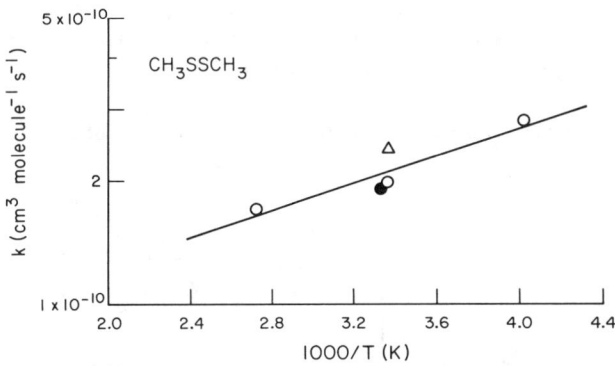

FIG. 88. Arrhenius plot of rate constants for the reaction of the OH radical with dimethyl disulfide. ($\triangle$) Cox and Sheppard;[2] ($\bigcirc$) Wine *et al.*;[3] ($\bullet$) Barnes *et al.*;[20,21] (———) recommendation (see text).

d. Thioethers

(1) Tetrahydrothiophene

The rate constant data of Wine and Thompson,[22] Martin *et al.*[13] and Witte and Zetzsch[19] (for which only the Arrhenius expression was available) are given in Table 13 and are plotted in Arrhenius form in Fig. 89. The rate constants obtained from these three studies[13,19,22] are in excellent agreement.

The recommended Arrhenius expression is based upon a unit-weighted least-squares analysis of the absolute rate constants of Wine and Thompson,[22] with

$k(\text{tetrahydrothiophene}) = (1.13^{+0.42}_{-0.31})$

$\times 10^{-11} e^{(166 \pm 97)/T} cm^3 \text{ molecule}^{-1} s^{-1}$

over the temperature range 255–377 K, where the indicated errors are two least-squares standard deviations, and

$k(\text{tetrahydrothiophene}) = 1.97$

$\times 10^{-11} cm^3 \text{ molecule}^{-1} s^{-1}$

at 298 K, with an estimated overall uncertainty of $\pm 30\%$ at 298 K.

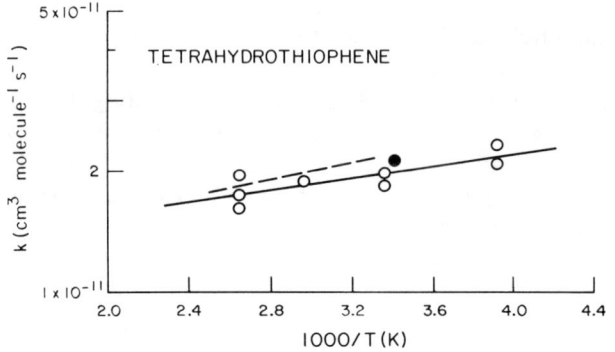

FIG. 89. Arrhenius plot of rate constants for the reaction of the OH radical with tetrahydrothiophene. (○) Wine and Thompson;[22] (●) Martin et al.;[13] (— — —) Witte and Zetzsch;[19] (———) recommendation (see text).

This OH radical reaction can proceed by the reaction pathways

$$OH + \text{[structure]} \longrightarrow$$

(a)

(b)

(c)

and it is expected that all three reaction channels will occur at around room temperature.[22] Based on estimation methods,[32] reaction pathway (b) is expected to dominate over pathway (a). Witte and Zetzsch[19] observed bi-exponential OH radical decays at temperatures $\geqslant 418$ K, showing the occurrence of the OH radical addition pathway (c), and obtained a rate constant of $k_{-c} = 3 \times 10^7 e^{-(6400 \pm 1000)/T} s^{-1}$ for the thermal decomposition of the OH-tetrahydrothiophene adduct formed in reaction

channel (c) back to reactants. However, the relative importance of pathways (b) and (c) are not known, although it is likely that the H-atom abstraction channel dominates.[19,32]

(2) Thiophene

The available rate constant data of Lee and Tang,[23] Atkinson et al.,[24] Mac Leod et al.,[5,25] Wine and Thompson,[22] Barnes et al.,[26] Martin et al.,[13] Wallington[27] and Witte and Zetzsch[19] are given in Table 13. The discharge flow measurements of Lee and Tang[23] and Mac Leod et al.[5,25] yield room temperature rate constants which are higher by a factor of ~ 5 than those determined from the flash photolysis,[19,22,27] the most recent discharge flow,[13] and the relative rate[24,26] studies, with the high rate constants obtained by Mac Leod et al.[5,25] being attributed to the occurrence of heterogeneous reactions on the flow tube walls.[13] The rate constants of Lee and Tang[23] and Mac Leod et al.[5,25] were thus not used in the rate constant evaluation.

At room temperature the rate constants obtained by Witte and Zetzsch[19] and Martin et al.[13] are $\sim 20\%$ higher than the data of Atkinson et al.,[24] Wine and Thompson,[22] Barnes et al.,[26] and Wallington[27] (all of which are plotted in Arrhenius form in Fig. 90), although at $\gtrsim 380$ K the data of Witte and Zetzsch[19] agree well with those of Wine and Thompson.[22]

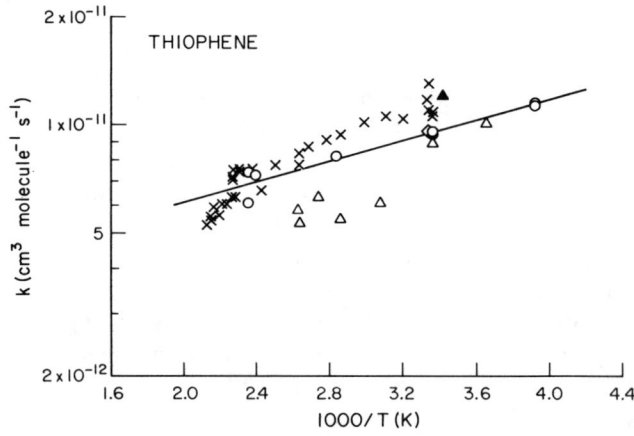

FIG. 90. Arrhenius plot of rate constants for the reaction of the OH radical with thiophene. (●) Atkinson et al.;[24] (○) Wine and Thompson;[22] (◇) Barnes et al.;[26] (▲) Martin et al.;[13] (△) Wallington;[27] (x) Witte and Zetzsch;[19] (———) recommendation (see text).

Furthermore, while the rate constants of Atkinson et al.,[24] Wine and Thompson,[22] Barnes et al.[26] and Wallington[27] are in excellent agreement at room temperature, significant discrepancies arise between the data of Wine and Thompson[22] and Wallington[27] at elevated temperatures.

A unit-weighted least-squares analysis of the rate constant data of Atkinson et al.,[24] Wine and Thompson[22] and Barnes et al.[26] leads to the recommendation of

k(thiophene) = $(3.19^{+0.68}_{-0.56})$

$\times \ 10^{-12} \ e^{(326 \ \pm \ 61)/T}$ cm^3 molecule^{-1} s^{-1}

over the temperature range 255–425 K, where the indicated errors are two least-squares standard deviations, and

k(thiophene) = 9.53×10^{-12} cm^3 molecule^{-1} s^{-1}

at 298 K, with an estimated overall uncertainty at 298 K of $\pm 30\%$. The rate constant is independent of the total pressure or, of the presence of O_2 at room temperature. This recommendation is essentially identical to that of

k(thiophene) = $3.20 \times 10^{-12} \ e^{324/T}$ cm^3 molecule^{-1} s^{-1}

of Atkinson.[30]

At around room temperature the OH radical reaction with thiophene is expected to proceed by OH radical addition to the $>C=C<$ bonds and to the S atom:

At the present time the relative contributions of these two reaction pathways (a) and (b) are not known. At elevated temperatures, these OH radical adducts are expected to undergo thermal decomposition back to the reactants (at least in part), and Witte and Zetzsch[19] have observed non-exponential OH radical decays at temperatures $\geqslant 434$ K, with a rate constant for the thermal decomposition of the OH-thiophene adduct [presumably that formed in reaction channel (b)] of $k = 3 \times 10^{10}$ $e^{-(10100 \ \pm \ 1400)/T}$ s^{-1} over the temperature range 434–471 K. (The OH-thiophene adduct formed in reaction channel (a) is expected to have a higher barrier to thermal decomposition.) Consequently, as for all OH radical addition reactions, the recommended rate expression given above cannot be used for temperatures outside of the range of the recommendation.

e. Other Organosulfur Compounds

Rate data are available for thiazole and dimethyl sulfoxide (Table 13), but since single studies have been carried out for each of these organosulfur compounds no recommendations are made.

References

[1] R. Atkinson, R. A. Perry, and J. N. Pitts, Jr., J. Chem. Phys. **66**, 1578 (1977).

[2] R. A. Cox and D. Sheppard, Nature **284**, 330 (1980).

[3] P. H. Wine, N. M. Kreutter, C. A. Gump, and A. R. Ravishankara, J. Phys. Chem. **85**, 2660 (1981).

[4] H. Mac Leod, G. Poulet, and G. Le Bras, J. Chim. Phys. **80**, 287 (1983).

[5] H. Mac Leod, J. L. Jourdain, G. Poulet, and G. Le Bras, Atmos. Environ. **18**, 2621 (1984).

[6] J. H. Lee and I. N. Tang, J. Chem. Phys. **78**, 6646 (1983).

[7] P. H. Wine, R. J. Thompson, and D. H. Semmes, Int. J. Chem. Kinet. **16**, 1623 (1984).

[8] I. Barnes, V. Bastian, K. H. Becker, E. H. Fink, and W. Nelsen, J. Atmos. Chem. **4**, 445 (1986).

[9] A. J. Hynes and P. H. Wine, J. Phys. Chem. **91**, 3672 (1987).

[10] R. Atkinson, R. A. Perry, and J. N. Pitts, Jr., Chem. Phys. Lett. **54**, 14 (1978).

[11] M. J. Kurylo, Chem. Phys. Lett. **58**, 233 (1978).

[12] R. Atkinson, J. N. Pitts, Jr., and S. M. Aschmann, J. Phys. Chem. **88**, 1584 (1984).

[13] D. Martin, J. L. Jourdain, and G. Le Bras, Int. J. Chem. Kinet. **17**, 1247 (1985).

[14] T. J. Wallington, R. Atkinson, E. C. Tuazon, and S. M. Aschmann, Int. J. Chem. Kinet. **18**, 837 (1986).

[15] A. J. Hynes, P. H. Wine, and D. H. Semmes, J. Phys. Chem. **90**, 4148 (1986).

[16] O. J. Nielsen, J. Treacy, L. Nelson, and H. Sidebottom, Proceedings, 4th European Symposium on the Physico-Chemical Behavior of Atmospheric Pollutants, 1986; D. Riedel Publishing Co., Dordrecht, Holland, 1987, p. 205.

[17] Y.-C. Hsu, D.-S. Chen, and Y.-P. Lee, Int. J. Chem. Kinet. **19**, 1073 (1987).

[18] I. Barnes, V. Bastian, and K. H. Becker, Int. J. Chem. Kinet. **20**, 415 (1988).

[19] F. Witte and C. Zetzsch, 9th International Symposium on Gas Kinetics, University of Bordeaux, Bordeaux, France, July 20–25, 1986; private communication, 1988.

[20] I. Barnes, V. Bastian, and K. H. Becker, 9th International Symposium on Gas Kinetics, University of Bordeaux, Bordeaux, France, July 20–25, 1986.

[21] I. Barnes, V. Bastian, and K. H. Becker, Proceedings, 4th European Symposium on the Physico-Chemical Behavior of Atmospheric Pollutants, 1986; D. Riedel Publishing Co., Dordrecht, Holland, 1987, p. 327.

[22] P. H. Wine and R. J. Thompson, Int. J. Chem. Kinet. **16**, 867 (1984).

[23] J. H. Lee and I. N. Tang, J. Chem. Phys. **77**, 4459 (1982).

[24] R. Atkinson, S. M. Aschmann, and W. P. L. Carter, Int. J. Chem. Kinet. **15**, 51 (1983).

[25] H. Mac Leod, J. L. Jourdain, and G. Le Bras, Chem. Phys. Lett. **98**, 381 (1983).

[26] I. Barnes, V. Bastian, K. H. Becker, and E. H. Fink, Proceedings, 3rd European Symposium on the Physico-Chemical Behavior of Atmospheric Pollutants, 1984; D. Riedel Publishing Co., Dordrecht, Holland, 1984, p. 149.

[27] T. J. Wallington, Int. J. Chem. Kinet. **18**, 487 (1986).

[28] W. B. DeMore, M. J. Molina, S. P. Sander, D. M. Golden, R. F. Hampson, M. J. Kurylo, C. J. Howard, and A. R. Ravishankara, "Chemical Kinetics and Photochemical Data for Use in Stratospheric Modeling," NASA Panel for Data Evaluation, Evaluation No. 8, Jet Propulsion Laboratory Publication 87–41, September 15, 1987.

[29] R. Atkinson, D. L. Baulch, R. A. Cox, R. F. Hampson, Jr., J. A. Kerr, and J. Troe, J. Phys. Chem. Ref. Data **18**, 881 (1989).

[30] R. Atkinson, Chem. Rev. **86**, 69 (1986).

[31] S. Hatakeyama and H. Akimoto, J. Phys. Chem. **87**, 2387 (1983).

[32] R. Atkinson, Int. J. Chem. Kinet. **19**, 799 (1987).

2.8. Nitrogen-Containing Organics

The available rate constant data obtained at, or close to, the high pressure second-order limit are given in Tables 15 (OH radical reactions) and 16 (OD radical reactions). To date, for those OH radical reactions which proceed by OH radical addition, rate constant data in the fall-off region between second- and third-order kinetics have been obtained only for HCN,[22,23,44,45] C_2N_2[25] and $CH_2=CHCN$.[26] As seen from Table 15, for most of these nitrogen-containing organic compounds kinetic studies have been carried out by only one research group. The kinetics and mechanisms of these OH (and OD) radical reactions are discussed by the various classes of the nitrogen-containing organics.

TABLE 15. Rate constants k and temperature-dependent parameters for the gas-phase reactions of the OH radical with nitrogen-containing organics at, or close to, the high pressure limits

Organic	$10^{12} \times A$ (cm^3 molecule^{-1} s^{-1})	B (K)	$10^{12} \times k$ (cm^3 molecule^{-1} s^{-1})	at T (K)	Technique	Reference	Temperature range covered (K)
Aliphatic Amines							
Methyl amine			22.0 ± 2.2	299.0	FP-RF	Atkinson et al.[1]	299–426
			19.4 ± 2.0	353.9			
	10.2	-229 ± 151	17.5 ± 1.8	426.1			
Ethyl amine			27.7 ± 2.8	299.6	FP-RF	Atkinson et al.[2]	300–426
			24.9 ± 2.5	354.1			
	14.7	-189 ± 151	23.0 ± 2.3	425.8			
Dimethyl amine			65.4 ± 6.6	298.5	FP-RF	Atkinson et al.[2]	298–425
			58.3 ± 5.9	354.5			
	28.9	-247 ± 151	51.1 ± 5.2	425.4			
Trimethyl amine			60.9 ± 6.1	298.7	FP-RF	Atkinson et al.[2]	299–425
			53.7 ± 5.4	352.5			
	26.2	-252 ± 151	47.4 ± 4.8	424.7			
Diethyl hydroxylamine			101	308	PR-RA	Gorse et al.[3]	
2-(Dimethyl-amino)-ethanol			47 ± 12	300 ± 2	FP-RF	Harris and Pitts[4]	
			63 ± 29	234	FP-RF	Anderson and Stephens[5]	234–364
			86 ± 29	269			
			103 ± 20	293			
			87 ± 15	333			
	90 ± 20	0	93 ± 30	364			
2-Amino-2-methyl-1-propanol			28 ± 5	300 ± 2	FP-RF	Harris and Pitts[4]	
N-Nitroso-dimethylamine			2.53 ± 0.21	298 ± 2	RR [relative to k(dimethyl ether) $= 2.98 \times 10^{-12}$][a]	Tuazon et al.[6]	
			3.6 ± 0.1	296 ± 2	LP-LIF	Zabarnick et al.[7]	
Dimethyl-nitramine			3.84 ± 0.15	298 ± 2	RR [relative to k(dimethyl ether) $= 2.98 \times 10^{-12}$][a]	Tuazon et al.[6]	
Hydrazines							
Hydrazine			22 ± 5	298	DF-EPR	Hack et al.[8]	
			65 ± 10	298 ± 1	FP-RF	Harris et al.[9]	298–424
			59 ± 9	355 ± 1			
	44	-116 ± 176	58 ± 9	424 ± 1			
	[61 ± 10	0]					

TABLE 15. Rate constants k and temperature-dependent parameters for the gas-phase reactions of the OH radical with nitrogen-containing organics at, or close to, the high pressure limits — Continued

Organic	$10^{12} \times A$ (cm^3 molecule^{-1} s^{-1})	B (K)	$10^{12} \times k$ (cm^3 molecule^{-1} s^{-1})	at T (K)	Technique	Reference	Temperature range covered (K)
Methyl-hydrazine	65 ± 13	0	65 ± 13	298—424	FP-RF	Harris et al.[9]	298–424
Nitrites							
Methyl nitrite			1.41 ± 0.19	292 ± 2	RR [relative to k(CO) $= 1.58 \times 10^{-13}$][b]	Campbell and Goodman[10]	
			1.09 ± 0.17	295 ± 3	RR [relative to k(n-butane) $= 2.50 \times 10^{-12}$][a]	Audley et al.[11]	
			0.21 ± 0.04	300 ± 3	RR [relative to k(n-hexane) $= 5.64 \times 10^{-12}$][a]	Tuazon et al.[12]	
			0.12 ± 0.03	300 ± 3	RR [relative to k(dimethyl ether) $= 3.01 \times 10^{-12}$][a]	Tuazon et al.[12]	
			1.00 ± 0.15	295 ± 2	DF-RF	Baulch et al.[13]	
Ethyl nitrite			1.77 ± 0.28	295 ± 3	RR [relative to k(n-butane) $= 2.50 \times 10^{-12}$][a]	Audley et al.[11]	
1-Propyl nitrite [CH$_3$(CH$_2$)$_2$ONO]			2.40 ± 0.45	295 ± 3	RR [relative to k(n-butane) $= 2.50 \times 10^{-12}$][a]	Audley et al.[11]	
			2.31 ± 0.34	295 ± 2	DF-RF	Baulch et al.[13]	
1-Butyl nitrite [CH$_3$(CH$_2$)$_3$ONO]			5.23 ± 1.76	295 ± 3	RR [relative to k(n-butane) $= 2.50 \times 10^{-12}$][a]	Audley et al.[11]	
			4.80 ± 0.72	295 ± 2	DF-RF	Baulch et al.[13]	
2-Butyl nitrite [CH$_3$CH$_2$CH(CH$_3$)ONO]			5.97 ± 0.71	295 ± 3	RR [relative to k(n-butane) $= 2.50 \times 10^{-12}$][a]	Audley et al.[11]	
2-Methyl-1-propyl nitrite [(CH$_3$)$_2$CHCH$_2$ONO]			5.35 ± 0.65	295 ± 3	RR [relative to k(n-butane) $= 2.50 \times 10^{-12}$][a]	Audley et al.[11]	
2-Methyl-2-propyl nitrite [(CH$_3$)$_3$CONO]			1.41 ± 0.20	295 ± 3	RR [relative to k(n-butane) $= 2.50 \times 10^{-12}$][a]	Audley et al.[11]	
Nitrates							
Methyl nitrate [CH$_3$ONO$_2$]			0.034 ± 0.004	298	DF-RF	Gaffney et al.[14]	
			0.38 ± 0.10	303 ± 2	RR [relative to k(ethene) $= 8.32 \times 10^{-12}$][a]	Kerr and Stocker[15]	
Ethyl nitrate [CH$_3$CH$_2$ONO$_2$]			0.49 ± 0.21	303 ± 2	RR [relative to k(ethene) $= 8.32 \times 10^{-12}$][a]	Kerr and Stocker[15]	

TABLE 15. Rate constants k and temperature-dependent parameters for the gas-phase reactions of the OH radical with nitrogen-containing organics at, or close to, the high pressure limits — Continued

Organic	$10^{12} \times A$ (cm^3 mole-cule^{-1} s^{-1})	B (K)	$10^{12} \times k$ (cm^3 molecule^{-1} s^{-1})	at T (K)	Technique	Reference	Tempera-ture range covered (K)
1-Propyl nitrate [CH$_3$CH$_2$CH$_2$ONO$_2$]			0.72 ± 0.23	303 ± 2	RR [relative to k(ethene) = 8.32×10^{-12}]^a	Kerr and Stocker[15]	
			0.62 ± 0.10	298 ± 2	RR [relative to k(cyclohexane) = 7.49×10^{-12}]^a	Atkinson and Aschmann[16]	
2-Propyl nitrate [(CH$_3$)$_2$CHONO$_2$]			0.18 ± 0.05	299 ± 2	RR [relative to k(cyclohexane) = 7.51×10^{-12}]^a	Atkinson et al.[17]	
			0.41 ± 0.06	298 ± 2	RR [relative to k(cyclohexane) = 7.49×10^{-12}]^a	Atkinson and Aschmann[16]	
1-Butyl nitrate [CH$_3$(CH$_2$)$_3$ONO$_2$]			1.40 ± 0.11	299 ± 2	RR [relative to k(cyclohexane) = 7.51×10^{-12}]^a	Atkinson et al.[17]	
			1.78 ± 0.19	298 ± 2	RR [relative to k(cyclohexane) = 7.49×10^{-12}]^a	Atkinson and Aschmann[16]	
2-Butyl nitrate [CH$_3$CH$_2$CH(CH$_3$)ONO$_2$]			0.68 ± 0.10	299 ± 2	RR [relative to k(cyclohexane) = 7.51×10^{-12}]^a	Atkinson et al.[17]	
			0.92 ± 0.16	298 ± 2	RR [relative to k(cyclohexane) = 7.49×10^{-12}]^a	Atkinson and Aschmann[16]	
2-Pentyl nitrate [CH$_3$(CH$_2$)$_2$CH(CH$_3$)ONO$_2$]			1.85 ± 0.13	299 ± 2	RR [relative to k(cyclohexane) = 7.51×10^{-12}]^a	Atkinson et al.[17]	
3-Pentyl nitrate [(C$_2$H$_5$)$_2$CHONO$_2$]			1.12 ± 0.20	299 ± 2	RR [relative to k(cyclohexane) = 7.51×10^{-12}]^a	Atkinson et al.[17]	
2-Methyl-3-butyl nitrate [(CH$_3$)$_2$CHCH(CH$_3$)ONO$_2$]			1.72 ± 0.06	298 ± 2	RR [relative to k(n-butane) = 2.54×10^{-12}]^a	Atkinson et al.[18]	
2,2-Dimethyl-1-propyl nitrate [(CH$_3$)$_3$CCH$_2$ONO$_2$]			0.85 ± 0.21	298 ± 2	RR [relative to k(n-butane) = 2.54×10^{-12}]^a	Atkinson et al.[18]	
2-Hexyl nitrate [CH$_3$(CH$_2$)$_3$CH(CH$_3$)ONO$_2$]			3.17 ± 0.16	299 ± 2	RR [relative to k(cyclohexane) = 7.51×10^{-12}]^a	Atkinson et al.[17]	
3-Hexyl nitrate [CH$_3$(CH$_2$)$_2$CH(ONO$_2$)CH$_2$CH$_3$]			2.70 ± 0.22	299 ± 2	RR [relative to k(cyclohexane) = 7.51×10^{-12}]^a	Atkinson et al.[17]	
Cyclohexyl nitrate			3.30 ± 0.36	298 ± 2	RR [relative to k(n-butane) = 2.54×10^{-12}]^a	Atkinson et al.[18]	
2-Methyl-2-pentyl nitrate [(CH$_3$)$_2$C(ONO$_2$)CH$_2$CH$_2$CH$_3$]			1.72 ± 0.22	298 ± 2	RR [relative to k(n-butane) = 2.54×10^{-12}]^a	Atkinson et al.[18]	

TABLE 15. Rate constants k and temperature-dependent parameters for the gas-phase reactions of the OH radical with nitrogen-containing organics at, or close to, the high pressure limits — Continued

Organic	$10^{12} \times A$ (cm^3 molecule^{-1} s^{-1})	B (K)	$10^{12} \times k$ (cm^3 molecule^{-1} s^{-1})	at T (K)	Technique	Reference	Temperature range covered (K)
3-Methyl-2-pentyl nitrate [CH$_3$CH(ONO$_2$)CH(CH$_3$)CH$_2$CH$_3$]			3.02 ± 0.08	298 ± 2	RR [relative to k(n-butane) = 2.54 × 10^{-12}]^a	Atkinson et al.[18]	
3-Heptyl nitrite [CH$_3$(CH$_2$)$_3$CH(ONO$_2$)CH$_2$CH$_3$]			3.69 ± 0.43	299 ± 2	RR [relative to k(cyclohexane) = 7.51 × 10^{-12}]^a	Atkinson et al.[17]	
3-Octyl nitrate [CH$_3$(CH$_2$)$_4$CH(ONO$_2$)CH$_2$CH$_3$]			3.88 ± 0.79	299 ± 2	RR [relative to k(cyclohexane) = 7.51 × 10^{-12}]^a	Atkinson et al.[17]	
Nitriles							
Hydrogen cyanide			0.33 ± 0.04	1950–2380	Flame	Haynes[19]	1950–2380
	42	5030		1318–2400	Flame	Fenimore[20]	1318–2400
			0.35	1790	Flame; OH, CN by LIF	Morley[21]	1790–2200
			0.33	1790			
			0.76	1790			
			0.67	2000			
			0.97	2130			
			1.25	2200			
	0.12 ± 0.05	400	0.03 ± 0.01	298	FP-RA	Fritz et al.[22,23]	296–433
Cyanogen [C$_2$N$_2$]			⩽0.03	298	FP-RF	Atkinson et al.[24]	298–424
			⩽0.05	424			
	0.311	1448	0.0025^c	300	DF-RF	Phillips[25]	300–555
Acetonitrile [CH$_3$CN]			0.0494 ± 0.006	297.2	FP-RF	Harris et al.[26]	297–424
			0.0620 ± 0.007	348.0			
	0.586	755 ± 126	0.105 ± 0.015	423.8			
			0.024 ± 0.003	295	FP-RF	Fritz et al.[22]	
			0.019 ± 0.002	296	FP-RF	Zetzsch[27,28]	
			0.0102 ± 0.0022	250	FP-RF	Kurylo and Knable[29]	250–363
			0.0146 ± 0.0015	273			
			0.0194 ± 0.0037	298			
	0.628	1030	0.0370 ± 0.0033	363			
			0.021 ± 0.003	295	DF-EPR	Poulet et al.[30]	295–393
			0.086 ± 0.01	393			
Propionitrile [C$_2$H$_5$CN]			0.194 ± 0.020	298.2	FP-RF	Harris et al.[26]	298–423
			0.233 ± 0.025	350.8			
			0.362 ± 0.036	384.0			
	2.69	800 ± 176	0.414 ± 0.040	423.0			
Acrylonitrile [CH$_2$=CHCN]			4.80 ± 0.50	298.7	FP-RF	Harris et al.[26]	299–423
			3.4 ± 0.5	296	FP-RF	Zetzsch[27]	
Nitrogen-Containing Heterocycles							
Aziridine^d			6.1 ± 0.5	295	FP-RF	Zetzsch[31]	

TABLE 15. Rate constants k and temperature-dependent parameters for the gas-phase reactions of the OH radical with nitrogen-containing organics at, or close to, the high pressure limits — Continued

Organic	$10^{12} \times A$ (cm^3 molecule^{-1} s^{-1})	B (K)	$10^{12} \times k$ (cm^3 molecule^{-1} s^{-1})	at T (K)	Technique	Reference	Temperature range covered (K)
Pyrrole[d]			122 ± 4	295 ± 1	RR [relative to k(propene) = 2.68×10^{-11}][a]	Atkinson et al.[32]	
			103 ± 6	298 ± 2	FP-RF	Wallington[33]	298–440
			98 ± 20	325			
			83 ± 13	355			
	27^{+8}_{-6}	-403 ± 93	68 ± 10	440			
			99.1 ± 2.0	298	FP-RF	Witte and Zetzsch[34]	298–442
			93.1 ± 7.3	298			
			93.1 ± 11.0	298			
			101.0 ± 21.8	298			
			83.9 ± 2.0	327			
			71.3 ± 2.0	347			
			68.7 ± 4.6	348			
	9.3 ± 3.8	-690 ± 120 (298–372 K)	59.4 ± 3.3	372			
			51.5 ± 2.0	392			
			48.5 ± 1.7	395			
			37.6 ± 2.0	422			
			24.4 ± 2.0	442			
Imidazole[d]			35.9 ± 3.3	297	FP-RF	Witte and Zetzsch[34]	297–440
			31.3 ± 0.1	316			
			27.3 ± 2.5	331			
	2.2 ± 0.2	-840 ± 30 (297–344 K)	25.2 ± 0.3	344			
			22.5 ± 1.0	353			
			19.8 ± 1.0	362			
			21.2 ± 1.6	363			
			16.3 ± 0.3	386			
			13.9 ± 0.5	402			
			8.6 ± 1.2	425			
			10.4 ± 0.7	425			
			9.1 ± 1.1	440			
1,2,4-Triazole[d]			<0.2	~ 298	FP-RF	Witte and Zetzsch[34]	
Pyridine[d]			0.494 ± 0.039	296 ± 2	RR [relative to k(dimethyl ether) = 2.96×10^{-12}][a]	Atkinson et al.[35]	
			0.159 ± 0.025	246	FP-RF	Witte and Zetzsch[34]	246–468
			0.190 ± 0.018	258			
			0.207 ± 0.021	268			
			0.221 ± 0.021	272			
	3.5 ± 1.4	760 ± 100 (246–286 K)	0.250 ± 0.020	286			
			0.256 ± 0.014	297			
			0.254 ± 0.014	308			
			0.245 ± 0.023	323			
			0.199 ± 0.014	348			
			0.119 ± 0.016	373			
			0.175 ± 0.012	398			
			0.191 ± 0.015	423			
			0.209 ± 0.015	448			
	12 ± 7	1800 ± 240 (423–468 K)	0.252 ± 0.012	468			
1,3,5-Triazine[d]			0.145 ± 0.027	296 ± 2	RR [relative to k(dimethyl ether) = 2.96×10^{-12}][a]	Atkinson et al.[35]	

TABLE 15. Rate constants k and temperature-dependent parameters for the gas-phase reactions of the OH radical with nitrogen-containing organics at, or close to, the high pressure limits — Continued

Organic	$10^{12} \times A$ (cm^3 molecule^{-1} s^{-1})	B (K)	$10^{12} \times k$ (cm^3 molecule^{-1} s^{-1})	at T (K)	Technique	Reference	Temperature range covered (K)
Miscellaneous							
Peroxyacetyl nitrate [CH$_3$C(O)OONO$_2$]			≤0.17	299 ± 1	FP-RF	Winer et al.[36]	
			0.113 ± 0.006	273 ± 2	FP-RF	Wallington et al.[37]	273–297
	1.23	651 ± 229	0.137 ± 0.005	297 ± 2			
			0.075 ± 0.014	298	DF-EPR	Tsalkani et al.[38]	
Nitromethane [CH$_3$NO$_2$]			1.00 ± 0.10	292 ± 2	RR [relative to k(CO) = 1.58 × 10^{-13}][b]	Campbell and Goodman[10]	
			0.0174 ± 0.0038	299	LP-LIF	Zabarnick et al.[39]	299–671
			0.0141 ± 0.0030	300			
			0.0273 ± 0.0085	372			
			0.0580 ± 0.0140	473			
			0.0735 ± 0.0071	572			
	0.81 ± 0.32	1208 ± 151	0.193 ± 0.030	671			
			0.156 ± 0.012	295 ± 2	PR-RA	Nielsen et al.[40]	
			0.272 ± 0.084	295 ± 2	RR [relative to k(ethane) = 2.58 × 10^{-13}][a]	Nielsen et al.[40]	
Nitromethane-d_3 [CD$_3$NO$_2$]			0.096 ± 0.011	295 ± 2	PR-RA	Nielsen et al.[40]	
			0.260 ± 0.046	295 ± 2	RR [relative to k(ethane) = 2.58 × 10^{-13}][a]	Nielsen et al.[40]	
Nitroethane [CH$_3$CH$_2$NO$_2$]			0.15 ± 0.05	295 ± 3	PR-RA	Nielsen et al.[41]	
1-Nitropropane [CH$_3$CH$_2$CH$_2$NO$_2$]			0.34 ± 0.08	295 ± 3	PR-RA	Nielsen et al.[41]	
1-Nitrobutane [CH$_3$CH$_2$CH$_2$CH$_2$NO$_2$]			1.35 ± 0.18	298 ± 2	RR [relative to k(propane) = 1.15 × 10^{-12}][a]	Atkinson and Aschmann[16]	
			1.55 ± 0.09	295 ± 3	PR-RA	Nielsen et al.[41]	
1-Nitropentane [CH$_3$CH$_2$CH$_2$CH$_2$CH$_2$NO$_2$]			3.30 ± 0.05	295 ± 3	PR-RA	Nielsen et al.[41]	
Isocyanic acid [HNCO]			0.0502 ± 0.0033	624	LP-LIF	Tully et al.[42]	624–875
			0.0723 ± 0.0048	676			
			0.100 ± 0.007	730			
			0.0942 ± 0.0075	731			
			0.133 ± 0.010	796			
	4.4 ± 0.9	2788 ± 141	0.182 ± 0.015	875			
CH$_2$=NOH			0.63 ± 0.31	300 ± 2	FP-KS	Horne and Norrish[43]	

TABLE 15. Rate constants k and temperature-dependent parameters for the gas-phase reactions of the OH radical with nitrogen-containing organics at, or close to, the high pressure limits — Continued

Organic	$10^{12} \times A$ (cm^3 mole-cule^{-1} s^{-1})	B (K)	$10^{12} \times k$ (cm^3 molecule^{-1} s^{-1})	at T (K)	Technique	Reference	Temperature range covered (K)
CH$_3$CH=NOH			2.2 ± 1.1	300 ± 2	FP-KS	Horne and Norrish[43]	

[a]From the present recommendations (see text).
[b]See Introduction.
[c]Calculated from cited Arrhenius expression.
[d]Structures:

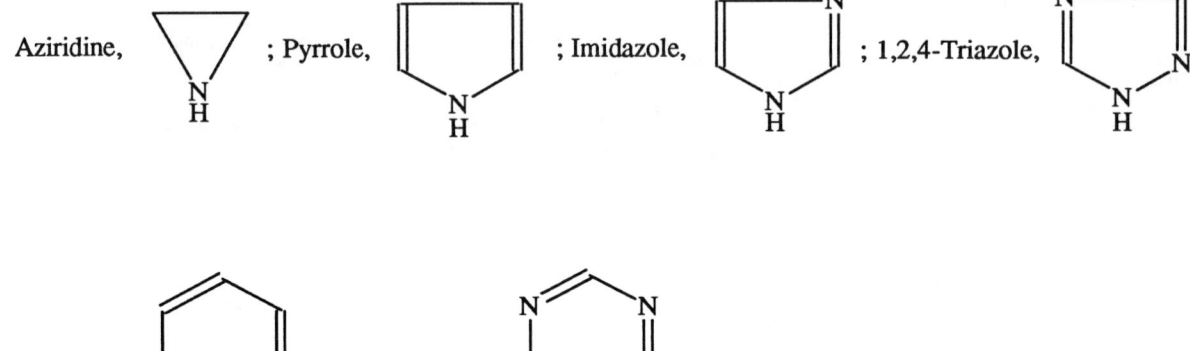

Aziridine,   ; Pyrrole,   ; Imidazole,   ; 1,2,4-Triazole,

Pyridine,   ; 1,3,5-Triazine,

TABLE 16. Rate constants k and temperature-dependent parameters for the gas-phase reactions of the OD radical with nitrogen-containing organic compounds

Organic	$10^{12} \times A$ (cm^3 mole-cule^{-1} s^{-1})	B (K)	$10^{12} \times k$ (cm^3 molecule^{-1} s^{-1})	at T (K)	Technique	Reference	Temperature range covered (K)
Nitromethane (CH$_3$NO$_2$)			0.0082 ± 0.0044	296	LP-LIF	Zabarnick et al.[39]	296–671
			0.0186 ± 0.0082	378			
			0.0443 ± 0.0004	474			
			0.0848 ± 0.0122	572			
	1.8 ± 1.0	1661 ± 201	0.218 ± 0.010	671			
Nitromethane-d_3 (CD$_3$NO$_2$)			0.0050 ± 0.0002	296	LP-LIF	Zabarnick et al.[39]	296–670
			0.0049 ± 0.0006	299			
			0.0074 ± 0.0013	376			
			0.0233 ± 0.0008	475			
			0.0382 ± 0.0013	566			
	0.53 ± 0.24	1439 ± 171	0.0846 ± 0.0006	670			

a. Aliphatic Amines

Only for 2-(dimethylamino)ethanol and N-nitrosodimethylamine have more than one study been carried out. For 2-(dimethylamino)ethanol the room temperature rate constants determined by Harris and Pitts[4] and Anderson and Stephens,[5] both using flash photolysis-resonance fluorescence methods, disagree by a factor of ~2. As discussed by Anderson and Stephens,[5] the rate constant measured by Harris and Pitts[4] may have been low due to adsorption of the 2-(dimethylamino)ethanol onto the reaction vessel walls [the 2-(dimethylamino)ethanol concentrations were measured before and after the reaction vessel in the study of Anderson and Stephens[5]]. Consequently, while no firm recommendation is made concerning the rate constant for this OH radical reaction, in the absence of further data the data of Anderson and Stephens[5] are preferred.

For N-nitrosodimethylamine, the room temperature rate constants of Tuazon et al.[6] and Zabarnick et al.[7] exhibit a discrepancy of a factor of ~1.4. However, since the rate constants measured by the laser photolysis-laser induced fluorescence technique of Zabarnick et al.[7] were observed to be dependent on the laser fluence, it is possible that the rate constant determined by Zabarnick et al.[7] is still an upper limit. Clearly, further kinetic data are needed for this reaction.

As shown in Table 15, the OH radical reactions with methyl amine, ethyl amine, dimethyl amine and trimethyl amine are rapid, with room temperature rate constants in the range $(2-6) \times 10^{-11}$ cm^3 molecule^{-1} s^{-1} and with negative temperature dependencies equivalent to $B \approx -230$ K. For the methyl-substituted amines, the trend of the room temperature rate constants suggests that these reactions proceed via abstraction from the C—H bonds and, where possible, the N—H bonds. From the rate constants measured by Atkinson et al.[1,2] and the bond dissociation energies for the C—H (93.3 ± 2, 87 ± 2 and 84 ± 2 kcal mol^{-1} in CH$_3$NH$_2$, (CH$_3$)$_2$NH and (CH$_3$)$_3$N, respectively[46]) and N—H (100.0 ± 2.5 and 91.5 ± 2 kcal mol^{-1} in CH$_3$NH$_2$ and (CH$_3$)$_2$NH, respectively[46]) bonds, it is expected that for CH$_3$NH$_2$ (and probably also C$_2$H$_5$NH$_2$) H-atom abstraction from the C—H bonds predominates, while for (CH$_3$)$_2$NH H-atom abstraction from the N—H bond is competitive with H-atom abstraction from the C—H bonds.[2]

Indeed, from a product study of irradiated HONO—(CH$_3$)$_2$NH—air mixtures, utilizing long pathlength FT-IR absorption spectroscopy, Lindley et al.[47] determined that at room temperature $k_a/(k_a + k_b) = 0.37 \pm 0.05$, where k_a and k_b are the rate constants for reaction pathways (a) and (b), respectively.

$$\text{OH} + (\text{CH}_3)_2\text{NH} \rightarrow \text{H}_2\text{O} + (\text{CH}_3)_2\dot{\text{N}} \qquad \text{(a)}$$

$$\rightarrow \text{H}_2\text{O} + \dot{\text{C}}\text{H}_2\text{NHCH}_3 \qquad \text{(b)}$$

However, the observations of negative temperature dependencies for these reactions[1,2] suggest that these OH radical reactions may proceed via the initial formation of an addition complex, which then rapidly decomposes to the observed products. Clearly, further experimental work concerning the reaction dynamics of these systems is needed.

The OH radical reactions with the substituted amines diethyl hydroxylamine, 2-(dimethylamino)ethanol, 2-amino-2-methyl-1-propanol, N-nitrosodimethylamine, dimethylnitramine and aziridine also almost certainly proceed via overall H-atom abstraction. However, the position of the H-atom abstracted cannot be predicted in all cases, since the C—H and N—H bond strengths are not known. Obviously, for N-nitrosodimethylamine and dimethylnitramine any H-atom abstraction must occur from the C—H bonds.

b. Hydrazines

To date, only for hydrazine and methylhydrazine are kinetic data available for the OH radical reactions (Table 15), and no unambiguous product data are available. The reactions of OH radicals with hydrazine and methylhydrazine are expected to occur via overall H-atom abstraction from the N—H bonds (of bond strength 87.5 kcal mol^{-1} in N$_2$H$_4$[48]). This is consistent with the magnitude of the rate constants observed,[9] although it is possible that the reaction proceeds via initial formation of an addition complex followed by rapid decomposition to the RNHNH or RNNH$_2$ radical and H$_2$O.

c. Nitrites

The available kinetic data, all obtained at room temperature, are given in Table 15. The relative rate studies of Campbell and co-workers,[10,11] using the dark heterogeneous reaction of H$_2$O$_2$ with NO$_2$ to generate OH radicals, are in excellent agreement with the discharge flow-resonance fluorescence data of Baulch et al.[13] for methyl nitrite, 1-propyl nitrite and 1-butyl nitrite. However, the rate constants obtained for methyl nitrite in the relative rate study of Tuazon et al.[12] (employing the dark reaction of N$_2$H$_4$ with O$_3$ to generate OH radicals) are lower by a factor of ~7. Jenkin et al.[49] have concluded that the products formed, and their yields, from the photolysis of CH$_3$ONO in CH$_3$ONO—NO—N$_2$ and CH$_3$ONO—O$_2$ mixtures at 298 K are consistent with the lower rate constant of Tuazon et al.[12] of ~1.2×10^{-13} cm^3 molecule^{-1} s^{-1}. While the reasons for these discrepancies are not known at present, it should be noted that the room temperature rate constant derived from the relative rate method of Campbell and Goodman[10] for nitromethane (CH$_3$NO$_2$) is 10 and 60 times higher than the absolute rate constants of Nielsen et al.[40] and Zabarnick et al.,[39] respectively. These observations suggest that fundamental problems are associated with the relative rate method of Campbell et al.[10,11] employing the dark heterogeneous reaction of H$_2$O$_2$ with NO$_2$ as a source of OH radicals.

Until the kinetics of these OH radical reactions with the alkyl nitrites are more fully understood and product studies carried out, the reaction mechanisms remain uncertain. These reactions may proceed via H-atom abstraction from the C—H bonds,

$$OH + CH_3ONO \rightarrow H_2O + \overset{\cdot}{C}H_2ONO$$
$$\downarrow$$
$$HCHO + NO,$$

and this is expected to be the major, if not only, reaction pathway if the room temperature rate constant for CH_3ONO is $\sim 1 \times 10^{-13}$ cm^3 molecule^{-1} s^{-1}.

d. Nitrates

The available kinetic data for the reactions of the OH radical with alkyl nitrates are given in Table 15. Data are available only at room temperature, and only for the C_1-C_4 alkyl nitrates have more than one kinetic study been carried out. For methyl nitrate the room temperature rate constants obtained by Gaffney et al.[14] and Kerr and Stocker[15] disagree by a factor of ~ 11. The recent rate constants of Atkinson and Aschmann,[16] employing a 6400-liter reaction chamber, supersedes the earlier data of Atkinson et al.[17] for the propyl and butyl nitrates. The room temperature rate constants obtained by Kerr and Stocker[15] and Atkinson and Aschmann[16] for 1-propyl nitrate are in good agreement. However, the room temperature rate constant recently determined by Atkinson and Aschmann[16] for 2-propyl nitrate is a factor of 2.3 higher than that of Atkinson et al.,[17] at least partially due to the slowness of this reaction and the presumably erroneous conclusion of Atkinson et al.[17] that wall losses of the alkyl nitrates, accounting for a significant fraction of the observed 2-propyl nitrate decays, were occurring in the ~ 60 liter reaction chamber used.[17]

As expected from the decreasing fractions of the overall alkyl nitrate decay rates attributed to wall losses in the study of Atkinson et al.[17] for the faster reacting alkyl nitrates, the room temperature rate constants of Atkinson and Aschmann[16] for 1- and 2-butyl nitrate are higher by only ~ 25-35% than those reported of Atkinson et al.[17] For the remaining C_5-C_8 alkyl nitrates studied by Atkinson et al.,[17] the reported rate constants are hence expected to be affected to an extent of $\lesssim 20$% by this assumption that wall losses were occurring. Based upon the estimation method developed by Atkinson,[50,51] the room temperature rate constant obtained by Kerr and Stocker[15] for ethyl nitrate is not consistent with those measured for the higher alkyl nitrates by Atkinson and Aschmann[16] and Atkinson et al.[17,18] This suggests that the rate constants obtained by Kerr and Stocker[15] for methyl and ethyl nitrate are systematically high, and that the absolute rate constant of Gaffney et al.[14] for methyl nitrate of 3.4×10^{-14} cm^3 molecule^{-1} s^{-1} at 298 K is to be preferred.

Since no direct product or mechanistic data are available for these compounds, mechanistic information can only be based upon the available kinetic data. Hydrogen atom abstraction from the C—H bonds appears to be the likely reaction pathway, with the —ONO$_2$ group

markedly decreasing the rate constant for H-atom abstraction from >CH— or —CH$_2$— groups bonded to the —ONO$_2$ group, and decreasing those for the β >CH—, —CH$_2$— or —CH$_3$ groups.[50,51] The distribution of nitratoalkyl radicals formed can be approximately calculated using the estimation method of Atkinson.[51]

e. Nitriles

The available rate constant data obtained at, or close to, the high-pressure second-order limit are given in Table 15. These kinetic data are discussed for the individual nitriles below.

(1) Hydrogen Cyanide

Rate constants have been derived at elevated temperatures (~ 1300-2400 K) by Haynes,[19] Fenimore[20] and Morley,[21] and at temperatures below ~ 850 K by Phillips[44,45] and Fritz et al.[22,23] At temperatures between 296 and 433 K the rate constants are in the fall-off region between second- and third-order kinetics at total pressures of N$_2$ below ~ 300 Torr,[22,23] with bimolecular rate constants (at 373 K) which extrapolate to zero as the total pressure approaches zero.[44] Thus, at temperatures $\lesssim 433$ K the OH radical reaction with HCN must proceed by initial OH radical addition to form an initially energy-rich adduct which can decompose back to reactants or be collisionally stabilized.

$$OH + HCN \rightleftharpoons \begin{bmatrix} OH \\ | \\ HC = \overset{\cdot}{N} \end{bmatrix}^{\ddagger} \text{ or } \begin{bmatrix} H\overset{\cdot}{C} = NOH \end{bmatrix}^{\ddagger}$$

$$\overset{OH}{\underset{HC = \overset{\cdot}{N}}{|}} \text{ or } H\overset{\cdot}{C} = NOH$$

Fritz et al.[22,23] also observed that the rate constants measured at total pressures of N$_2$ of 10–40 Torr (which are in the fall-off region for temperatures $\leqslant 433$ K) over the temperature range 298–850 K showed marked non-Arrhenius behavior. These rate constants increased rapidly with increasing temperature for temperatures $\gtrsim 500$ K, and had an extrapolated temperature dependence of $B \sim 4500$ K for temperatures $\gtrsim 850$ K. Fritz et al.[23] attributed this increase in rate constant at temperatures $\gtrsim 500$ K to the occurrence of either the H-atom abstraction process

$$OH + HCN \rightarrow H_2O + CN \qquad (a)$$

or a "transfer" process

$$OH + HCN \rightarrow HOCN + H \qquad (b)$$

The rate constants obtained by Haynes[19] were derived on the assumption that the reaction of OH radicals with

HCN proceeds by channel (b) and not by channel (a) [which would be in equilibrium with the reverse reaction under the conditions employed]. As discussed in detail by Miller and Melius,[52] reaction pathway (b) involves initial OH radical addition,

$$OH + HCN \rightarrow [HOHCN]^{\ddagger} \rightarrow HOCN + H$$

and hence the rate constant k_b must be less or equal to the high-pressure OH radical addition rate constant k_∞. However, as shown in Fig. 91, in which the rate constant data of Haynes,[19] Fenimore,[20] Morley[21] and Fritz et al.[22,23] are plotted in Arrhenius form, the limiting high pressure rate constant k_∞ reported by Fritz et al.[22,23] does not extrapolate linearly to the reported rate constants at flame temperatures.[19-21]

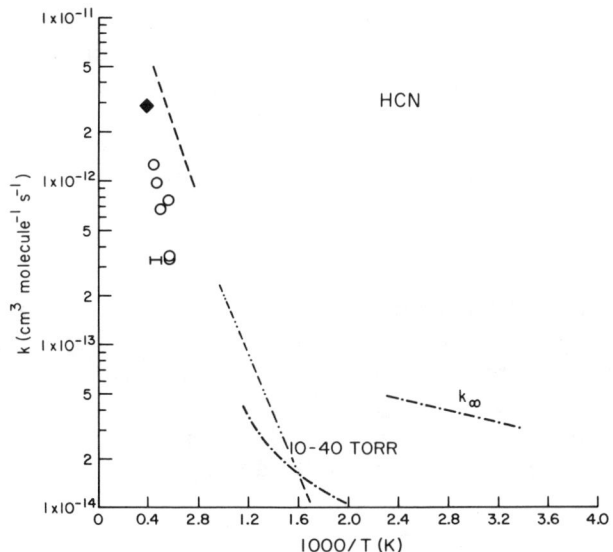

FIG. 91. Arrhenius plot of rate constants for the reaction of the OH radical with HCN. (⊢) Haynes;[19] (— — —) Fenimore;[20] (○) Morley;[21] (⋅—⋅⋅—⋅) Fritz et al.;[22,23] (◆) calculated by Szekely et al.[53] from the reverse reaction (—⋅⋅⋅—) calculated by Jacobs et al.[54] from the reverse reaction.

It is possible, however, that the limiting high-pressure rate constant k_∞ exhibits non-Arrhenius behavior and/or that the reported values of k_∞ of Fritz et al.[22,23] were still in the fall-off regime, with k_∞ having a steeper temperature dependence than reported.

From shock tube and LP-LIF studies of the kinetics of the reaction,

$$CN + H_2O \rightarrow HCN + OH \qquad (-a)$$

Szekely et al.[53] and Jacobs et al.[54] calculated the rate of the reverse reaction (a) to be

$$k_a = (2.9 \pm 1.4) \times 10^{-12} \text{ cm}^3 \text{ molecule}^{-1} \text{ s}^{-1}$$

over the temperature range 2460–2840 K[53] and

$$k_a = (1.28 \pm 0.17) \times 10^{-11} e^{-4162/T} \text{ cm}^3 \text{ molecule}^{-1} \text{ s}^{-1}$$

over the temperature range 518–1027 K,[54] and these calculated rate constants are also plotted in Fig. 91. These calculated values of k_a[53,54] are in reasonably good agreement with extrapolation of the rate constants obtained by Fritz et al.[22,23] at total pressures of 10–40 Torr and 298–850 K to higher temperatures.

Thus, in the absence of kinetic and product data carried out as a function of total pressure at temperatures >400 K, no firm recommendations can be made. However, it is clear that at temperatures $\lesssim 500$ K the reaction of the OH radical with HCN proceeds by addition to form an adduct which can back dissociate or be collisionally stabilized. At elevated temperatures characteristic of flame conditions this addition reaction will be of no importance because of fall-off effects and thermal decomposition of the OH—HCN adduct, as for the OH radical reaction with C_2H_2. The kinetic data of Fenimore[20] and Morley[21] are reasonably consistent with the reaction proceeding by H-atom abstraction at these elevated temperatures of $\geqslant 1300$ K,

$$OH + HCN \rightarrow H_2O + CN \qquad (a)$$

with a rate constant of[53,54]

$$k_a = 1.3 \times 10^{-11} e^{-4160/T} \text{ cm}^3 \text{ molecule}^{-1} \text{ s}^{-1}$$

For recent theoretical calculations of the rate constants for the four channels

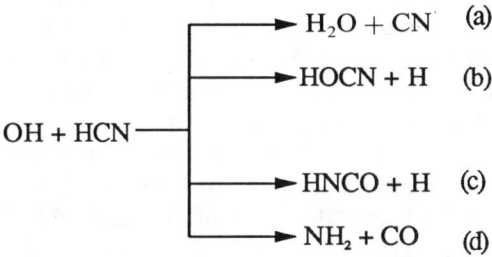

and their comparison with experimental data, the article of Miller and Melius[52] should be consulted.

(2) Acetonitrile (CH₃CN)

The available rate constant data of Harris et al.,[26] Fritz et al.,[22] Zetzsch,[27,28] Kurylo and Knable[29] and Poulet et al.[30] are given in Table 15 and are plotted in Arrhenius form in Fig. 92. In the flash photolysis-resonance fluorescence study of Zetzsch,[28] the room temperature rate constant was observed to be pressure dependent, increasing from $\sim 8 \times 10^{-15}$ cm³ molecule⁻¹ s⁻¹ at ~ 5 Torr total pressure of argon to $\sim 1.1 \times 10^{-14}$ cm³ molecule⁻¹ s⁻¹ at 10 Torr total pressure of argon to 1.8×10^{-14} cm³ molecule⁻¹ s⁻¹ at 100–300 Torr total pressure of argon.

J. Phys. Chem. Ref. Data, Monograph 1 (1989)

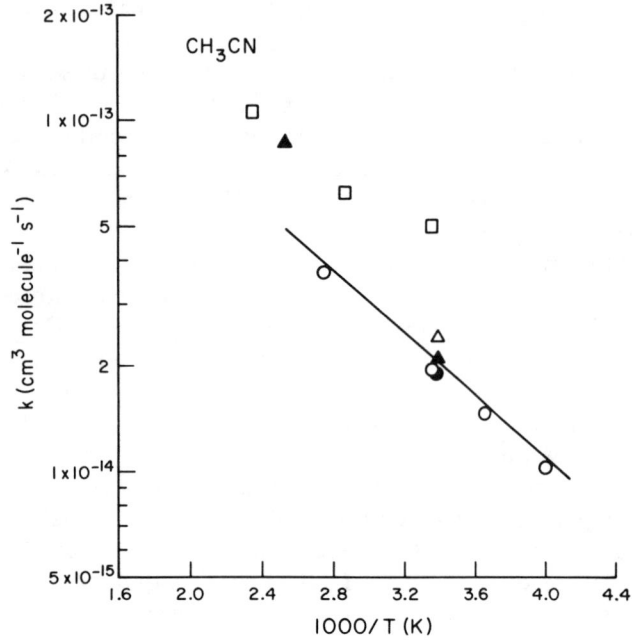

FIG. 92. Arrhenius plot of rate constants for the reaction of the OH radical with CH₃CN. (▢) Harris *et al.*;[26] (△) Fritz *et al.*;[22] (●) Zetzsch;[27,28] (○) Kurylo and Knable;[29] (▲) Poulet *et al.*;[30] (———) recommendation (see text).

However, the absolute room temperature rate constants of Poulet *et al.*,[30] Fritz *et al.*[22] and Kurylo and Knable,[29] obtained at total pressures of 1.2 Torr, 7 Torr and 20–50 Torr, respectively, are in good agreement, suggesting that the lower pressure data of Zetzsch[28] may be in error.

From a comparison of the room temperature rate constants, it is clear (Fig. 92) that the rate constants of Fritz *et al.*,[22] Kurylo and Knable[29] and Poulet *et al.*[30] and the 100–300 Torr total pressure rate constant of Zetzsch[28] are in good agreement, but that these rate constants are a factor of ~2–2.5 lower than that of Harris *et al.*[26] This observation suggests that the rate constants of Harris *et al.*[26] were erroneously high due to the occurrence of secondary reactions or the presence of a reactive impurity. Neglecting these data of Harris *et al.*,[26] the only temperature-dependent data then arise from the study of Kurylo and Knable[29] and rate constants measured by Poulet *et al.*[30] at 393 K (cited as a footnote in their publication) and 295 K. Unfortunately, the temperature dependencies obtained from these data of Kurylo and Knable[29] and Poulet *et al.*[30] do not agree. Since Le Bras and co-workers have experienced problems associated with heterogeneous wall reactions in previous studies,[55] the flash photolysis data of Kurylo and Knable[29] are used to derive the temperature dependence of this reaction.

A unit-weighted average of the room temperature rate constants of Fritz *et al.*,[22] Zetzsch,[27,28] Kurylo and Knable[29] and Poulet *et al.*,[30] combined with the temperature dependence of Kurylo and Knable,[30] leads to the recommended Arrhenius expression of

$$k(CH_3CN) = 6.77 \times 10^{-13} e^{-1030/T} \text{ cm}^3 \text{ molecule}^{-1} \text{ s}^{-1}$$

over the temperature range 250–363 K, and

$$k(CH_3CN) = 2.14 \times 10^{-14} \text{ cm}^3 \text{ molecule}^{-1} \text{ s}^{-1}$$

at 298 K, with an estimated overall uncertainty of ±40% at 298 K, with the uncertainties increasing at both lower and higher temperatures. Although an Arrhenius expression has been recommended, non-Arrhenius behavior is expected at higher and lower temperatures. Clearly, further kinetic studies are necessary, especially at higher temperatures (≳350 K) and as a function of pressure to resolve the above-mentioned discrepancies.

Based upon the apparent lack of a pressure dependence over the range ~1–50 Torr at room temperature (but see above discussion) and the (qualitative) observation that the room temperature rate constants increase by a factor of ~4–10 in going from acetonitrile to propionitrile (Table 15), it is expected that the OH radical reaction with acetonitrile proceeds mainly by H-atom abstraction from the —CH₃ group:

$$OH + CH_3CN \rightarrow H_2O + \dot{C}H_2CN$$

(3) Other Nitriles

The available kinetic data are given in Table 15. For cyanogen, the rate constants are pressure dependent over the temperature range 300–555 K,[25] with the rate constants extrapolating to zero at zero pressure.[25] Thus the OH radical reaction with C_2N_2 proceeds by initial addition over this temperature range:

$$OH + C_2N_2 \rightleftharpoons HOC_2N_2^*$$
$$\downarrow M$$
$$HOC_2N_2$$

It is also possible that the limiting high pressure rate constants were not attained at the highest total pressures employed by Phillips[25] of ~15 Torr, and hence that the Arrhenius expression cited in Table 15 is still in the falloff region.

For acrylonitrile, Harris *et al.*[26] observed that the measured rate constants were pressure dependent at room temperature, increasing from (4.06 ± 0.41) × 10⁻¹³ cm³ molecule⁻¹ s⁻¹ at 50 Torr total pressure of argon to (4.80 ± 0.50) × 10⁻¹³ cm³ molecule⁻¹ s⁻¹ at 500 Torr total pressure of argon. At 50 Torr total pressure of argon, the rate constant was independent of temperature over the range 299–423 K.[26] These data, together with the product data of Hashimoto *et al.*,[56] show that the OH radical reaction with CH₂=CHCN proceeds by initial OH radical addition to the >C=C< bond.

$$OH + CH_2=CHCN \xrightarrow{M} HOCH_2\dot{C}HCN$$

$$\text{and } \dot{C}H_2CHOHCN$$

f. Nitrogen-Containing Heterocycles

The available kinetic data are given in Table 15.

(1) Pyrrole

The rate constants of Atkinson et al.,[32] Wallington[33] and Witte and Zetzsch[34] are given in Table 15 and are plotted in Arrhenius form in Fig. 93.

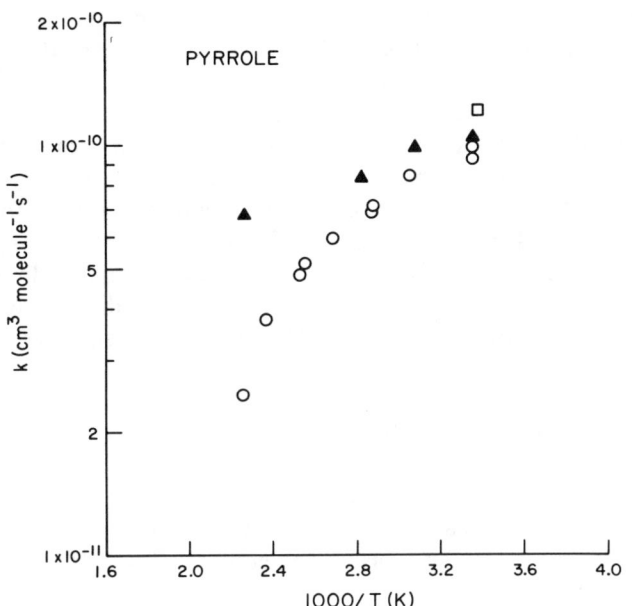

FIG. 93. Arrhenius plot of rate constants for the reaction of the OH radical with pyrrole. (□) Atkinson et al.;[32] (▲) Wallington;[33] (○) Witte and Zetzsch.[34]

There are substantial disagreements concerning the temperature and pressure effects on the rate constants for this reaction. At 298 K, Witte and Zetzsch[34] observed that the rate constants were pressure dependent, increasing from 6.3×10^{-11} cm^3 molecule^{-1} s^{-1} at 22 Torr total pressure of argon to $(9.3-10.1) \times 10^{-11}$ cm^3 molecule^{-1} s^{-1} at 100–500 Torr total pressure of argon. In contrast, Wallington[33] observed no pressure dependence of the rate constant at 298 ± 2 K over the range 25–100 Torr total pressure (of argon diluent). Furthermore, at 100 Torr total pressure, distinct fall-off behavior was observed by Witte and Zetzsch[34] for temperatures $\geqslant 392$ K.

However, at room temperature the (supposedly) limiting high-pressure rate constants of Wallington[33] and Witte and Zetzsch[34] and that obtained by Atkinson et al.[32] in the presence of ~740 Torr total pressure of air are in good agreement. The mean of the high pressure rate constants of Atkinson et al.,[32] Wallington[33] and Witte and Zetzsch[34] leads to the recommendation of

$$k(\text{pyrrole}) = 1.1 \times 10^{-10} \text{ cm}^3 \text{ molecule}^{-1} \text{ s}^{-1}$$

at 298 K, with an estimated overall uncertainty of ±30%.

Based upon the observed effects of pressure and temperature on the rate constant for this reaction, it is clear[34] that at temperatures $\lesssim 400$ K the reaction proceeds, at least in part, by OH radical addition. This OH radical reaction can proceed by the pathways

with channel (b) also possibly involving initial OH radical addition to the N atom followed by decomposition.

(2) Other Nitrogen-Containing Heterocycles

The available kinetic data are given in Table 15. Only for pyridine has more than a single study been carried out, and the room temperature rate constants of Atkinson et al.[35] and Witte and Zetzsch[34] disagree by a factor of ~2. For pyridine, the absolute rate constant data obtained by Witte and Zetzsch[34] exhibit non-Arrhenius behavior, with two distinct temperature regimes ($\leqslant 290$ K and > 420 K) which suggest predominantly an OH radical addition process in the lower temperature region and an H-atom abstraction (from the aromatic ring C—H bonds) process at the higher temperatures.

As noted above, the OH radical reaction with aziridine is expected to proceed by overall H-atom abstraction from the N—H bond and (less likely) the C—H bonds of the three-membered ring. For imidazole and 1,2,4-triazole, the OH radical reactions are expected to be analogous to that for pyrrole, with the ring N-atom(s) decreasing the reactivity of the ring towards OH radical addition. This is observed.[34] Similarly, based upon the relative rate data of Atkinson et al.,[35] 1,3,5-triazine is less reactive than pyridine towards OH radical addition to the ring, as expected.

g. Miscellaneous

(1) Peroxyacetyl Nitrate

While the room temperature rate constants of Wallington et al.[37] and Tsalkani et al.[38] are consistent with the upper limit determined by Winer et al.,[36] the rate constants of Wallington et al.[37] and Tsalkani et al.[38] disagree by a factor of almost 2 (Table 15). The magnitude of this rate constant suggests that the reaction proceeds by H-atom abstraction from the —CH$_3$ group

$$OH + CH_3C(O)OONO_2 \rightarrow H_2O + \dot{C}H_2C(O)OONO_2$$

(2) Nitromethane

The available rate constant data of Campbell and Goodman,[10] Zabarnick et al.[39] and Nielsen et al.[40] are given in Table 15 and are plotted in Arrhenius form in Fig. 94. At room temperature, the reported rate constants for CH_3NO_2 span two orders of magnitude.

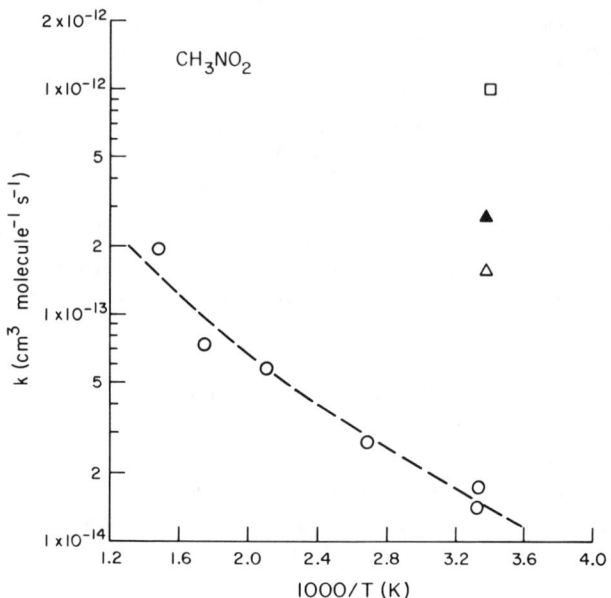

FIG. 94.   Arrhenius plot of rate constants for the reaction of the OH radical with CH_3NO_2. (□) Campbell and Goodman;[10] (○) Zabarnick et al.;[39] (△) Nielsen et al.;[40] absolute rate constant; (▲) Nielsen et al.,[40] relative rate constant; (— — —) three-parameter expression fit to the data of Zabarnick et al.[39]

Despite the discrepancies between the absolute and relative rate studies of Zabarnick et al.[39] and Nielsen et al.[40] (and the factor of 2 disagreement between the absolute and relative rate data of Nielsen et al.[40]), it appears that the rate constant derived at 292 ± 2 K by Campbell and Goodman[10] is in error. The study of Zabarnick et al.[39] utilized the laser photolysis of CH_3NO_2 itself to generate OH radicals, and the measured rate constants were observed to be dependent on the laser photolysis energy (presumably due to secondary reactions involving a photolytically formed species). This dependence of the measured rate constant on the laser fluence necessitated a significant extrapolation to zero laser energy to obtain the supposedly secondary-reaction free rate constant.[39] Thus, as the cited error limits indicate, the rate constants obtained by Zabarnick et al.[39] are subject to significant uncertainties. Zabarnick et al.[39] observed no effect, within the experimental errors, on the measured rate constant at room temperature when the total pressure was varied from 100–300 Torr of argon, indicating that the rate constant is not pressure-dependent.

However, based upon the rate constants obtained by Nielsen et al.[41] and Atkinson and Aschmann[16] for 1-nitrobutane from absolute and relative rate studies (which are in reasonable agreement), it is apparent that the —NO_2 group markedly decreases the reactivity of the α and β —CH_2— groups [since the observed rate constant[16] is consistent with substituent factors[50,51] of $F(-NO_2) \sim 0$ and $F(-CH_2NO_2) \lesssim 0.5$ at 298 K]. This implies that the OH radical reaction rate constant for CH_3NO_2 should be $\ll 1.5 \times 10^{-13}$ cm^3 molecule^{-1} s^{-1} at 298 K. The nitroalkanes are expected to photolyze,[57] and it is not clear whether photolysis occurred under the experimental conditions of the relative rate study of Nielsen et al.,[40] or whether or not photolysis (if it occurred) was taken into account.

In the absence of further experimental data, the rate constants obtained by Zabarnick et al.[39] for the reactions of the OH radical with CH_3NO_2 (Table 15) and of the OD radical with CH_3NO_2 and CD_3NO_2 (Table 16) are preferred. The magnitude of these rate constants and the observation of a deuterium isotope effect[39] indicates that these reactions proceed by H- or D-atom abstraction from the —CH_3 or —CD_3 groups

$$OH + CH_3NO_2 \rightarrow H_2O + \dot{C}H_2NO_2$$

A unit-weighted least-squares analysis of the rate constant data of Zabarnick et al.[39] for the reaction of the OH radical with CH_3NO_2, using the equation $k = CT^2e^{-D/T}$, leads to

$$k(CH_3NO_2) = 5.6$$
$$\times 10^{-19} T^2 e^{-360/T} \text{ cm}^3 \text{ molecule}^{-1} \text{ s}^{-1}$$

over the temperature range 299–671 K, and this expression is shown in Fig. 94 as the dashed line.

(3) Isocyanic Acid (HNCO)

The only kinetic study carried out to date is that of Tully et al.[42] Attempts to determine rate constants for the reaction of the OH radical with DNCO were rendered difficult by the ready D-atom exchange with the OH radical precursor H_2O. However, the magnitude of the rate constants obtained for HNCO suggests that the reaction proceeds mainly by H-atom abstraction.

$$OH + HNCO \rightarrow H_2O + NCO$$

References

[1] R. Atkinson, R. A. Perry, and J. N. Pitts, Jr., J. Chem. Phys. **66**, 1578 (1977).

[2] R. Atkinson, R. A. Perry, and J. N. Pitts, Jr., J. Chem. Phys. **68**, 1850 (1978).

[3] R. A. Gorse, Jr., R. R. Lii, and B. B. Saunders, Science **197**, 1365 (1977).

[4] G. W. Harris and J. N. Pitts, Jr., Environ. Sci. Technol. **17**, 50 (1983).

[5] L. G. Anderson and R. D. Stephens, Int. J. Chem. Kinet. **20**, 103 (1988).

[6]E. C. Tuazon, W. P. L. Carter, R. Atkinson, A. M. Winer, and J. N. Pitts, Jr., Environ. Sci. Technol. **18**, 49 (1984).

[7]S. S. Zabarnick, J. W. Fleming, A. P. Baronavski, and M. C. Lin, Proceedings, 17th International Symposium on Free Radicals, 1985; National Bureau of Standards Special Publication 716, April 1986; p. 731.

[8]W. Hack, K. Hoyermann, and H. Gg. Wagner, Ber. Bunsenges Phys. Chem. **78**, 386 (1974).

[9]G. W. Harris, R. Atkinson, and J. N. Pitts, Jr., J. Phys. Chem. **83**, 2557 (1979).

[10]I. M. Campbell and K. Goodman, Chem. Phys. Lett. **36**, 382 (1975).

[11]G. J. Audley, D. L. Baulch, I. M. Campbell, D. J. Waters, and G. Watling, J. Chem. Soc. Faraday Trans. 1, **78**, 611 (1982).

[12]E. C. Tuazon, W. P. L. Carter, R. Atkinson, and J. N. Pitts, Jr., Int. J. Chem. Kinet. **15**, 619 (1983).

[13]D. L. Baulch, I. M. Campbell, and S. M. Saunders, Int. J. Chem. Kinet. **17**, 355 (1985).

[14]J. S. Gaffney, R. Fajer, G. I. Senum, and J. H. Lee, Int. J. Chem. Kinet. **18**, 399 (1986).

[15]J. A. Kerr and D. W. Stocker, J. Atmos. Chem. **4**, 253 (1986).

[16]R. Atkinson and S. M. Aschmann, Int. J. Chem. Kinet., in press (1989).

[17]R. Atkinson, S. M. Aschmann, W. P. L. Carter, and A. M. Winer, Int. J. Chem. Kinet. **14**, 919 (1982).

[18]R. Atkinson, S. M. Aschmann, W. P. L. Carter, A. M. Winer, and J. N. Pitts, Jr., Int. J. Chem. Kinet. **16**, 1085 (1984).

[19]B. S. Haynes, Combust. Flame, **28**, 113 (1977).

[20]C. P. Fenimore, 17th International Symposium on Combustion, 1978; The Combustion Institute, Pittsburgh, PA, 1979, p. 661.

[21]C. Morley, 18th International Symposium on Combustion, 1980; The Combustion Institute, Pittsburgh, PA, 1981, p. 23.

[22]B. Fritz, K. Lorenz, W. Steinert, and R. Zellner, Proceedings, 2nd European Symposium on the Physico-Chemical Behavior of Atmospheric Pollutants, 1981; D. Riedel Publishing Co., Dordrecht, Holland, 1982, p. 192.

[23]B. Fritz, K. Lorenz, W. Steinert, and R. Zellner, Oxid. Commun. **6**, 363 (1984).

[24]R. Atkinson, R. A. Perry, and J. N. Pitts, Jr., Combust. Flame **31**, 213 (1978).

[25]L. F. Phillips, Combust. Flame **35**, 233 (1979).

[26]G. W. Harris, T. E. Kleindienst, and J. N. Pitts, Jr., Chem. Phys. Lett. **80**, 479 (1981).

[27]C. Zetzsch, report to Bundeminister für Forschung und Technologie, Projektträger für Umweltchemikalien, 1982.

[28]C. Zetzsch, presented at Bunsen-Kolloquium, Frankfurt am Main, March 1983.

[29]M. J. Kurylo and G. L. Knable, J. Phys. Chem. **88**, 3305 (1984).

[30]G. Poulet, G. Laverdet, J. L. Jourdain, and G. Le Bras, J. Phys. Chem. **88**, 6259 (1984).

[31]C. Zetzsch, presented at Bunsen-Kolloquium, Göttingen, October 1980.

[32]R. Atkinson, S. M. Aschmann, A. M. Winer, and W. P. L. Carter, Atmos. Environ. **18**, 2105 (1984).

[33]T. J. Wallington, Int. J. Chem. Kinet. **18**, 487 (1986).

[34]F. Witte and C. Zetzsch, 9th International Symposium on Gas Kinetics, University of Bordeaux, Bordeaux, France, July 20–25, 1986, private communication (1988).

[35]R. Atkinson, E. C. Tuazon, T. J. Wallington, S. M. Aschmann, J. Arey, A. M. Winer, and J. N. Pitts, Jr., Environ. Sci. Technol. **21**, 64 (1987).

[36]A. M. Winer, A. C. Lloyd, K. R. Darnall, R. Atkinson, and J. N. Pitts, Jr., Chem. Phys. Lett. **51**, 221 (1977).

[37]T. J. Wallington, R. Atkinson, and A. M. Winer, Geophys. Res. Lett. **11**, 861 (1984).

[38]N. Tsalkani, A. Mellouki, G. Poulet, G. Toupance, and G. Le Bras, J. Atmos. Chem. **7**, 409 (1988).

[39]S. Zabarnick, J. W. Fleming, and M. C. Lin, Chem. Phys. **120**, 319 (1988).

[40]O. J. Nielsen, H. W. Sidebottom, D. J. O'Farrell, M. Donlon, and J. Treacy, Chem. Phys. Lett. **146**, 197 (1988).

[41]O. J. Nielsen, H. W. Sidebottom, D. J. O'Farrell, M. Donlon, and J. Treacy, Chem. Phys. Lett. **156**, 312 (1989).

[42]F. P. Tully, R. A. Perry, L. R. Thorne, and M. D. Allendorf, presented at 22nd International Symposium on Combustion, Seattle, August 14–19, 1988.

[43]D. G. Horne and R. G. W. Norrish, Proc. Roy. Soc. (London) **A315**, 287 (1970).

[44]L. F. Phillips, Chem. Phys. Lett. **57**, 538 (1978).

[45]L. F. Phillips, Aust. J. Chem. **32**, 2571 (1979).

[46]D. F. McMillen and D. M. Golden, Ann. Rev. Phys. Chem. **33**, 493 (1982).

[47]C. R. C. Lindley, J. G. Calvert, and J. H. Shaw, Chem. Phys. Lett. **67**, 57 (1979).

[48]M. A. Grela and A. J. Colussi, Int. J. Chem. Kinet. **20**, 713 (1988).

[49]M. E. Jenkin, G. D. Hayman, and R. A. Cox, J. Photochem. Photobiol. A: Chem. **42**, 187 (1988).

[50]R. Atkinson, Chem. Rev. **86**, 69 (1986).

[51]R. Atkinson, Int. J. Chem. Kinet. **19**, 799 (1987).

[52]J. A. Miller and C. F. Melius, 21st International Symposium on Combustion, 1986; The Combustion Institute, Pittsburgh, PA, 1988; p. 919.

[53]A. Szekely, R. K. Hanson, and C. T. Bowman, Int. J. Chem. Kinet. **16**, 1609 (1984).

[54]A. Jacobs, M. Wahl, R. Weller, and J. Wolfrum, Chem. Phys. Lett. **144**, 203 (1988).

[55]D. Martin, J. L. Jourdain, and G. Le Bras, Int. J. Chem. Kinet. **17**, 1247 (1985) and references therein.

[56]S. Hashimoto, H. Bandow, H. Akimoto, J.-H. Weng, and X.-Y. Tang, Int. J. Chem. Kinet. **16**, 1385 (1984).

[57]W. D. Taylor, T. D. Allston, M. J. Moscato, G. B. Fazekas, R. Kozlowski, and G. A. Takacs, Int. J. Chem. Kinet. **12**, 231 (1980).

2.9. Phosphorus-Containing Organics

The available kinetic data for the gas-phase reactions of the OH radical with organophosphorus compounds are given in Table 17. In addition, the available data for the OH radical reaction with phosphine (PH_3) are also included. For each of the organophosphorus compounds, as for phosphine, only a single study has been carried out, and hence no recommendations are given.

From a product study carried out at atmospheric pressure of air and room temperature, Tuazon et al.[2] concluded that the reaction of the OH radical with trimethyl phosphate does not proceed by the displacement mechanism,

$$OH + (CH_3O)_3PO \rightarrow CH_3\dot{O} + (CH_3O)_2P(O)OH$$

since the expected HCHO product formed from reaction of the CH_3O radical with O_2 was not observed. This observation then suggests that this reaction proceeds by H-atom abstraction from the —OCH_3 groups.

$$OH + (CH_3O)_3PO \rightarrow H_2O + (CH_3O)_2P(O)O\dot{C}H_2$$

Based upon these product data for trimethyl phosphate[2] and the kinetic data for triethyl phosphate[3] and a series of trimethyl phosphorothioates,[4] Goodman et al.,[4] Atkinson et al.[3] and Atkinson[6] proposed that at room temperature the OH radical reactions with these organophosphorus compounds proceed by H-atom abstraction from the —OCH_3 and —SCH_3 groups and by initial OH radical addition to the —$>P=S$ bond. For example, for $(CH_3O)_2P(S)SCH_3$,

$$
OH + (CH_3O)_2P(S)SCH_3 \longrightarrow
\begin{cases}
\overset{\cdot}{O}CH_2 \\
| \\
CH_3OP(S)SCH_3 & \text{(a)} \\
\\
(CH_3O)_2P(S)S\overset{\cdot}{C}H_2 & \text{(b)} \\
\\
\overset{SOH}{|} \\
(CH_3O)_2PSCH_3 & \text{(c)} \\
\downarrow \\
\downarrow \\
products
\end{cases}
$$

with the rate constant for H-atom abstraction from either a —OCH_3 or —SCH_3 group being $\sim 2.9 \times 10^{-12}$ cm^3 molecule^{-1} s^{-1} at 298 K and the rate constant at ~ 298 K for addition to the —$>P=S$ bond being $\sim 5.5 \times 10^{-11}$ cm^3 molecule^{-1} s^{-1}. The occurrence of reaction pathway (c) is further supported by the observed formation of $(CH_3O)_3PO$ and $(CH_3O)_2P(O)SCH_3$ from the OH radical-initiated reactions of $(CH_3O)_3PS$ and $(CH_3O)_2P(S)SCH_3$, respectively, under simulated atmospheric conditions.[7]

The kinetic data for the dimethyl phosphoroamidates and dimethyl phosphorothioamidates suggest that for these compounds the OH radical reactions also proceed, in part, by reaction with the —$N(CH_3)_2$, —$NHCH_3$ or —NH_2 groups,[5] presumably by a process involving overall H-atom abstraction from these groups.[5]

References

[1] B. Fritz, K. Lorenz, W. Steinert, and R. Zellner, Proceedings, 2nd European Symposium on the Physico-Chemical Behavior of Atmospheric Pollutants, 1981; D. Riedel Publishing Co., Dordrecht, Holland, 1982, p. 192; "Methods of the Ecotoxicological Evaluation of Chemicals, Photochemical Degradation in the Gas Phase," Vol. 6, *OH Reaction Rate Constants and Tropospheric Lifetimes of Selected Environmental Chemicals.* Report 1980–1983, K. H. Becker, H. M. Biehl, P. Bruckmann, E. H. Fink, F. Führ, W. Klöpffer, R. Zellner, and C. Zetzsch, Eds., Kernforschungsanlage Jülich GmbH, November 1984.

[2] E. C. Tuazon, R. Atkinson, S. M. Aschmann, J. Arey, A. M. Winer, and J. N. Pitts, Jr., Environ. Sci. Technol. 20, 1043 (1986).

[3] R. Atkinson, S. M. Aschmann, M. A. Goodman, and A. M. Winer, Int. J. Chem. Kinet. 20, 273 (1988).

[4] M. A. Goodman, S. M. Aschmann, R. Atkinson, and A. M. Winer, Arch. Environ. Contam. Toxicol. 17, 281 (1988).

[5] M. A. Goodman, S. M. Aschmann, R. Atkinson, and A. M. Winer, Environ. Sci. Technol. 22, 578 (1988).

[6] R. Atkinson, Environ. Toxicol. Chem. 7, 435 (1988).

[7] R. Atkinson, S. M. Aschmann, J. Arey, P. A. McElroy, and A. M. Winer, Environ. Sci. Technol. 23, 243 (1989).

TABLE 17. Rate constants k and temperature-dependent parameters for the gas-phase reactions of the OH radical with phosphorus-containing organics

Organic	$10^{12} \times A$ (cm^3 molecule^{-1} s^{-1})	B (K)	$10^{12} \times k$ (cm^3 molecule^{-1} s^{-1})	at T (K)	Technique	Reference
Phosphine [PH_3]			14 ± 3	249	LP-RF	Fritz et al.[1]
			17 ± 3	249		
			13 ± 3	256		
			18 ± 3	296		
			12 ± 3	296		
			13 ± 3	296		
			18 ± 3	370		
	27 ± 6	155	18 ± 3	438		
Trimethyl phosphate [$(CH_3O)_3PO$]			7.37 ± 0.74	296 ± 2	RR [relative to k(dimethyl ether) $= 2.96 \times 10^{-12}$][a]	Tuazon et al.[2]
Triethyl phosphate [$(C_2H_5O)_3PO$]			55.3 ± 3.5	296 ± 2	RR [relative to k(propene) $= 2.66 \times 10^{-11}$][a]	Atkinson et al.[3]
O,O,S-Trimethyl-phosphorothioate [$(CH_3O)_2P(O)SCH_3$]			9.29 ± 0.68	298 ± 2	RR [relative to k(cyclohexane) $= 7.49 \times 10^{-12}$][a]	Goodman et al.[4]
O,S,S-Trimethyl-phosphorodithioate [$(CH_3S)_2P(O)OCH_3$]			9.59 ± 0.75	298 ± 2	RR [relative to k(cyclohexane) $= 7.49 \times 10^{-12}$][a]	Goodman et al.[4]

TABLE 17. Rate constants k and temperature-dependent parameters for the gas-phase reactions of the OH radical with phosphorus-containing organics — Continued

Organic	$10^{12} \times A$ (cm^3 molecule^{-1} s^{-1})	B (K)	$10^{12} \times k$ (cm^3 molecule^{-1} s^{-1})	at T (K)	Technique	Reference
O,O,O-Trimethyl-phosphorothioate [(CH$_3$O)$_3$PS]			69.7 ± 3.9	298 ± 2	RR [relative to k(2-methyl-1,3-butadiene) = 1.01 × 10^{-10}][a]	Goodman et al.[4]
O,O,S-Trimethyl-phosphorodithioate [(CH$_3$O)$_2$P(S)SCH$_3$]			56.0 ± 1.8	298 ± 2	RR [relative to k(2-methyl-1,3-butadiene) = 1.01 × 10^{-10}][a]	Goodman et al.[4]
(CH$_3$O)$_2$P(O)N(CH$_3$)$_2$			31.9 ± 2.4	296 ± 2	RR [relative to k(propene) = 2.66 × 10^{-11}][a]	Goodman et al.[5]
(CH$_3$O)$_2$P(S)N(CH$_3$)$_2$			46.8 ± 1.4	296 ± 2	RR [relative to k(propene) = 2.66 × 10^{-11}][a]	Goodman et al.[5]
(CH$_3$O)$_2$P(S)NHCH$_3$			233 ± 15	296 ± 2	RR [relative to k(propene) = 2.66 × 10^{-11}][a]	Goodman et al.[5]
			232 ± 13	296 ± 2	RR [relative to k(2,3-dimethyl-2-butene) = 1.11 × 10^{-10}][a]	Goodman et al.[5]
(CH$_3$O)$_2$P(S)NH$_2$			244 ± 9	296 ± 2	RR [relative to k(2,3-dimethyl-2-butene) = 1.11 × 10^{-10}][a]	Goodman et al.[5]
Dimethyl chloro-phosphorothioate [(CH$_3$O)$_2$P(S)Cl]			59.0 ± 3.8	296 ± 2	RR [relative to k(2-methyl-1,3-butadiene) = 1.01 × 10^{-10}][a]	Atkinson et al.[3]

[a]From present recommendations (see text).

2.10. Silicon-Containing Compounds

The available kinetic data for the reactions of the OH radical with organosilicon compounds are given in Table 18. The rate constant data for the reaction of the OH radical with silane (SiH$_4$) are also included for completeness. To date, the only silicon-containing compounds for which OH radical reaction rate data are available are silane[1] and tetraethylsilane.[2] Both of these reactions have been postulated to occur by overall H-atom abstraction:[1,2]

$$OH + SiH_4 \rightarrow H_2O + SiH_3$$

$$OH + (C_2H_5)_4Si \begin{cases} \rightarrow H_2O + CH_3\dot{C}HSi(C_2H_5)_3 & \text{(a)} \\ \rightarrow H_2O + \dot{C}H_2CH_2Si(C_2H_5)_3 & \text{(b)} \end{cases}$$

with reaction pathway (a) being expected to dominate for the tetraethylsilane reaction since the Si atom must activate the neighboring —CH$_2$— groups towards reaction if the measured rate constant[2] at 793 K is correct (although it should be noted that the reported rate constant[2] was not subsequently revised to take into account the effects of self-heating of the reaction mixtures, as was carried out for a series of alkanes[3]). Clearly, further kinetic and mechanistic data for this class of compounds are required.

References

[1]R. Atkinson and J. N. Pitts, Jr., Int. J. Chem. Kinet. **10**, 1151 (1978).

[2]R. R. Baldwin, C. J. Everett, and R. W. Walker, Trans. Faraday Soc. **64**, 2708 (1968).

[3]R. R. Baldwin and R. W. Walker, J. Chem. Soc. Faraday Trans. 1, **75**, 140 (1979).

TABLE 18. Rate constants k and temperature-dependent parameters for the gas-phase reactions of the OH radical with silicon-containing compounds

Compound	$10^{12} \times A$ (cm^3 molecule^{-1} s^{-1})	B (K)	$10^{12} \times k$ (cm^3 molecule^{-1} s^{-1})	at T (K)	Technique	Reference	Temperature range covered
Silane [SiH$_4$]			12.4 ± 1.9	299.6	FP-RF	Atkinson and Pitts[1]	300–426
			12.4 ± 1.9	355.6			
	14.4	48 ± 201	13.0 ± 2.0	425.8			
Tetraethylsilane [(C$_2$H$_5$)$_4$Si]			75	793	RR [relative to $k(H_2)$ = 1.02 × 10^{-12}][a]	Baldwin et al.[2]	

[a]See Introduction.

2.11. Aromatic Compounds

a. Kinetics

The available kinetic data are listed in Table 19. In general, these rate constant data, obtained from both absolute and rate constant studies, are in reasonably good agreement. Room temperature rate constants are available for a wide variety of aromatic hydrocarbons and substituted aromatics. Additionally, temperature dependence studies have been carried out for benzene,[4,6,8,9,13,14,16] benzene-d_6,[6,8] toluene,[4,6] toluene-d_3,[6] toluene-d_5,[6] toluene-d_8,[4,6] the xylene isomers,[4,21] the trimethylbenzene isomers,[4] phenol,[29,30] methoxybenzene,[31] o-cresol,[31] fluorobenzene,[16] chlorobenzene,[16] bromobenzene,[14,16] iodobenzene,[16] aniline,[14,40] N,N-dimethylaniline,[40] nitrobenzene,[14] o-nitrophenol,[36] 1,2,4-trichlorobenzene,[11] hexafluorobenzene,[16] naphthalene[8,9] and phenanthrene.[9,48]

For a number of the aromatic compounds, three distinct temperature regimes have been observed with the flash or laser photolysis techniques employed to date:[4,6,8,9,14,16,21,29,31,40] (a) at low temperatures, i.e., $\lesssim 325$ K for the monocyclic aromatic hydrocarbons and $\lesssim 410$ K for naphthalene and phenanthrene, exponential OH radical decays are observed, and the rate constants change only slightly with temperature, with negative temperature dependencies being obtained in many cases; (b) at elevated temperatures, $\gtrsim 400\text{-}450$ K for the monocyclic aromatic hydrocarbons and $\gtrsim 600$ K for naphthalene and phenanthrene, exponential OH radical decays are also observed. For the aromatic hydrocarbons, at elevated temperatures the measured rate constants increase rapidly with increasing temperature, with the rate constants at $\sim 400\text{-}450$ K (or ~ 600 K for naphthalene and phenanthrene) being typically a factor of 5–10 lower than those at ~ 325 K; and (c) at intermediate temperatures of $\sim 325\text{-}400$ K for the monocyclic aromatics, and $\sim 410\text{-}600$ K for naphthalene and phenanthrene, non-exponential decays of OH radicals are observed,[4,6,10,14,21] with the decay rate decreasing with the reaction time. In this temperature regime any rate data obtained are a combination of the forward and reverse reaction steps (see below) and are dependent on the experimental conditions (for example, the observation time) employed.

As discussed below, for the aromatic hydrocarbons, phenol, methoxybenzene, o-cresol, fluorobenzene, chlorobenzene, bromobenzene, iodobenzene, aniline and N,N-dimethylaniline, the available kinetic and mechanistic data show that in the low temperature regime OH radical addition to the aromatic ring is the dominant reaction pathway, while at elevated temperatures H-atom abstraction occurs. The intermediate temperature regime where non-exponential OH radical decays occur is characterized by formation and redissociation of the OH-aromatic adducts. Furthermore, this precise intermediate temperature range where non-exponential decays are observed is dependent to some extent on the time-resolution of the experimental technique. For these reasons the reported kinetic data in this intermediate temperature regime are not discussed, apart from tabulating these data, as reported, in Table 19.

Furthermore, although exponential OH radical decays were observed by Perry et al.[4] for the monocyclic aromatics and by Wallington et al.[16] for benzene for temperature $\gtrsim 380$ K, Tully et al.[6] report that the OH radical addition process continues to contribute to the high temperature reaction pathway up to ~ 450 K. Hence in the discussion and derivation of temperature dependent rate constants for the individual aromatic hydrocarbons in the sections below, only rate constants in the temperature regimes $\leqslant 325$ K and $\geqslant 450$ K for the monocyclic aromatic hydrocarbons, and $\leqslant 410$ K and $\geqslant 600$ K for naphthalene and phenanthrene, have been utilized. The kinetic data for the individual compounds are discussed below.

(1) Benzene

The available rate constant data obtained at, or close to, the high pressure limit are listed in Table 19. The kinetic data of Davis et al.,[1] Lorenz and Zellner,[8] Wahner and Zetzsch,[10] Witte et al.,[14] Baulch et al.[52] and Bourmada et al.[53] show that at around room temperature the rate constant is in the fall-off regime between second and third-order kinetics below $\sim 25\text{-}50$ Torr total pressure of argon or helium diluent, with the room temperature rate constant being essentially independent of total pressure above this pressure range[3,6,8,10,11,14] (although Davis et al.[1] observed a pressure-dependence extending to $\geqslant 100$ Torr total pressure of helium, in contrast to more recent studies[6]). Based upon the room temperature pressure-dependent data of Lorenz and Zellner[8] and Witte *et al.*,[14] the limiting low pressure third-order rate constant k_o for argon diluent is

$$k_o^{Ar}(\text{benzene}) = 3 \times 10^{-29} \text{ cm}^6 \text{ molecule}^{-2} \text{ s}^{-1}$$

at 298 K, with a lower value (by a factor of ~ 2) for helium diluent.[52] With a value of $F = 0.6$[14] in the Troe fall-off expression and $k_\infty \approx 1.3 \times 10^{-12}$ cm^3 molecule^{-1} s^{-1} at 298 K, this shows that at 298 K and 760 and 100 Torr total pressure of argon diluent the measured rate constant is ~ 5–6% and $\sim 12\%$, respectively, below the high pressure limiting value, k_∞.

TABLE 19. Rate constants k and temperature-dependent parameters for the gas-phase reactions of the OH radical with aromatic compounds at, or close to, the high pressure limit

Aromatic	$10^{12} \times A$ (cm^3 molecule^{-1} s^{-1})	B (K)	$10^{12} \times k$ (cm^3 molecule^{-1} s^{-1})	at T (K)	Technique	Reference	Temperature range covered (K)
Benzene			1.59 ± 0.12	298	FP-RF	Davis *et al.*[1]	
			$\leqslant 2.6$	304 ± 1	RR [relative to $k(n\text{-butane}) = 2.61 \times 10^{-12}$][a]	Doyle *et al.*[2]	
			1.24 ± 0.12	298	FP-RF	Hansen *et al.*[3]	
			1.20 ± 0.15	297.6	FP-RF	Perry *et al.*[4]	298–422
			1.32 ± 0.30	304.4			
			1.33 ± 0.25	305.8			
			1.66 ± 0.25	322.7			
			1.37 ± 0.20[b]	331.9			
			1.66[b]	333.2			
			1.04[b]	350.6			
			0.63[b]	354.7			
			1.00[b]	355.2			
			1.00[b]	361.2			
			0.31[b]	364.8			
			0.31[b]	380.8			
			0.26 ± 0.15	396.2			
			0.34 ± 0.07	396.4			
			0.34 ± 0.12	405.8			
			0.45 ± 0.07	422.0			
			0.84	300	RR [relative to $k(\text{ethene}) = 8.44 \times 10^{-12}$][a]	Cox *et al.*[5]	
			1.04 ± 0.08	250	FP-RF	Tully *et al.*[6]	250–1017
			1.20 ± 0.09	270			
	3.1 ± 2.6	270 ± 220 (250–298 K)	1.24 ± 0.09	298			
			0.7[b]	352			
			0.3[b]	390			
			0.4[b]	442			
			0.543 ± 0.023	542			
			0.639 ± 0.029	621			
			0.682 ± 0.074	630			
			0.606 ± 0.034	653			
			1.02 ± 0.04	715			
			1.20 ± 0.16	742			
			1.59 ± 0.09	817			
			1.90 ± 0.20	895			
			2.26 ± 0.13	917			
			2.35 ± 0.23	981			
	24 ± 9	2260 ± 300 (621–1017 K)	2.20 ± 0.34	1017			

TABLE 19. Rate constants k and temperature-dependent parameters for the gas-phase reactions of the OH radical with aromatic compounds at, or close to, the high pressure limit — Continued

Aromatic	$10^{12} \times A$ (cm³ molecule⁻¹ s⁻¹)	B (K)	$10^{12} \times k$ (cm³ molecule⁻¹ s⁻¹)	at T (K)	Technique	Reference	Temperature range covered (K)
			0.93	300	RR [relative to k(ethene) = 8.44×10^{-12}][a]	Barnes et al.[7]	
			0.76 ± 0.15	244	LP-RF	Lorenz and Zellner[8,9]	244–870
	6.3 ± 1.7	500 ± 50 (244–298 K)	1.15 ± 0.25	298 ± 2			
			1.26 ± 0.25	336			
			0.83 ± 0.17	373			
			0.50 ± 0.10	384			
			0.40 ± 0.09	453			
			0.48 ± 0.16	522			
			0.43 ± 0.09	523			
			0.75 ± 0.20	567			
			0.46 ± 0.10	604			
			0.76 ± 0.25	665			
			1.28 ± 0.16	720			
			1.13 ± 0.25	803			
	20 ± 10	2100 ± 400 (522–870 K)	2.16 ± 0.50	870			
			0.88 ± 0.04	295	FP-RF	Wahner and Zetzsch[10]	
			1.02 ± 0.2	296	FP-RF	Rinke and Zetzsch[11]	
			1.46 ± 0.06	c	RR [relative to k(n-hexane) = 5.61×10^{-12}][a]	Ohta and Ohyama[12]	
			2.09 ± 0.06	787	FP-RF	Madronich and Felder[13]	787–1409
			1.78 ± 0.06	805			
			2.67 ± 0.18	865			
			3.04 ± 0.29	1019			
			3.52 ± 0.25	1196			
			6.66 ± 0.23	1309			
	35 ± 3	2300 ± 100	7.36 ± 1.12	1409			
			1.00 ± 0.04	239	FP-RF	Witte et al.[14]	239–354
			1.11 ± 0.12	239			
			1.04 ± 0.03	245			
			1.06 ± 0.03	253			
			1.13 ± 0.04	253			
			1.04 ± 0.05	259			
			1.11 ± 0.02	274			
			1.12 ± 0.08	274			
			1.11 ± 0.04	283			
			1.05 ± 0.05	299			
			1.18 ± 0.17	299			
			1.23 ± 0.09	312			
			1.37 ± 0.09	312			
			1.21 ± 0.12	325			
			1.28 ± 0.05	328			
			1.26 ± 0.08	331			
			1.39 ± 0.22	331			
			1.48 ± 0.12	334			
			1.50 ± 0.29	342			
			1.25 ± 0.22	352			
	2.3	190 ± 60	1.33 ± 0.36	354			
			1.22 ± 0.45	296	RR [relative to k(propane) = 1.13×10^{-12}][a]	Edney et al.[15]	

TABLE 19. Rate constants k and temperature-dependent parameters for the gas-phase reactions of the OH radical with aromatic compounds at, or close to, the high pressure limit — Continued

Aromatic	$10^{12} \times A$ (cm^3 molecule^{-1} s^{-1})	B (K)	$10^{12} \times k$ (cm^3 molecule^{-1} s^{-1})	at T (K)	Technique	Reference	Temperature range covered (K)
			1.40 ± 0.23	234	FP-RF	Wallington et al.[16]	234–438
			1.30 ± 0.19	263			
			1.29 ± 0.14	296			
			0.193 ± 0.037	393			
			0.258 ± 0.034	438			
Benzene-d_6			1.08 ± 0.05	250	FP-RF	Tully et al.[6]	250–1150
			1.19 ± 0.05	298			
			0.4^b	498			
			0.227 ± 0.03	568			
			0.424 ± 0.045	630			
			0.300 ± 0.032	653			
			0.430 ± 0.023	675			
			0.481 ± 0.019	734			
			0.720 ± 0.046	830			
			1.04 ± 0.03	917			
			1.08 ± 0.10	981			
			1.47 ± 0.07	1002			
	13 ± 6	2300 ± 340 (568–1150 K)	1.91 ± 0.28	1150			
			1.10 ± 0.22	298	LP-RF	Lorenz and Zellner[8]	298–524
			1.00 ± 0.20	336			
			0.76 ± 0.15	380			
			0.48 ± 0.10	398			
			0.28 ± 0.05	436			
			0.25 ± 0.05	524			
Toluene			6.11 ± 0.40	298	FP-RF	Davis et al.[1]	
			3.7 ± 1.6	304 ± 1	RR [relative to $k(n\text{-butane}) = 2.61 \times 10^{-12}$]^a	Doyle et al.[2]	
			5.78 ± 0.58	298	FP-RF	Hansen et al.[3]	
			6.40 ± 0.64	297.9	FP-RF	Perry et al.[4]	298–473
			4.90 ± 0.6	323.7			
			4.99 ± 0.6^b	325.3			
			4.04^b	334.6			
			5.36 ± 0.9^b	338.5			
			1.51^b	339.7			
			1.66 ± 0.25^b	352.6			
			1.19^b	354.2			
			1.38 ± 0.17^b	364.0			
			1.35^b	366.0			
			1.22 ± 0.14^b	378.4			
			1.49 ± 0.22	379.3			
			1.58 ± 0.24	394.2			
			1.69 ± 0.25	408.7			
			1.76 ± 0.18	424.4			
			1.71 ± 0.20	472.7			
			7.6	300	RR [relative to $k(\text{ethene}) = 8.44 \times 10^{-12}$]^a	Cox et al.[5]	
			8.20 ± 0.54	213	FP-RF	Tully et al.[6]	213–1046
			8.73 ± 0.39	231			
			7.97 ± 0.56	250			
			8.53 ± 0.37	260			
			7.44 ± 0.55	270			

TABLE 19. Rate constants k and temperature-dependent parameters for the gas-phase reactions of the OH radical with aromatic compounds at, or close to, the high pressure limit — Continued

Aromatic	$10^{12} \times A$ (cm^3 molecule^{-1} s^{-1})	B (K)	$10^{12} \times k$ (cm^3 molecule^{-1} s^{-1})	at T (K)	Technique	Reference	Temperature range covered (K)
	3.8 ± 2.5	-180 ± 170 (213–298 K)	6.36 ± 0.69	298			
			6.3 ± 0.6^b	320			
			5.4 ± 1.1^b	332			
			3.6^b	352			
			1.4^b	397			
			1.7^b	442			
			2.16 ± 0.08	504			
			2.45 ± 0.05	568			
			2.49 ± 0.12	568			
			3.26 ± 0.29	666			
			3.58 ± 0.16	694			
			4.67 ± 0.19	793			
			5.54 ± 0.27	868			
			6.87 ± 0.23	958			
			9.5 ± 1.0	1046			
			6.06 ± 0.17	c	RR [relative to $k(n\text{-hexane}) = 5.61 \times 10^{-12}]^a$	Ohta and Ohyama[12]	
			5.44 ± 0.55	297	RR [relative to $k(\text{cyclohexane}) = 7.47 \times 10^{-12}]^a$	Edney et al.[15]	
			6.45 ± 0.74	773	RR [relative to $k(\text{H}_2) = 9.22 \times 10^{-13}]^d$	Baldwin et al.[17]	
			5.48 ± 0.16	296 ± 2	RR [relative to $k(\text{propene}) = 2.66 \times 10^{-11}]^a$	Atkinson and Aschmann[18]	
Toluene-d_3 [C$_6$H$_5$CD$_3$]			5.62 ± 0.52	250	FP-RF	Tully et al.[6]	250–1002
			5.97 ± 0.17	270			
			5.63 ± 0.30	298			
			0.8^b	383			
			1.23 ± 0.09	518			
			1.32 ± 0.08	568			
			1.41 ± 0.06	568			
			1.46 ± 0.09	568			
			1.40 ± 0.08	568			
			2.10 ± 0.07	653			
			3.01 ± 0.10	742			
			3.59 ± 0.13	817			
			4.55 ± 0.34	895			
			5.92 ± 0.42	966			
			8.54 ± 1.21	1002			
Toluene-d_5 [C$_6$D$_5$CH$_3$]			6.11 ± 0.40	250	FP-RF	Tully et al.[6]	250–1002
			6.02 ± 1.68	270			
			6.47 ± 0.65	298			
			3.0^b	358			
			1.1^b	412			
			1.66 ± 0.10	470			
			2.04 ± 0.14	518			
			2.52 ± 0.14	568			
			2.69 ± 0.28	630			
			3.29 ± 0.25	653			
			4.53 ± 0.52	742			
			5.08 ± 0.32	793			
			6.48 ± 0.41	895			
			6.52 ± 0.91	996			

TABLE 19. Rate constants k and temperature-dependent parameters for the gas-phase reactions of the OH radical with aromatic compounds at, or close to, the high pressure limit — Continued

Aromatic	$10^{12} \times A$ (cm³ molecule⁻¹ s⁻¹)	B (K)	$10^{12} \times k$ (cm³ molecule⁻¹ s⁻¹)	at T (K)	Technique	Reference	Temperature range covered (K)
			7.97 ± 0.73	1002			
Toluene-d_8 [$C_6D_5CD_3$]			6.13 ± 0.63	298.1	FP-RF	Lloyd et al.[4]	298–432
			4.78^b	323.6			
			3.56^b	324.2			
			0.38 ± 0.06	385.2			
			0.51 ± 0.07	397.0			
			0.70 ± 0.07	432.2			
			6.04 ± 0.48	250	FP-RF	Tully et al.[6]	250–1150
			6.36 ± 0.52	270			
			6.40 ± 0.20	298			
			0.5^b	390			
			0.73 ± 0.07	470			
			1.17 ± 0.09	498			
			1.27 ± 0.03	542			
			1.15 ± 0.05	568			
			1.97 ± 0.12	621			
			2.35 ± 0.16	700			
			2.18 ± 0.10	715			
			3.53 ± 0.28	793			
			2.76 ± 0.14	842			
			3.18 ± 0.30	842			
			4.05 ± 0.30	868			
			4.52 ± 0.25	966			
			6.91 ± 1.32	1017			
			6.51 ± 1.50	1150			
Ethylbenzene			6.94 ± 1.39	305 ± 2	RR [relative to k(n-butane) $= 2.62 \times 10^{-12}$][a]	Lloyd et al.[19]	
			7.95 ± 0.50^e	298	FP-RF	Ravishankara et al.[20]	
			6.51 ± 0.29	c	RR [relative to k(n-hexane) $= 5.61 \times 10^{-12}$][a]	Ohta and Ohyama[12]	
o-Xylene			11.2 ± 3.4	304 ± 1	RR [relative to k(n-butane) $= 2.61 \times 10^{-12}$][a]	Doyle et al.[2]	
			15.3 ± 1.5	298.0	FP-RF	Hansen et al.[3]	
			14.3 ± 1.5	298.5	FP-RF	Perry et al.[4]	298–432
			12.9^b	313.5			
			14.0 ± 2.0	319.0			
			12.3^b	332.1			
			9.27^b	348.2			
			4.98^b	367.7			
			3.76^b	372.8			
			3.25 ± 0.45	379.5			
			3.35 ± 0.46	395.6			
			3.27 ± 0.46	414.6			
			3.63 ± 0.43	425.3			
			3.34 ± 0.35	432.4			
			12.4 ± 1.2^e	298	FP-RF	Ravishankara et al.[20]	
			14.0	300	RR [relative to k(ethene) $= 8.44 \times 10^{-12}$][a]	Cox et al.[5]	

TABLE 19. Rate constants k and temperature-dependent parameters for the gas-phase reactions of the OH radical with aromatic compounds at, or close to, the high pressure limit — Continued

Aromatic	$10^{12} \times A$ (cm³ molecule⁻¹ s⁻¹)	B (K)	$10^{12} \times k$ (cm³ molecule⁻¹ s⁻¹)	at T (K)	Technique	Reference	Temperature range covered (K)
			14.2 ± 1.7	298	FP-RF	Nicovich et al.[21]	298–970
			15.8 ± 1.8	320			
			5.1[b]	357			
			2.39 ± 0.20	400			
			4.19 ± 0.48	508			
			5.42 ± 0.45	576			
			6.87 ± 0.91	647			
			10.20 ± 0.91	757			
			12.8 ± 1.1	886			
	65 ± 11	1420 ± 120 (508–970 K)	15.7 ± 1.3	970			
			12.6 ± 0.6	c	RR [relative to $k(n\text{-hexane}) = 5.61 \times 10^{-12}$][a]	Ohta and Ohyama[12]	
			13.6	300	RR [relative to $k(1,3,5\text{-trimethylbenzene}) = 5.75 \times 10^{-11}$][a]	Klöpffer et al.[22]	
			12.6	297	RR [relative to $k(\text{cyclohexane}) = 7.47 \times 10^{-12}$][a]	Edney et al.[15]	
			12.2 ± 0.6	296 ± 2	RR [relative to $k(\text{propene}) = 2.66 \times 10^{-11}$][a]	Atkinson and Aschmann[18]	
m-Xylene			19.6 ± 1.4	304 ± 1	RR [relative to $k(n\text{-butane}) = 2.61 \times 10^{-12}$][a]	Doyle et al.[2]	
			23.6 ± 2.4	297.3	FP-RF	Hansen et al.[3]	
			18.8 ± 3.8	305 ± 2	RR [relative to $k(n\text{-butane}) = 2.62 \times 10^{-12}$][a]	Lloyd et al.[19]	
			24.0 ± 2.5	298.3	FP-RF	Perry et al.[4]	298–427
			24.4 ± 3.6	314.5			
			20.5[b]	320.0			
			13.1[b]	327.8			
			2.81[b]	354.9			
			1.68[b]	365.2			
			2.19[b]	373.8			
			2.21 ± 0.33	379.1			
			2.23 ± 0.33	390.9			
			2.49 ± 0.36	403.5			
			2.86 ± 0.38	414.0			
			3.02 ± 0.30	427.0			
			20.6 ± 1.3[e]	298	FP-RF	Ravishankara et al.[20]	
			19.6	300	RR [relative to $k(\text{ethene}) = 8.44 \times 10^{-12}$][a]	Cox et al.[5]	
			26.5 ± 2.5	250	FP-RF	Nicovich et al.[21]	250–960
			25.6 ± 4.3	269			
			25.4 ± 3.5	298			
			5.2[b]	330			
			2.47 ± 0.41	400			

TABLE 19. Rate constants k and temperature-dependent parameters for the gas-phase reactions of the OH radical with aromatic compounds at, or close to, the high pressure limit — Continued

Aromatic	$10^{12} \times A$ (cm^3 molecule^{-1} s^{-1})	B (K)	$10^{12} \times k$ (cm^3 molecule^{-1} s^{-1})	at T (K)	Technique	Reference	Temperature range covered (K)
			3.44 ± 0.34	508			
			4.60 ± 0.54	576			
			6.2 ± 1.1	684			
			9.3 ± 1.1	757			
			10.1 ± 1.5	875			
	68 ± 23	1540 ± 240 (508–960 K)	14.6 ± 3.1	960			
			21.4 ± 1.4	299 ± 2	RR [relative to k(cyclohexane) $= 7.51 \times 10^{-12}$][a]	Atkinson et al.[23]	
			22.3 ± 0.7	c	RR [relative to k(n-hexane) $= 5.61 \times 10^{-12}$][a]	Ohta and Ohyama[12]	
			23.1	297	RR [relative to k(cyclohexane) $= 7.47 \times 10^{-12}$][a]	Edney et al.[15]	
			23.0 ± 0.6	296 ± 2	RR [relative to k(propene) $= 2.66 \times 10^{-11}$][a]	Atkinson and Aschmann[18]	
p-Xylene			10.7 ± 2.4	304 ± 1	RR [relative to k(n-butane) $= 2.61 \times 10^{-12}$][a]	Doyle et al.[2]	
			12.2 ± 1.2	297.3	FP-RF	Hansen et al.[3]	
			15.3 ± 1.7	298.0	FP-RF	Perry et al.[4]	298–428
			18.2 ± 2.2	306.3			
			$18.2 \pm 2,2$	310.7			
			17.3 ± 2.2	313.2			
			16.7^b	315.0			
			14.9 ± 2.0	324.2			
			15.3^b	330.2			
			11.7^b	352.7			
			5.49^b	358.7			
			3.39^b	369.6			
			2.50^b	372.1			
			2.66 ± 0.40	383.8			
			2.43 ± 0.32	385.3			
			2.67 ± 0.36	387.1			
			2.96 ± 0.40	392.8			
			3.17 ± 0.43	400.0			
			3.29 ± 0.40	404.3			
			3.68 ± 0.45	412.6			
			3.56 ± 0.55	422.4			
			3.29 ± 0.33	428.4			
			10.5 ± 1.0^e	298	FP-RF	Ravishankara et al.[20]	
			13.5 ± 1.4	298	FP-RF	Nicovich et al.[21]	298–960
			13.8 ± 1.1	320			
			12.5 ± 1.3	335			
			4.3^b	357			
			1.71 ± 0.28	400			
			3.70 ± 0.64	484			
			3.40 ± 0.48	526			
			5.03 ± 0.88	576			
			6.01 ± 0.59	647			

TABLE 19. Rate constants k and temperature-dependent parameters for the gas-phase reactions of the OH radical with aromatic compounds at, or close to, the high pressure limit — Continued

Aromatic	$10^{12} \times A$ (cm^3 molecule^{-1} s^{-1})	B (K)	$10^{12} \times k$ (cm^3 molecule^{-1} s^{-1})	at T (K)	Technique	Reference	Temperature range covered (K)
			9.66 ± 0.85	757			
			11.6 ± 1.6	886			
	64 ± 24	1440 ± 250 (526–960 K)	14.6 ± 1.9	960			
			13.0 ± 0.6	c	RR [relative to $k(n\text{-hexane}) = 5.61 \times 10^{-12}$]^a	Ohta and Ohyama[12]	
			13.6	300	RR [relative to $k(1,3,5\text{-trimethylbenzene}) = 5.75 \times 10^{-11}$]^a	Klöpffer et al.[22]	
			13.9	296	RR [relative to $k(\text{cyclohexane}) = 7.45 \times 10^{-12}$]^a	Edney et al.[15]	
			14.2 ± 4.1	298 ± 2	RR [relative to $k(\text{cyclohexane}) = 7.49 \times 10^{-12}$]^a	Becker and Klein[24]	
			13.0 ± 0.5	296 ± 2	RR [relative to $k(\text{propene}) = 2.66 \times 10^{-11}$]^a	Atkinson and Aschmann[18]	
Xylenes (mixture of isomers)			18.7	298	DF-MS	Morris and Niki[25]	
n-Propyl-benzene			5.42 ± 1.09	305 ± 2	RR [relative to $k(n\text{-butane}) = 2.62 \times 10^{-12}$]^a	Lloyd et al.[19]	
			5.86 ± 0.50^e	298	FP-RF	Ravishankara et al.[20]	
			6.62 ± 0.23	c	RR [relative to $k(n\text{-hexane}) = 5.61 \times 10^{-12}$]^a	Ohta and Ohyama[12]	
Isopropyl-benzene			5.32 ± 1.07	305 ± 2	RR [relative to $k(n\text{-butane}) = 2.62 \times 10^{-12}$]^a	Lloyd et al.[19]	
			7.79 ± 0.50	298	FP-RF	Ravishankara et al.[20]	
			6.28 ± 0.34	c	RR [relative to $k(n\text{-hexane}) = 5.61 \times 10^{-12}$]^a	Ohta and Ohyama[12]	
o-Ethyl-toluene			12.0 ± 2.4	305 ± 2	RR [relative to $k(n\text{-butane}) = 2.62 \times 10^{-12}$]^a	Lloyd et al.[19]	
			12.5 ± 1.3	c	RR [relative to $k(n\text{-hexane}) = 5.61 \times 10^{-12}$]^a	Ohta and Ohyama[12]	
m-Ethyl-toluene			17.0 ± 3.4	305 ± 2	RR [relative to $k(n\text{-butane}) = 2.62 \times 10^{-12}$]^a	Lloyd et al.[19]	

TABLE 19. Rate constants k and temperature-dependent parameters for the gas-phase reactions of the OH radical with aromatic compounds at, or close to, the high pressure limit — Continued

Aromatic	$10^{12} \times A$ (cm³ molecule⁻¹ s⁻¹)	B (K)	$10^{12} \times k$ (cm³ molecule⁻¹ s⁻¹)	at T (K)	Technique	Reference	Temperature range covered (K)
			21.3 ± 1.1	c	RR [relative to $k(n\text{-hexane}) = 5.61 \times 10^{-12}$]ᵃ	Ohta and Ohyama[12]	
p-Ethyl-toluene			11.3 ± 2.3	305 ± 2	RR [relative to $k(n\text{-butane}) = 2.62 \times 10^{-12}$]ᵃ	Lloyd et al.[19]	
			12.9 ± 1.3	c	RR [relative to $k(n\text{-hexane}) = 5.61 \times 10^{-12}$]ᵃ	Ohta and Ohyama[12]	
t-Butyl-benzene			4.60 ± 0.45	c	RR [relative to $k(n\text{-hexane}) = 5.61 \times 10^{-12}$]ᵃ	Ohta and Ohyama[12]	
1,2,3-Tri-methylbenzene			19.8 ± 4.2	304 ± 1	RR [relative to $k(n\text{-butane}) = 2.61 \times 10^{-12}$]ᵃ	Doyle et al.[2]	
			26.4 ± 2.6	297.1	FP-RF	Hansen et al.[3]	
			33.3 ± 4.5	296.9	FP-RF	Perry et al.[4]	297–421
			27.6^b	317.5			
			24.5 ± 3.7	325.1			
			19.4^b	338.6			
			4.32^b	374.4			
			3.54^b	377.4			
			4.28 ± 0.64	388.6			
			4.89 ± 0.74	396.8			
			6.00 ± 0.80	420.7			
			29.7 ± 4.1	c	RR [relative to $k(n\text{-hexane}) = 5.61 \times 10^{-12}$]ᵃ	Ohta and Ohyama[12]	
			32.7 ± 1.9	296 ± 2	RR [relative to $k(\text{propene}) = 2.66 \times 10^{-11}$]ᵃ	Atkinson and Aschmann[18]	
1,2,4-Tri-methylbenzene			28.7 ± 5.3	304 ± 1	RR [relative to $k(n\text{-butane}) = 2.61 \times 10^{-12}$]ᵃ	Doyle et al.[2]	
			33.5 ± 3.4	296.9	FP-RF	Hansen et al.[3]	
			40.0 ± 4.5	298.2	FP-RF	Perry et al.[4]	298–430
			37.3 ± 4.8	314.3			
			32.4 ± 4.8	323.4			
			15.2^b	340.2			
			4.56^b	370.9			
			2.47^b	374.1			
			3.34 ± 0.44	383.7			
			4.82 ± 0.63	397.7			
			4.75 ± 0.62	400.3			
			5.31 ± 0.69	423.5			
			5.24 ± 0.60	429.5			
			31.7 ± 1.3	c	RR [relative to $k(n\text{-hexane}) = 5.61 \times 10^{-12}$]ᵃ	Ohta and Ohyama[12]	

TABLE 19. Rate constants k and temperature-dependent parameters for the gas-phase reactions of the OH radical with aromatic compounds at, or close to, the high pressure limit — Continued

Aromatic	$10^{12} \times A$ (cm^3 molecule^{-1} s^{-1})	B (K)	$10^{12} \times k$ (cm^3 molecule^{-1} s^{-1})	at T (K)	Technique	Reference	Temperature range covered (K)
			32.5 ± 1.1	296 ± 2	RR [relative to k(propene) = 2.66×10^{-11}][a]	Atkinson and Aschmann[18]	
1,3,5-Tri-methylbenzene			44.4 ± 5.3	304 ± 1	RR [relative to k(n-butane) = 2.61×10^{-12}][a]	Doyle et al.[2]	
			47.2 ± 4.8	297.1	FP-RF	Hansen et al.[3]	
			62.4 ± 7.5	298.3	FP-RF	Perry et al.[4]	298–420
			51.9 ± 6.3	318.4			
			52.1^b	322.5			
			3.16^b	368.0			
			3.38 ± 0.45	372.4			
			3.45 ± 0.45	381.0			
			3.82 ± 0.50	390.1			
			5.03 ± 0.60	420.1			
			38.9 ± 5.3	c	RR [relative to k(n-hexane) = 5.61×10^{-12}][a]	Ohta and Ohyama[12]	
			57.5 ± 3.0	296 ± 2	RR [relative to k(propene) = 2.66×10^{-11}][a]	Atkinson and Aschmann[18]	
Styrene [C$_6$H$_5$CH=CH$_2$]			52 ± 5	298 ± 2	RR [relative to k(2,2,4-trimethyl-pentane) = 3.68×10^{-12}][a]	Bignozzi et al.[26]	
			58.1 ± 1.5	296 ± 2	RR [relative to k(2-methyl-1,3-buta-diene) = 1.01×10^{-10}][a]	Atkinson and Aschmann[27]	
α-Methylstyrene [C$_6$H$_5$C(CH$_3$)=CH$_2$]			52 ± 6	298 ± 2	RR [relative to k(2,2,4-trimethylpentane) = 3.68×10^{-12}][a]	Bignozzi et al.[26]	
β-Methylstyrene [C$_6$H$_5$CH=CHCH$_3$]			59 ± 6	298 ± 2	RR [relative to k(2,2,4-trimethylpentane) = 3.68×10^{-12}][a]	Bignozzi et al.[26]	
β-Dimethylstyrene [C$_6$H$_5$CH=C(CH$_3$)$_2$]			33 ± 5	298	RR [relative to k(2,2,4-trimethylpentane) = 3.68×10^{-12}][a]	Chiorboli et al.[28]	
Phenol			28.3 ± 5.7	296	FP-RF	Rinke and Zetzsch[11]	
			35.2 ± 1.7	245	FP-RF	Witte and Zetzsch[29]	245–470
			33.2 ± 1.0	257			
			29.4 ± 2.9	272			
			27.7 ± 0.9	286			
			26.0 ± 1.8	296			
	5.3 ± 0.9	-470 ± 50 (245–296 K)	25.7 ± 0.9	296			
			22.0 ± 0.8	310			
			20.5 ± 0.6	319			
			19.8 ± 0.5	320			
			17.6 ± 0.9	324			
			17.0 ± 0.3	330			
			14.4 ± 0.5	335			

TABLE 19. Rate constants k and temperature-dependent parameters for the gas-phase reactions of the OH radical with aromatic compounds at, or close to, the high pressure limit — Continued

Aromatic	$10^{12} \times A$ (cm³ mole-cule⁻¹ s⁻¹)	B (K)	$10^{12} \times k$ (cm³ molecule⁻¹ s⁻¹)	at T (K)	Technique	Reference	Temperature range covered (K)
			14.8 ± 0.7	339			
			13.7 ± 0.5	343			
			10.6 ± 0.6	346			
			10.6 ± 1.0	349			
			10.3 ± 0.6	351			
			7.85 ± 1.00	356			
			6.98 ± 0.74	359			
			3.23 ± 0.19	393			
			3.56 ± 0.39	394			
			3.08 ± 0.16	394			
			3.46 ± 0.11	423			
			3.23 ± 0.07	445			
			2.90 ± 0.23	466			
			3.10 ± 0.17	470			
			9.5	1000–1150	RR [relative to $k(\text{CO}) = 1.12 \times 10^{-13}$ $e^{0.0009077 T}]^d$	He et al.[30]	1000–1150
Methoxybenzene [$C_6H_5OCH_3$]			19.6 ± 2.4	299.9	FP-RF	Perry et al.[31]	300–435
			17.3 ± 2.6	309.0			
			17.5 ± 2.6	309.7			
			17.5 ± 2.6	318.5			
	3.7	−403 (300–322 K)	17.8 ± 2.7	321.7			
			12.7^b	329.5			
			6.7^b	357.4			
			3.0^b	370.4			
			3.33 ± 0.50	385.5			
			3.25 ± 0.50	392.3			
			3.31 ± 0.50	404.1			
			3.90 ± 0.56	413.2			
			2.85 ± 0.43	417.9			
			3.30 ± 0.40	422.0			
			2.76 ± 0.41	428.7			
	1.7	−252 (386–435 K)	2.72 ± 0.41	435.3			
			14.1 ± 0.6	c	RR [relative to $k(n\text{-hexane}) = 5.61 \times 10^{-12}]^a$	Ohta and Ohyama[12]	
o-Cresol			34.1 ± 6.8	299.4	FP-RF	Perry et al.[31]	299–423
			29.3 ± 5.9	310.6			
			29.8 ± 6.0	322.0			
			26.8 ± 5.4	330.7			
	1.6	−906 (299–335 K)	25.5 ± 5.1	335.4			
			18.5^b	344.4			
			15.0^b	356.5			
			6.4^b	385.8			
			6.0^b	392.8			
			5.6 ± 1.1	400.6			
			5.4 ± 1.1	407.8			
	50	906 (401–423 K)	6.2 ± 1.2	423.1			
			42.5 ± 3.7	300 ± 1	RR [relative to $k(n\text{-butane}) - k(\text{neopentane}) = 1.70 \times 10^{-12}]^a$	Atkinson et al.[32]	

TABLE 19. Rate constants k and temperature-dependent parameters for the gas-phase reactions of the OH radical with aromatic compounds at, or close to, the high pressure limit — Continued

Aromatic	$10^{12} \times A$ (cm³ molecule⁻¹ s⁻¹)	B (K)	$10^{12} \times k$ (cm³ molecule⁻¹ s⁻¹)	at T (K)	Technique	Reference	Temperature range covered (K)
			42.6 ± 2.2	296 ± 2	RR [relative to k(propene) = 2.66×10^{-11}][a]	Atkinson and Aschmann[33]	
m-Cresol			59.6 ± 3.4	300 ± 1	RR [relative to k(o-cresol) = 4.2×10^{-11}][a]	Atkinson et al.[32]	
			67.8 ± 4.0	296 ± 2	RR [relative to k(propene) = 2.66×10^{-11}][a]	Atkinson and Aschmann[33]	
p-Cresol			46.2 ± 2.1	300 ± 1	RR [relative to k(o-cresol) = 4.2×10^{-11}][a]	Atkinson et al.[32]	
			48.4 ± 5.1	296 ± 2	RR [relative to k(propene) = 2.66×10^{-11}][a]	Atkinson and Aschmann[33]	
Thiophenol [C_6H_5SH]			11.2 ± 1.4	300	RR [relative to k(n-hexane) = 5.64×10^{-12}][a]	Barnes et al.[34]	
Acetophenone [$C_6H_5COCH_3$]			2.74 ± 0.15	298	FP-RF	Nolting et al.[35]	
Fluoro-benzene			0.54 ± 0.05	296	FP-RF	Zetzsch[36]	
			0.90 ± 0.12	c	RR [relative to k(n-hexane) = 5.61×10^{-12}][a]	Ohta and Ohyama[12]	
			0.524 ± 0.088	234	FP-RF	Wallington et al.[16]	234–438
			0.632 ± 0.103	253			
			0.649 ± 0.099	263			
			0.610 ± 0.080	277			
			0.631 ± 0.081	296			
			0.656 ± 0.074	303			
			0.196 ± 0.047	393			
			0.209 ± 0.038	438			
Chloro-benzene			0.67 ± 0.05	296	FP-RF	Zetzsch[36,37]	
			0.91 ± 0.12	299 ± 2	RR [relative to k(benzene) = 1.24×10^{-12}][a]	Atkinson et al.[38]	
			0.55 ± 0.44	297	RR [relative to k(n-butane) = 2.53×10^{-12}][a]	Edney et al.[15]	
			0.707 ± 0.084	234	FP-RF	Wallington et al.[16]	234–438
			0.624 ± 0.062	263			
			0.741 ± 0.094	296			
			0.214 ± 0.046	393			
			0.191 ± 0.033	438			
Bromobenzene			0.70 ± 0.07	296	FP-RF	Zetzsch[36]	
			0.57 ± 0.07	245	FP-RF	Witte et al.[14]	245–362
			0.64 ± 0.04	253			

TABLE 19. Rate constants k and temperature-dependent parameters for the gas-phase reactions of the OH radical with aromatic compounds at, or close to, the high pressure limit — Continued

Aromatic	$10^{12} \times A$ (cm³ molecule⁻¹ s⁻¹)	B (K)	$10^{12} \times k$ (cm³ molecule⁻¹ s⁻¹)	at T (K)	Technique	Reference	Temperature range covered (K)
			0.68 ± 0.06	259			
			0.72 ± 0.04	265			
			0.71 ± 0.04	274			
			0.66 ± 0.05	283			
			0.71 ± 0.04	299			
			0.72 ± 0.03	312			
			0.69 ± 0.04	316			
			0.76 ± 0.02	316			
			0.75 ± 0.04	325			
			0.77 ± 0.03	334			
			0.92 ± 0.12	339			
			0.79 ± 0.06	343			
			0.76 ± 0.07	343			
			0.78 ± 0.10	346			
			0.92 ± 0.15	354			
	1.3	180 ± 60	0.91 ± 0.19	362			
			0.915 ± 0.187	234	FP-RF	Wallington et al.[16]	234–438
			1.02 ± 0.16	263			
			0.915 ± 0.097	296			
			0.763 ± 0.053	353			
			0.219 ± 0.039	438			
Iodobenzene			0.93 ± 0.05	296	FP-RF	Zetzsch[36]	
			1.25 ± 0.15	263	FP-RF	Wallington et al.[16]	263–438
			1.32 ± 0.16	296			
			1.32 ± 0.19	353			
			1.03 ± 0.12	393			
Benzyl chloride			2.97 ± 0.16	298	RR [relative to k(n-butane) = 2.54×10^{-12}][a]	Edney et al.[15]	
			2.80 ± 0.19	298 ± 2	RR [relative to k(dimethyl ether) = 2.98×10^{-12}][a]	Tuazon et al.[39]	
Benzyl alcohol			22.9 ± 2.5	298	FP-RF	Nolting et al.[35]	
Benzotrifluoride [$C_6H_5CF_3$]			0.46 ± 0.12	299 ± 2	RR [relative to k(benzene) = $= 1.24 \times 10^{-12}$][a]	Atkinson et al.[38]	
Aniline			119 ± 24	296	FP-RF	Rinke and Zetzsch[11]	
			173 ± 8	239	FP-RF	Witte et al.[14]	239–359
			164 ± 11	245			
			158 ± 4	253			
			133 ± 9	265			
			136 ± 7	274			
			128 ± 10	283			
			114 ± 3	299			
			106 ± 3	312			
			98 ± 6	316			
			93 ± 3	325			
			92 ± 7	336			
			86 ± 6	342			
			86 ± 3	346			
			83 ± 28	352			
	17	-560 ± 30	83 ± 15	359			

TABLE 19. Rate constants k and temperature-dependent parameters for the gas-phase reactions of the OH radical with aromatic compounds at, or close to, the high pressure limit — Continued

Aromatic	$10^{12} \times A$ (cm³ molecule⁻¹ s⁻¹)	B (K)	$10^{12} \times k$ (cm³ molecule⁻¹ s⁻¹)	at T (K)	Technique	Reference	Temperature range covered (K)
			112 ± 12	265	FP-RF	Atkinson et al.[40]	265–455
			118 ± 10	283			
			118 ± 11	298			
			92^b	310			
			67^b	325			
			47.8 ± 5.3	342			
			36.3 ± 6.1	382			
			39.6 ± 7.5	391			
			29.7 ± 1.2	426			
			29.7 ± 5.4	455			
N,N-Dimethyl-aniline			151 ± 31	278	FP-RF	Atkinson et al.[40]	278–464
			148 ± 11	298			
			119 ± 6	303			
			57^b	318			
			29^b	329			
			5.8^b	361			
			1.71 ± 0.25	421			
			2.20 ± 0.51	425			
			2.09 ± 0.30	437			
			2.85 ± 0.24	460			
			3.12 ± 0.34	464			
Benzonitrile			0.33 ± 0.03	296	FP-RF	Zetzsch[36,37]	
Nitrobenzene			0.16 ± 0.05	296	FP-RF	Zetzsch[14,36,37]	
			<0.9	296 ± 2	RR [relative to k(dimethyl ether) $= 2.96 \times 10^{-12}$]^a	Atkinson et al.[40]	
			<0.7	296 ± 2	RR [relative to k(benzene) $= 1.23 \times 10^{-12}$]^a	Atkinson et al.[40]	
			0.120 ± 0.007	259	FP-RF	Witte et al.[14]	259–362
			0.110 ± 0.012	265			
			0.119 ± 0.007	274			
			0.126 ± 0.009	283			
			0.132 ± 0.007	288			
			0.137 ± 0.014	299			
			0.146 ± 0.015	312			
			0.169 ± 0.022	316			
			0.137 ± 0.025	316			
			0.181 ± 0.028	331			
			0.154 ± 0.021	339			
			0.169 ± 0.011	342			
			0.163 ± 0.065	352			
			0.136 ± 0.013	358			
			0.133 ± 0.021	358			
	0.6	440 ± 80 (259–342 K)	0.158 ± 0.040	362			
4-Chloro-benzotri-fluoride			0.24 ± 0.08	299 ± 2	RR [relative to k(benzene) $= 1.24 \times 10^{-12}$]^a	Atkinson et al.[38]	
o-Dichloro-benzene			0.42 ± 0.02	295	FP-RF	Wahner and Zetzsch[10]	
m-Dichloro-benzene			0.72 ± 0.02	295	FP-RF	Wahner and Zetzsch[10]	

TABLE 19. Rate constants k and temperature-dependent parameters for the gas-phase reactions of the OH radical with aromatic compounds at, or close to, the high pressure limit — Continued

Aromatic	$10^{12} \times A$ (cm³ molecule⁻¹ s⁻¹)	B (K)	$10^{12} \times k$ (cm³ molecule⁻¹ s⁻¹)	at T (K)	Technique	Reference	Temperature range covered (K)
p-Dichloro-benzene			0.32 ± 0.02	295	FP-RF	Wahner and Zetzsch[10]	
			0.52	300	RR [relative to k(ethene) = 8.44 × 10⁻¹²]ᵃ	Klöpffer et al.[22]	
p-Chloro-aniline			83.0 ± 4.2	295	FP-RF	Wahner and Zetzsch[10]	
			~43	296 ± 5	RR [relative to k(ethene) = 8.61 × 10⁻¹²]ᵃ	Klöpffer et al.[22]	
o-Nitro-phenol	0.42 ± 0.05	−217 ± 60	0.90 ± 0.02	294	FP-RF	Zetzsch[36]	273–353
2,3-Dimethyl-phenol			80.2 ± 3.0	296 ± 2	RR [relative to k(2-methyl-1,3-butadiene) = 1.01 × 10⁻¹⁰]ᵃ	Atkinson and Aschmann[33]	
2,4-Dimethyl-phenol			71.5 ± 4.1	296 ± 2	RR [relative to k(2-methyl-1,3-butadiene) = 1.01 × 10⁻¹⁰]ᵃ	Atkinson and Aschmann[33]	
2,5-Dimethyl-phenol			80.0 ± 11.0	296 ± 2	RR [relative to k(2-methyl-1,3-butadiene) = 1.01 × 10⁻¹⁰]ᵃ	Atkinson and Aschmann[33]	
2,6-Dimethyl-phenol			65.9 ± 5.0	296 ± 2	RR [relative to k(2-methyl-1,3-butadiene) = 1.01 × 10⁻¹⁰]ᵃ	Atkinson and Aschmann[33]	
3,4-Dimethyl-phenol			81.4 ± 5.8	296 ± 2	RR [relative to k(2-methyl-1,3-butadiene) = 1.01 × 10⁻¹⁰]ᵃ	Atkinson and Aschmann[33]	
3,5-Dimethyl-phenol			113 ± 8	296 ± 2	RR [relative to k(2-methyl-1,3-butadiene) = 1.01 × 10⁻¹⁰]ᵃ	Atkinson and Aschmann[33]	
1,2,4-Tri-chloro-benzene	2.3 ± 1.0	429 ± 125	0.497 ± 0.036 0.532 ± 0.050 0.631 ± 0.082 0.706 ± 0.054 0.712 ± 0.083	273 296 323 348 368	FP-RF	Rinke and Zetzsch[11]	273–368
			0.58	300	RR [relative to k(toluene) = 5.91 × 10⁻¹²]ᵃ	Klöpffer et al.[22]	
2,3-Dichloro-phenol			1.66 ± 0.15	298	FP-RF	Nolting et al.[35]	
2,4-Dichloro-phenol			1.06 ± 0.06	298	FP-RF	Nolting et al.[35]	
2,4- + 2,6-Toluene diisocyanate			7.09 ± 0.24	298 ± 2	RR [relative to k(toluene) = 5.96 × 10⁻¹²]ᵃ	Becker et al.[41]	
2,4-Toluene-diamine			192 ± 71	298 ± 2	RR [relative to k(cyclohexene) = 6.77 × 10⁻¹¹]ᵃ	Becker et al.[41]	

TABLE 19. Rate constants k and temperature-dependent parameters for the gas-phase reactions of the OH radical with aromatic compounds at, or close to, the high pressure limit — Continued

Aromatic	$10^{12} \times A$ (cm^3 molecule^{-1} s^{-1})	B (K)	$10^{12} \times k$ (cm^3 molecule^{-1} s^{-1})	at T (K)	Technique	Reference	Temperature range covered (K)
2,6-Toluene-diamine			$\geqslant 101$	298 ± 2	RR [relative to k(cyclohexene) = 6.77×10^{-11}]^a	Becker et al.[41]	
Hexafluoro-benzene			0.219 ± 0.016^e	298	FP-RF	Ravishankara et al.[20]	
			0.093 ± 0.013	234	FP-RF	Wallington et al.[16]	234–438
			0.122 ± 0.013	263			
			0.161 ± 0.024	296			
			0.222 ± 0.029	353			
			0.266 ± 0.030	393			
	1.3 ± 0.3	610 ± 80	0.358 ± 0.059	438			
o-Nitro-toluene			0.70 ± 0.05	298	FP-RF	Nolting et al.[35]	
m-Nitro-toluene			0.95 ± 0.05	298	FP-RF	Nolting et al.[35]	
			1.3 ± 0.9	298 ± 2	RR [relative to k(benzene) = 1.23×10^{-12}]^a	Atkinson et al.[42]	
n-Propyl-pentafluoro-benzene			3.06 ± 0.24^e	298	FP-RF	Ravishankara et al.[20]	
Biphenyl			5.8 ± 0.8	296	FP-RF	Zetzsch[36,37]	
			7.61 ± 0.67	294 ± 1	RR [relative to k(n-nonane) = 1.01×10^{-11}]^a	Atkinson et al.[43]	
			8.32 ± 0.75	295 ± 1	RR [relative to k(cyclohexane) = 7.43×10^{-12}]^a	Atkinson and Aschmann[44]	
			8.0	300	RR [relative to k(ethene) = 8.44×10^{-12}]^a	Klöpffer et al.[22]	
2-Chloro-biphenyl			2.82 ± 0.38	295 ± 1	RR [relative to k(cyclohexane) = 7.43×10^{-12}]^a	Atkinson and Aschmann[44]	
3-Chloro-biphenyl			5.28 ± 0.82	295 ± 1	RR [relative to k(cyclohexane) = 7.43×10^{-12}]^a	Atkinson and Aschmann[44]	
4-Chloro-biphenyl			3.86 ± 0.67	295 ± 1	RR [relative to k(cyclohexane) = 7.43×10^{-12}]^a	Atkinson and Aschmann[44]	
Methylene-dianiline			30 ± 10	298 ± 2	RR [relative to k(cyclohexene) = 6.77×10^{-11}]^a	Becker et al.[41]	
1,4-Naphtho-quinone			3.1 ± 1.2	298 ± 2	RR [relative to k(cyclohexane) = 7.49×10^{-12}]^a	Atkinson et al.[45]	
Tetralin^f			34.3 ± 0.6	296 ± 2	RR [relative to k(propene) = 2.66×10^{-11}]^a	Atkinson and Aschmann[27]	

TABLE 19. Rate constants k and temperature-dependent parameters for the gas-phase reactions of the OH radical with aromatic compounds at, or close to, the high pressure limit — Continued

Aromatic	$10^{12} \times A$ (cm^3 molecule^{-1} s^{-1})	B (K)	$10^{12} \times k$ (cm^3 molecule^{-1} s^{-1})	at T (K)	Technique	Reference	Temperature range covered (K)
Indane[f]			9.2	295	DF-RF	Baulch et al.[46]	
Indene[f]			>51	295	DF-RF	Baulch et al.[46]	
2,3-Dihydro-benzofuran[f]			36.6 ± 1.1	298 ± 2	RR [relative to k(propene) = 2.63×10^{-11}][a]	Atkinson et al.[47]	
1,4-Benzo-dioxan[f]			25.2 ± 0.4	298 ± 2	RR [relative to k(propene) = 2.63×10^{-11}][a]	Atkinson et al.[47]	
2,3-Benzo-furan[f]			37.3 ± 4.8	298 ± 2	RR [relative to k(propene) = 2.63×10^{-11}][a]	Atkinson et al.[47]	
Fluorene[f]			13.0	300	RR [relative to k(ethene) = 8.44×10^{-12}][a]	Klöpffer et al.[22]	
Naphthalene			18.6 ± 1.0	300	LP-RF	Lorenz and Zellner[8,9]	300–873
			14.6 ± 5.0	337			
			11.0 ± 4.4	358			
			10.1 ± 4.0	378 ± 2			
			11.6 ± 3.0	404			
	2.3 ± 1.5	−640 ± 300 (300–407 K)	10.5 ± 4.0	407			
			6.3 ± 2.0	452			
			4.3 ± 1.5	476			
			1.3 ± 0.5	502			
			1.2 ± 0.4	525 ± 1			
			0.7 ± 0.2	528			
			0.6 ± 0.1	531			
			1.1 ± 0.1	636			
			1.1 ± 0.2	665			
			1.4 ± 0.2	727			
	50	2500 (636–873 K)	3.0 ± 0.5	873			
			22.8 ± 1.6	294 ± 1	RR [relative to k(n-nonane) = 1.01×10^{-11}][a]	Atkinson et al.[43]	
			23.5 ± 0.6	298 ± 1	RR [relative to k(propene) = 2.63×10^{-11}][a]	Biermann et al.[48]	
			25.9 ± 2.4	295 ± 1	RR [relative to k(2-methyl-1,3-butadiene) = 1.02×10^{-10}][a]	Atkinson and Aschmann[49]	
			21.6	300	RR [relative to k(ethene) = 8.44×10^{-12}][a]	Klöpffer et al.[22]	
1-Methyl-naphthalene			53.0 ± 4.8	298 ± 2	RR [relative to k(2-methyl-1,3-butadiene) = 1.01×10^{-10}][a]	Atkinson and Aschmann[50]	
2-Methyl-naphthalene			52.3 ± 4.2	295 ± 1	RR [relative to k(2-methyl-1,3-butadiene) = 1.02×10^{-10}][a]	Atkinson and Aschmann[49]	

TABLE 19. Rate constants k and temperature-dependent parameters for the gas-phase reactions of the OH radical with aromatic compounds at, or close to, the high pressure limit — Continued

Aromatic	$10^{12} \times A$ (cm^3 molecule^{-1} s^{-1})	B (K)	$10^{12} \times k$ (cm^3 molecule^{-1} s^{-1})	at T (K)	Technique	Reference	Temperature range covered (K)
1-Nitro-naphthalene			5.4 ± 1.8	298 ± 2	RR [relative to k(cyclohexane) = 7.49 × 10^{-12}][a]	Atkinson et al.[45]	
2-Nitro-naphthalene			5.6 ± 0.9	298 ± 2	RR [relative to k(cyclohexane) = 7.49 × 10^{-12}][a]	Atkinson et al.[45]	
2,3-Dimethyl-naphthalene			76.8 ± 4.8	295 ± 1	RR [relative to k(2-methyl-1,3-butadiene) = 1.02 × 10^{-10}][a]	Atkinson and Aschmann[49]	
2-Methyl-1-nitronaphthalene			<8.3	298 ± 2	RR [relative to k(cyclohexane) = 7.49 × 10^{-12}][a]	Arey et al.[51]	
1,4-Dichloro-naphthalene			5.8	300	RR [relative to k(toluene) = 5.91 × 10^{-12}][a]	Klöpffer et al.[22]	
Acenaphthene[f]			58.4	300	RR [relative to k(ethene) = 8.44 × 10^{-12}][a]	Klöpffer et al.[22]	
			103 ± 13	296 ± 2	RR [relative to k(2,3-dimethyl-2-butene) = 1.11 × 10^{-10}][a]	Atkinson and Aschmann[27]	
Acenaph-thylene[f]			110 ± 11	296 ± 2	RR [relative to k(2,3-dimethyl-2-butene) = 1.11 × 10^{-10}][a]	Atkinson and Aschmann[27]	
Phenan-threne			15.6 ± 2.0	338	LP-RF	Lorenz and Zellner[9]	338–748
			16.1 ± 2.0	355			
			19.1 ± 2.5	387			
			12.0 ± 1.7	399			
			8.3 ± 0.8	431			
			4.0 ± 0.7	492			
			2.8 ± 0.7	526			
			1.2 ± 0.2	597			
			1.2 ± 0.4	648			
			2.2 ± 0.5	748			
			34 ± 12	298 ± 1	RR [relative to k(propene) = 4.85 × 10^{-12}e$^{504/T}$][a]	Biermann et al.[48]	298–319
			28 ± 6	319 ± 1			
Anthracene			112 ± 9	325 ± 1	RR [relative to k(propene) = 2.29 × 10^{-11}][a]	Biermann et al.[48]	

[a]From the present recommendations (see text).
[b]Non-exponential OH radical decays observed (see text).
[c]Room temperature, not specified.
[d]See Introduction.
[e]At 200 Torr total pressure of helium diluent, data also obtained at other total pressures.

TABLE 19. Footnotes — Continued

[f]Structures:

Tetralin, ; Indane, ; Indene, ; 2,3-Dihydrobenzofuran, ;

1,4-Benzodioxan, ; 2,3-Benzofuran, ; Fluorene, ;

Acenaphthene, ; Acenaphthylene, .

The absolute rate constants obtained at, or close to, the high pressure limit of Davis et al.,[1] Perry et al.,[4] Tully et al.,[6] Lorenz and Zellner,[8,9] Wahner and Zetzsch,[10] Rinke and Zetzsch,[11] Madronich and Felder,[13] Witte et al.[14] and Wallington et al.[16] are plotted in Arrhenius form in Fig. 95 (the 298 K rate constant of Hansen et al.[3] is identical to that of Tully et al.[6] and is hence not shown). At room temperature these absolute rate constants exhibit a significant degree of scatter, of a factor of ~1.8. The reasons for these discrepancies are not known, but may, at least in part, be due to the relatively low magnitude of this rate constant. The recommendations are based on the flash and laser photolysis-resonance fluorescence studies of Hansen et al.,[3] Perry et al.,[4] Tully et al.,[6] Lorenz and Zellner,[8,9] Madronich and Felder,[13] Witte et al.[14] (which is judged to supersede the earlier room temperature studies of Wahner and Zetzsch[10] and Rinke and Zetzsch[11]), and Wallington et al.[16] For temperatures ⩽355 K, a unit-weighted least-squares analysis of the rate constants of Hansen et al.,[3] Perry et al.,[4] Tully et al.,[6] Lorenz and Zellner,[8,9] Witte et al.[14] and Wallington et al.[16] yields the Arrhenius expression of

$$k(\text{benzene}; \ T \leqslant 355 \text{ K}) = (2.47^{+0.90}_{-0.66})$$
$$\times \ 10^{-12} \ e^{-(207 \pm 89)/T} \ \text{cm}^3 \ \text{molecule}^{-1} \ \text{s}^{-1}$$

over the temperature range 234–354 K, where the indicated error limits are two least-squares standard deviations, and

$$k(\text{benzene}) = 1.23 \times 10^{-12} \ \text{cm}^3 \ \text{molecule}^{-1} \ \text{s}^{-1} \text{ at 298 K,}$$

with an estimated overall uncertainty at 298 K of ±30%.

This rate constant and Arrhenius expression are applicable for ~100 Torr total pressure of argon diluent and are expected to be somewhat below the limiting high-pressure rate constant k_∞. Based upon the estimation discussed above, at 298 K the rate constant k_∞ is expected to be:

$$k_\infty(\text{benzene}) = 1.40 \times 10^{-12} \ \text{cm}^3 \ \text{molecule}^{-1} \ \text{s}^{-1}$$

and hence

$$k_\infty(\text{benzene}) \approx 3.58 \times 10^{-12} \ e^{-280/T} \ \text{cm}^3 \ \text{molecule}^{-1} \ \text{s}^{-1}$$

over the temperature range ~235–355 K.

The rate constants derived from the relative rate studies of Doyle et al.,[2] Cox et al.,[5] Barnes et al.,[7] Ohta and Ohyama[12] and Edney et al.[15] are in reasonably good agreement with this recommendation.

At temperatures ⩾450 K the only reported rate constants are those of Tully et al.,[6] Lorenz and Zellner,[8,9] Madronich and Felder[13] and Felder and Madronich[54] [who extended the study of Madronich and Felder[13] to obtain $k(\text{benzene}) = (2.5 \pm 0.3) \times 10^{-11} \ e^{-(2050 \pm 125)/T}$ cm^3 molecule^{-1} s^{-1} over the same temperature range of 787–1409 K, but did not tabulate the additional rate constants measured], and these data are in reasonably good agreement. Consistent with the recommendations for the alkanes, haloalkanes and alkenes, a unit-weighted least-squares analysis of the data of Tully *et al.*,[6] Lorenz and Zellner[8,9] and Madronich and Felder,[13] using the expression $k = CT^2e^{-D/T}$, yields the recommendation of

$$k(\text{benzene}, T \geqslant 450 \text{ K}) = (4.67^{+1.73}_{-1.27})$$

$$\times 10^{-18} T^2 e^{-(543 \pm 226)/T} \text{ cm}^3 \text{ molecule}^{-1} \text{ s}^{-1}$$

over the temperature range 453–1409 K, where the error limits are two least-squares standard deviations. The rate constants reported by Perry *et al.*[4] between 396 and 422 K are, as expected,[6] somewhat higher than predicted from this recommendation, though in agreement within the experimental errors.

over the temperature range 250–298 K, where the indicated error limits are two least-squares standard deviations, and

$$k(\text{benzene-}d_6) = 1.14 \times 10^{-12} \text{ cm}^3 \text{ molecule}^{-1} \text{ s}^{-1}$$

at 298 K, with an estimated overall uncertainty at 298 K of ±30%. The significant uncertainties in the above Arrhenius parameters are largely due to the small temperature range (250–298 K) covered. Again, this rate expression is applicable for a total pressure of ~100 Torr of argon diluent, and is expected to be slightly ($\lesssim$10% at 298 K) below the limiting high-pressure rate constant k_∞.

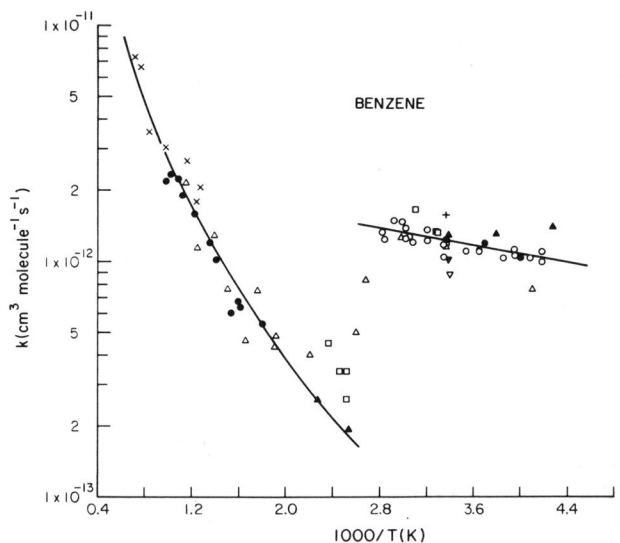

FIG. 95. Arrhenius plot of rate constants obtained at, or close to, the high-pressure limit for the reaction of the OH radical with benzene. (+) Davis *et al.*;[1] (□) Perry *et al.*;[4] (●) Tully *et al.*;[6] (△) Lorenz and Zellner;[8,9] (▽) Wahner and Zetzsch;[10] (▼) Rinke and Zetzsch;[11] (x) Madronich and Felder;[13] (○) Witte *et al.*;[14] (▲) Wallington *et al.*;[16] (———) recommendations (see text).

(2) Benzene-d_6

The rate constants reported by Tully *et al.*[6] and Lorenz and Zellner[8] are listed in Table 19 and are plotted in Arrhenius form in Fig. 96. It can be seen that the rate constants obtained by Lorenz and Zellner[8] at 298 and 524 K are in excellent agreement with those of Tully *et al.*[6]

For temperatures $\lesssim$325 K, a unit-weighted least-squares analysis of the rate constants of Tully *et al.*[6] and Lorenz and Zellner[8] yields the recommended Arrhenius expression of

$$k(\text{benzene-}d_6; T \leqslant 325 \text{ K}) = (1.54^{+1.81}_{-0.83})$$

$$\times 10^{-12} e^{-(90 \pm 216)/T} \text{ cm}^3 \text{ molecule}^{-1} \text{ s}^{-1}$$

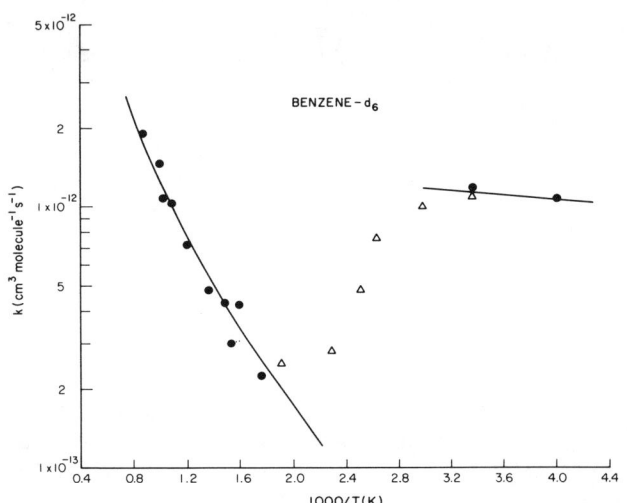

FIG. 96. Arrhenius plot of rate constants for the reaction of the OH radical with benzene-d_6. (●) Tully *et al.*;[6] (△) Lorenz and Zellner;[8] (———) recommendations (see text).

At elevated temperatures, using the criteria discussed above, the recommendation is based on the rate constants obtained at temperatures $\geqslant$450 K. A unit-weighted least-squares analysis of the data of Tully *et al.*[6] and Lorenz and Zellner,[8] using the expression $k = CT^2e^{-D/T}$, yields the recommendation of

$$k(\text{benzene-}d_6; T \geqslant 450 \text{ K}) = (2.23^{+1.15}_{-0.75})$$

$$\times 10^{-18} T^2 e^{-(582 \pm 298)/T} \text{ cm}^3 \text{ molecule}^{-1} \text{ s}^{-1}$$

over the temperature range 524–1150 K, where the indicated error limits are two least-squares standard deviations.

It can be seen from Table 19 and Figs. 95 and 96 that at temperatures $\lesssim$325 K the rate constants for benzene and benzene-d_6 are essentially identical, within the experimental errors. However, for temperatures $\geqslant$450 K the rate constants for benzene-d_6 are significantly lower than those for benzene-h_6. As discussed below, these observations are totally consistent with OH radical addition to the aromatic ring dominating for temperatures $\leqslant 325$ K, while H-atom abstraction dominates for temperatures $\geqslant 450$ K (as shown by the study of Madronich and Felder[13]), with the corresponding expected kinetic isotope effect.

(3) Toluene

The available rate constants obtained at, or close to, the high pressure limit are given in Table 19, and those of Davis et al.,[1] Hansen et al.,[3] Perry et al.,[4] Cox et al.,[5] Tully et al.,[6] Edney et al.,[15] Baldwin et al.[17] and Atkinson and Aschmann[18] are plotted in Arrhenius form in Fig. 97 for the temperature regimes for which exponential OH radical decays have been observed in the two temperature-dependent flash photolysis studies.[4,6]

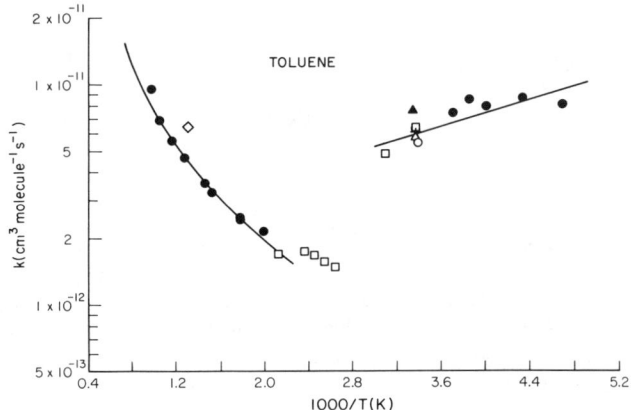

FIG. 97. Arrhenius plot of rate constants obtained at, or close to, the high-pressure limit for the reaction of the OH radical with toluene. (+) Davis et al.;[1] (Δ) Hansen et al.;[3] (□) Perry et al.;[4] (▲) Cox et al.;[5] (●) Tully et al.;[6] (○) Edney et al.,[15] Atkinson and Aschmann;[18] (◇) Baldwin et al.;[17] (———) recommendations (see text).

Davis et al.,[1] Tully et al.[6] and Bourmada et al.[53,55] have reported that at room temperature the rate constant for this reaction is in the fall-off regime between second- and third-order kinetics below ~100 Torr total pressure of helium or argon diluent.

Based upon the discharge flow-resonance fluorescence study of Bourmada et al.,[53,55] the limiting low pressure third order rate constant k_o at 295 ± 2 K for helium diluent is

$$k_o^{He}(\text{toluene}) = (4.0 \pm 0.5) \times 10^{-28} \text{ cm}^6 \text{ molecule}^{-2} \text{ s}^{-1}$$

Combined with a limiting high-pressure rate constant of $k_\infty = 6.0 \times 10^{-12}$ cm^3 molecule^{-1} s^{-1} [53,55] and $F = 0.6$ and assuming $k_o^{Ar} \geqslant k_o^{He}$, this leads to the expectation that at ~100 Torr total pressure of argon diluent and

room temperature the measured rate constants are within ~5% of k_∞.

For temperatures $\leqslant 325$ K, a unit-weighted least-squares analysis of the absolute rate constants of Hansen et al.,[3] Perry et al.[4] and Tully et al.[6] (the rate constant of Davis et al.[1] has not been included since the corresponding rate constant for benzene appears to be anomalously high; see above) and the relative rate data of Edney et al.[15] and Atkinson and Aschmann[18] yields the recommended Arrhenius expression of

$$k(\text{toluene}, T \leqslant 325) = (1.81^{+1.27}_{-0.74})$$
$$\times 10^{-12} \text{ e}^{(355 \pm 143)/T} \text{ cm}^3 \text{ molecule}^{-1} \text{ s}^{-1}$$

over the temperature range 213–324 K, where the indicated errors are two least-squares standard deviations, and

$$k(\text{toluene}) = 5.96 \times 10^{-12} \text{ cm}^3 \text{ molecule}^{-1} \text{ s}^{-1} \text{ at 298 K,}$$

with an estimated overall uncertainty at 298 K of ±25%.

For temperatures $\geqslant 450$ K, a unit-weighted least-squares analysis of the rate constants obtained by Perry et al.[4] and Tully et al.,[6] using the expression $k = CT^2 e^{-D/T}$, yields the recommendation of

$$k(\text{toluene}, T \geqslant 450 \text{ K}) = (7.58^{+1.35}_{-1.14})$$
$$\times 10^{-18} T^2 \text{ e}^{(11 \pm 106)/T} \text{ cm}^3 \text{ molecule}^{-1} \text{ s}^{-1}$$

over the temperature range 473–1046 K, where the indicated errors are two least-squares standard deviations. While the rate constant obtained by Perry et al.[4] at 473 K is in good agreement with those of Tully et al.,[6] the rate constants of Perry et al.[4] at temperatures between 378 and 424 K are ~25% higher than those predicted from the above expression. This may well be due to a continuing (but decreasing with increasing temperature) contribution of the addition process to the observed overall rate constant, as discussed by Tully et al.[6]

The rate constants obtained from the relative rate studies of Cox et al.,[5] Ohta and Ohyama[12] and Baldwin et al.[17] are in good agreement with the above recommendations for the two temperature regions.

(4) Toluene-d_8

Rate constants have been obtained for toluene-d_8 at, or close to, the high-pressure limit by Perry et al.[4] and Tully et al.[6] These data are given in Table 19, and the rate constants obtained in the temperature regimes corresponding to exponential OH radical decays are plotted in Arrhenius form in Fig. 98. The rate constants from these two studies[4,6] are in excellent agreement. At temperatures $\leqslant 325$ K, a unit-weighted least-squares analysis of these rate constants[4,6] yields the recommended Arrhenius expression of

$$k(\text{toluene-}d_8, T \leqslant 325 \text{ K}) = (7.31^{+4.25}_{-2.70})$$

$$\times 10^{-12} e^{-(44 \pm 127)/T} \text{ cm}^3 \text{ molecule}^{-1} \text{ s}^{-1}$$

over the temperature range 250–298 K, where the indicated errors are two least-squares standard deviations, and

$$k(\text{toluene-}d_8) = 6.31 \times 10^{-12} \text{ cm}^3 \text{ molecule}^{-1} \text{ s}^{-1}$$

at 298 K, with an estimated overall uncertainty of ±20% at 298 K.

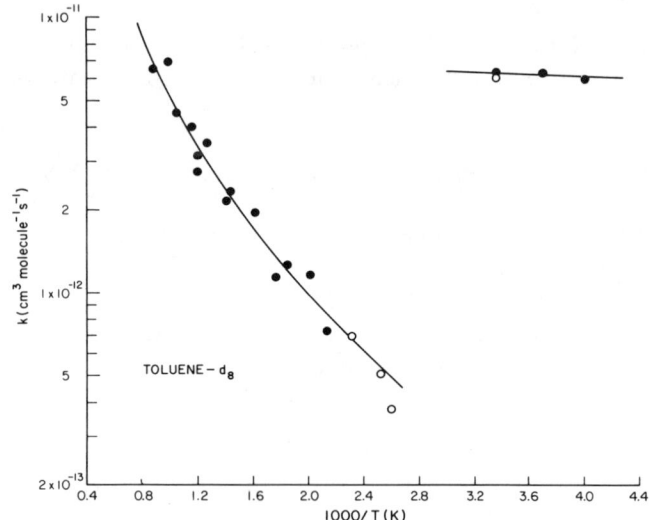

FIG. 98. Arrhenius plot of rate constants for the reaction of the OH radical reaction with toluene-d_8. (○) Perry et al.;[4] (●) Tully et al.;[6] (——) recommendations (see text).

For temperatures ⩾450 K, a unit-weighted least-squares analysis of the rate constants of Tully et al.,[6] using the expression $k = CT^2 e^{-D/T}$, yields the recommendation of

$$k(\text{toluene-}d_8, T \geqslant 450 \text{ K}) = (6.85^{+2.55}_{-1.86})$$

$$\times 10^{-18} T^2 e^{-(276 \pm 216)/T} \text{ cm}^3 \text{ molecule}^{-1} \text{ s}^{-1}$$

over the temperature range 470–1150 K, where the indicated errors are two least-squares standard deviations. These recommendations are identical to those of Atkinson,[56] using the same data set.

As for benzene and benzene-d_6, the rate constants at ⩽325 K for toluene and toluene-d_8 are very similar, consistent with the dominance of OH radical addition to the aromatic ring. However, for temperatures ⩾450 K the OH radical rate constant for toluene-d_8 is significantly lower than that for toluene-h_8. This is also shown in Fig. 99, in which the reported elevated temperature (⩾450 K) rate constants for toluene,[4,6] toluene-d_3 (C$_6$H$_5$CD$_3$),[6] toluene-d_5 (C$_6$D$_5$CH$_3$)[6] and toluene-d_8[6] are plotted in Arrhenius form.

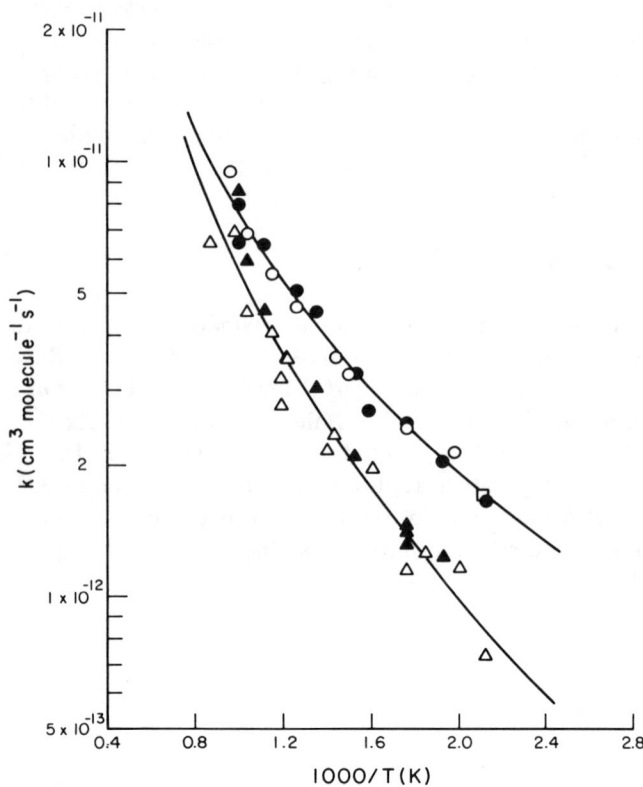

FIG. 99. Arrhenius plot of rate constants at temperatures ⩾450 K for the reactions of the OH radical with toluene and partially- and fully-deuterated toluenes. Toluene-h_8: (□) Perry et al.;[4] (○) Tully et al.[6] Toluene-d_5 (C$_6$D$_5$CH$_3$): (●) Tully et al.[6] Toluene-d_3 (C$_6$H$_5$CD$_3$): (▲) Tully et al.[6] Toluene-d_8: (△) Tully et al.[6] (——) recommendations (see text).

It can be seen that to a good approximation these data fall into two sets, namely those for C$_6$H$_5$CH$_3$ and C$_6$D$_5$CH$_3$ and those for C$_6$H$_5$CD$_3$ and C$_6$D$_5$CD$_3$, with the rate constants for toluene and toluene-d_5 being significantly higher than those for toluene-d_3 and toluene-d_8 (at least up to 1000 K). The lines shown in Fig. 99 are those calculated from unit-weighted least-squares analyses of the data for C$_6$H$_5$CH$_3$[4,6] and C$_6$D$_5$CH$_3$[6] [k(C$_6$X$_5$CH$_3$)] and C$_6$H$_5$CD$_3$[6] and C$_6$D$_5$CD$_3$[6] [k(C$_6$X$_5$CD$_3$)], respectively, with

$$k(\text{C}_6\text{X}_5\text{CH}_3) = (7.63^{+1.01}_{-0.90})$$

$$\times 10^{-18} T^2 e^{(3 \pm 82)/T} \text{ cm}^3 \text{ molecule}^{-1} \text{ s}^{-1}$$

and

$$k(\text{C}_6\text{X}_5\text{CD}_3) = (8.19^{+2.39}_{-1.85})$$

$$\times 10^{-18} T^2 e^{-(361 \pm 172)/T} \text{ cm}^3 \text{ molecule}^{-1} \text{ s}^{-1}$$

over the temperature range 470–1150 K, where the indicated errors are two least-squares standard deviations. This leads to a deuterium isotope effect for H- or D-atom abstraction from the —CH$_3$ or —CD$_3$ groups in the toluenes of $k^D/k^H = 1.07e^{-364/T}$, which is in between the isotope effects observed for primary and secondary C—H bonds in the alkanes (Sec. 2.1). These H- or D-atom abstraction rate constants are significantly higher than those for benzene or benzene-d_6, again showing that the major process involves H- or D-atom abstraction from the —CH$_3$ or —CD$_3$ substituent groups. While there may be consistent differences between the rate constants for toluene and toluene-d_5, and between those for toluene-d_3 and toluene-d_8, due to H- or D-atom abstraction from the aromatic ring C—H or C—D bonds, these are minor and are probably within the experimental errors.

(5) Ethylbenzene

The available rate constants of Lloyd et al.,[19] Ravishankara et al.[20] and Ohta and Ohyama,[12] all obtained at room temperature, are given in Table 19. Within the likely overall experimental error limits, these rate constants are in agreement and, from a unit-weighted average of these data,[12,19,20] it is recommended that

$$k(\text{ethylbenzene}) = 7.1 \times 10^{-12} \text{ cm}^3 \text{ molecule}^{-1} \text{ s}^{-1}$$

at ~ 298 K, with an estimated overall uncertainty of $\sim \pm 35\%$. The temperature dependence at temperatures $\lesssim 320$ K is expected to be close to zero.

(6) o-Xylene

The available rate constants obtained at, or close to, the high-pressure limit are given in Table 19, and those of Hansen et al.,[3] Perry et al.,[4] Ravishankara et al.,[20] Cox et al.,[5] Nicovich et al.[21] and Atkinson and Aschmann[18] are plotted in Arrhenius form in Fig. 100. In general, the agreement between these studies and those of Doyle et al.,[2] Ohta and Ohyama,[12] Klöpffer et al.[22] and Edney et al.[15] is good.

For temperatures $\leqslant 325$ K rate constants have been reported only over the very limited temperature range 296–320 K with, within the experimental error limits, no obvious temperature dependence. Hence, a unit-weighted average of the absolute rate constants of Hansen et al.,[3] Perry et al.,[4] Ravishankara et al.[20] and Nicovich et al.[21] and the recent relative rate data of Ohta and Ohyama,[12] Edney et al.[15] and Atkinson and Aschmann[18] (the rate constant of Klöpffer et al.[22] was not used due to a lack of details) yields the recommendation of

$$k(o\text{-xylene}; T \leqslant 325 \text{ K}) = 1.37$$

$$\times 10^{-11} \text{ cm}^3 \text{ molecule}^{-1} \text{ s}^{-1},$$

independent of temperature over the range 296–320 K, with an estimated overall uncertainty over this temperature range of $\pm 25\%$. At room temperature, the rate constant for this reaction is close to the limiting high pressure value for total pressures of helium or argon diluent of $\gtrsim 20$ Torr.[20]

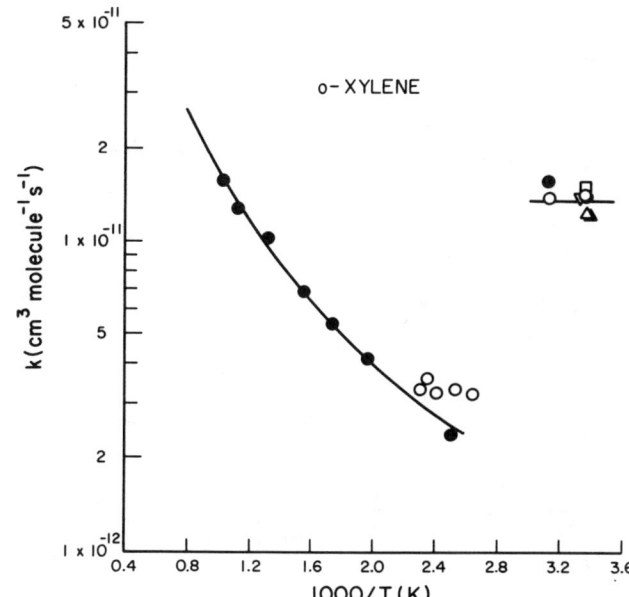

FIG. 100. Arrhenius plot of rate constants for the reaction of the OH radical with o-xylene. (□) Hansen et al.;[3] (○) Perry et al.;[4] (△) Ravishankara et al.;[20] (▽) Cox et al.;[5] (●) Nicovich et al.;[21] (▲) Atkinson and Aschmann;[18] (———) recommendations (see text).

For temperatures $\geqslant 450$ K, a unit-weighted least-squares analysis of the rate constants of Nicovich et al.,[21] using the expression $k = CT^2e^{-D/T}$, yields the recommendation of

$$k(o\text{-xylene}, T \geqslant 450 \text{ K}) = (1.75^{+0.25}_{-0.23})$$

$$\times 10^{-17} T^2 e^{-(35 \pm 90)/T} \text{ cm}^3 \text{ molecule}^{-1} \text{ s}^{-1}$$

over the temperature range 508–970 K, where the indicated errors are two least-squares standard deviations. Again, as is the case for m- and p-xylene (see below), the rate constants determined by Perry et al.[4] over the small temperature range ~ 379–432 K are somewhat higher (by up to $\sim 50\%$) than predicted from the recommended $\geqslant 450$ K expression.

(7) m-Xylene

The available rate constants obtained at, or close to, the high-pressure limit are given in Table 19, and those of Hansen et al.,[3] Lloyd et al.,[19] Perry et al.,[4] Ravishankara et al.,[20] Cox et al.,[5] Nicovich et al.,[21] Atkinson et al.,[23] Edney et al.[15] and Atkinson and Aschmann[18] are plotted in Arrhenius form in Fig. 101. Ravishankara et al.[20] have shown that at 298 K this reaction is in the fall-off regime between second- and third-order kinetics at 3 Torr total pressure of argon, with the limiting high pressure value being approached at ~20 Torr total pressure of helium or argon.[20]

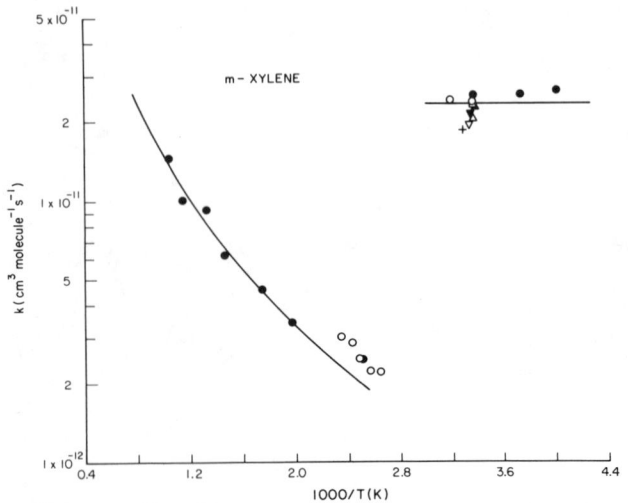

FIG. 101. Arrhenius plot of rate constants for the reaction of the OH radical with m-xylene. (☐) Hansen et al.;[3] (+) Lloyd et al.;[19] (○) Perry et al.;[4] (△) Ravishankara et al.;[20] (▽) Cox et al.;[5] (●) Nicovich et al.;[21] (▼) Atkinson et al.;[23] (▲) Edney et al.,[15] Atkinson and Aschmann;[18] (_____) recommendations (see text).

The rate constant data of Perry et al.[4] and Nicovich et al.[21] suggest that the temperature dependence is essentially zero over the range 250–315 K. Thus, for temperatures ⩽325 K a unit-weighted average of the absolute rate constants of Hansen et al.,[3] Perry et al.,[4] Ravishankara et al.[20] and Nicovich et al.[21] and the recent relative rate data of Atkinson et al.,[23] Ohta and Ohyama,[12] Edney et al.[15] and Atkinson and Aschmann[18] yields the recommendation of

$$k(m\text{-xylene, } T \leqslant 325 \text{ K}) = 2.36$$

$$\times \, 10^{-11} \text{ cm}^3 \text{ molecule}^{-1} \text{ s}^{-1}$$

over the temperature range 250–315 K, with an estimated overall uncertainty of ±25%.

For temperatures ⩾450 K, a unit-weighted least-squares analysis of the data of Nicovich et al.,[21] using the expression $k = CT^2 e^{-D/T}$, yields the recommendation of

$$k(m\text{-xylene, } T \geqslant 450 \text{ K}) = (1.71^{+0.71}_{-0.51})$$

$$\times \, 10^{-17} \, T^2 \, e^{-(127 \pm 235)/T} \text{ cm}^3 \text{ molecule}^{-1} \text{ s}^{-1}$$

over the temperature range 508–960 K, where the indicated errors are two least-squares standard deviations.

(8) p-Xylene

The available rate constants obtained at, or close to, the high pressure limit are listed in Table 19, and those of Hansen et al.,[3] Perry et al.,[4] Ravishankara et al.,[20] Nicovich et al.,[21] Edney et al.[15] and Atkinson and Aschmann[18] are plotted in Arrhenius form in Fig. 102.

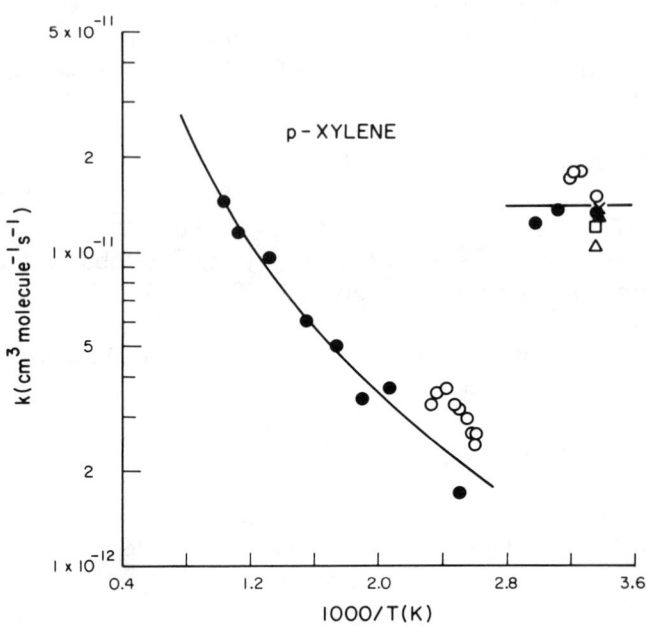

FIG. 102. Arrhenius plot of rate constants for the reaction of the OH radical with p-xylene. (☐) Hansen et al.;[3] (○) Perry et al.;[4] (△) Ravishankara et al.;[20] (●) Nicovich et al.;[21] (x) Edney et al.;[15] (▲) Atkinson and Aschmann;[18] (_____) recommendations (see text).

As for m-xylene, Ravishankara[20] have reported that at 298 K the rate constant for this reaction is in the fall-off regime between second- and third-order kinetics at 3 Torr total pressure of argon, with the rate constants at 20 Torr total pressure of helium or argon being close to the high-pressure kinetic regime. The rate constants obtained at around room temperature exhibit a significant degree of scatter, and a unit-weighted average of the rate constants determined by Hansen et al.,[3] Perry et al.,[4] Ravishankara et al.,[20] Nicovich et al.,[21] Ohta and Ohyama,[12] Edney et al.[15] and Atkinson and Aschmann[18] at temperatures ⩽335 K yields the recommendation of

$$k(p\text{-xylene; } T \leqslant 335 \text{ K}) = 1.43$$

$$\times \, 10^{-11} \text{ cm}^3 \text{ molecule}^{-1} \text{ s}^{-1},$$

independent of temperature over the range 296–335 K, with an estimated overall uncertainty of ±40%.

At temperatures ⩾450 K, a unit-weighted least-squares analysis of the rate constants determined by Nicovich et

$al.,$[21] using the expression $k = CT^2 e^{-D/T}$, yields the recommendation of

$$k(p\text{-xylene}; \ T \geqslant 450 \ K) = (1.74^{+0.70}_{-0.50})$$

$$\times \ 10^{-17} \ T^2 \ e^{-(99 \pm 215)/T} \ \text{cm}^3 \ \text{molecule}^{-1} \ \text{s}^{-1}$$

over the temperature range 484–960 K, where the indicated errors are two least-squares standard deviations.

As noted previously,[21,56] at any given temperature $\geqslant 450$ K the rate constants for the H-atom abstraction pathway from o-, m- and p-xylene are similar, and are close to a factor of 2 higher than those for toluene and toluene-d_5 ($C_6D_5CH_3$). This suggests that the OH radical reaction rate constant for H-atom abstraction from the —CH_3 groups on the toluenes and xylenes is independent of the aromatic hydrocarbon and depends only on the number of —CH_3 substituent groups. This is shown in Fig. 103, in which the H-atom abstraction rate constants per —CH_3 group are plotted in Arrhenius form for toluene, toluene-d_5 and o-, m- and p-xylene for temperatures $\geqslant 450$ K.

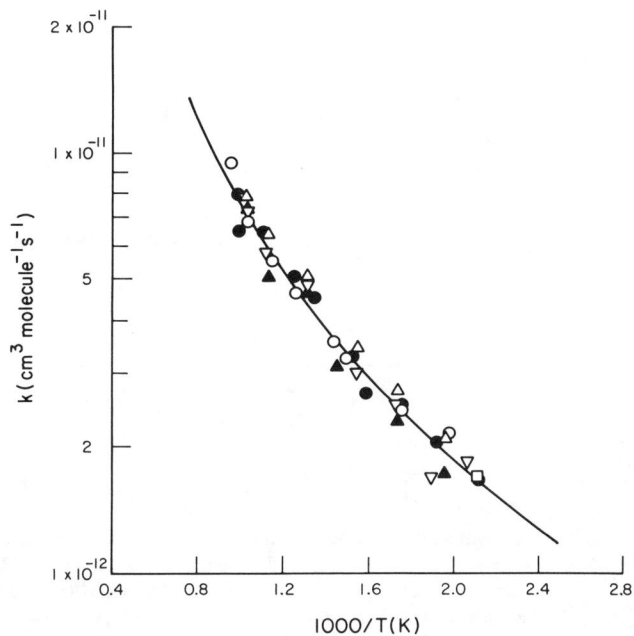

FIG. 103. Arrhenius plot of rate constants at temperatures $\geqslant 450$ K per —CH_3 group for the reaction of the OH radical with toluene, toluene-d_5 ($C_6D_5CH_3$) and o-, m- and p-xylene. Toluene: (□) Perry et $al.$;[4] (○) Tully et $al.$[6] Toluene-d_5: (●) Tully et $al.$[6] o-Xylene: (△) Nicovich et $al.$[21] m-Xylene: (▲) Nicovich et $al.$[21] p-Xylene: (▽) Nicovich et $al.$[21] (———) recommendation (see text).

A unit-weighted least-squares analysis of these data,[4,6,21] using the expression $k = CT^2 e^{-D/T}$, yields

$$k(\text{H-atom abstraction per}-CH_3 \text{ group}) = (8.07^{+0.99}_{-0.88})$$

$$\times \ 10^{-18} \ T^2 \ e^{-(38 \pm 76)/T} \ \text{cm}^3 \ \text{molecule}^{-1} \ \text{s}^{-1}$$

over the temperature range 470–1046 K, where the indicated errors are two least-squares standard deviations. Extrapolation to 298 K yields an H-atom abstraction rate constant per —CH_3 group of 6.3×10^{-13} cm^3 molecule^{-1} s^{-1}.

(9) *n*-Propylbenzene

The available room temperature rate constants of Lloyd et $al.$,[19] Ravishankara et $al.$[20] and Ohta and Ohyama[12] are in good agreement (Table 19), and a unit-weighted average of these data[12,19,20] leads to the recommendation of

$$k(n\text{-propylbenzene}) = 6.0 \times 10^{-12} \ \text{cm}^3 \ \text{molecule}^{-1} \ \text{s}^{-1}$$

at ~ 298 K, with an estimated overall uncertainty of $\pm 30\%$.

(10) Isopropylbenzene

The available rate constants of Lloyd et $al.$,[19] Ravishankara et $al.$[20] and Ohta and Ohyama,[12] all obtained at room temperature, exhibit a spread of $\sim 50\%$ (Table 19). A unit-weighted average of these data[12,19,20] leads to the recommendation of

$$k(\text{isopropylbenzene}) = 6.5 \times 10^{-12} \ \text{cm}^3 \ \text{molecule}^{-1} \ \text{s}^{-1}$$

at ~ 298 K, with an estimated overall uncertainty of $\pm 35\%$.

These room temperature rate constants for n-propylbenzene and isopropylbenzene (and ethylbenzene) are similar to that for toluene, and indicate that the rate constants for toluene are reasonably applicable to the higher monoalkylbenzenes.

(11) *o*-, *m*- and *p*-Ethyltoluene

Two room temperature relative rate constant studies have been carried out for each of these isomers,[12,19] with the measured rate constants being in good agreement (Table 19). Unit-weighted averages of these rate data[12,19] lead to the recommendations of

$$k(o\text{-ethyltoluene}) = 1.23 \times 10^{-11} \ \text{cm}^3 \ \text{molecule}^{-1} \ \text{s}^{-1},$$

$$k(m\text{-ethyltoluene}) = 1.92 \times 10^{-11} \ \text{cm}^3 \ \text{molecule}^{-1} \ \text{s}^{-1},$$

and

$$k(p\text{-ethyltoluene}) = 1.21 \times 10^{-11} \ \text{cm}^3 \ \text{molecule}^{-1} \ \text{s}^{-1},$$

all at ~ 298 K, with estimated overall uncertainties of $\pm 40\%$. These rate constants are similar to those for the corresponding xylene isomers, again showing that the reactions proceed by OH radical addition to the aromatic ring, with the rate constants depending on the number and positions of the alkyl substituents.

(12) 1,2,3-, 1,2,4- and 1,3,5-Trimethylbenzene

The available rate constants of Doyle *et al.*,[2] Hansen *et al.*,[3] Perry *et al.*,[4] Ohta and Ohyama[12] (carried out at room temperature, which was not specified) and Atkinson and Aschmann[18] are given in Table 19 and are plotted in Arrhenius form in Figs. 104 through 106 (assuming 298 K as the temperature of the Ohta and Ohyama[12] study). In all cases, there is a significant amount of scatter in the reported rate constants for temperatures ≲325 K. The rate constants derived from the relative rate study of Doyle *et al.*[2] are subject to significant uncertainties since dilution, which was of a similar magnitude to the OH radical reaction as a loss process for the *n*-butane reference compound, had to be taken into account.

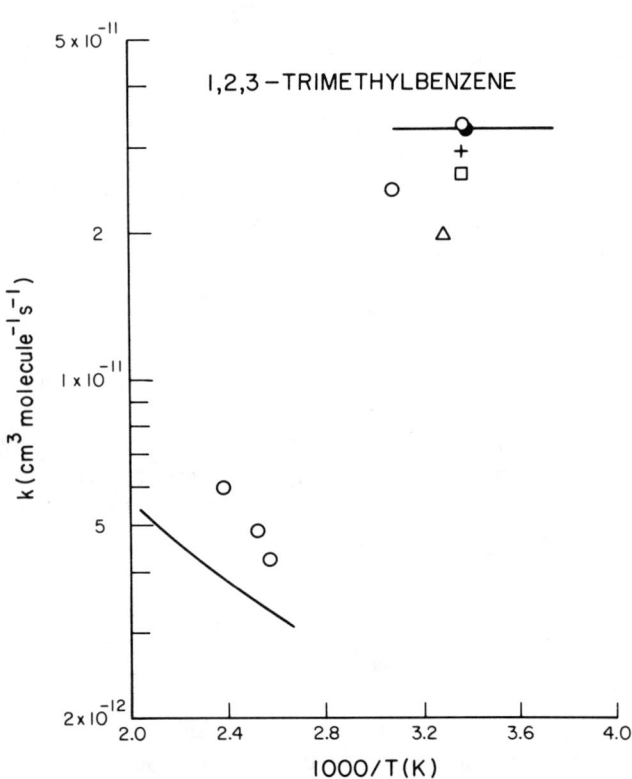

FIG. 104. Arrhenius plot of rate constants for the reaction of the OH radical with 1,2,3-trimethylbenzene. (△) Doyle *et al.*;[2] (□) Hansen *et al.*;[3] (○) Perry *et al.*;[4] (+) Ohta and Ohyama,[12] assuming a temperature of 298 K; (●) Atkinson and Aschmann;[18] (———) recommendations (see text).

The recommendations utilize the room temperature rate constants obtained from the most recent relative rate study of Atkinson and Aschmann,[18] with an assumed zero temperature dependence over the narrow temperature range ~295–325 K. Thus, it is recommended that

$$k(1,2,3\text{-trimethylbenzene}) = 3.27$$

$$\times 10^{-11} \text{ cm}^3 \text{ molecule}^{-1} \text{ s}^{-1},$$

$$k(1,2,4\text{-trimethylbenzene}) = 3.25$$

$$\times 10^{-11} \text{ cm}^3 \text{ molecule}^{-1} \text{ s}^{-1}$$

and

$$k(1,3,5\text{-trimethylbenzene}) = 5.75$$

$$\times 10^{-11} \text{ cm}^3 \text{ molecule}^{-1} \text{ s}^{-1},$$

all independent of temperature over the range ~295–325 K and with estimated overall uncertainties of ±35%.

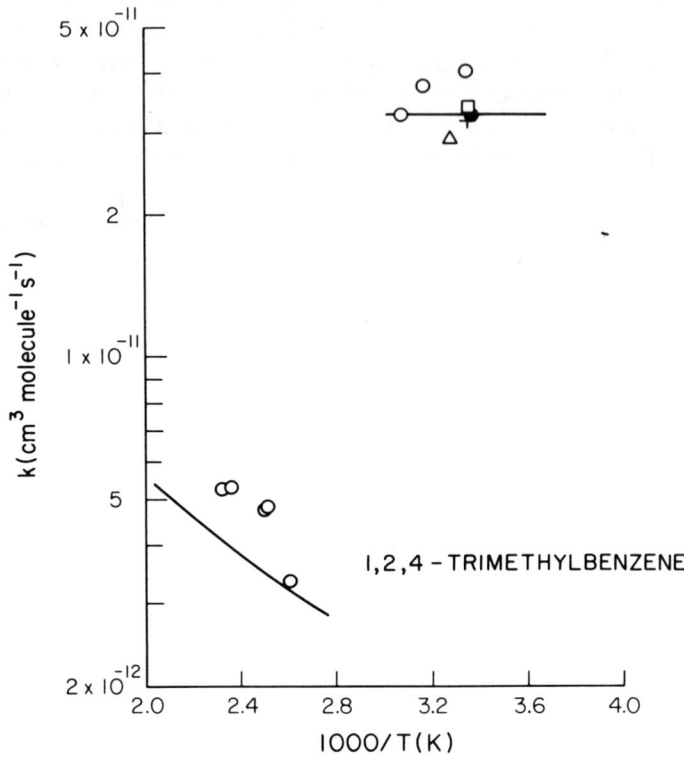

FIG. 105. Arrhenius plot of rate constants for the reaction of the OH radical with 1,2,4-trimethylbenzene. (△) Doyle *et al.*;[2] (□) Hansen *et al.*;[3] (○) Perry *et al.*;[4] (+) Ohta and Ohyama,[12] assuming a temperature of 298 K, (●) Atkinson and Aschmann;[18] (———) recommendations (see text).

At elevated temperatures (≳370–380 K), this OH radical addition process is no longer observed experimentally, and the H-atom abstraction pathway is measured. As noted above and by Tully *et al.*,[6] at temperatures up to ~450 K the OH radical addition process still appears to contribute slightly to the measured rate constants, which are thus greater than those for the abstraction reaction. Since the rate constant for H-atom abstraction from —CH₃ groups in the methyl-substituted benzenes depends only on the number of —CH₃ groups (see above and Fig. 103), for temperatures ≳450 K it is recommended that

k(trimethylbenzenes) = 2.42

$$\times \ 10^{-17} \ T^2 \ e^{-38/T} \ cm^3 \ molecule^{-1} \ s^{-1},$$

and the lines in Figs. 104 through 106 for this temperature range reflect this recommendation. As anticipated, the rate data of Perry et al.[4] at temperatures ~370–430 K are somewhat higher than these recommendations.

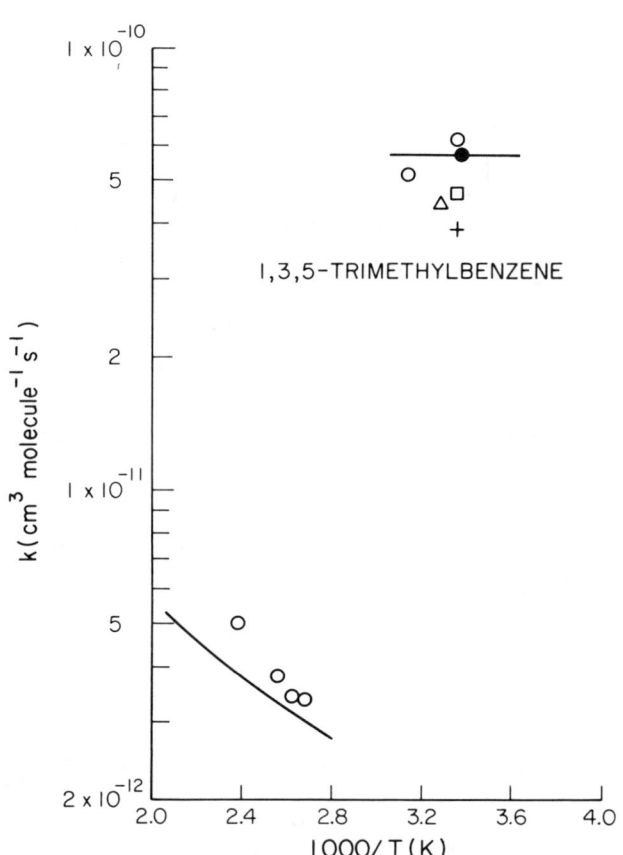

FIG. 106. Arrhenius plot of rate constants for the reaction of the OH radical with 1,3,5-trimethylbenzene. (Δ) Doyle et al.;[2] (□) Hansen et al.;[3] (○) Perry et al.;[4] (+) Ohta and Ohyama,[12] assuming a temperature of 298 K; (●) Atkinson and Aschmann;[18] (———) recommendations (see text).

(13) Styrene

The available rate constants of Bignozzi et al.[26] and Atkinson and Aschmann[27] are given in Table 19. Both rate constants were obtained from relative rate studies carried out at room temperature. The agreement is good. Since in the study of Bignozzi et al.[26] the styrene was a factor of 14 more reactive than the 2,2,4-trimethylpentane reference compound, the rate constant of Atkinson and Aschmann[27] is preferred, leading to the recommendation of

k(styrene) = 5.8 × 10^{-11} cm^3 molecule^{-1} s^{-1} at 298 K,

with an estimated overall uncertainty of ±25%.

The room temperature product studies of Bignozzi et al.[26] and Chiorboli et al.[28] show that the OH radical reactions with styrene and the other aromatic alkenes investigated proceed by OH radical addition to the alkene moiety, for example

$$OH + C_6H_5CH{=}CH_2 \rightarrow C_6H_5CHOH\dot{C}H_2$$

$$and \ C_6H_5\dot{C}HCH_2OH$$

At elevated temperatures, ≳550–650 K, this process is expected to become rapidly reversible, with H-atom abstraction from the various C—H bonds then becoming the important process.

(14) Phenol

The available rate constant data of Rinke and Zetzsch,[11] Witte and Zetzsch[29] and He et al.[30] are given in Table 19 and are plotted in Arrhenius form in Fig. 107.

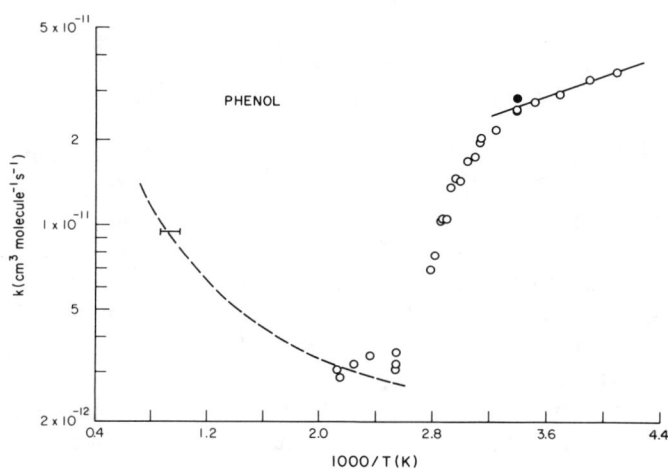

FIG. 107. Arrhenius plot of rate constants obtained at, or close to, the high-pressure limit for the reaction of the OH radical with phenol. (●) Rinke and Zetzsch,[11] (○) Witte and Zetzsch;[29] (⊢) He et al.;[30] (— — —, ———) recommendations (see text).

Rinke and Zetzsch[11] observed that at room temperature the rate constant is pressure dependent below ~30 Torr total pressure of helium diluent, while Witte and Zetzsch[29] observed non-exponential OH radical decays in the temperature range 320–359 K, indicating thermal decomposition of the OH-phenol addition adduct back to reactants over this temperature range.

The room temperature rate constants of Rinke and Zetzsch[11] and Witte and Zetzsch[29] are in good agreement, and a unit-weighted least-squares analysis of the rate constants of Rinke and Zetzsch[11] and Witte and Zetzsch[29] for temperatures <300 K leads to the recommended Arrhenius expression of

J. Phys. Chem. Ref. Data, Monograph 1 (1989)

$$k(\text{phenol}; \ T < 300 \ \text{K}) = (6.75^{+2.93}_{-2.04})$$

$$\times \ 10^{-12} \ e^{(405 \pm 100)/T} \ \text{cm}^3 \ \text{molecule}^{-1} \ \text{s}^{-1}$$

over the temperature range 245–296 K, where the indicated errors are two least-squares standard deviations, and

$$k(\text{phenol}) = 2.63 \times 10^{-11} \ \text{cm}^3 \ \text{molecule}^{-1} \ \text{s}^{-1} \ \text{at 298 K,}$$

with an estimated overall uncertainty at 298 K of ±30%.

At temperatures $\gtrsim 390$ K, it is expected that the reaction process observed is H-atom abstraction from the C—H bonds of the aromatic ring and/or the substituent O—H bond.

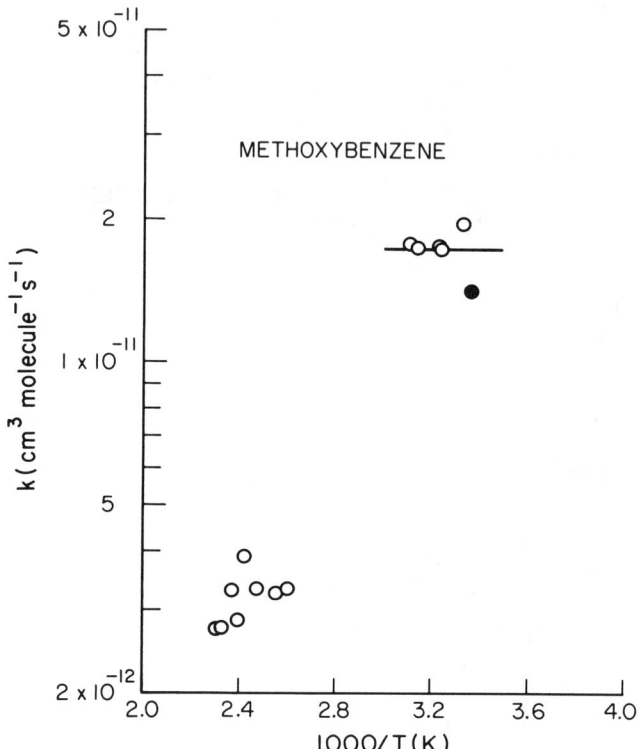

From the recommended rate expression for benzene, and assuming that the rate constant for pathway (a) will be 0.83 k(benzene) [2.26 × 10^{-12} cm³ molecule⁻¹ s⁻¹ at 1000 K and 1.6 × 10^{-13} cm³ molecule⁻¹ s⁻¹ at 400 K], it appears that H-atom abstraction from the substituent O—H bond to form the phenoxy radical [reaction pathway (b)] dominates below ∼1200 K, as also concluded by He et al.[30]

Assuming the data of Witte and Zetzsch[29] and He et al.[30] at 390–1150 K to be reasonably correct, these data can be fitted by the expression,

$$k(\text{phenol}; \ T \gtrsim 390 \ \text{K}) = 5$$

$$\times \ 10^{-18} \ T^2 \ e^{500/T} \ \text{cm}^3 \ \text{molecule}^{-1} \ \text{s}^{-1}$$

which is shown as the dashed line in Fig. 107. Clearly, further kinetic data are needed at elevated temperatures, ⩾400 K, before any firm recommendation can be made.

(15) Methoxybenzene

The available data of Perry et al.[31] and Ohta and Ohyama[12] (carried out at an unspecified room temperature) are given in Table 19 and are plotted in Arrhenius form in Fig. 108 (assuming 298 K for the temperature of the Ohta and Ohyama[12] study). The agreement between these two studies[12,31] at ∼299 K is reasonable, especially since wall-adsorption of the methoxybenzene would be expected to occur in the static reaction vessel used by Ohta and Ohyama.[12] A unit-weighted average of the rate

constants obtained by Perry et al.[31] and Ohta and Ohyama[12] at temperatures <325 K leads to the recommendation of

$$k(\text{methoxybenzene}; \ T \leqslant 325 \ \text{K}) = 1.73$$

$$\times \ 10^{-11} \ \text{cm}^3 \ \text{molecule}^{-1} \ \text{s}^{-1},$$

independent of temperature over the range 298–322 K, with an estimated overall uncertainty of ±35%.

FIG. 108. Arrhenius plot of rate constants for the reaction of the OH radical with methoxybenzene. (○) Perry et al.;[31] (●) Ohta and Ohyama,[12] assuming a temperature of 298 K; (———) recommendation (see text).

At temperatures $\gtrsim 325$ K, the OH-methoxybenzene addition adduct thermally decomposes[31] and at temperatures $\gtrsim 380$ K only an H-atom abstraction process is observed. As for phenol, the magnitude of the measured rate constants at ∼400 K indicates that H-atom abstraction from the —OCH₃ group

$$OH + C_6H_5OCH_3 \rightarrow H_2O + C_6H_5O\dot{C}H_2$$

dominates over H-atom abstraction from the C—H bonds of the aromatic ring.

(16) o-, m- and p-Cresol

The available rate constant data of Perry et al.[31] (for o-cresol only), Atkinson et al.[32] and Atkinson and

Aschmann[33] are given in Table 19. The rate constant data of Perry *et al.*,[31] Atkinson *et al.*[32] and Atkinson and Aschmann[33] for *o*-cresol are plotted in Arrhenius form in Fig. 109.

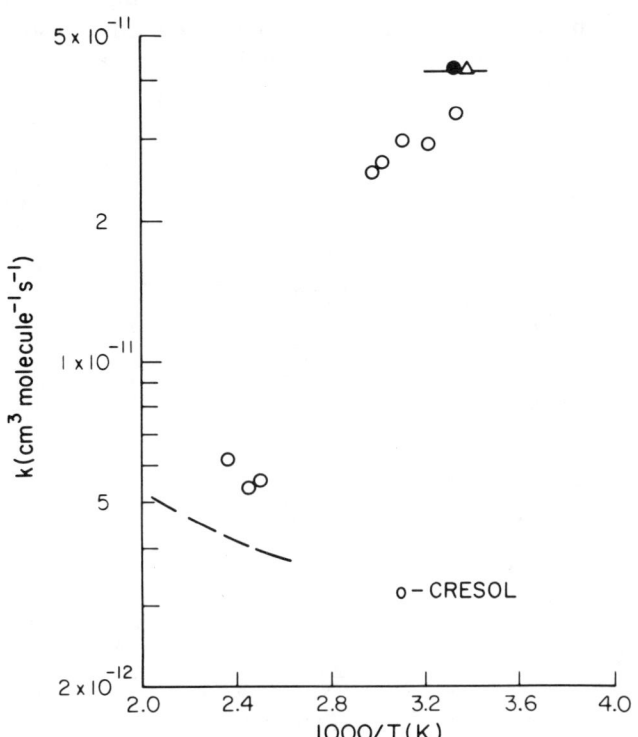

FIG. 109.   Arrhenius plot of rate constants for the reaction of the OH radical with *o*-cresol. (○) Perry *et al.*;[31] (●) Atkinson *et al.*;[32] (△) Atkinson and Aschmann;[33] (— — —, ——) recommendations (see text).

The room temperature rate constants of Perry *et al.*,[31] Atkinson *et al.*[32] and Atkinson and Aschmann[33] for this isomer show a discrepancy of ~25%, although they agree within the combined experimental error limits. As noted by Atkinson *et al.*,[32] the rate constants determined by Perry *et al.*[31] may have been somewhat low due to wall adsorption problems (especially in the small optical calibration cells used). Since the higher overall error limits assigned by Perry *et al.*[31] take into account (at least in part) such adsorption problems, a weighted average of these room temperature rate constants[31–33] yields the recommendation of

$$k(o\text{-cresol}) = 4.2 \times 10^{-11} \text{ cm}^3 \text{ molecule}^{-1} \text{ s}^{-1}$$

at 300 K, with an estimated overall uncertainty of ±30%. No recommendation is made concerning the temperature dependence of the rate constant at around room temperature.

The OH-*o*-cresol addition adduct thermally decomposes at temperatures ≳335 K, and at ≳400 K the measured OH radical reaction rate constants reflect H-atom abstraction from the C—H or O—H bonds

The magnitude of the measured rate constants at 400–425 K[31] shows that H-atom abstraction from the C—H bonds of the aromatic ring [pathway (c)] is of minor (~3%) significance. In fact, the experimental data are consistent with the occurrence of pathways (a) and (b) with rate constants at 400 K (derived from the discussions and/or recommendations for toluene, toluene-*d*₅, the xylenes and phenol) of

$$k_a = 1.2 \times 10^{-12} \text{ cm}^3 \text{ molecule}^{-1} \text{ s}^{-1}$$

and

$$k_b \approx 2.8 \times 10^{-12} \text{ cm}^3 \text{ molecule}^{-1} \text{ s}^{-1}.$$

The rate constant for the sum of pathways (a) and (b)

$$[k_a = 8.07 \times 10^{-18} \ T^2 \ e^{-38/T} \text{ cm}^3 \text{ molecule}^{-1} \text{ s}^{-1},$$

$$k_b \approx 5 \times 10^{-18} \ T^2 \ e^{500/T} \text{ cm}^3 \text{ molecule}^{-1} \text{ s}^{-1}]$$

is shown in Fig. 109 as the dashed line.

From unit-weighted averages of the rate constant data of Atkinson *et al.*[32] and Atkinson and Aschmann,[33] the recommended rate constants for *m*- and *p*-cresol at 298 K are then

$$k(m\text{-cresol}) = 6.4 \times 10^{-11} \text{ cm}^3 \text{ molecule}^{-1} \text{ s}^{-1}$$

and

$$k(p\text{-cresol}) = 4.7 \times 10^{-11} \text{ cm}^3 \text{ molecule}^{-1} \text{ s}^{-1},$$

both with estimated overall uncertainties of ±35%.

(17) Fluorobenzene

The available rate constants of Zetzsch,[36] Ohta and Ohyama[12] and Wallington *et al.*[16] are given in Table 19 and (assuming a temperature of 298 K for the Ohta and

Ohyama[12] study) are plotted in Arrhenius form in Fig. 110. The room temperature rate constants show a significant degree of scatter. Over the temperature range 234–303 K the data of Wallington *et al.*[16] are, within the experimental errors, independent of temperature. From a unit-weighted average of the room temperature rate constants of Zetzsch,[36] Ohta and Ohyama[12] and Wallington *et al.*[16] and assuming a zero temperature dependence over this temperature range, it is recommended that

$$k(\text{fluorobenzene; } T \lesssim 310 \text{ K}) = 6.9$$

$$\times 10^{-13} \text{ cm}^3 \text{ molecule}^{-1} \text{ s}^{-1},$$

independent of temperature over the range 234–303 K, with an estimated overall uncertainty over this temperature range of ± a factor of 2.

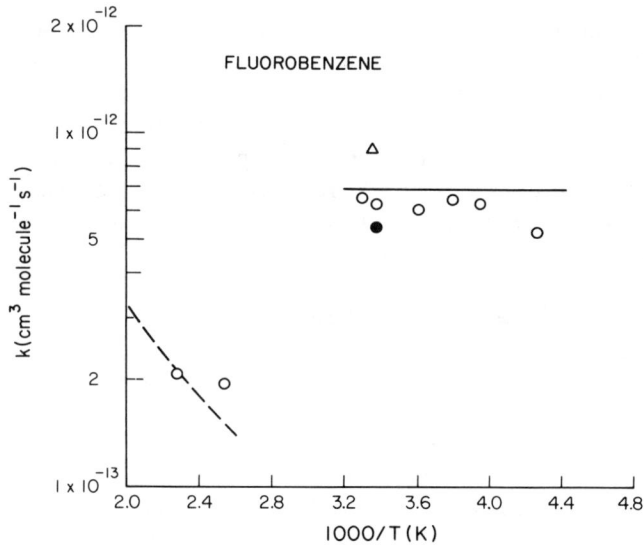

FIG. 110. Arrhenius plot of rate constants for the reaction of the OH radical with fluorobenzene. (●) Zetzsch;[36] (Δ) Ohta and Ohyama,[12] assuming a temperature of 298 K; (○) Wallington *et al.*;[16] (— — —, ———) recommendations (see text).

Above ~310 K the OH-fluorobenzene addition adduct thermally decomposes[16] and above ~390 K the measured rate constants presumably reflect an H-atom abstraction process (from the C—H bonds of the aromatic ring). Moreover, as shown in Fig. 110 by the dashed line, the measured rate constants at 393 and 438 K[16] are consistent with the H-atom abstraction rate constant recommended for benzene, scaled by a factor of 0.83 to take into account the fact that fluorobenzene has only five C—H bonds. This leads to

$$k_{\text{abstraction}} \approx 3.9 \times 10^{-18} T^2 e^{-543/T} \text{ cm}^3 \text{ molecule}^{-1} \text{ s}^{-1}.$$

(18) Chlorobenzene

The available rate constants of Zetzsch,[36,37] Atkinson *et al.*,[38] Edney *et al.*[15] and Wallington *et al.*[16] are given in Table 19, and those of Zetzsch,[36,37] Atkinson *et al.*[38] and Wallington *et al.*[16] are plotted in Arrhenius form in Fig. 111 (the rate constant of Edney *et al.*[15] has a high associated uncertainty which encompasses the other room temperature rate constants,[16,36–38] and is hence not used in the evaluation).

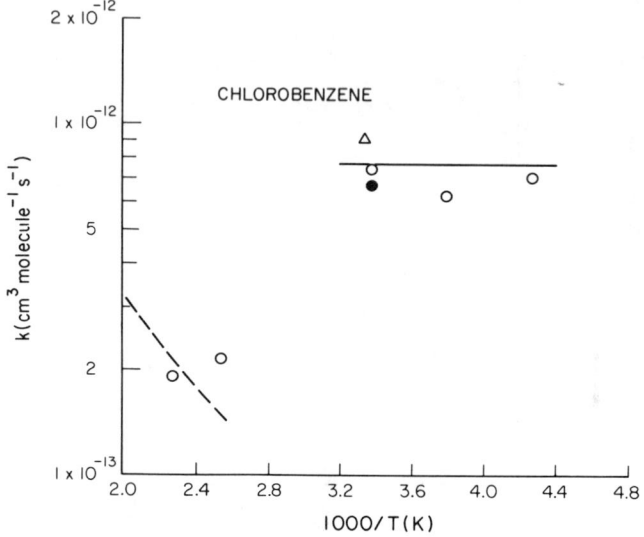

FIG. 111. Arrhenius plot of rate constants for the reaction of the OH radical with chlorobenzene. (●) Zetzsch;[36,37] (Δ) Atkinson *et al.*;[38] (○) Wallington *et al.*;[16] (— — —, ———) recommendations (see text).

The only temperature dependent study is that of Wallington *et al.*,[16] which indicates no appreciable temperature dependence over the range 234–296 K, within the experimental error limits. A unit-weighted average of the room temperature rate constants of Zetzsch,[36,37] Atkinson *et al.*[38] and Wallington *et al.*,[16] with an assumed zero temperature dependence below ~300 K, leads to the recommendation of

$$k(\text{chlorobenzene}) = 7.7 \times 10^{-13} \text{ cm}^3 \text{ molecule}^{-1} \text{ s}^{-1}$$

over the temperature range 234–299 K, with an estimated overall uncertainty of ±40% over this temperature range. As for fluorobenzene, the OH radical addition adduct thermally decomposes above ~300 K,[16] and the measured rate constants at >390 K presumably reflect H-atom abstraction from the C—H bonds of the aromatic ring. Again, the H-atom abstraction rate constant for benzene, scaled by a factor of 0.83, fits the measurements reasonably well, as shown by the dashed line in Fig. 111.

(19) Bromobenzene

The rate constants of Zetzsch,[36] Witte *et al.*[14] and Wallington *et al.*[16] are given in Table 19 and are plotted in Arrhenius form in Fig. 112. The absolute studies of Witte *et al.*[14] and Wallington *et al.*[16] show differences with respect to both the room temperature rate constant and the sign of the temperature dependence. A unit-weighted average of the entire data set[14,16,36] below 365 K leads to the recommendation of

$$k(\text{bromobenzene}) = 7.7 \times 10^{-13} \text{ cm}^3 \text{ molecule}^{-1} \text{ s}^{-1},$$

independent of temperature over the range 234–362 K, with an estimated overall uncertainty of $\pm 40\%$ over this temperature range.

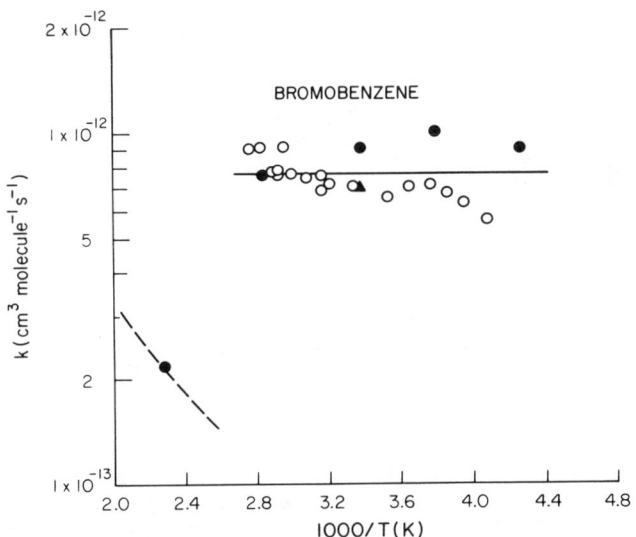

FIG. 112. Arrhenius plot of rate constants for the reaction of the OH radical with bromobenzene. (▲) Zetzsch;[36] (○) Witte *et al.*;[14] (●) Wallington *et al.*;[16] (— — —, ———) recommendations (see text).

Thermal decomposition of the OH-bromobenzene addition adduct has been observed to occur at temperatures $\geqslant 316$ K[14] (being rapid above 353 K[16]). Hence the rate constant of Wallington *et al.*[16] at 438 K is presumably that for H-atom abstraction from the C—H bonds of the aromatic ring. As for fluorobenzene and chlorobenzene, the magnitude of this rate constant is totally consistent with that derived from the recommendation for benzene, scaled by a factor of 0.83 (which is shown as the dashed line in Fig. 112).

(20) Iodobenzene

The rate constant data of Zetzsch[36] and Wallington *et al.*[16] are given in Table 19. At room temperature, the rate constant of Wallington *et al.*[16] is ~50% higher than that of Zetzsch.[36] A unit-weighted average of these rate con-

stants, together with the lack of a temperature dependence observed by Wallington *et al.*[16] for $T \leqslant 393$ K, leads to

$$k(\text{iodobenzene}) = 1.1 \times 10^{-12} \text{ cm}^3 \text{ molecule}^{-1} \text{ s}^{-1},$$

independent of temperature over the range 263–393 K. Above ~400 K, thermal decomposition of the OH-iodobenzene adduct occurs,[16] and H-atom abstraction is then expected to be the observed reaction pathway for temperatures $\gtrsim 400$–500 K.

(21) Benzyl chloride

The rate constants of Edney *et al.*[15] and Tuazon *et al.*,[39] both obtained at room temperature from relative rate studies, are in excellent agreement (Table 19). A unit-weighted average of these data[15,39] leads to the recommendation of

$$k(\text{benzyl chloride}) = 2.9 \times 10^{-12} \text{ cm}^3 \text{ molecule}^{-1} \text{ s}^{-1}$$

at 298 K, with an estimated overall uncertainty of $\pm 30\%$.

(22) Aniline

The available rate constant data of Rinke and Zetzsch,[11] Witte *et al.*[14] and Atkinson *et al.*[40] are given in Table 19 and are plotted in Arrhenius form in Fig. 113.

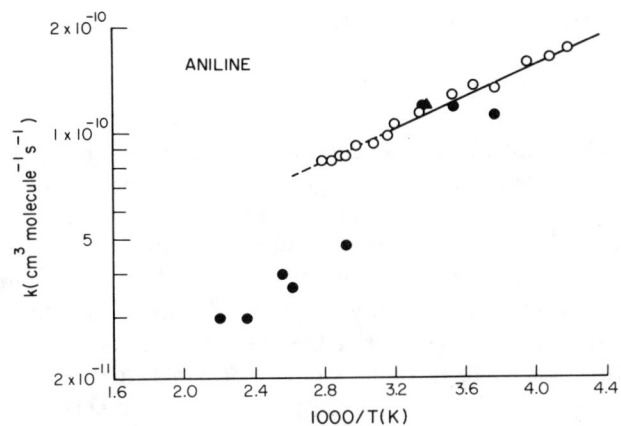

FIG. 113. Arrhenius plot of rate constants for the reaction of the OH radical with aniline. (▲) Rinke and Zetzsch;[11] (○) Witte and Zetzsch;[14] (●) Atkinson *et al.*;[40] (———) recommendation (see text).

Rinke and Zetzsch[11] observed that at room temperature the rate constant was independent of total pressure of helium diluent down to the lowest pressure studied (~15 Torr). Witte *et al.*[14] observed biexponential OH radical decays for temperatures $\geqslant 336$ K, showing that thermal decomposition of the OH-aniline addition adduct was occurring. This observation[14] is in reasonable agreement with the observation by Atkinson et al.[40] of non-exponential OH radical decays at temperatures $\geqslant 310$ K.

The rate constants measured by Atkinson et al.[40] at 265–298 K are in agreement, within the experimental errors, with the much more extensive data set of Witte et al.[14] (note that for the temperature regime where the OH-aniline adduct decomposes the OH radical addition rate constants were obtained[14] from the biexponential OH radical decay curves).

Accordingly, for the temperature region $T \leqslant 360$ K a unit-weighted least-squares analysis of the rate constants of Rinke and Zetzsch,[11] Witte et al.[14] and the 265, 283 and 298 K data points of Atkinson et al.[40] yields the recommended Arrhenius expression of

$$k(\text{aniline; } T \leqslant 360 \text{ K}) = (1.94^{+0.52}_{-0.41})$$

$$\times 10^{-11} e^{(519 \pm 69)/T} \text{ cm}^3 \text{ molecule}^{-1} \text{ s}^{-1}$$

over the temperature range 239–359 K (but see below), where the indicated errors are two least-squares standard deviations, and

$$k(\text{aniline}) = 1.11 \times 10^{-10} \text{ cm}^3 \text{ molecule}^{-1} \text{ s}^{-1}$$

at 298 K, with an estimated overall uncertainty at 298 K of $\pm 25\%$. It must be noted, however, that thermal decomposition of the OH-aniline addition adduct begins to occur at ~ 310–335 K,[14,40] and hence the above recommendation is only valid for the temperature range $\lesssim 320$ K unless thermal decomposition of the addition adduct is taken into account.

No recommendation is made concerning the OH radical reaction rate constants at temperatures $\gtrsim 380$ K.

(23) Nitrobenzene

The available kinetic data of Zetzsch,[14,36,37] Atkinson et al.[40] and Witte et al.[14] are given in Table 19. The upper limits to the room temperature rate constants measured by Atkinson et al.[40] are consistent with the data of Zetzsch[14,36,37] (as revised[14]) and Witte et al.[14] The sole temperature-dependence study is that of Witte et al.,[14] who observed non-exponential OH radical decays over the entire temperature range studied, possibly due to regeneration of OH radicals from photolysis fragments.[14]

(24) Hexafluorobenzene

The available rate constants of Ravishankara et al.[20] and Wallington et al.[16] are given in Table 19, and are plotted in Arrhenius form in Fig. 114. The room temperature rate constant of Ravishankara et al.[20] is $\sim 35\%$ higher than that of Wallington et al.[16] A unit-weighted least-squares analysis of the rate constants of Ravishankara et al.[20] and Wallington et al.[16] leads to the recommendation of

$$k(\text{hexafluorobenzene}) = (1.46^{+0.91}_{-0.56})$$

$$\times 10^{-12} e^{-(638 \pm 148)/T} \text{ cm}^3 \text{ molecule}^{-1} \text{ s}^{-1}$$

over the temperature range 234–438 K, where the indicated errors are two least-squares standard deviations, and

$$k(\text{hexafluorobenzene}) = 1.72$$

$$\times 10^{-13} \text{ cm}^3 \text{ molecule}^{-1} \text{ s}^{-1}$$

at 298 K, with an estimated overall uncertainty at 298 K of $\pm 40\%$.

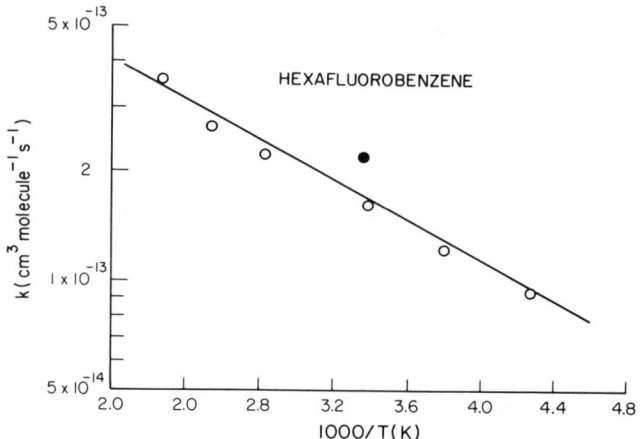

FIG. 114. Arrhenius plot of rate constants for the reaction of the OH radical with hexafluorobenzene. ($\bullet$) Ravishankara et al.;[20] ($\bigcirc$) Wallington et al.;[16] (——) recommendation (see text).

Wallington et al.[16] observed no evidence for thermal decomposition of the OH-hexafluorobenzene addition adduct over the temperature range (234–438 K) studied. Clearly, kinetic data are needed at elevated temperatures of $\gtrsim 500$ K.

(25) Biphenyl

The rate constants of Zetzsch,[36,37] Atkinson et al.,[43] Atkinson and Aschmann[44] and Klöpffer et al.,[22] all obtained at room temperature, are given in Table 19. These studies[22,36,37,43,44] are in generally good agreement, although the rate constant determined by Zetzsch[36,37] (which required knowledge of the vapor pressure of biphenyl) is lower by $\sim 25\%$ than the other values. Since no details are available concerning the study of Klöpffer et al.,[22] the rate constant from that study is not used in the evaluation. A unit-weighted average of the room temperature rate constants of Zetzsch,[36,37] Atkinson et al.[43] and Atkinson and Aschmann[44] leads to the recommendation of

$$k(\text{biphenyl}) = 7.2 \times 10^{-12} \text{ cm}^3 \text{ molecule}^{-1} \text{ s}^{-1} \text{ at } 298 \text{ K,}$$

with an estimated overall uncertainty of ±30%. This rate constant is that for OH radical addition to the aromatic rings. At temperatures $\gtrsim 400$ K the OH-biphenyl adduct will rapidly thermally decompose and only H-atom abstraction from the C—H bonds of the aromatic rings will be observed. Based upon the recommendation for benzene at temperatures $\gtrsim 450$ K, it is expected that the rate constant for H-atom abstraction from biphenyl will be

$$k_{\text{abstraction}} \approx 7.8 \times 10^{-18}\ T^2\ e^{-543/T}\ \text{cm}^3\ \text{molecule}^{-1}\ \text{s}^{-1}.$$

(26) Naphthalene

The available rate constants obtained at, or close to, the high-pressure limit are given in Table 19 and those of Lorenz and Zellner,[8,9] Atkinson et al.,[43] Biermann et al.[48] and Atkinson and Aschmann[49] are plotted in Arrhenius form in Fig. 115. Lorenz and Zellner[8] have shown that at 378 ± 2 K the rate constant for this reaction is in the fall-off region between second- and third-order kinetics below ~50 Torr total pressure of helium, but that no such fall-off behavior is observed at 525 ± 1 K.

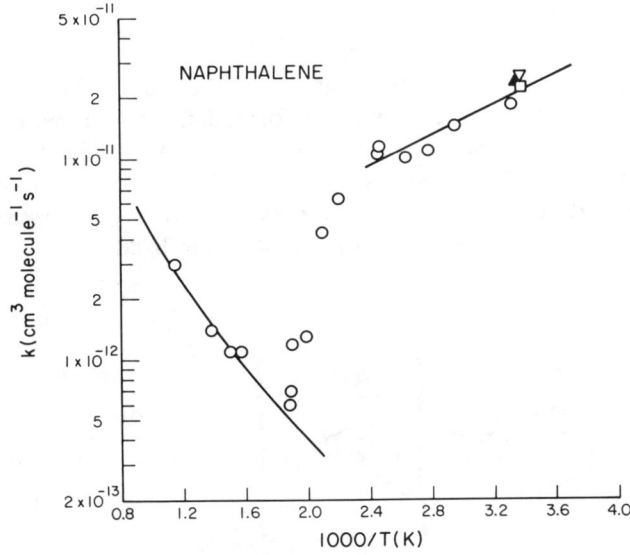

FIG. 115. Arrhenius plot of rate constants obtained at, or close to, the high-pressure limit for the reaction of the OH radical with naphthalene. (○) Lorenz and Zellner;[8,9] (□) Atkinson et al.;[43] (▲) Biermann et al.;[48] (▽) Atkinson and Aschmann;[49] (———) recommendations (see text).

At temperatures $\leqslant 410$ K the rate constants obtained by Lorenz and Zellner,[8,9] Atkinson and co-workers[43,48,49] and Klöpffer et al.[22] are in good agreement (the rate constant of Klöpffer et al.[22] is not used in the evaluation because of a lack of details). A unit-weighted least-squares analysis of the data of Lorenz and Zellner,[8,9]

Atkinson et al.,[43] Biermann et al.[48] and Atkinson and Aschmann[49] yields the recommended Arrhenius expression of

$$k(\text{naphthalene},\ T \leqslant 410\ \text{K}) = (1.07^{+1.14}_{-0.55})$$

$$\times\ 10^{-12}\ e^{(895\ \pm\ 239)/T}\ \text{cm}^3\ \text{molecule}^{-1}\ \text{s}^{-1}$$

over the temperature range 294–407 K, where the indicated errors are two least-squares standard deviations, and

$$k(\text{naphthalene}) = 2.16 \times 10^{-11}\ \text{cm}^3\ \text{molecule}^{-1}\ \text{s}^{-1}$$

at 298 K, with an estimated overall uncertainty of ±30% at 298 K.

At elevated temperatures, $\geqslant 600$ K for this particular aromatic hydrocarbon,[8,9] the only rate constants available are those of Lorenz and Zellner.[9] A unit-weighted least-squares fit of these data,[9] using the expression $k = CT^2 e^{-D/T}$, yields the recommendation of

$$k(\text{naphthalene};\ T \geqslant 600\ \text{K}) = (1.12^{+2.12}_{-0.73})$$

$$\times\ 10^{-17}\ T^2\ e^{-(969\ \pm\ 752)/T}\ \text{cm}^3\ \text{molecule}^{-1}\ \text{s}^{-1}$$

over the temperature range 636–873 K, where the indicated errors are two least-squares standard deviations.

(27) Phenanthrene

The available rate constants of Lorenz and Zellner[9] and Biermann et al.[48] are given in Table 19 and are plotted in Arrhenius form in Fig. 116.

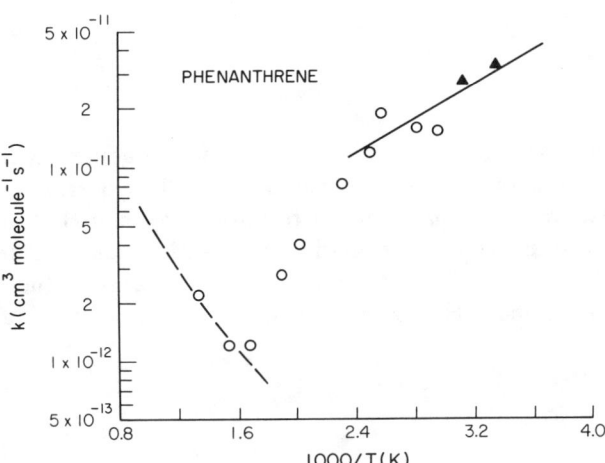

FIG. 116. Arrhenius plot of rate constants for the reaction of the OH radical with phenanthrene. (○) Lorenz and Zellner;[9] (▲) Biermann et al.;[48] (— — —, ———) recommendations (see text).

The rate constants obtained by Biermann *et al.*[48] from a relative rate study are consistent with the higher temperature ($T \geqslant 338$ K) data of Lorenz and Zellner.[9] For the temperature range $\leqslant 410$ K (the same as for naphthalene) a unit-weighted least-squares analysis of the data of Lorenz and Zellner[9] and Biermann *et al.*[48] yields the recommendation of

$$k(\text{phenanthrene}; T \leqslant 410 \text{ K}) = (1.02^{+5.41}_{-0.86})$$

$$\times 10^{-12} e^{(1021 \pm 634)/T} \text{ cm}^3 \text{ molecule}^{-1} \text{ s}^{-1}$$

over the temperature range 298–399 K, where the indicated errors are two least-squares standard deviations, and

$$k(\text{phenanthrene}) = 3.1 \times 10^{-11} \text{ cm}^3 \text{ molecule}^{-1} \text{ s}^{-1}$$

at 298 K, with an estimated overall uncertainty at 298 K of $\pm$ a factor of 2.

At elevated temperatures, where thermal decomposition of the OH-phenanthrene adduct is sufficiently rapid that only the H-atom abstraction process is observed, data are available only at 648 and 748 K.[9] Since there are ten aromatic ring C—H bonds on phenanthrene versus eight on naphthalene, it is expected that the H-atom abstraction rate constant for phenanthrene will be 1.25 that for naphthalene,

$$k(\text{phenanthrene}; T \geqslant 600 \text{ K}) = 1.40$$

$$\times 10^{-17} T^2 e^{-969/T} \text{ cm}^3 \text{ molecule}^{-1} \text{ s}^{-1}$$

and this expression, shown as the dashed line in Fig. 116, is in good agreement with the limited data available in this temperature regime.

b. Mechanism

The available data discussed above, together with product data,[42,56-73] show that in general two reaction pathways can occur: a direct reaction involving H-atom abstraction from the aromatic ring C—H bonds or from X—H bonds (X = C, O, N, S) of the substituent group(s), and OH radical addition to the aromatic ring.

For example, for toluene

Reaction leading to substituent group (or H-atom) elimination, for example

does not appear to be of any significance.[13,30] The hydroxycyclohexadienyl radical formed from OH radical addition to benzene has been observed and its reactions with O_2, NO and NO_2 studied.[66,70] The initially energy-rich OH radical addition adducts can either decompose back to reactants or be collisionally stabilized.

A further reaction step involves the unimolecular decomposition of this thermalized OH-aromatic adduct back to the reactants

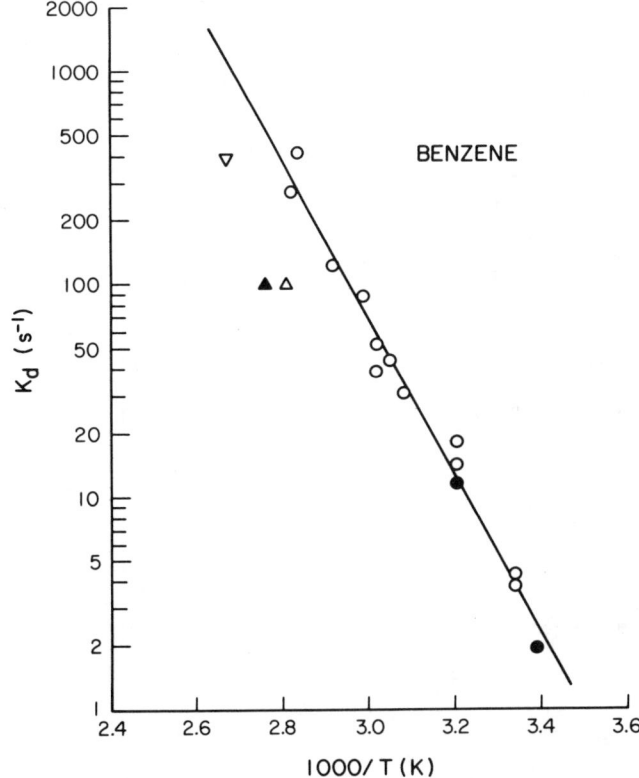

and this reaction pathway obviously becomes more rapid as the temperature increases.

Indeed, it is this thermal back-decomposition of the OH-aromatic adduct which gives rise to the non-exponential OH radical decays observed in the flash or laser photolysis kinetic studies and to the occurrence of distinct temperature regimes with differing kinetic behavior. The thermal decomposition rate constant of the OH-aromatic adduct, k_d, can be estimated from the temperature region over which non-exponential OH radical decays are observed[4,8,16,31] and, more accurately, from numerical analysis of the time-dependent behavior of the OH radical decays in this temperature region.[10,14,29] The available data (or estimates) for the Arrhenius parameters of these thermal decomposition rate constants are given in Table 20. (These data, and the recommended Arrhenius expression calculated for the hydroxycyclohexadienyl radical, are assumed to be at, or close to, the high-pressure limit, although it is likely that at temperatures $\gtrsim 350$ K this rate constant k_d will be significantly into the fall-off region at total pressures below approximately one atmosphere.)

For benzene, the available values of k_d are plotted in Arrhenius form in Fig. 117, and the recommended thermal decomposition rate of the hydroxycyclohexadienyl radical, obtained from a unit-weighted least-squares analysis of the data of Wahner and Zetzsch[10] and Witte et al.,[14] is given in Table 20. The thermal decomposition rates of the methyl-substituted benzenes and other monocyclic aromatics studied are reasonably similar at ~ 300–400 K. Thus, for benzene the thermalized hydroxycyclohexadienyl radical has a lifetime with respect to thermal decomposition of ~ 0.3 sec at 298 K, ~ 0.03 sec at 325 K, ~ 0.6 ms at 380 K and ~ 0.2 ms at 400 K, and these lifetimes are reasonably representative of those for the methyl-substituted benzenes and other monocyclic aromatics. These lifetimes are then totally consistent with the above discussion of the reaction dynamics of these OH radical reactions.

Thus, for most of the aromatic compounds studied, at around room temperature, i.e., $\leqslant 325$ K, OH radical addition to the aromatic ring dominates, while for temperatures $\gtrsim 450$ K ($\gtrsim 600$ K for the OH-naphthalene adduct[8,9]) back-dissociation of the OH-aromatic adducts becomes so rapid that on the time scale of the flash or laser photolysis studies carried out to date only the direct reaction involving H-atom abstraction is observed.

Apart from benzaldehyde (and presumably other aromatic aldehydes) and the aromatic alkenes such as styrene, where at room temperature the reactions proceed by H-atom abstraction from the —CHO group and by OH radical addition to the alkene $>C=C<$ bond, respectively,

$$OH + C_6H_5CHO \rightarrow H_2O + C_6H_5\dot{C}O$$

$$OH + C_6H_5CH=CH_2 \rightarrow C_6H_5\dot{C}HCH_2OH$$

$$\text{and } C_6H_5CHOH\dot{C}H_2$$

the major fraction of the OH radical reactions with the aromatic compounds studied to date proceeds by OH radical addition to the aromatic ring at around room temperature.

FIG. 117. Arrhenius plot of rate constants k_d for the thermal decomposition of the hydroxycyclohexadienyl radical formed from OH radical addition to benzene. ($\triangle$) Perry et al.;[4] (∇) Lorenz and Zellner;[8] ($\bullet$) Wahner and Zetzsch;[10] ($\bigcirc$) Witte and Zetzsch;[14] ($\blacktriangle$) Wallington et al.;[16] (———) recommendation (see text and Table 20).

The fractions of the overall reactions proceeding by H-atom abstraction from the C—H bonds of the aromatic ring or from X—H bonds of the substituent groups (X = C or O) can be estimated from the kinetic recommendations given above. Table 21 gives estimated rate constant ratios k_{abs}/k_{total} at 298 K for benzene, naphthalene and phenanthrene and the substituted benzenes for which estimates can be made, where k_{abs} and k_{total} are the rate constants for the H-atom abstraction reaction and the overall reaction, respectively. For toluene and the xylenes, these estimates, derived from extrapolations of the elevated temperature (generally $\geqslant 450$ K) rate constants, are in agreement, within the likely uncertainties, with recent product data.[42,59,60,62,64,65,67-69,72]

TABLE 20. Thermal decomposition rate constants, $k_d = A_d\,e^{-B_d/T}$, for OH aromatic addition adducts

Aromatic	A_d (s^{-1})	B_d (K)	Reference
Benzene	$3 \times 10^{13\,a}$	9410 ± 1000	Perry et al.[4]
	$4 \times 10^{13\,a}$	9500 ± 720	Lorenz and Zellner[8]
	$3 \times 10^{13\,a}$	8960 ± 690	Wahner and Zetzsch[10]
	3×10^{12}	8180 ± 720	Witte et al.[14]
	$3 \times 10^{13\,a}$	9560	Wallington et al.[16]
	9.4×10^{12}	8540 ± 750	Recommended[b]
Toluene	$3 \times 10^{13\,a}$	9110 ± 1000	Perry et al.[4]
o-Xylene	$3 \times 10^{13\,a}$	9260 ± 1000	Perry et al.[4]
m-Xylene	$3 \times 10^{13\,a}$	9010 ± 1000	Perry et al.[4]
p-Xylene	$3 \times 10^{13\,a}$	9410 ± 1000	Perry et al.[4]
1,2,3-Trimethylbenzene	$3 \times 10^{13\,a}$	9360 ± 1000	Perry et al.[4]
1,2,4-Trimethylbenzene	$3 \times 10^{13\,a}$	9210 ± 1000	Perry et al.[4]
1,3,5-Trimethylbenzene	$3 \times 10^{13\,a}$	9110 ± 1000	Perry et al.[4]
Phenol	1×10^{13}	8900 ± 1300	Witte and Zetzsch[29]
Methoxybenzene	$3 \times 10^{13\,a}$	9110 ± 1000	Perry et al.[31]
o-Cresol	$3 \times 10^{13\,a}$	9610 ± 1000	Perry et al.[31]
Fluorobenzene	$3 \times 10^{13\,a}$	9560	Wallington et al.[16]
Chlorobenzene	$3 \times 10^{13\,a}$	10070	Wallington et al.[16]
Bromobenzene	2×10^{10}	6740 ± 600	Witte et al.[14]
	$3 \times 10^{13\,a}$	10570	Wallington et al.[16]
Iodobenzene	$3 \times 10^{13\,a}$	11580	Wallington et al.[16]
Aniline	6×10^{11}	8420 ± 1080	Witte et al.[14]
Naphthalene	$4 \times 10^{13\,a}$	11430 ± 720	Lorenz and Zellner[8]

[a]Value of A_d assumed or estimated.
[b]Calculated from unit-weighted least-squares analysis of the thermal decomposition rate constants reported by Wahner and Zetzsch[10] and Witte et al.[14]

TABLE 21. Rate constant ratios k_{abs}/k_{total} at 298 K for the gas-phase reactions of the OH radical with a series of aromatic compounds

Aromatic	k_{abs}/k_{total} at 298 K[a]
Benzene	0.05
Benzene-d_6	0.02
Toluene	0.12
Toluene-d_8	0.04
o-Xylene	0.10
m-Xylene	0.04
p-Xylene	0.08
1,2,3-Trimethylbenzene	0.06
1,2,4-Trimethylbenzene	0.06
1,3,5-Trimethylbenzene	0.03
Phenol	~0.09
Methoxybenzene	~0.14
o-Cresol	~0.07
Naphthalene	0.0018
Phenanthrene	0.0015

[a]From extrapolation of the elevated temperature rate constant data to 298 K, using the recommendations (see text) for the rate constants k_{abs} and k_{total}. These extrapolated values are expected to be subject to uncertainties of the order of $\pm 50\%$.

References

[1]D. D. Davis, W. Bollinger, and S. Fischer, J. Phys. Chem. **79**, 293 (1975).

[2]G. J. Doyle, A. C. Lloyd, K. R. Darnall, A. M. Winer, and J. N. Pitts, Jr., Environ. Sci. Technol. **9**, 237 (1975).

[3]D. A. Hansen, R. Atkinson, and J. N. Pitts, Jr., J. Phys. Chem. **79**, 1763 (1975).

[4]R. A. Perry, R. Atkinson, and J. N. Pitts, Jr., J. Phys. Chem. **81**, 296 (1977).

[5]R. A. Cox, R. G. Derwent, and M. R. Williams, Environ. Sci. Technol. **14**, 57 (1980).

[6]F. P. Tully, A. R. Ravishankara, R. L. Thompson, J. M. Nicovich, R. C. Shah, N. M. Kreutter, and P. H. Wine, J. Phys. Chem. **85**, 2262 (1981).

[7]I. Barnes, V. Bastian, K. H. Becker, E. H. Fink, and F. Zabel, Atmos. Environ. **16**, 545 (1982).

[8]K. Lorenz and R. Zellner, Ber. Bunsenges Phys. Chem. **87**, 629 (1983).

[9]K. Lorenz and R. Zellner, 8th International Symposium on Gas Kinetics, University of Nottingham, Nottingham, UK, July 15–20, 1984; private communication 1985.

[10]A. Wahner and C. Zetzsch, J. Phys. Chem. **87**, 4945 (1983).

[11]M. Rinke and C. Zetzsch, Ber. Bunsenges Phys. Chem. **88**, 55 (1984).

[12]T. Ohta and T. Ohyama, Bull. Chem. Soc. Jpn. **58**, 3029 (1985).

[13]S. Madronich and W. Felder, J. Phys. Chem. **89**, 3556 (1985).

[14]F. Witte, E. Urbanik, and C. Zetzsch, J. Phys. Chem. **90**, 3251 (1986).

[15]E. O. Edney, T. E. Kleindienst, and E. W. Corse, Int. J. Chem. Kinet. **18**, 1355 (1986).

[16]T. J. Wallington, D. M. Neuman, and M. J. Kurylo, Int. J. Chem. Kinet. **19**, 725 (1987).

[17]R. R. Baldwin, M. Scott, and R. W. Walker, 21st International Symposium on Combustion, 1986; The Combustion Institute, Pittsburgh, PA, 1988; p. 991.

[18]R. Atkinson and S. M. Aschmann, Int. J. Chem. Kinet. **21**, 355 (1989).

[19]A. C. Lloyd, K. R. Darnall, A. M. Winer, and J. N. Pitts, Jr., J. Phys. Chem. **80**, 789 (1976).

[20]A. R. Ravishankara, S. Wagner, S. Fischer, G. Smith, R. Schiff, R. T. Watson, G. Tesi, and D. D. Davis, Int. J. Chem. Kinet. **10**, 783 (1978).

[21]J. M. Nicovich, R. L. Thompson, and A. R. Ravishankara, J. Phys. Chem. **85**, 2913 (1981).

[22]W. Klöpffer, R. Frank, E.-G. Kohl, and F. Haag, Chemiker-Zeitung, **110**, 57 (1986); "Methods of the Ecotoxicological Evaluation of Chemicals, Photochemical Degradation in the Gas Phase," Vol. 6, *OH Reaction Rate Constants and Tropospheric Lifetimes of Selected Environmental Chemicals*. Report 1980–1983, K. H. Becker, H. M. Biehl, P. Bruckmann, E. H. Fink, F. Führ, W. Klöpffer, R. Zellner, and C. Zetzsch, Editors, Kernforschungsanlage Jülich GmbH, November 1984.

[23]R. Atkinson, S. M. Aschmann, and W. P. L. Carter, Int. J. Chem. Kinet. **15**, 37 (1983).

[24]K. H. Becker and Th. Klein, Proceedings, 4th European Symposium on the Physico-Chemical Behavior of Atmospheric Pollutants; D. Riedel Publishing Co., Dordrecht, Holland, 1987, p. 320.

[25]E. D. Morris, Jr. and H. Niki, J. Phys. Chem. **75**, 3640 (1971).

[26]C. A. Bignozzi, A. Maldotti, C. Chiorboli, C. Bartocci, and V. Carassiti, Int. J. Chem. Kinet. **13**, 1235 (1981).

[27]R. Atkinson and S. M. Aschmann, Int. J. Chem. Kinet. **20**, 513 (1988).

[28]C. Chiorboli, C. A. Bignozzi, A. Maldotti, P. F. Giardini, A. Rossi, and V. Carassiti, Int. J. Chem. Kinet. **15**, 579 (1983).

[29]F. Witte and C. Zetzsch, private communication, 1988.

[30]Y. Z. He, W. G. Mallard, and W. Tsang, J. Phys. Chem. **92**, 2196 (1988).

[31]R. A. Perry, R. Atkinson, and J. N. Pitts, Jr., J. Phys. Chem. **81**, 1607 (1977).

[32]R. Atkinson, K. R. Darnall, and J. N. Pitts, Jr., J. Phys. Chem. **82**, 2759 (1978).

[33]R. Atkinson and S. M. Aschmann, Int. J. Chem. Kinet., in press (1989).

[34]I. Barnes, V. Bastian, K. H. Becker, E. H. Fink, and W. Nelsen, J. Atmos. Chem. **4**, 445 (1986).

[35]F. Nolting, F. Witte, and C. Zetzsch, report to the Umweltbundesamt, December 1987; private communication, 1988.

[36]C. Zetzsch, report to the Bundesminister für Forschung und Technologie, Projektträger für Umweltchemikalein, 1982; private communication, 1985.

[37]C. Zetzsch, 15th Informal Conference on Photochemistry, Stanford University, Stanford, CA, June 27–July 1, 1982.

[38]R. Atkinson, S. M. Aschmann, A. M. Winer, and J. N. Pitts, Jr., Arch. Environ. Contamin. Toxicol. **14**, 417 (1985).

[39]E. C. Tuazon, R. Atkinson, and S. M. Aschmann, Int. J. Chem. Kinet., to be submitted for publication (1989).

[40]R. Atkinson, E. C. Tuazon, T. J. Wallington, S. M. Aschmann, J. Arey, A. M. Winer, and J. N. Pitts, Jr., Environ. Sci. Technol. **21**, 64 (1987).

[41]K. H. Becker, V. Bastian, and Th. Klein, J. Photochem. Photobiol., A: Chemistry **45**, 195 (1988).

[42]R. Atkinson, S. M. Aschmann, J. Arey, and W. P. L. Carter, Int. J. Chem. Kinet. **21**, 801 (1989).

[43]R. Atkinson, S. M. Aschmann, and J. N. Pitts, Jr., Environ. Sci. Technol. **18**, 110 (1984).

[44]R. Atkinson and S. M. Aschmann, Environ. Sci. Technol. **19**, 462 (1985).

[45]R. Atkinson, S. M. Aschmann, J. Arey, B. Zielinska, and D. Schuetzle, Atmos. Environ., in press (1989).

[46]D. L. Baulch, I. M. Campbell, and S. M. Saunders, 9th International Symposium on Gas Kinetics, University of Bordeaux, Bordeaux, France, July 20–25, 1986.

[47]R. Atkinson, J. Arey, E. C. Tuazon, and S. M. Aschmann, Int. J. Chem. Kinet., to be submitted for publication (1989).

[48]H. W. Biermann, H. Mac Leod, R. Atkinson, A. M. Winer, and J. N. Pitts, Jr., Environ. Sci. Technol. **19**, 244 (1985).

[49]R. Atkinson and S. M. Aschmann, Int. J. Chem. Kinet. **18**, 569 (1986).

[50]R. Atkinson and S. M. Aschmann, Atmos. Environ. **21**, 2323 (1987).

[51]J. Arey, R. Atkinson, S. M. Aschmann, and D. Schuetzle, Polycyclic Arom. Compounds, in press (1989).

[52]D. L. Baulch, I. M. Campbell, and S. M. Saunders, J. Chem. Soc. Faraday Trans. 2, **84**, 377 (1988).

[53]N. Bourmada, P. Devolder, J.-F. Pauwels, and J.-P. Sawerysyn, 10th International Symposium on Gas Kinetics, University College of Swansea, Swansea, UK, July 24–29, 1988.

[54]W. Felder and S. Madronich, Combust. Sci. Technol. **50**, 135 (1986).

[55]N. Bourmada, P. Devolder, and L.-R. Sochet, Chem. Phys. Lett. **149**, 339 (1988); N. Bourmada, M. Carlier, J.-F. Pauwels, and P. Devolder, J. Chim. Phys. **85**, 881 (1988).

[56]R. Atkinson, Chem. Rev. **86**, 69 (1986).

[57]M. Hoshino, H. Akimoto, and M. Okuda, Bull. Chem. Soc. Jpn. **51**, 718 (1978).

[58]K. R. Darnall, R. Atkinson, and J. N. Pitts, Jr., J. Phys. Chem. **83**, 1943 (1979).

[59]H. Takagi, N. Washida, H. Akimoto, K. Nagasawa, Y. Usui, and M. Okuda, J. Phys. Chem. **84**, 478 (1980).

[60]R. Atkinson, W. P. L. Carter, K. R. Darnall, A. M. Winer, and J. N. Pitts, Jr., Int. J. Chem. Kinet. **12**, 779 (1980).

[61]R. A. Kenley, J. E. Davenport, and D. G. Hendry, J. Phys. Chem. **85**, 2740 (1981).

[62]R. Atkinson, W. P. L. Carter, and A. M. Winer, J. Phys. Chem. **87**, 1605 (1983).

[63]B. E. Dumdei and R. J. O'Brien, Nature **311**, 248 (1984).

[64]P. B. Shepson, E. O. Edney, and E. W. Corse, J. Phys. Chem. **88**, 4122 (1984).

[65]J. A. Leone, R. C. Flagan, D. Grosjean, and J. H. Seinfeld, Int. J. Chem. Kinet. **17**, 177 (1985).

[66]B. Fritz, V. Handwerk, M. Preidel, and R. Zellner, Ber. Bunsenges Phys. Chem. **89**, 343 (1985).

[67]M. W. Gery, D. L. Fox, H. E. Jeffries, L. Stockburger, and W. S. Weathers, Int. J. Chem. Kinet. **17**, 931 (1985).

[68]H. Bandow, N. Washida, and H. Akimoto, Bull. Chem. Soc. Jpn. **58**, 2531 (1985).

[69]H. Bandow and N. Washida, Bull. Chem. Soc. Jpn. **58**, 2541 (1985).

[70]R. Zellner, B. Fritz, and M. Preidel, Chem. Phys. Lett. **121**, 412 (1985).

[71]E. C. Tuazon, H. Mac Leod, R. Atkinson, and W. P. L. Carter, Environ. Sci. Technol. **20**, 383 (1986).

[72]M. W. Gery, D. L. Fox, R. M. Kamens, and L. Stockburger, Environ. Sci. Technol. **21**, 339 (1987).

[73]R. Atkinson, J. Arey, B. Zielinska, and S. M. Aschmann, Environ. Sci. Technol. **21**, 1014 (1987).

2.12. Organometallic Compounds

a. Kinetics

The available rate constant data are listed in Table 22. Only three organometallic compounds have been studied to date, and for tetramethyl- and tetraethyl lead two kinetic studies have been carried out at room temperature by Harrison and Laxen[2] and Nielsen et al.[3] However, the rate constants reported for tetraethyl lead[2,3] disagree by a factor of ~7. Thus, although the two rate constants for tetramethyl lead[2,3] (obtained from the same studies as those for tetraethyl lead[2,3]) are in reasonable agreement, no recommendations are made.

b. Mechanisms

The sole product study carried out concerning the reactions of OH radicals with organometallic compounds under atmospheric conditions is that of Niki et al.[1] for CH_3HgCH_3. It was concluded from this FT-IR absorption spectroscopy study[1] that the initial reaction proceeds via a displacement process,

$$OH + CH_3HgCH_3 \rightarrow CH_3HgOH + CH_3$$

followed by subsequent oxidation of the CH_3 radical to formaldehyde and other minor products, and by further homogeneous and/or heterogeneous reactions of CH_3HgOH to yield compounds such as $[(CH_3Hg)_3O]NO_3$.[1] The occurrence of such a displacement reaction is consistent with the magnitude of the rate constant observed.[1]

For the tetraalkyl lead compounds studied, neither the initial reaction pathways nor the products under atmospheric conditions are known, although again displacement mechanisms leading to the initial formation of $(CH_3)_3PbOH$ and $(C_2H_5)_3PbOH$ are possible.

TABLE 22. Rate constants k for the gas-phase reactions of the OH radical with organometallic compounds

Organometallic	$10^{12} \times k$ (cm³ molecule^{-1} s^{-1})	at T (K)	Technique	Reference
Organomercury Compounds				
Dimethyl mercury	19.7 ± 1.6	~ 300	RR [relative to k(ethene) $= 8.44 \times 10^{-12}$]ª	Niki *et al.*[1]
	18.4 ± 1.5	~ 300	RR [relative to k(propene) $= 2.60 \times 10^{-11}$]ª	Niki *et al.*[1]
Organolead Compounds				
Tetramethyl lead	9.0	295 ± 3	RR [relative to k(toluene) $= 6.03 \times 10^{-12}$]ª	Harrison and Laxen[2]
	6.3 ± 1.4	296	PR-RA	Nielsen *et al.*[3]
Tetraethyl lead	80	295 ± 3	RR [relative to k(*m*-xylene) $= 2.36 \times 10^{-11}$]ª	Harrison and Laxen[2]
	11.6 ± 1.7	296	PR-RA	Nielsen *et al.*[3]

ªFrom the present recommendations (see text).

References

[1] H. Niki, P. D. Maker, C. M. Savage, and L. P. Breitenbach, J. Phys. Chem. **87**, 4978 (1983).

[2] R. M. Harrison and D. P. H. Laxen, Environ. Sci. Technol. **12**, 1384 (1978).

[3] O. J. Nielsen, T. Nielsen, and P. Pagsberg, "Direct Spectrokinetic Investigation of the Reactivity of OH with Tetralkyllead Compounds in Gas Phase. Estimate of Lifetimes of Tetraalkyllead Compounds in Ambient Air." Report Risø-R-463; Risø National Laboratory, Roskilde, Denmark, May 1982.

2.13. Organic Radicals

The available kinetic data are given in Table 23. As may be expected, few direct studies have been carried out due to the difficulties of investigating radical-radical reactions.

a. CH₃

In addition to the rate constants given in Table 23, Roth and Just[8] have derived, from computer modeling of CH₄—O₂ and CH₃—O₂ systems, a rate constant of

$$k(CH_3) = 3.3 \times 10^{-9} e^{-9580/T} \text{ cm}^3 \text{ molecule}^{-1} \text{ s}^{-1}$$

for the reaction

$$OH + CH_3 \rightarrow CH_3O + H$$

over the temperature range 1800–2300 K, with

$$k(CH_3) = (1.3\text{–}1.8)$$

$$\times 10^{-11} \text{ cm}^3 \text{ molecule}^{-1} \text{ s}^{-1} \text{ at } 1800 \text{ K}$$

and

$$k(CH_3) = (4.2\text{–}5.8)$$

$$\times 10^{-11} \text{ cm}^3 \text{ molecule}^{-1} \text{ s}^{-1} \text{ at } 2300 \text{ K}.$$

At around room temperature, the reaction of OH radicals with CH₃ radicals probably proceeds mainly by addition.[3] However, at elevated temperatures the rate constant for the addition reaction will be far into the fall-off regime and only direct reaction pathways, such as

$$OH + CH_3 \rightarrow CH_3O + H \qquad (a)$$

$$\rightarrow HCHO + H_2 \qquad (b)$$

$$\rightarrow H_2O + CH_2 \qquad (c)$$

will be observed. Roth and Just[8] conclude from their computer modeling study that reaction (c) cannot be the sole reaction pathway occurring at ~ 1800–2300 K.

b. HCO

The only reasonably direct measurement is that of Temps and Wagner.[4] Seery[9] has inferred a rate constant for this reaction of $(8 \pm 8) \times 10^{-11}$ cm³ molecule^{-1} s^{-1}

at 1700–2000 K from induction time measurements. Clearly, this reaction is rapid and proceeds by

$$OH + HCO \rightarrow H_2O + CO$$

The meager data available[4,9] can be fitted by the expressions

$$k(HCO) = 7 \times 10^{-11}\, e^{350/T}\, cm^3\, molecule^{-1}\, s^{-1}$$

or

$$k(HCO) = 2.2 \times 10^{-10}\, (T/298)^{-0.5}\, cm^3\, molecule^{-1}\, s^{-1}$$

c. $C_2(X^3\Pi_u)$

The data of Bulewicz et al.[5] refer to the overall rate constant for the reactions

$$OH + C_2(X^3\Pi_u) \rightarrow CH(A^2\Delta) + CO(X^1\Sigma^+) \qquad (a)$$

and

$$OH + C_2(X^3\Pi_u) \rightarrow CH(^2\Sigma) + CO(X^1\Sigma^+) \qquad (b)$$

From earlier flame studies, Porter et al.[10] estimated that $k_b/k_a \sim 0.1$ over the temperature range $\sim$900–1700 K, and hence it appears that reaction (a) dominates.

d. CN

The rate constants derived by Morley[6] and Haynes[7] for the reaction

$$OH + CN \rightarrow H + NCO$$

are in good agreement.

TABLE 23. Rate constants k for the gas-phase reactions of the OH radical with organic radicals

Organic Radical	$10^{12} \times k$ (cm^3 molecule^{-1} s^{-1})	at T (K)	Technique	Reference
CH$_3$	4.3 ± 0.5	1970–2185	Flame-MS	Fenimore[1]
	10	1800–1958	Flame-MS	Jones and Fenimore[2]
	93 ± 25^a	296	FP-RA; Computer modeling	Sworski et al.[3]
HCO	220 ± 80	296	RR [relative to $k(HCHO) = 9.78 \times 10^{-12}]^b$	Temps and Wagner[4]
$C_2(X^3\Pi_u)$	8 ± 4	2200	Flame-optical absorption	Bulewicz et al.[5]
CN	100	2300–2560	Flame-MS	Morley[6]
	93 ± 12	1950–2380	Flame-product study	Haynes[7]

^{a}At atmospheric pressure.
^{b}From the present recommendation (see text).

References

[1]C. P. Fenimore, 12th International Symposium on Combustion, 1968; The Combustion Institute, Pittsburgh, PA, 1969, p. 463.

[2]G. W. Jones and C. P. Fenimore, unpublished data, cited in reference 1.

[3]T. J. Sworski, C. J. Hochanadel, and P. J. Ogren, J. Phys. Chem. **84**, 129 (1980).

[4]F. Temps and H. Gg. Wagner, Ber. Bunsenges Phys. Chem. **88**, 415 (1984).

[5]E. M. Bulewicz, P. J. Padley, and R. E. Smith, Proc. Roy. Soc. (London) **A315**, 129 (1970).

[6]C. Morley, Combust. Flame **27**, 189 (1976).

[7]B. S. Haynes, Combust. Flame **28**, 113 (1977).

[8]P. Roth and Th. Just, 20th International Symposium on Combustion, 1984; The Combustion Institute, Pittsburgh, PA, 1985; p. 807.

[9]D. J. Seery, 12th International Symposium on Combustion, 1968; The Combustion Institute, Pittsburgh, PA, 1969, p. 588.

[10]R. P. Porter, A. H. Clark, W. E. Kaskan, and W. E. Browne, 11th International Symposium on Combustion, 1966; The Combustion Institute, Pittsburgh, PA, 1967; p. 907.

2.14. Addendum

Two 1988 publications[1,2] which included rate constant data for the gas-phase reactions of the OH radical with oxygen-containing organic compounds were inadvertently overlooked and the data omitted from Sec. 2.6. Both studies used absolute techniques and the rate constants obtained are given below.

Organic	$10^{12} \times k$ (cm^3 molecule^{-1} s^{-1})	at T (K)	Technique	Reference
Acetone-d_6	0.0358 ± 0.0029	298	FP-RF	Wallington et al.[1]
1,1,1-Trifluoroacetone	0.0151 ± 0.0013	298	FP-RF	Wallington et al.[1]
Methanol	0.88 ± 0.18	298	PR-RA	Pagsberg et al.[2]
Methanol-d_4	0.323 ± 0.002	298	FP-RF	Wallington et al.[1]
Ethanol-d_6	1.15 ± 0.09	298	FP-RF	Wallington et al.[1]
2-Chloroethanol	1.28 ± 0.09	298	FP-RF	Wallington et al.[1]
2,2,2-Trichloroethanol	0.245 ± 0.024	298	FP-RF	Wallington et al.[1]
2,2,2-Trifluoroethanol	0.0955 ± 0.0071	298	FP-RF	Wallington et al.[1]
1,2-Epoxy-butane	1.91 ± 0.08	298	FP-RF	Wallington et al.[1]

The rate constant of Pagsberg et al.[2] for methanol is in agreement with the recommendation (Sec. 2.6) and those of Wallington et al.[1] for 2-chloroethanol and 1,2-epoxybutane are in agreement with the rate constants given in Table 11. However, the rate constant determined by Wallington et al.[1] for methanol-d_4 is 60% higher than that of McCaulley et al.[3] (Table 11).

References

[1]T. J. Wallington, P. Dagaut, and M. J. Kurylo, J. Phys. Chem. **92**, 5024 (1988).

[2]P. Pagsberg, J. Munk, A. Sillesen, and C. Anastasi, Chem. Phys. Lett. **146**, 375 (1988).

[3]J. A. McCaulley, N. Kelly, M. F. Golde, and F. Kaufman, J. Phys. Chem. **93**, 1014 (1989).

3. Conclusions

The available (through 1988) kinetic and mechanistic data for the gas-phase reactions of the OH radical with organic compounds have been compiled and evaluated in the above sections. For a large number of compounds, temperature dependent rate expressions have been recommended, often over large temperature ranges which extend from room temperature or below to around 1000 K. However, there is still a paucity of reliable kinetic data at the elevated temperatures characteristic of combustion conditions. Just as important, there is a serious lack of knowledge concerning the reaction mechanisms and products formed under combustion conditions, and it must be recognized by combustion chemists and modelers that for many organic compounds the reaction mechanisms and products observed at around room temperature are not applicable at temperatures >600 K. This is clearly true for those reactions which proceed by an OH radical addition process at "low" (<300 K) temperatures, since fall-off effects and thermal decomposition of the addition adducts result in the addition reaction pathways being of generally negligible importance at temperatures >1000 K and only direct reactions, often involving H-atom abstraction, are operable. Obvious examples are the OH radical reactions with the alkenes, alkynes and aromatic hydrocarbons, which proceed by OH radical addition at room temperature and atmospheric pressure, but by H-atom abstraction under combustion conditions.

It is also clear that experimental data are only available for a small number of the organic compounds encountered as a result of biogenic and anthropogenic activities. Although not dealt with here in any detailed manner, estimation procedures are available for the calculation of rate constants for the reactions of the OH radical with organic compounds of low-to-moderate complexity,[1-4] and these references can be consulted for details. Hopefully, future experimental studies will continue to expand the present kinetic and mechanistic data base to more complex organics and to both higher and lower temperatures.

References

[1]R. Atkinson, Chem. Rev. **86**, 69 (1986).
[2]R. Atkinson, Int. J. Chem. Kinet. **18**, 555 (1986).
[3]R. Atkinson, Int. J. Chem. Kinet. **19**, 799 (1987).
[4]R. Atkinson, Environ. Toxicol. Chem. **7**, 435 (1988).

4. Acknowledgments

I especially thank Ms. Christy J. LaClaire for her excellent work in the preparation of this manuscript, and gratefully acknowledge the financial support of the Office of Standard Reference Data, National Institute of Standards and Technology through Award No. 60NANB7D0747. I thank Drs. Phillipe Dagaut, Charles D. Jonah, J. Alistair Kerr, Walter Klöpffer, Michael J. Kurylo, M. C. Lin, Ole John Nielsen, Gregory P. Smith, Frank P. Tully, Timothy J. Wallington and Cornelius Zetzsch and their co-workers for communicating their data prior to publication and for helpful discussions.

Journal of Physical and Chemical Reference Data
Cumulative Listing of Reprints and Supplements

Reprints from Volume 1

1. Gaseous Diffusion Coefficients, *T.R. Marrero and E.A. Mason,* Vol. 1, No. 1, pp. 1–118 (1972) $7.00

2. Selected Values of Critical Supersaturation for Nucleation of Liquids from the Vapor, *G.M. Pound,* Vol. 1, No. 1, pp. 119–134 (1972) $3.00

3. Selected Values of Evaporation and Condensation Coefficients for Simple Substances, *G.M. Pound,* Vol. 1, No. 1, pp. 135–146 (1972) $3.00

4. Atlas of the Observed Absorption Spectrum of Carbon Monoxide between 1060 and 1900 Å, *S.G. Tilford and J.D. Simmons,* Vol. 1, No. 1, pp. 147–188 (1972) $4.50

5. Tables of Molecular Vibrational Frequencies, Part 5, *T. Shimanouchi,* Vol. 1, No. 1, pp. 189–216 (1972) (superseded by No.103) $4.00

6. Selected Values of Heats of Combustion and Heats of Formation of Organic Compounds Containing the Elements C, H, N, O, P, and S, *Eugene S. Domalski,* Vol. 1, No. 2, pp. 221–278 (1972) $5.00

7. Thermal Conductivity of the Elements, *C.Y. Ho, R.W. Powell, and P.E. Liley,* Vol. 1, No. 2, pp. 279–422 (1972) $7.50

8. The Spectrum of Molecular Oxygen, *Paul H. Krupenie,* Vol. 1, No. 2, pp. 423–534 (1972) $6.50

9. A Critical Review of the Gas-Phase Reaction Kinetics of the Hydroxyl Radical, *Wm. E. Wilson, Jr.,* Vol. 1, No. 2, pp. 535–574 (1972) $4.50

10. Molten Salts: Volume 3, Nitrates, Nitrites, and Mixtures, Electrical Conductance, Density, Viscosity, and Surface Tension Data, *G.J. Janz, Ursula Krebs, H.F. Siegenthaler, and R.P.T. Tomkins,* Vol. 1, No. 3, pp. 581–746 (1972) $8.50

11. High Temperature Properties and Decomposition of Inorganic Salts—Part 3. Nitrates and Nitrites, *Kurt H. Stern,* Vol. 1, No. 3, pp. 747–772 (1972) $4.00

12. High-Pressure Calibration: A Critical Review, *D.L. Decker, W.A. Bassett, L. Merrill, H.T. Hall, and J.D. Barnett,* Vol. 1, No. 3, pp. 773–836 (1972) $5.00

13. The Surface Tension of Pure Liquid Compounds, *Joseph J. Jasper,* Vol. 1, No. 4, pp. 841–1009 (1972) $8.50

14. Microwave Spectra of Molecules of Astrophysical Interest, I. Formaldehyde, Formamide, and Thioformaldehyde, *Donald R. Johnson, Frank J. Lovas, and William H. Kirchhoff,* Vol. 1, No. 4, pp. 1011–1046 (1972) $4.50

15. Osmotic Coefficients and Mean Activity Coefficients of Uni-univalent Electrolytes in Water at 25° C, *Walter J. Hamer and Yung-Chi Wu,* Vol. 1, No. 4, pp. 1047–1099 (1972) $5.00

16. The Viscosity and Thermal Conductivity Coefficients of Gaseous and Liquid Fluorine, *H.J.M. Hanley and R. Prydz,* Vol. 1, No. 4, pp. 1101–1113 (1972) $3.00

Reprints from Volume 2

17. Microwave Spectra of Molecules of Astrophysical Interest, II. Methylenimine, *William H. Kirchhoff, Donald R. Johnson, and Frank J. Lovas,* Vol. 2, No. 1, pp. 1–10 (1973) $3.00

18. Analysis of Specific Heat Data in the Critical Region of Magnetic Solids, *F.J. Cook,* Vol. 2, No. 1, pp. 11–24 (1973) $3.00

19. Evaluated Chemical Kinetic Rate Constants for Various Gas Phase Reactions, *Keith Schofield,* Vol. 2, No. 1, pp. 25–84 (1973) $5.00

20. Atomic Transition Probabilities for Forbidden Lines of the Iron Group Elements. (A Critical Data Compilation for Selected Lines), *M.W. Smith and W.L. Wiese,* Vol. 2, No. 1, pp. 85–120 (1973) $4.50

21. Tables of Molecular Vibrational Frequencies, Part 6, *T. Shimanouchi,* Vol. 2, No. 1, pp. 121–162 (1973) (superseded by No. 103) $4.50

22. Compilation of Energy Band Gaps in Elemental and Binary Compound Semiconductors and Insulators, *W.H. Strehlow and E.L. Cook,* Vol. 2, No. 1, pp. 163–200 (1973) $4.50

23. Microwave Spectra of Molecules of Astrophysical Interest, III. Methanol, *R.M. Lees, F.J. Lovas, W.H. Kirchhoff, and D.R. Johnson,* Vol. 2, No. 2, pp. 205–214 (1973) $3.00

24. Microwave Spectra of Molecules of Astrophysical Interest, IV. Hydrogen Sulfide, *Paul Helminger, Frank C. De Lucia, and William H. Kirchhoff,* Vol. 2, No. 2, pp. 215–224 (1973) $3.00

- -

**Journal of Physical and Chemical Reference Data
Reprint and Supplement Orders**

To: American Chemical Society
    Distribution Office
    1155 Sixteenth Street, N.W.
    Washington, DC 20036

Name:_____

Title: _____

Organization:_____

Address:_____

City:_____ State: _____

Country: _____ Zip: _____

I am a member of _____
       (ACS, AIP, or Affiliated Society)

ORDERS FOR REPRINTS AND SUPPLEMENTS MUST BE PREPAID.
*Foreign orders for Reprints, add $2.50 for each reprint for postage and handling. Foreign orders for Reprint Packages, add $5.00 for each Reprint Package for postage and handling. Make checks payable to the American Chemical Society.

BULK RATES: Subtract 20% from the listed price for orders of 50 or more of any one item.

Please ship the following reprints and supplements:

Reprint No./Package _____, _____ copies $ _____

Reprint No./Package _____, _____ copies $ _____

Reprint No./Package _____, _____ copies $ _____

Vol. 6, Suppl. 1   ☐ Hardcover _____ copies $ _____
            ☐ Softcover

Vol. 10, Suppl. 1   ☐ Hardcover _____ copies $ _____

Vol. 11, Suppl. 1   ☐ Hardcover _____ copies $ _____

Vol. 11, Suppl. 2   ☐ Hardcover _____ copies $ _____

Vol. 13, Suppl. 1   ☐ Hardcover _____ copies $ _____

Vol. 14, Suppl. 1   ☐ Hardcover _____ copies $ _____

Vol. 14, Suppl. 2   ☐ Hardcover _____ copies $ _____

Vol. 16, Suppl. 1   ☐ Hardcover _____ copies $ _____

Vol. 17, Suppl. 1   ☐ Hardcover _____ copies $ _____

Vol. 17, Suppl. 2   ☐ Hardcover _____ copies $ _____

Vol. 17, Suppl. 3   ☐ Hardcover _____ copies $ _____

Vol. 17, Suppl. 4   ☐ Hardcover _____ copies $ _____

Other Suppl.:   _____ _____ copies $ _____

          Total Enclosed $ _____

Reprints from Volume 3

Reprints from Volume 4

Reprints from Volume 5

Reprints from Volume 6

(Continuation of Cumulative Listing of Reprints)

Reprints from Volume 13

Reprints from Volume 16

(Continuation of Cumulative Listing of Reprints)

Special Reprints Packages

These special reprints packages offer selected articles in specific subject areas from the JOURNAL OF PHYSICAL AND CHEMICAL REFERENCE DATA, and they are offered at a better rate than when purchased individually. You will have available a complete library of literature for your specific requirements at a fraction of the cost of purchasing back issues of the journal.

Look over the reprints packages available—they are listed by subject area. In the Cumulative Listing of Reprints you will find the titles corresponding to the reprint numbers. You are sure to find building your information bank in this manner to be thorough and economical.

Package C1 (5 Parts) MOLECULAR VIBRATIONAL FREQUENCIES. Consisting of Reprint Nos. 103, 129, 170, 257, NSRD 39.
If purchased individually: $ 33.00
Special package price: **$ 26.00**

Package C2 (22 Parts) ATOMIC ENERGY LEVELS. Consisting of Reprint Nos. 26, 54, 64, 68, 94, 100, 109, 125, 126, 131, 132, 149, 150, 154, 156, 160, 179, 180, 192, 200, 222, 278.
If purchased individually: $121.00
Special package price: **$ 96.00**

Package C3 (6 Parts) ATOMIC SPECTRA. Consisting of Reprint Nos. 33, 56, 77, 78, 110, 132.
If purchased individually: $ 33.00
Special package price: **$ 27.00**

Package C4 (5 Parts) ATOMIC TRANSITION PROBABILITIES. Consisting of Reprint Nos. 20, 63, 82, 118, 182.
If purchased individually: $ 35.00
Special package price: **$ 28.00**

Package C5 (7 Parts) MOLECULAR SPECTRA. Consisting of Reprint Nos. 4, 8, 53, 79, 93, 130, 146.
If purchased individually: $ 51.50
Special package price: **$ 41.00**

Package C6 (9 Parts) THERMODYNAMIC PROPERTIES OF ELECTROLYTE SOLUTIONS. Consisting of Reprint Nos. 15, 95, 111, 151, 152, 174, 184, 185, 186.
If purchased individually: $ 46.00
Special package price: **$ 37.00**

Package C7 (12 Parts) IDEAL GAS THERMODYNAMIC PROPERTIES. Consisting of Reprint Nos. 30, 42, 43, 62, 65, 66, 70, 80, 83, 113, 115, 141.
If purchased individually: $ 38.00
Special package price: **$ 31.00**

Package C8 (7 Parts) RESISTIVITY. Consisting of Reprint Nos. 138, 139, 155, 221, 258, 259, 260.
If purchased individually: $ 47.50
Special package price: **$ 39.00**

Package C9 (7 Parts) MOLTEN SALTS. Consisting of Reprint Nos. 10, 41, 71, 96, 135, 167, 168.
If purchased individually: $ 62.50
Special package price: **$ 44.00**

Package C10 (4 Parts) REFRACTIVE INDEX. Consisting of Reprint Nos. 81, 158, 162, 240.
If purchased individually: $ 32.50
Special package price: **$ 26.00**

Supplements to JPCRD

When the topic demands it, and the quality of the data justifies it, the JOURNAL OF PHYSICAL AND CHEMICAL REFERENCE DATA issues a special Supplement. Each Supplement is a monograph—collected tables of highly significant physical or chemical property data in one complete volume. Listed below are the special Supplements to JPCRD that have been published. Each is a valuable resource for the physical chemist and chemical physicist.

ATOMIC AND IONIC SPECTRUM LINES BELOW 2000 ANGSTROMS: HYDROGEN THROUGH KRYPTON, by Raymond L. Kelly. (Supplement No. 1 to Volume 16) 1987, 1689 pages, 3 volumes. Hardcover.
U.S. & Canada: $75.00
Abroad: $90.00

ATOMIC ENERGY LEVELS OF THE IRON-PERIOD ELEMENTS: POTASSIUM THROUGH NICKEL, by J. Sugar and C. Corliss. (Supplement No. 2 to Volume 14) 1985, 664 pages. Hardcover.
U.S. & Canada: $50.00
Abroad: $58.00

JANAF THERMOCHEMICAL TABLES, Third Edition, by M. W. Chase, Jr., C. A. Davies, J. R. Downey, Jr., D. J. Frurip, R. A. McDonald, and A. N. Syverud. (Supplement No. 1 to Volume 14) 1985, 1896 pages, 2 volumes. Hardcover.
U.S. & Canada: $130.00
Abroad: $156.00

HEAT CAPACITIES AND ENTROPIES OF ORGANIC COMPOUNDS IN THE CONDENSED PHASE, by E.S. Domalski, W.H. Evans, and E.D. Hearing. (Supplement No. 1 to Volume 13) 1984, 288 pages. Hardcover.
U.S. & Canada: $40.00
Abroad: $48.00

THE NBS TABLES OF CHEMICAL THERMODYNAMIC PROPERTIES. SELECTED VALUES FOR INORGANIC AND C_1 AND C_2 ORGANIC SUBSTANCES IN SI UNITS, by

D.D. Wagman, W.H. Evans, V.B. Parker, R.H. Schumm, I. Halow, S.M. Bailey, K.L. Churney, and R.L. Nuttall. (Supplement No. 2 to Volume 11) 1982, 394 pages. Hardcover.
U.S. & Canada: $40.00
Abroad: $48.00

GAS-PHASE ION AND NEUTRAL THERMOCHEMISTRY, by S. G. Lias, J. E. Bartmess, J. F. Liebman, J. L. Holmes, R. D. Levin, and W. G. Mallard. (Supplement No. 1 to Volume 17) 1988, 872 pages. Hardcover.
U.S. & Canada: $70.00
Abroad: $84.00

THERMODYNAMIC AND TRANSPORT PROPERTIES FOR MOLTEN SALTS: CORRELATED EQUATIONS FOR CRITICALLY EVALUATED DENSITY, SURFACE TENSION ELECTRICAL CONDUCTANCE, AND VISCOSITY DATA, by George J. Janz. (Supplement No. 2 to Volume 17) 1988, 320 pages. Hardcover.
U.S. & Canada: $25.00
Abroad: $30.00

ATOMIC TRANSITION PROBABILITIES SCANDIUM THROUGH MANGANESE, by G. A. Martin, J. R. Fuhr, and W. L. Wiese. (Supplement No. 3 to Volume 17) 1988, 523 pages.
U.S. & Canada: $65.00
Abroad: $78.00

ATOMIC TRANSITION PROBABILITIES IRON THROUGH NICKEL, by J. R. Fuhr, G. A. Martin, and W. L. Wiese. (Supplement No. 4 to Volume 17) 1988, 504 pages.
U.S. & Canada: $65.00
Abroad: $78.00